left until the pain is relieved. Then the skin should be dried, and either ointment or antiseptic talcum powder should be applied.

When the burn is large and severe, call a physician at once. If emergency treatment is required and a physician is not available within fifteen minutes, place the patient on a cot and cover him with blankets, but be careful to avoid touching the burned parts. Remove adhering clothing from the burned parts and cover with one of the jellies mentioned above, pending the physician's arrival. In order to relieve the severe pain, apply compresses with aluminum acetate or a gauze moistened with 1 to 2 per cent aqueous solution of a local anesthetic.

Chemicals in the Eye

Wash the eye immediately with a large amount of water. If the chemical is an acid, flushing with water should be followed by application of a 1 per cent solution of sodium bicarbonate. An eye cup is very convenient for this purpose. If the chemical is an alkali, flushing with water should be followed by application of a 1 per cent solution of boric acid. After washing with water and dilute neutralizing solution, a drop of sterile olive oil should be applied and the patient should be taken to a physician. The authors use olive oil that contains 1-2 g of ethyl p-aminobenzoate per 100 ml and that has been heated to 110° for a few minutes. For the same purpose an ophthalmic ointment containing 2 per cent butyn sulfate may be used. Injuries to the eye require a specialist's care.

Acids and Alkalies on the Skin

Wash with large amounts of water. In the case of acids, follow by application of a paste of sodium bicarbonate and allow to remain for 15-20 minutes. Then remove the excess, dry, and cover the skin with an ointment. In the case of alkalies, after washing with water, rinse the affected parts with a saturated boric acid solution or a 1 per cent solution of acetic acid. Then dry the skin and cover with tannic acid jelly.

Bromine Burns

Wash at once with water and a 2 per cent solution of sodium thiosulfate. Cover immediately with glycerin and follow with an ointment.

Organic Substances

Rinse at once with water and remove any insoluble part by washing with ethyl alcohol. This may be followed by washing with soap and water. Dry the skin and apply an ointment.

Cuts

If the cut is a minor one, allow it to bleed for a few seconds, wash it with water, and then apply a prepared bandage with an antiseptic center. If a ready-made bandage is not available, apply an antiseptic from the first-aid cabinet and then cover the wound with a sterile 1- or 2-inch bandage. In case of profuse bleeding, wash the cut and apply a bandage, using pressure about four inches above and below the cut to diminish circulation and to aid clotting. Continuous pressure should not be maintained for more than five minutes. When a clot has formed, apply a mild antiseptic and call a physician.

Identification
of Organic
Compounds

A Student's Text
Using Semimicro Techniques

Identification
of Organic Compounds

A Student's Text Using Semimicro Techniques

NICHOLAS D. CHERONIS

Formerly Professor of Chemistry
Brooklyn College
Brooklyn, New York

JOHN B. ENTRIKIN

Head, Department of Chemistry
Centenary College of Louisiana
Shreveport, Louisiana

INTERSCIENCE PUBLISHERS
a Division of John Wiley & Sons, Inc.   New York • London • Sydney

Printed in the United States of America

Preface

The preparation of this preface was to have been the final written contribution to this book by Dr. Nicholas D. Cheronis. His untimely death on July 2, 1962, as a result of an automobile accident near Chicago saddens all who knew him as a true friend and causes a feeling of real loss to the host of people who recognized him as a great chemist and a stimulating teacher.

This text, although it is essentially an abridgment of a larger work by the same authors (Cheronis and Entrikin, *Semimicro Qualitative Organic Analysis*, 2nd ed., Interscience Publishers, New York, 1957), has been completely rewritten. All of the material from the larger text that had been found essential or particularly useful to students of elementary or intermediate organic chemistry has also been covered in the present book. Some less frequently used techniques, tests, and derivatives have been omitted in order to present a text of moderate size and cost.

As compared to the larger work by these authors, the following characteristics of the present text may be noted:

1. The recent literature has been considered, and, wherever possible, new techniques, tests, and library references have been included.

2. The discussions and procedures for the classification of an unknown substance by its solubility behavior and its acid-base characteristics have been combined into a single chapter (Chapter 5). An expanded table presents a classification of the common types of organic compounds on the basis of: (*a*) the elements present, (*b*) their solubility in selected solvents, and (*c*) the acid-base character of the substance.

3. The general tests for structure and functional groups have been incorporated with the specific tests for chemical classes into a single chapter (Chapter 6).

4. An entirely new chapter (Chapter 7) is devoted to the use of paper chromatography and infrared spectroscopy in the identification of organic substances. An extensive bibliography is provided for this chapter as well as for several other chapters.

5. The number and variety of exercises provided for student study have been considerably increased in this text since it has been specifically written as a laboratory textbook.

6. Although the tables of organic compounds (pages 345–426) have been reduced in that a small number of derivatives is included, it is be-

lieved that adequate information is provided for the identification of compounds which the student is likely to encounter.

7. Directions for the preparation of derivatives have been limited to the more commonly used derivatives, and these procedures have been consolidated into two chapters (Chapters 10 and 11).

8. Useful information for the application of paper chromatographic techniques has been included in the Appendix.

The authors wish to express their sincere appreciation to all those who have assisted in any way the preparation of this book. Obviously, all those who contributed to the preparation of the larger book by these authors also indirectly contributed to this text. Special mention should be made of the contributions by Dr. Robert C. Gore of the Perkin-Elmer Corporation, who materially assisted in the section on infrared spectra, and by Dr. A. L. McClellan of the California Research Corporation, who critically reviewed the section on the classification by solubility. The authors are particularly indebted to their students and assistants who have made library searches and who have tested the procedures used in this book.

J.B.E.

Shreveport, Louisiana
August 1962

Contents

1

Basis for Identification

Identification

The term *identity* as applied to chemical compounds means the assignment of a molecular or ionic structure to a chemical substance so that it can be distinguished as a single substance, always the same and different from other chemical substances. Qualitative organic analysis comprises the methods of identifying an organic substance. There are three general approaches for identification: (*a*) the systematic approach, (*b*) microscopic methods, and (*c*) the use of physical methods.

The Systematic Approach

The systematic scheme involves a series of steps by which one proves that the unknown compound *A* has identical physical and chemical properties with a known compound *B*, that is, a compound which has been described in the literature.[1] These steps may be briefly summarized:

1. The sample is first fractionated to isolate relatively pure components. A component is regarded as pure when further fractionation results in no variation of its physical constants. Melting points are usually employed as criteria of purity of solids, and boiling points and refractive indices as criteria of purity for liquids. Therefore, in this step one or more physical constants of the unknown are determined.

2. The pure unknown is then subjected to gross examination, analysis of the elements present, tests of solubility in a few selected solvents, and tests of reaction to a few pH indicators. On the basis of these tests the un-

---

[1] For the characterization of new compounds, that is, compounds which have not been described in the literature, see Cheronis and Entrikin, *Semimicro Qualitative Organic Analysis* (2nd ed., N.Y.: Interscience, 1957), p. 3.

known is assigned to a large division, which contains several classes of organic compounds.

3. Small samples of the unknown substance are subjected to tests with a number of reagents in order to detect the presence or absence of functional groups, such as the carboxyl (COOH), carbonyl (CO), hydroxyl (OH), nitro (NO_2), amino (NH_2), and the like. By means of these tests the classification of the unknown is restricted to a *single class* of organic compounds.

4. All the data obtained in the above tests are carefully coordinated; the literature is then consulted and a list of compounds is prepared. The compound that best fits the experimental data is selected as the most probable one to be identical with the unknown.

5. By means of chemical reactions one or more suitable solid derivatives of the unknown are prepared. If these derivatives melt within 1–2°C[2] of the melting points recorded in the literature for the same derivatives of the compound tentatively identified as the unknown, the identification is regarded as complete.

To illustrate the steps used in the identification, let us assume that the unknown is a liquid. On distillation 90 per cent boils at 116–118°; analysis of a sample from this fraction for elements shows the absence of nitrogen, halogens, and sulfur. The substance is found to be soluble in water and ether and therefore is classified (page 93) in Solubility Division S_1, which includes *carboxylic acids, alcohols, carbonyl compounds, anhydrides, esters, ethers,* and some *phenols.* Since the aqueous solution of the unknown is neutral, all acidic substances are excluded. Tests of the unknown with a mixture of potassium alkoxide and carbon disulfide (xanthate test, page 119) indicate the presence of an alcohol group, and hence the unknown is restricted to the class of alcohols. The presence of an alcohol is confirmed by the hydroxamate test (page 119). Reference to Table 6A (pages 362–64),[3] which lists the alcohols with their boiling points, indicates that, since the unknown has a boiling point of 116–118°, it is probably *n*-butyl alcohol (b.p. 117.7°). One of recommended derivatives for alcohols is the 3,5-dinitrobenzoyl ester, prepared by the reaction of the alcohol with 3,5-dinitrobenzoyl chloride. When the 3,5-dinitrobenzoyl ester of the unknown is prepared, it shows a melting point of 63° after one crystallization; the melting point of the 3,5-dinitrobenzoate of butyl alcohol is given in the literature as 64°. The 3,5-dinitrobenzoate prepared from a known sample of *n*-butyl alcohol is found to melt at 63°, and a mixture of the two dinitrobenzoates

---

[2] All temperatures in this text are Centigrade.

[3] The nomenclature used in the tables follows, as far as possible, that of *Chemical Abstracts.* In some instances the synonym of a compound appears in parentheses after the name chosen, but inclusion of a separate column of synonyms was abandoned as too ambitious for a book of this type.

(from unknown and known) shows no alteration in the melting point. Another derivative is prepared by the reaction of a sample of the unknown with α-naphthyl isocyanate; the resulting α-naphthylurethan (α-naphthylcarbamate) melts at 71°, which is the melting point listed in Table 6A (page 363) for the α-naphthylurethan of *n*-butyl alcohol. Therefore the unknown is definitely identified as *n*-butyl alcohol.

In the present text this approach will be used almost exclusively, although Chapter 7 is devoted to chromatographic and infrared methods. The reasons for choosing the systematic approach is to extend the student's knowledge of organic chemistry and provide experience in reflective thinking through judicious problem solving.

Microscopic and Physical Methods

Table 1.1 gives a summary of the microscopic methods. Procedure I is based on the reaction of the unknown compound with a definite number of reagents under the microscope and the identification of the resulting products by their crystal structure. Procedure II employs the accurate determination of the crystallographic properties of the unknown and their derivatives. Procedure III depends on the determination of the melting points of the unknown and their behavior when they are "fused" with known standards. The microscopic methods are applicable to a limited number of alkaloids and other biologically active compounds such as drugs.[3a]

TABLE 1.1

Microscopic Methods for Identification of Organic Compounds

Procedure	Basis of procedure
I. Reactions under the microscope	Comparison of reactions of known and unknown with the same reagent
II. Crystallographic properties	Comparison of known and unknown and their derivatives
III. Fusion techniques	Comparison of melting points, eutectics, and refractive indices of melts of known and unknown

Table 1.2 summarizes the various physical methods which are used either extensively or in restricted applications to the identification of organic substances. Of those listed, paper chromatography and infrared spectra are the most important and most widely used; they are discussed in Chapter 7.

[3a] McCrone, *Fusion Methods in Chemical Microscopy—A Textbook and Laboratory Manual* (N.Y. Interscience, 1957).

TABLE 1.2

Summary of Physical Methods Employed for the Identification
of Organic Compounds

Method
Differential migration
Adsorption (column)
Paper chromatography
Electrophoresis
Spectroscopic
Infrared and Raman
Ultraviolet
X-ray
Mass
Nuclear magnetic resonance
Fluorometric
Radiochemical

Analytical Procedures

In all qualitative analytical procedures *small portions* of the unknown
are used for the various tests described in the preceding paragraph. The
word *small* as used in this text refers to milligram quantities. With solubility and functional group tests samples of 25–50 mg are employed. As the
beginner acquires experience, most tests can be performed with quantities of
1–5 mg. However, for the preparation of derivatives which require purification, the procedures described use 100–200 mg of the sample. Once experience has been acquired it is possible to complete the identification of an
unknown compound starting with a few milligrams. For procedures employed in the identification of less than 1 mg of organic compounds, the advanced student is referred to the literature.[4] Although quantities of less

TABLE 1.3

Microchemical Units

Units	Explanation	Symbol
Gram	1×10^{-3} kg	g
Milligram	1×10^{-3} g	mg
Microgram	1×10^{-6} g	μg
Nanogram	1×10^{-9} g	ng
Picogram	1×10^{-12} g	pg
Microliter	1×10^{-6} liter or 1×10^{-3} ml	μl

[4] Cheronis, *Micro and Semimicro Methods,* Vol. VI of Weissberger (ed.), *Technique
of Organic Chemistry* (2nd ed., N.Y.: Interscience, 1957), pp. 413–582.

than 1 mg will seldom be referred to in this elementary text, it is well for the student to be acquainted with microchemical units. Table 1.3 lists the most common units for small masses and volumes.

The terms *macro, semimicro, micro,* and *submicro,* as usually applied to analytical procedures, are differentiated according to the size of the sample taken for analysis. Macro refers to methods and samples which are commonly employed in the laboratory. Micro procedures employ a sample usually one-tenth as large as in the macro procedure, and semimicro methods use samples which are intermediate between the two.

Approach to Student "Unknowns" and Method of Using the Text

The instructor usually gives detailed directions as to the best approach to the identification of an "unknown," and the outline given here is merely for general information, for the systematic steps to be followed by beginners.

One may start with gross observations (page 106) and then determine whether the unknown is a relatively pure organic compound, or an impure "practical" or "commercial" grade of an organic substance or a mixture (pages 9–11). This step, that is, the ascertaining of purity through fractionation procedures (Chapter 2), involves, of course, the determination of boiling point (page 52) and refractive index (page 54) if the sample is a liquid and of the melting point (page 41) if the sample is a solid. If the results of the first step in the purification procedure (page 10) indicate that the sample is a mixture, reference should be made to Chapter 8 on the separation of mixtures. The purified compound or each pure component separated from a mixture is then examined as suggested in the following chapters: Chapter 4, determination of elements present; Chapter 5, classification of the unknown by solubility and acid-base character. At this point, preliminary tests should be selected from the first part of Chapter 6 that would guide in the logical selection of specific class tests. The performance of several functional group tests, as outlined in Chapter 6, enable one to classify the unknown into a single class of organic compounds. All the available data are now used and correlated and the literature or tables are consulted, as outlined in Chapter 9, in order to arrive at a tentative assumption as to the nature of the unknown.

The tentative identification now enables one to consult Chapters 10 and 11, in order to select suitable derivatives for final and conclusive identification.

The text also may be used as a laboratory manual in the regular one-year course of organic chemistry, and one of the authors[5] has used it suc-

[5] Entrikin, *J. Chem. Educ.*, **24**, 604 (1947).

cessfully for many years. While the facts and theories concerning the various classes of organic compounds are being studied, known members of these classes may be used as samples in the performance of those tests pertinent to the class being studied. Experience may also be gained in making preparations and in recognizing typical organic reactions by preparing derivatives of these known compounds by several methods.

To illustrate, let it be assumed that alcohols are being studied in the lecture course. Three representative alcohols may be selected for laboratory investigation. Their properties may be studied by the various tests for alcohols (pages 119–22) as, for example, the hydroxamate and xanthate tests. Further experience with alcohols may be gained by preparing derivatives for selected alcohols (pages 247–56), purifying them, and then determining their melting points. A comparison may then be made between the data from the experiments and those given in the tables (Tables 6A, 6B, page 362). If an unknown compound is secured and identified as soon as the preliminary work has been completed with knowns of any one class, the work will generally prove more interesting and the preliminary work will be done with more care and attention to detail.

Periodically the student may be given an unknown that may belong to any chemical class studied up to that time. The identification of such a compound is made according to the outline described in the preceding paragraphs. Such work is mentally stimulating and helps the correlation of facts about the properties of the important groups of organic compounds.

Advanced students who have had one or more courses in organic chemistry may follow the methods of this text systematically. Either of two plans may be adopted: one is systematically to study the theory involved and acquire experience in all the methods suggested in the text before undertaking the identification of general unknowns; the other is to start the identification of unknowns early in the course and to study the theory and perform the experiments that seem necessary for each unknown as the need arises. The first plan affords a more complete knowledge of the whole subject; the second sustains interest better, but often omits many of the functional tests and methods of making derivatives. Physical methods should be considered and used for the determination of functional groups. After determining physical constants and elements and performing other preliminary tests, the advanced student may choose to use infrared spectra, and thus arrive at a tentative assumption of the probable nature of the unknown; he may then proceed directly to the preparation of derivatives to prove the assumption. A considerable portion of the time spent in identifying unknowns should be devoted to mixtures. Chromatographic procedures for the fractionation and identification of the components of mixtures should be tried. Numerous problems should be solved.

Laboratory Notebook and Reports

Although the exact type of notebook to be kept and the form of report to be made differ with each laboratory, the basic principles involved are the same: (a) the notebook entries should record observations and conclusions as they are made so that the notebook may give a clear summary of the work accomplished; (b) the report should be an abstract of the entries made in the notebook with particular emphasis on the final conclusions or the solution of the problem. For beginners, in particular, the authors recommend that the following items be covered in the notebook entries for each known or unknown that is studied:

1. The date received and the nature of the material (single class, "general" unknown, or mixture).

2. Results of gross examination, determination of purity, and fractionation.

3. Determination of elements present.

4. Solubility and indicator data and a list of the probable class or classes.

5. A list of the functional-group tests that are to be made, together with the results of those tests.

6. A summarized argument, prepared with care as to both grammatical form and logic, regarding the probable nature of the compound, and based on the data collected (including one or more physical constants and a comparison of these constants with the data found in the tables).

7. A list of the types of reactions that can be used for the preparation of derivatives of the probable compound.

8. A summary of the method used in the preparation of the derivative(s), including the equations for the reactions, the exact methods of purification of the derivatives, and the reasons for using the methods chosen.

9. A comparison of the data on physical constants of the original compound and the prepared derivative with data for known compounds taken from the tables.

10. A statement of the conclusions as to final identity.

2

Fractionation and
Purification Procedures

Introduction

An unknown organic material may be (*a*) a relatively pure organic substance, (*b*) an impure organic compound (that is, one containing small amounts of impurities), or (*c*) a mixture of organic compounds.[1] A *pure organic substance* is one that *on repeated fractionation yields fractions with the same melting point, boiling point, refractive index, density, solubility, and other physical constants.* Thus if 1 g of a solid unknown organic substance is crystallized three times and the crystals from the second and third crystallizations show substantially the same melting point and solubility, it is usually assumed that the compound is relatively pure. Similarly, if a 5-ml sample of a liquid, which was separated from a mixture, is distilled and two 2-ml fractions are collected that boil within 1° of each other and show substantially the same refractive index, for most operational purposes one can assume that the liquid is relatively pure.

The methods usually employed for the purification of organic compounds involve fractionation by *crystallization, distillation,* or *sublimation.* In a number of cases *chromatographic* procedures and separation by *ion*

---

[1] The difference between the last two categories is a matter of degree; for example, a sample of an amino acid containing 1 per cent of other amino acids may be regarded as a chemical of "practical" grade for some purposes even though it is a mixture. The term *mixture* is usually applied to materials in which no single component constitutes more than 90 per cent of the total. Although the term *pure* is applied to many organic substances containing less than 1 per cent of an impurity, the presence of 0.01 per cent of an impurity may materially alter the activity of compounds possessing physiological activity.

9

exchange resins and *extraction* yield best results, but the methods most commonly employed are crystallization for the purification of solids and distillation for the purification of liquids.

If the unknown is a mixture of two or more substances the procedures described in Chapter 8 should be followed for fractionation and separation into relatively pure components.

If the unknown is an impure organic compound—as is the case with most of the so-called "practical" and "commercial" grades of organic compounds —it should be purified by fractionation. Even when preliminary tests indicate that the unknown is a relatively pure organic substance, it is advisable to perform a simple fractionation before beginning systematic analysis.

Selection of Fractionation Procedures

It is assumed that the student is familiar with the elementary aspects of crystallization and distillation. Moreover, it is advisable to review the theory of these fractionation procedures in an elementary laboratory text of organic chemistry.[2] In the present text only a brief review of all fractionation procedures will be given, particularly as they are applicable to the masses and volumes which the student may encounter. For a more extensive treatment of micro and semimicro fractionation procedures the student is referred to the more advanced texts by the authors.[3,4]

If the sample is a solid and the amount available is 1 g or more, crystallization should be the fractionation procedure selected. A small amount of the unknown is used to determine the best solvent (see pages 11–12) and then about 50–75 per cent of the available material is purified.

If the amount of solid unknown sample is far below 500 mg, consideration should be given to vacuum sublimation, the procedures for which are described on pages 35–37.

If the sample is a liquid and the volume is 5–10 ml, the procedures outlined on pages 30–33 for semimicro fractional distillation give relatively good results provided the rate of distillation is not more than 0.3 ml per minute. However, fractional distillation of 2 ml or less of a liquid sample should not be undertaken by beginners unless experience is first gained through trials with known mixtures. A beginner should practice distillation and fractionation of small volumes of liquids, at either atmospheric or reduced pressure, by making known mixtures and then separating them by

[2] Cheronis, *Semimicro Experimental Organic Chemistry* (N.Y.: Hadrian Press, 1960), pp. 31–50 and 54–88.

[3] Cheronis and Entrikin, *Semimicro Qualitative Organic Analysis* (2nd ed., N.Y.: Interscience, 1957), pp. 29–111.

[4] Cheronis, *Micro and Semimicro Methods,* Vol. VI of Weissberger (ed.), *Technique of Organic Chemistry* (2nd ed., N.Y.: Interscience, 1957), pp. 13–133.

fractional distillation as described on pages 31–33. A good mixture, from which several trial runs can be made, is 1.9 ml of chlorobenzene and 0.1 ml of bromobenzene. The efficiency of the separation can be ascertained by determining the boiling point and the refractive index of each fraction. Liquid samples of 1 ml or less are best fractionated by gas chromatographic apparatus.

CRYSTALLIZATION[5]

Selection of Solvents

The choice of the solvent in the crystallization of solids is of extreme importance. In most cases the solvent cannot be selected on the basis of rules or theoretical considerations, but must be experimentally determined; therefore, the solvent is specified for recommended derivatives.

The literature describing derivatives contains a reference to the solvent used but very seldom to the solubility data. In such cases the beginner should proceed cautiously to determine roughly the solubility as described on page 13. The following general rules with reference to solvents will be valuable to the beginner.

1. A solid usually dissolves best in a liquid that it resembles in structure. For solid esters such solvents as methanol, ethanol, and ethyl acetate should be among the first tried.

2. It is advisable to select a solvent that will dissolve the crude solid when hot, but only sparingly when cold.

3. When a solvent dissolves a solid very readily in the cold, it may be useful as a crystallizing medium if it is mixed with another solvent in which the compound is sparingly soluble. Thus, if a substance is very soluble in alcohol in the cold and sparingly soluble in water, it is dissolved in a *small amount of hot alcohol*, the solution is filtered, and water is added until cloudiness develops. The solution is then heated until clear, filtered if necessary, and cooled. The use of *solvent pairs* is very helpful in many crystallizations.

4. The impurities present should either be very readily soluble or as sparingly soluble as possible in the solvent. For example, in the 3,5-dinitrobenzoates ($C_6H_3(NO_2)_2COOR$) used for the identification of hydroxy compounds, the impurity (dinitrobenzoic acid) is completely soluble in methanol or ethanol used as a solvent for the crystallization. Another derivative for hydroxy compounds may be prepared by allowing the alcohol to react with

---

[5] For an extensive discussion of crystallization, recrystallization, and solvents, see Tipson, "Crystallization and Recrystallization," in Vol. III of Weissberger (ed.), *Technique of Organic Chemistry* (2nd ed., N.Y.: Interscience, 1956), and Weissberger and Proskauer, *Organic Solvents,* 2nd rev. ed. by Riddick and Toops, Vol. VII of Weissberger (ed.), *Technique of Organic Chemistry* (N.Y.: Interscience, 1955).

an isocyanate, for example, phenyl isocyanate (C_6H_5NCO). The desired derivative is known as the *urethan*, $C_6H_5NHCOOR$, and the impurity is diphenylurea, $C_6H_5NHCONHC_6H_5$, produced by the action of moisture present in the hydroxy compound or on the walls of the vessel. Hot petroleum ether dissolves the urethan but not the diphenylurea.

5. The solvent should be chemically inert to the compound to be crystallized. In some cases, however, a solvent may be chosen because it reacts chemically with the compound. For example, some aromatic acids may be purified by being dissolved in dilute sodium hydroxide and then, after filtration of the solution, precipitated by neutralization with dilute hydrochloric or sulfuric acid.

6. In the event that a solvent is not given in the literature, it is recommended that the procedure outlined on page 13 be tried with solvents in the following order: methanol or ethanol, mixture of lower alcohols and water, acetone or mixture of acetone and alcohol, benzene or mixtures of benzene and toluene, petroleum ether or benzene and petroleum ether, glacial acetic acid or aqueous acetic acid. Table 2.1 shows a number of solvents and solvent pairs used for crystallization of derivatives. The general pro-

TABLE 2.1

Common Solvents for Crystallization of Derivatives

Solvent or solvent pair	Useful derivatives
Water	Carboxylic acids, amides, and substituted amides
Methanol	Most derivatives: benzoates, 3,5-dinitrobenzoates, amides, *p*-toluidides, nitro and bromo compounds, etc.
Methanol–water	*p*-Nitrobenzyl esters, sulfonamides, anilides, picrates, semicarbazones, hydrazones, substituted hydrazones, etc.
Ethanol	Same as methanol and methanol–water mixtures, molecular complexes
Dioxane–water	Xanthylamides
Petroleum ether[a]	Phenylurethans, α-naphthylurethans
Petroleum ether–benzene	*p*-Nitrophenylurethans, 3,5-dinitrophenylurethans
Acetone–alcohol	Osazones, bromo compounds, nitro compounds
Isopropyl ether	Quaternary ammonium salts
Ethyl acetate	Quaternary ammonium salts, esters
Benzene	Picrates, molecular complexes
Chloroform and carbon tetrachloride	Sulfonchlorides, acid chlorides, anhydrides

[a] The original fractions of petroleum distillates such as ligroin or petroleum ether (b.p. 30–60°, 60–80°, or 90–110°, etc.) have been displaced in recent years by the commercial availability of hydrocarbons, such as pentane; hexane, and heptane, with boiling ranges within 5–10° from the boiling points of the pure hydrocarbons.

cedure for the preparation of solvent pairs is to use a solvent in which the derivative is most soluble and then, after the solution has been prepared, to add cautiously a solvent in which it is least soluble. The two solvents must be miscible and should be used hot.

7. If several solvents are found to be applicable, additional factors, such as volatility, cost, and availability, should be considered. A solvent with a high boiling point is usually an advantage because evaporation is minimal and because there is a larger interval between the solution and crystallization temperatures. This generalization is valid only for solvents boiling between 60 and 150°, since beyond this range the difficulty of removing the solvent and the danger of decomposition of the substance offset the advantages.

Some empiric knowledge is useful in restricting the number of solvents to be tried in a given case. Substances with hydroxyl groups are likely to be soluble in methanol or water. Substances soluble in methanol are slightly less soluble in ethanol and isopropyl alcohol, and even less soluble in higher alcohols. Polyhydroxy compounds may be more soluble in water than in methanol. Alcohol–water is usually a good solvent pair for compounds containing hydroxyl groups. Substances with aromatic structures are likely to be soluble in benzene, toluene, and ether. Substances soluble in aliphatic hydrocarbons (ligroin, petroleum ether, and the like) have low solubilities in water. Heterocyclic compounds soluble in alcohols may often be precipitated by the addition of ether. When a substance is found to be sparingly soluble in the common organic solvents, glacial acetic acid or pyridine should be tried.

When it is impossible to determine through the literature what solvent to use, or if the solvent is known but not the amount to be used, the solubility of the unknown substance or derivative should be determined roughly, both at room temperature and at the boiling point of the solvent. About 10–50 mg of the solid is placed in a 3-inch test tube (preferably with tapered end) and, by means of a graduated pipet dropper, 0.1–0.5 ml of the solvent is added and the contents of the tube are stirred with a small glass rod to bring the solvent in contact with all the solid in the tube and then allowed to stand 1–2 minutes. The ratio of solid to solvent in the initial test should be about 1:10. If, at this ratio, the solid dissolves completely at room temperature, the solvent is not useful for crystallization unless it is miscible with another solvent in which the solute is sparingly soluble. This can be determined by a separate test in which the same quantities of solute and first solvent are used and between one fifth and one tenth as much of the second solvent is added.

If the solid does not dissolve, successive portions of 0.1–0.5 ml of solvent[6]

---

[6] For measuring small volumes see page 15.

may be added, the tube being shaken after the addition of each portion, until the solid dissolves. The number of milliliters of solvent used divided into the number of milligrams of solid taken for the solubility test gives the number of milligrams of solid dissolved by each milliliter of solvent at room temperature.

The solubility of the solid at or near the boiling point of the solvent is determined by a similar procedure. If, for example, the solubility tests are performed on 50 mg of the sample, after the addition of the first 0.5 ml of solvent, the tube is cautiously heated until the solvent just begins to boil. If the solid dissolves completely, the test is repeated, 0.3 ml of solvent being used. If some of the solid remains undissolved, more solvent is added in portions of 0.1–0.2 ml, the tube being heated after the addition of each portion.

In this manner the solubility of the solid at or near the boiling point of the solvent and also at room temperature is determined. This information is used to determine the solvent and the amount to be used. Let it be assumed, for example, that 8 ml of methanol was used to dissolve 50 mg of a compound at 20° and 0.9 ml at the boiling point of alcohol. The solubility, therefore, at 20° is about 6 mg/ml and, at or near the boiling point of methanol, 55 mg/ml. If the amount of derivative to be crystallized is 200 mg, the amount of methanol to be used is 4–4.5 ml. After crystallization about 25 mg of the derivative will remain in the mother liquor. Generally the solvent is not particularly useful unless the solubility near the boiling point of the solvent is *at least 5 times the solubility at room temperature.* If, in the above example, the solubility at room temperature had been 12 mg/ml, the loss would have been at least 50 mg in the first crystallization.

The solvent used most extensively in the experimental part of this work is methanol. Wherever possible, the crude derivative is dissolved in methanol and precipitated after filtration by cautious addition of water. Although methanol is toxic when absorbed in the tissues in appreciable quantities, the handling of small amounts by beginners has been found entirely safe—provided caution is used. A number of other factors make the use of methanol more desirable than ethanol. In general the solubilities of organic compounds are not greatly different in two homologs. The commercial grade of methanol is almost anhydrous and of greater purity than the commercial grade of ethanol. In addition, the ease with which methanol is obtained in the market, plus its low price, makes its use desirable wherever possible.

Measurement of Solvents and Other Liquids

Virtually all solvents and liquid reagents used in small-scale experimentation are dispensed from reagent bottles having a capacity of 15, 30, or 60 ml and provided with plastic caps holding a glass dropper and rubber bulb. The glass droppers usually have a tip of about 2.5–3.0 mm in outer diameter and deliver approximately 0.5 ml of water if the bulb is pressed on the upper part. However, it is best to check the dropper by withdrawing several dropperfuls and emptying in a 10-ml graduate. A dropper calibrated to 0.25, 0.5, 0.75, and 1 ml is commercially available[7] and has been extensively used for small-scale work.

When graduate droppers are not available, ordinary medicine droppers may be employed and the volume estimated by counting the number of drops. With droppers having tips of 3 mm in outside diameter, 20 drops of water is about 1 ml; hence 1 drop = 0.05 ml. Since the surface tension of a liquid and the density must be taken into consideration for measuring other liquids, the following figures are approximations: methanol and ethanol = 40 drops/ml; acetone = 50 drops/ml; benzene = 55 drops/ml; and ether = 60 drops/ml.

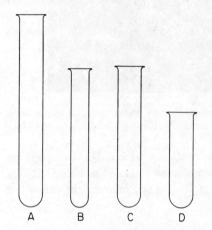

Figure 2.1 A. Pyrex tube 25 × 200 mm (8-inch) employed in crystallization of 100–500 mg or more of organic compounds. The 8-inch tube with a side arm is employed as a filter vessel in most semimicro crystallizations (see Figure 2.6). B and C. Pyrex tubes 18 × 150 mm and 20 × 150 mm (common 6-inch tubes) used for crystallization of 50–100 mg of material. D. Pyrex tube 25 × 100 mm, often employed as reaction vessel.

A   B   C   D

In general, for most preparative work the measurement need only be approximate. For example, if it is desired to wash the crystals of a derivative with 1 ml of 50 per cent methanol, no serious error is introduced by making a mixture with one dropperful (from the semimicro reagent bottle) of methanol and one dropperful of water. Calculation[8] of the alcohol–water mixture volume thus produced shows that about 1.3 ml of methanol–water containing 50 per cent of alcohol by weight has been mixed.

Preparation of Solutions for Crystallization

Figure 2.1 shows four types of test tubes that have been found most useful in the preparation of solutions for the crystallization of solid organic

[7] Microware, Inc., Vineland, N.J.
[8] Cheronis and Entrikin, *op. cit.*, p. 19.

substances. In each case both the size of the tube and the range of quantities for which it is suitable are given. For example, in the preparation and purification of 100–500 mg of derivatives, 8-inch tubes are used throughout. If the amount of solvent to be used in the preparation of the nearly saturated hot solution is less than 5 ml, a 6-inch tube (20 × 150 mm) is preferable. For amounts of 10–30 ml, 8-inch tubes (25 × 200 mm) are used.

Figures 2.2 and 2.3 show a number of heating arrangements for

Figure 2.2 *Left:* Microburner. *Center:* Semimicro burner. *Right:* Bunsen burner with barrel removed. (Microware Inc., Vineland, N.J.; A. H. Thomas Co., Philadelphia, Pa.)

small-scale work.[9] Since most organic solvents are flammable, the question of whether the beginner should use a flame is of some importance. It has been found by the authors that in dealing with small amounts of solvent it

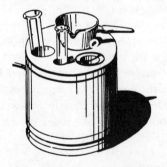

Figure 2.3 Heating bath for small-scale experimentation. (Microware Inc., Vineland, N.J.; A. H. Thomas Co., Philadelphia, Pa.)

is possible with a little care to avoid accidents. The vessel should never be more than one-third full; the flame should be directed first at the top of the liquid layer and then slowly moved downward. If complete solution of the

[9] Microware, Inc., Vineland, N.J.; A. H. Thomas Co., Philadelphia, Pa.

solid is not effected when the liquid just begins to boil, more solvent is added. Finally, when the solid dissolves slowly or the amount of solvent exceeds 10 ml, the apparatus shown in Figure 2.4 for effecting solution by

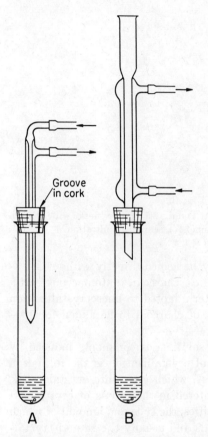

Groove in cork

A   B

Figure 2.4 A. Apparatus for heating under reflux using tube with finger micro-condenser. B. Apparatus for heating under reflux using a Liebig microcondenser. (Microware Inc., Vineland, N.J.; A. H. Thomas Co., Philadelphia, Pa.)

heating under reflux is used. If benzene or ether is used as a solvent, heating with a direct flame should be avoided and the water bath shown in Figure 2.3 employed.

Filtration of the Hot Solution and Formation of Crystals

The hot solution for crystallization should be saturated (or nearly so) at a temperature 10° below the boiling point of the solvent. If the solution is saturated at the boiling point of the solvent, crystallization will occur during filtration; the filter will become clogged and the troublesome transfer and preparation of the hot solution must be done over again. Crystallization during filtration may sometimes be prevented by preheating the filtering apparatus. If the hot solution is free from suspended or insolu-

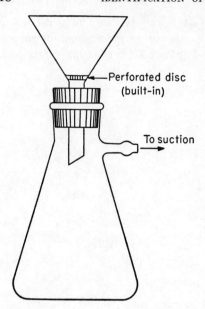

Figure 2.5 Filter flask with Hirsch funnel for semimicro filtration.

ble particles or colored impurities, it may be cooled directly without filtration. For the removal of colored impurities, charcoal, diatomaceous earth, and silica gel have been found to be better adapted to microcrystallization than inorganic precipitates. The amount of charcoal to be added for 100–200 mg of solid should not exceed 20 mg.

The separation of crystals from hot solutions is commonly induced by cooling and stirring. The first evidence of crystallization is the formation of minute crystalline particles (nucleates), which grow into crystals.

The most important problem encountered in the course of crystallization procedure is the elimination of impurities, the type and amount of which influence the shape and size of the crystals and usually the rate of crystallization as well. Some impurities, particularly when present in minute amounts, may be removed by treatment of the hot solution with such adsorbing media as activated carbon or diatomaceous earths (Filter-cel and the like). The bulk of the impurities, however, remain in the solution. Therefore, conditions must be chosen that allow the formation of pure crystals and their separation from the mother liquor without adherent impurities and with minimum loss of the substance being crystallized.

In the following section procedures and apparatus are described for the crystallization of quantities above 25 mg. With proper selection of a solvent the procedure may be used for smaller quantities. For quantities of a few milligrams the reader is referred to the more extensive treatise of the authors.

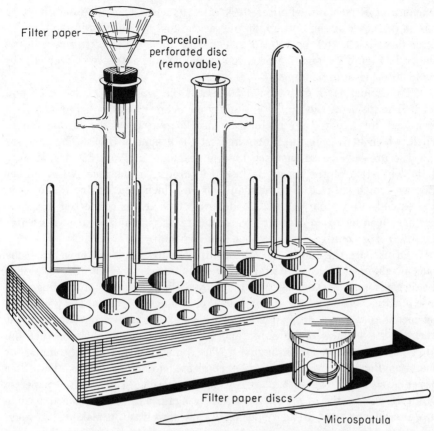

Figure 2.6   Apparatus for semimicro filteration consisting of frosted funnel with removable porcelain disc and 8-inch side arm.

Regardless of which procedure is used, the steps in the sequence of purification are the same: solution, crystallization, separation of the mother liquor from the crystals, removal of a small quantity for determination of the melting point, and repetition of the cycle until samples from two successive crystallizations exhibit essentially the same melting temperature (within 0.5°).

Apparatus and Procedure for Quantities of 25 mg or More[10]

The hot solution, prepared according to the directions given in the preceding section, is filtered by means of the apparatus shown in Figures 2.5 and 2.6. Two setups may be used for filtration. One is the traditional

[10] For quantities less than 25 mg see Cheronis and Entrikin, *op. cit.,* pp. 44–49.

Buchner or Hirsch funnel–filter–flask arrangement. The funnel is about 15 mm in diameter at the bottom, 35 mm at the top, and 10 mm in depth. The filter flask is 50, 100, or 150 ml in capacity. The other filtration setup is shown in Figure 2.6 and has been found by the authors more convenient for rapid filtering and cleaning.

The funnel has a diameter of 50 mm at the top, and its stem is 55 mm long. The inside of the funnel is slightly etched or ground so as to provide a firm seat for a porcelain perforated disc 20 mm in diameter and 5 mm thick, which fits inside the funnel against the ground surface. The edges of the disc are beveled so that the bottom diameter is about 15 mm (Figure 2.6). The stem of the funnel is inserted through a one-hole No. 4 rubber stopper, which is fitted into the mouth of an 8-inch tube having a side arm, and extends 10–12 mm below the side arm. This arm is connected to a rubber hose leading to an ordinary water aspirator (in the absence of a water aspirator it is connected to a rubber aspirator bulb).

To prepare the funnel for filtration, the tube is placed on a test-tube rack or clamped to a small stand. The perforated porcelain disc is placed inside the funnel and arranged in place by a slight pressure of the finger. A disc of filter paper[11] 24–25 mm in diameter is placed on top of the porcelain; by means of a pipet dropper 2 drops of water are placed on two different parts of the surface of the paper and the funnel tilted slightly so that the filter paper is moistened throughout. The funnel is fitted into the mouth of the receiving tube, which in turn is connected to the aspirator. The filter paper is sucked down into place by gentle suction. If the filter paper is properly placed, the edges are held tightly against the funnel and protrude upward. If any part of the paper protrudes downward, a leak will develop during filtration; in such a case the suction is discontinued, another drop of water is added, and the filter paper is adjusted with a spatula. After the filter paper has been made to adhere to the sides of the funnel, the suction is momentarily discontinued, 2 drops of methanol is added to the paper, and suction is applied again. The receiving tube is changed, and the funnel is ready for filtration if the solvent is an alcohol, ether, ester, or an organic acid. When a hydrocarbon (benzene, heptane, or the like) is used as a solvent, the moistening of the paper by water is followed by washing with acetone and then with 5–6 drops of the hydrocarbon; then suction is applied.

The funnel prepared as described in the preceding paragraph is fitted into the mouth of a clean tube, and gentle suction is applied through the side arm. The tube containing the hot solution is held in one hand and its mouth is lowered over the funnel so that the solution is poured through the center of the disc. It is advisable to use a glass rod in pouring the hot solution into the funnel. The tube containing the solution is lowered so that it

[11] Microware, Inc., Vineland, N.J.

just touches a rod held vertically with one end touching the filter disc. The filtration is completed within a few seconds. If the amount of the solid being crystallized is very small, 0.5 ml of fresh solvent is added to the tube from which the solution was poured out and heated until all the crystalline solid adhering to the sides is dissolved; the hot solvent is then poured into the filter so as to wash down any solid adhering to it.

The funnel is removed and cleaned immediately and then set aside, upside down, to drain and be ready for the next filtration. The tube containing the filtered hot solution is immersed in a beaker or a small jar in which running tap water circulates; the solution is stirred with a glass rod from time to time. For most derivatives 5–10 minutes of cooling is sufficient; others require 15 minutes or more for complete crystallization.

If mixed solvents, such as alcohol (methanol or ethanol) and water, are used, it is advisable to effect solution in alcohol and after filtration to add warm water dropwise by means of the pipet dropper until a permanent cloudiness is obtained on shaking. The tube is heated until the cloudiness disappears and is then cooled.

The crystals and mother liquor are thoroughly stirred by means of the rod so as to loosen most of the solid adhering to the walls of the tube. A clean 8-inch tube with side arm is fitted with the filter funnel prepared as described above. The tube is shaken two or three times and the contents are poured into the funnel. The mother liquor is poured back into the crystallizing tube, and the process is repeated until practically all the adhering crystals have been transferred into the funnel. The draining is complete within a minute or two. About 0.5–1 ml of solvent is added to the tube in which the crystallization took place, and the tube is shaken so that washing of the adhering crystals takes place, since the tube is to be used directly for the second crystallization. The amount of solvent added should be insufficient to dissolve the crystals remaining in the tube. The suction is discontinued, and the washings are added slowly over the crystals in the filter so that the entire mass is moistened; after a minute the washing is repeated. The filtrates are saved until the crystallization is complete, and the melting point determined. A small amount (5–10 mg) of the crystals is removed from the filter and dried as directed on pages 24–25 and labeled as a sample from the first crystallization. The balance is then used in the next crystallization.

To perform a recrystallization loosen with the spatula the filter paper containing the crystals and transfer it directly into the 8-inch tube in which the previous crystallization took place (Figure 2.7). If this manipulation is difficult for the beginner, the funnel may be placed upside down on a small piece of clean paper (Figure 2.8), and, by a slight pressure with the spatula, the porcelain disc and the crystals made to fall on the paper. The porcelain

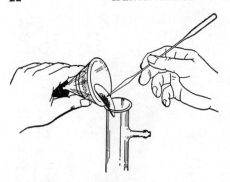

Figure 2.7 Transfer of filter paper and crystals directly into the solution vessel.

disc is pushed away with the spatula, and the crystals, together with the filter paper, are transferred into the crystallizing tube.

The amount of solvent to be added varies between 50 and 80 per cent of the amount used in the first crystallization. Assume that 7 ml of methanol

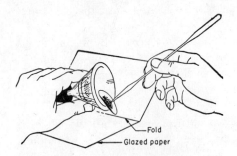

Figure 2.8 Transfer of filter paper and crystals onto glazed paper for subsequent transfer into the solution vessel.

was used in the first crystallization. Since a certain amount of the solid remains in the mother liquor, depending on the solubility of the derivative, it is advisable to begin with 4–5 ml of methanol for the second crystallization. The tube is heated until the solvent begins to boil; by means of a spatula or a glass rod the filter paper is pulled up on the sides about 50 mm from the bottom and the heating is resumed, so that the hot vapor condenses on the region of the filter paper and washes down any adhering solid. The filter paper is pulled out and discarded. Additional solvent is added until the solid dissolves at a temperature near the boiling point of the solvent. If proper care has been taken, the solution will be clear and without shreds of paper, so that filtration of the solution may not be necessary; if the solution is not clear, it should be filtered by the same procedure as before. The cooling of the hot solution and filtration of the crystals are accomplished in the manner already described. A sample of the crystals (5 mg) is removed, dried as directed on page 24, and properly labeled. The bulk of the crystals from the second crystallization is saved until the identification of the compound is complete.

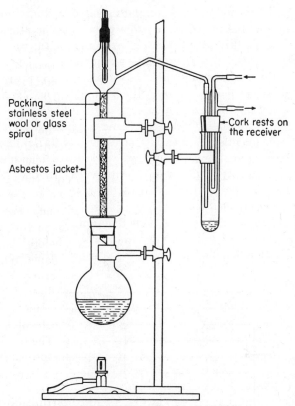

Packing—stainless steel wool or glass spiral

Asbestos jacket→

←Cork rests on the receiver

Figure 2.15 Apparatus for semimicro fractionation. The column may be secured with a wider top for insertion of a cork to hold the thermometer instead of the glass thermometer sleeve.

efficient than either the steel wool or glass helix type. For insulation an asbestos jacket is used. The asbestos jacket is 100–110 mm long and is made by winding several layers of asbestos paper, which have been painted with a sodium silicate solution and allowed to dry, around an 8-mm tube to a thickness of 4–5 mm. The side arm of the column is connected to the condensing-receiving system as described on page 28.

For a detailed discussion of the factors that determine the efficiency of fractionation the reader is referred to the literature given at the end of this chapter.[16] The most important of these are: (a) height of column; (b) nature of packing; (c) insulation; and (d) rate of distillation. Even with a column so designed as to meet the requirements in the first three categories, the efficiency of the fractionation rapidly diminishes if the withdrawal of distillate is at the rate of more than 0.3 ml per minute.

[16] See also Cheronis, *Micro Methods,* pp. 62–70.

It is advisable to wet the column with some of the mixture to be fractionated before distillation is begun; this is best accomplished by adding the liquid to be distilled slowly through the top of the column into the boiling vessel, from which the cork has been loosened to permit the escape of air. The various connections are inspected and adjusted, and heat is slowly applied to the boiling vessel with a small flame. In the beginning there will be a certain amount of refluxing within the column; when flooding appears on the top of the column, the flame is removed and the vessel is allowed to cool momentarily. Heating is resumed with the flame moved to and fro until the vapor begins to enter the side tube. The heating is then adjusted so that the rate of distillate withdrawal is 0.2–0.3 ml per minute. The receiving tubes may be calibrated at the 0.5-, 1-, 2-, and 5-ml marks with thin strips of gummed paper. If care is used, 5–10 ml of mixtures may be separated efficiently with a single fractionation.

The effect of the rate of distillate withdrawal on a routine fractionation of a 50 per cent methanol–water mixture is shown graphically in Figure 2.16.

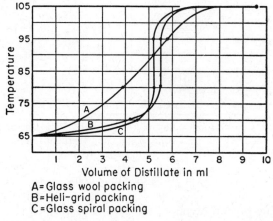

A = Glass wool packing
B = Heli-grid packing
C = Glass spiral packing

Figure 2.16   Effect of packing and rate of distillation on semimicro fractionation. Distillation curves of three 10 ml methanol–water (equal parts) mixtures. A. Glass wool packing, 0.5 ml/min. B. Heli-grid packing, 0.25 ml/min. C. Glass spiral packing, 0.25 ml/min.

In order to reduce the holdup of the column and apparatus, which often is 0.5–1.0 ml, a high-boiling compound or "chaser" is added to the mixture to be fractionated. The boiling point of the "chaser" should be at least 20° higher than the boiling point of the last fraction; in addition, the "chaser" should be relatively inert and not have a tendency to form azeotropes. Among the "chasers" commonly employed are: cymene (b.p. 175°); biphenyl (b.p. 254°); acenaphthene (b.p. 277°); and phenanthrene (b.p.

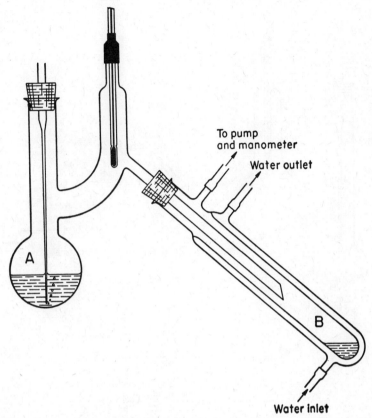

Figure 2.17 Apparatus for semimicro distillation at reduced pressure. The receiver has a built-in water jacket.

340°). The amount of "chaser" added should be a little more than the total estimated holdup.

Apparatus and Procedure for Distillation under Reduced Pressure

Figure 2.17 shows a simple arrangement for semimicro distillation under reduced pressure. It consists of a small (25–50 ml) Claisen flask, A, which has two necks, one for a capillary tube (which should reach almost to the bottom of the flask), and the other for the thermometer. The side arm of the distilling flask is 10–12 mm in diameter and 140 mm in length and is bent at an angle of 135° and fits through a one-hole rubber stopper into the receiver. The specially constructed receiver, B, is a 40 × 100 mm tube with a built-in condenser.[17] When this special tube is not

[17] Microware, Inc., Vineland, N.J.

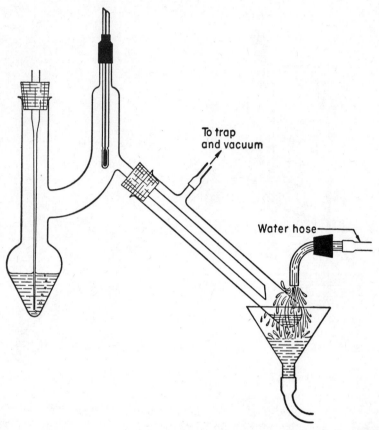

Figure 2.18   Apparatus for semimicro distillation at reduced pressure. The receiver is an 8-inch tube with side arm, cooled either as shown or by immersion in a cold bath.

available, the arrangement shown in Figure 2.18 is employed. This arrangement uses an ordinary Pyrex 8-inch tube with a side arm which is cooled by immersion in a cold bath or by a stream of cold water.

The diameter of the side arm tubing in most distilling flasks is 6–8 mm. In some semimicro and micro distilling tubes the side arm has been constructed of 5–6 mm tubing. This practice is not recommended, since at low pressures the volume of vapor passing through the side arm into the receiver is tremendously greater than the volume passing through during distillation at atmospheric pressure. Hence the side arm of distilling flasks used for vacuum distillations (for any scale: macro, semimicro, or micro) should be 8-, 10-, or 12-mm diameter.

It is assumed that the student is familiar with the procedure for vacuum

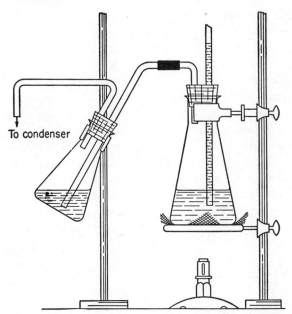

To condenser

Figure 2.19   Apparatus for semimicro steam distillation using two Erlenmeyer flasks.

distillation and use of manometers. In case of doubt it is advisable to review the directions in a text of experimental organic chemistry.[18]

Apparatus and Procedure for Steam Distillation

Figure 2.19 shows an assembly for steam distillation which is readily arranged from apparatus available in all laboratories since it employs two small Erlenmeyer flasks. The liquid to be steam-distilled is placed in the smaller flask and the rubber tubing that joins the steam inlet and steam injector is disconnected. Heat is applied to the large flask until the water rises in the gauge and steam issues from the steam-outlet tube. The flame is then removed momentarily, and, after about 30 seconds, the steam inlet is adjusted to the injector and heating is resumed. The flame is so adjusted that the splashing does not reach the middle part of the vapor outlet tube. An 8-inch tube is used in the receiving system, and, if the microcondenser is not sufficient to cool the vapor, a beaker of cold water is raised so as to surround the lower part of the receiving tube.

SUBLIMATION

In many cases sublimation is superior to crystallization as a method for the purification of very small quantities of solid substances, since it

[18] Cheronis, *Semimicro Chemistry*, pp. 78–87.

entails a minimum loss of material. It can be employed either for the purification of a single compound or for the separation of several components from a mixture. However, fractional sublimation is limited in application as a process of fractionation because there is difficulty in obtaining a series of multiple resublimations in the same vessel. Although, theoretically, any solid organic compound that can be distilled at atmospheric or reduced pressure without decomposition may also be sublimed, for practical purposes a solid that fails to give a sublimate at a pressure of a few microns when heated at a temperature 25–50° below its melting point for several hours is not considered to sublime operationally.

Fractional sublimation under reduced pressure (1–5 mm) is often more suitable for the purification of small quantities of many organic solids (1–50 mg) than crystallization. The apparatus shown in Figure 2.20 con-

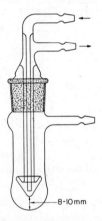

8-10mm

Figure 2.20 Microsublimator tube with ground-glass joint.

sists of a 25 × 100-mm tube which has a bulb of about 30 mm in diameter. The condenser has a conelike end (the upper part of which forms a well) and fits into the tube by a glass joint.

The sample to be sublimed is spread at the bottom of the bulb; the glass joint is greased lightly,[19] the condenser is adjusted in place and water is run through very slowly; heat is applied from a shallow bath into which the sublimator dips about 5–6 mm. For ordinary work a pressure of 5–20 mm is satisfactory. The temperature is raised gradually to 45–50° and kept in this range for about 30 minutes. If no cloudy film forms at the lower part of the condenser, the temperature is raised stepwise 10–15° and allowed to remain for 0.5–1 hour at each interval until a film of sublimate is obtained. After about 1 mg of sublimate has been formed, the vacuum is released gradually, the condenser is lifted over a glass slide, and (with the sharp end of the spatula blade) a few crystals are detached from the

[19] Celvasene Light, Distillation Products Industries, Rochester, N.Y.

center of the glass slide for microscopic examination, which is followed by determination of the melting point. The condenser is now washed with a thin stream of solvent into a watch glass or dish so that the material may be recovered after evaporation of the solvent. In this manner 6–8 fractions are obtained, and it is possible, on the basis of the melting point data, to decide upon a plan of fractionation. An illustration of fractional sublimation for the purification of derivatives is described in some detail on pages 377–81 of the larger work of the authors.

If the nature of the impure sublimand is known, and if it melts below 200°, the temperature is adjusted 10–15° below its melting point and fractions of about 1 mg are sublimed until the sublimate gives the desired melting point. In the purification of derivatives a rapid procedure is to fractionate one third of the sublimand, and afterward to collect a second fraction that is usually relatively pure.

If the desired compound melts above 200° and no information is available regarding the appropriate temperature and pressure for efficient sublimation, it is best to start with a pressure of 10–50 microns and a temperature of 100–150°. If the vacuum system gives a pressure of only 1–10 mm, it is advisable to raise the temperature to 150–180° and allow the sublimation process to proceed for 6–12 hours or more.

EXTRACTION[20]

Extraction for the separation of organic compounds from mixtures is usually accomplished by shaking the mixture in a separatory funnel with a pair of immiscible solvents. Ether, benzene, and chloroform are the most commonly used solvents to extract organic substances from aqueous solutions or dispersions. For compounds with appreciable water solubility, esters (such as ethyl acetate) or higher alcohols (1-butanol or pentanols) are more efficient. The choice of the solvent depends on considerations of solubility and structural relations between solvent and solute. For example, chloroform is a better "extraction solvent" than carbon tetrachloride for alcohols, esters, ketones, and aldehydes because it forms hydrogen bonds.

It is possible to effect a fractionation of small quantities of several organic compounds contained in a mixture through (a) adjustment of pH, (b) selection of the solvents, and (c) addition of other solutes that will alter the solubilities of one or more components. Assume that it is desired to separate a small amount of fatty acids from a mixture that contains mostly sterols. If the mixture is shaken with dilute alkali, the fatty acids are converted to their sodium salts, which are very soluble in water, and

[20] For an extensive discussion of extraction techniques, see Craig and Craig, "Extraction and Distribution," in Vol. III of Weissberger (ed.), *Technique of Organic Chemistry* (2nd ed., N.Y.: Interscience, 1956).

thus separated from the sterols, which now can be removed by one or two extractions with benzene. The alkaline aqueous phase is adjusted to a pH below 7.0 and subjected to extraction either with chloroform or with ether. Similarly, a mixture of aldehydes can be separated from other organic compounds by converting them first to bisulfite addition products, which are soluble in water, and then by raising the pH of the aqueous phase to decompose the addition product and subjecting it to extraction.

Separatory funnels of 25- or 50-ml capacity are used for extraction of small amounts of aqueous solutions. Extraction of 1–2 ml of solutions may be accomplished with minimum losses without the use of a separatory funnel. The solution is placed in a 3-inch tube and, after shaking with 1–2 ml of solvent, the separation of the two layers is accomplished by means of a pipet having a long capillary tip (see Appendix, Figure A-1). Ether, benzene, or other solvent is added to the solution and the tube is stoppered with a solid rubber stopper. The thumb of the hand holding the tube is placed on top of the rubber stopper so as to hold it down firmly, and the contents of the tube are shaken. Then the pressure of the thumb is gradually released with the tube held at an angle and with the opening toward the hood, the stopper is removed, and the tube is placed on a rack or stand, until the two layers separate. The tube is then inclined slightly and the desired layer is removed by several withdrawals with the capillary pipet dropper and transferred into another tube (Figure 2.21). The bulb of

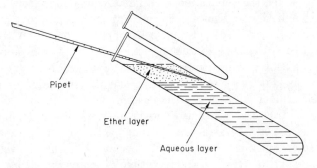

Figure 2.21 Extraction and separation by means of a capillary pipet.

the dropper is pressed and the capillary tip is inserted into the tube until it reaches 1–2 mm above the junction of the two liquids. The pressure on the bulb is then released gradually until the desired layer has been withdrawn. Several extractions of the solution can be performed in this manner. In some extractions, when it is desirable to increase the contact between the two phases, the pipet dropper is used to suck up one phase. Then place the tip at the bottom of the tube and expel the liquid vigorously. This is repeated several times for a period of 3–5 minutes in order to ensure complete extraction.

Exercises

1. Outline a practical procedure by which the purity of a sample assumed to be an organic compound can be ascertained: (a) if it is a solid; (b) if it is a liquid.

2. Outline a procedure by which it is possible to arrive at a conclusion as to whether an unknown is a relatively pure substance or a mixture: (a) if the unknown is a solid and the available quantity of material is 3 g; (b) if the unknown is a liquid and the volume is 5 ml; (c) if the unknown sample is a solid and its weight is 300 mg.

3. Consult a text on physical chemistry and derive from the Clapeyron equation an expression which shows that the vapor pressure of a liquid is an exponential function of the absolute temperature.

4. An unknown impure liquid compound is found to be thermolabile. Outline a procedure by which it can be purified.

5. Assume that 5 g of commercial p-dichlorobenzene is to be purified. Indicate (a) selection of a suitable solvent; (b) procedure to be followed so as to obtain crystals of high purity with minimum loss of material.

6. A 3-ml sample of a liquid organic base with K_b 8×10^{-5} is to be purified. Indicate a procedure by which purification can be effected with the minimum loss.

7. An organic compound is very soluble in methanol and acetone but sparingly soluble in water, benzene, and ether. Outline two procedures for its purification.

8. Which of the following conditions should be chosen so as to obtain the most pure crystals: (a) use of a solvent in which the impure compound is very soluble and subsequent addition of a solvent in which the compound is sparingly soluble until crystals appear with rapid cooling; (b) same conditions as in (a) except that after the appearance of crystals more of the solvent is added and the mixture is heated, then rapidly cooled; (c) same conditions as in (b) but after heating the solution is covered and allowed to cool very slowly. Give reasons for the answer.

9. What factor must be carefully controlled in a given distillation by means of a fractionating column assembly in order to obtain efficient fractionation?

10. What compounds are best removed from a mixture by means of steam distillation?

11. Assume that 1 g of an organic substance is dissolved in 12 ml of water and the distribution constant of the substance between water and ether is 3.8. Calculate how much of the substance will be removed by (a) one extraction with 10 ml of ether. (b) by two extractions with two 5-ml portions of ether.

3

Determination of Physical Constants in Analysis of Organic Substances

Melting Points

The melting point of a pure solid organic compound is used as a criterion of its purity. In the systematic identification of organic compounds the melting point of the derivative that is prepared serves as the most important constant in the proof of identity.

For the determination of melting points the *capillary-tube* method is commonly used. About a milligram (or less) of the solid is placed in a thin glass capillary tube having a diameter close to 1 mm. The capillary tube is attached to a thermometer, then placed in a liquid bath, and heated slowly. The interval of temperature in which the solid within the capillary tube begins to liquefy and the temperature at which the liquid is clear is recorded as the *observed melting range*. When the values are corrected they are called *corrected melting points*. It should be noted that melting points determined by this method are not *true melting points* but *capillary melting points;* the latter are slightly higher than the true melting points, which are determined by cooling or heating curves;[1] these require larger samples but give more exact information as to the purity of the compound. For all ordinary purposes the capillary-tube method is used. The amount of substance required for a single determination by the capillary-tube method is usually 1–2 mg, although a fraction of this amount may be used.

Other methods for the determination of melting temperatures of solids

[1] Skau and Wakeham in *Physical Methods of Organic Chemistry,* Vol. I of Weissberger (ed.), *Technique of Organic Chemistry* (2nd ed., N.Y.: Interscience, 1949), p. 70.

are *heating bars* and the *microscope heating stage*.[2] In the former, the crystals are heated on a metal bar whose temperature is determined either by a thermometer or thermocouple; in the latter, a few crystals weighing a fraction of a milligram are placed on an electrically heated stage and the temperature at which the crystals melt is observed. With this method it is possible, in most cases, to observe the temperature at which liquid and fragments of crystals coexist and, for this reason, *corrected micromelting points* should be differentiated from corrected *capillary melting points*.

Several factors determine the precision and accuracy of the measurement of melting temperatures. Although it is possible to obtain a precision of 0.5° with ordinary thermometers, and with specially constructed thermometers a precision of 0.1°, the accuracy depends primarily on the calibration of the device by which the temperature is measured at the region where the crystals are situated.

Construction and Filling of Melting-Point Tubes

Glass capillaries of uniform diameter (1–1.2 mm) and 70–75 or 100 mm length are commerically available packed in vials; the authors prefer capillary tubes of about 100–120 mm length.[3] It is recommended, however, that beginners learn to prepare glass capillaries by heating and drawing out glass tubing that has been thoroughly *cleaned*. A soft glass test tube 12–16 mm in diameter, or any thin-walled glass tube 6–8 mm in diameter and 16–20 cm in length, is rotated in the hottest part of the Bunsen flame. When soft, it is removed from the flame and drawn out slowly in such a manner as to insure a uniform capillary bore; this bore should be about 1 mm in diameter. The capillary is cut into lengths of 90–120 mm, and one end is sealed by heating for a few seconds in the outside tip of the flame. Care should be taken not to make the sealed end thick. The capillaries are placed in a dry test tube, which is tightly corked to keep out moisture and other impurities. Consideration should be given to the possibility that alkali-sensitive compounds are affected by the alkalinity of the soft glass and may give low melting temperatures.[4] In such a case capillaries from Pyrex glass should be employed.

To load the capillary tube, a few milligrams of the crystalline material

---

[2] Cheronis and Entrikin, *Semimicro Qualitative Organic Analysis* (2nd ed., N. Y.: Interscience, 1957), pp. 118–24; Cheronis, *Micro and Semimicro Methods*, Vol. VI of Weissberger (ed.), *Technique of Organic Chemistry* (2nd ed., N.Y.: Interscience, 1957), pp. 157–80.

[3] Microware, Inc., Vineland, N.J.

[4] Jones and Wood, *J. Am. Chem. Soc.* **63**, 1760 (1941); Jones, *Ind. Eng. Chem., Anal. Ed.* **13**, 819 (1941); Norton and Hansberry, *J. Am. Chem. Soc.* **67**, 1610 (1945); George, *Helv. Chim. Acta* **15**, 924 (1932).

are placed on a watch glass or piece of clean paper and crushed to fine powder by drawing the spatula over them. The open end of the capillary tube is pressed into the fine powder; then the closed end is tapped on the desk, or the tube is lightly scratched with the flat part of a file, in order to force the sample to the bottom. The tube is filled to a height of about 1–2 mm and is then attached to the thermometer so that the end of the capillary tube reaches the middle of the mercury bulb. If an oil is used as a bath, the capillary tube is attached to the thermometer by means of a small rubber band cut from ordinary $\frac{3}{16}$-inch tubing. The rubber band is so adjusted that it is near the top of the capillary tube and does not come in contact with the liquid bath. The rubber band has the disadvantage that on heating it may come in contact with the bath liquid and color it; therefore, it is advisable to use new rubber bands for every determination.[5] A fine copper wire wound several times around the thermometer and capillary has been successfully used by the authors. Another method suggested in the literature is to omit the rubber band and attach the capillary by placing a glass rod against the entire length of the thermometer above the bulb; the capillary is fitted in the groove formed by the rod and thermometer. If sulfuric acid is used as a bath the rubber band is unnecessary, as capillary attraction will hold the melting-point tube to the thermometer.

Procedure for Observing Melting-Point Temperatures

Figures 3.1 and 3.2 show a variety of liquid heating baths for observing the melting temperatures of solids placed in capillary tubes. Two of these

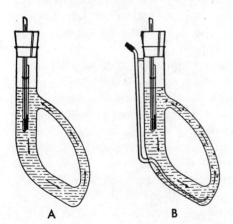

A   B

Figure 3.1 A. Thiele tube for determination of melting points. B. Modified Thiele tube for melting-point determinations.

(Figs. 3.1B and 3.2B) can be connected to an air-supply line or to an aspirator bulb; thus air can be blown at a very slow rate (so that bubbles

[5] Silicone fluids do not affect rubber or metal bands.

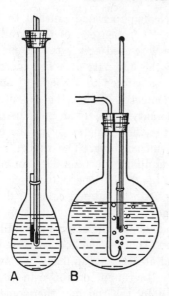

Figure 3.2 A. Kjeldahl flask (25-ml) assembly for melting-point determinations. B. Round-bottom flask assembly for melting-point determinations.

can be counted) to provide better circulation. The liquid commonly used for this type of bath (up to 230°) is a grade of heavy petroleum oil (mineral oil). If a sulfuric acid–potassium sulfate mixture is to be used, special instruction must be obtained from the instructor. The thermometer is arranged in the apparatus so that the lower end of the capillary is clearly visible. If a rubber band is used to hold the capillary to the thermometer, it is so adjusted that it is out of the liquid. In the Thiele apparatus the oil level is about 10–15 mm above the circular side tube and the thermometer 15 mm below it, so that the latter is near the mid-point between the upper and lower side arms.

When the thermometer bearing the capillary tube has been adjusted, the tube or flask is heated rapidly to about 10–15° below the known melting point of the substance. If the substance is an unknown, the approximate melting point is first determined by heating fairly rapidly until the substance has melted. The bath is then allowed to cool to about 20° below the observed melting point; the thermometer is carefully removed and held until it has acquired the temperature of the room; then a new loaded capillary tube is inserted. The thermometer is replaced and the bath heated until the temperature rises to within 10–15° of the melting point. The flame is removed until the temperature begins to drop. The heating is then resumed at such a rate that the temperature rises 2–3° per minute, the liquid bath being stirred so that the temperature in the various parts of the apparatus will be as uniform as possible. When the temperature comes to within 2–4° of the melting point, a rise of 1° per minute is desirable. It should be stressed that the slower the rate of heating the greater is the precision and the rate for the last 1° should be 1° each 2–3 minutes. The temperature at which the substance begins to liquefy and the temperature at which the liquid is clear are noted. This interval of temperature is recorded as the melting-point range of the substance. If the compound melts without decomposition, it is suggested that a second and a third observation be made by removing the thermometer from the bath, holding it in air until the liquid in the capillary solidifies, then repeating the melting-point determination.

A number of compounds which have bonds of partial ionic character (amino acids, quaternary ammonium salts, and the like) melt with decomposition, and the melting points depend on the rate of heating. In such cases it is advisable after the first exploratory determination to make a second determination by first preheating the bath to within 10° of the first observed melting point and then inserting the thermometer with the capillary; then the bath is heated rapidly so that the rise of temperature is 8–10° per minute. The decomposition temperature is noted, and the initial temperature of the bath and the rate of temperature rise are reported.

Calibration of Thermometers

The accuracy of melting-point determinations by the capillary-tube method, and by almost every method in which a thermometer is used, depends to a large extent on the calibration of the instrument by which the temperature is measured. The procedure recommended for the calibration of thermometers is to use a thermometer calibrated by the manufacturer

TABLE 3.1

Tentative List of Primary Standards for Thermometer Calibration
Employed in Melting Point Determinations

Substance	Melting point (°C)	Substance	Melting point (°C)
Water-ice	0	Urea	132.8
Cyclohexanol	25.45	Salicylic acid	158.3
Menthol	42.5	Succinic acid	182.8
Benzophenone	48.1	Anthracene	216.18
p-Nitrotoluene	51.65	Phthalimide	233.5
Naphthalene	80.25	p-Nitrobenzoic acid	241.0
Acetanilide	114.2	Phenolphthalein	265.0
Benzoic acid	122.36	Anthraquinone	286.0

Source: Cheronis, *Micro and Semimicro Methods*, Vol. VI of Weissberger (ed.), *Technique of Organic Chemistry* (N.Y.: Interscience, 1954), p. 156.

by partial immersion, and then to calibrate it by means of reference standards in the apparatus that is to be employed. A thermometer is selected in which 1° is equivalent to 1–1.2 mm and calibrated by partial immersion. If greater precision is desired, two thermometers, one reading from 0–180° and the other from 150–320°, with subdivision in 0.5°, are employed.

The thermometer to be calibrated is first heated in an oven for 6–8 hours at about 250–300°, which is higher than the temperatures to which they are commonly exposed. The thermometer is allowed to stand at room temperature for 2–3 days and is then calibrated by the reference standards listed in Table 3.1, *in the same apparatus and with the same technique that is employed for the melting-point determinations*. Three determinations or more are made for each fixed point; the deviation from the average value should not exceed 0.5°. Average values are used to plot the calibration curve from which the correction to be applied to the observed melting points may be read directly. Figure 3.3 shows calibration curves for three thermometers.

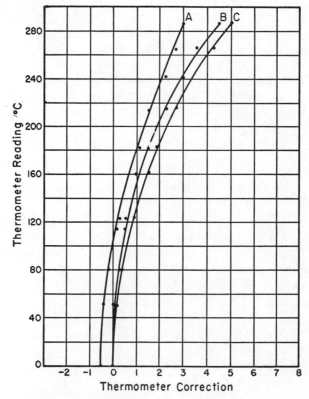

Figure 3.3 Calibration curves of three thermometers.

Evaluation of Melting-Point Data

Melting-point data constitute one of the main criteria employed in the determination of the purity of solid organic compounds and, also, the sole criterion in the final systematic identification of an unknown organic substance. The last step in any systematic scheme of characterization involves the preparation of one or two derivatives and comparison of the melting-point data of these derivatives with the values listed in the literature. For a critical discussion of the compilation of melting points of derivatives listed in Tables 1–29 (page 345 ff.) the reader is referred to the larger work of the authors.[6] It should be pointed out that in some cases the melting points of derivatives prepared by the student will not agree with the melting points listed in the literature. Differences of 1–2° between observed melting temperatures of pure derivatives as prepared by research workers and the values listed in the literature are common. Hence it is a common practice among experienced workers in the identification of organic substances to use the method of mixed melting points. In many schools, however, this practice is not encouraged or even permitted when "student unknowns" are being identified, since it is desirable that the beginner prepare more than one derivative. Therefore, for students the procedure described in this section should be modified in accordance with directions by the instructor. On the other hand, the experienced analyst relies on this procedure to eliminate possibilities in "identity" or to introduce short cuts in routine characterizations.

Mixed Melting-Point Method

To illustrate the use of mixed melting points in characterization work assume that an organic liquid under investigation boiling at 106–108° is provisionally identified as isobutyl alcohol. One of the derivatives prepared for the final proof of the identity is the 3,5-dinitrobenzoate; the melting points of the crystals of this derivative were 84–85° after the first crystallization and 85–86° for the second crystallization. The melting point of the 3,5-dinitrobenzoate of isobutyl alcohol is listed in Table 6A (Appendix) as 87°. For a mixed melting point the 3,5-dinitrobenzoate of a known sample (100–200 mg) of isobutyl alcohol is prepared, and the melting point of the crystals is determined. Approximately equal amounts of the crystals of the dinitrobenzoate derived from the "known" and the "unknown" samples are thoroughly mixed by crushing in a mortar or a watch glass and the melting point of the mixture is determined. If all three melting points are essentially the same, or if the melting point of the mixture lies between

[6] Cheronis and Entrikin, op. cit., pp. 127–30.

that of the two dinitrobenzoates, the "unknown" is identified as isobutyl alcohol. If the unknown is not isobutyl alcohol, the melting point of the mixture of the two dinitrobenzoates will be at least 10° or more below that of the components and the melting will not be "sharp" but will soften and melt gradually over a range of several degrees.

The basis of the above method is that a mixture of two unlike crystalline substances will melt at a considerably lower temperature than either of the individual components alone, owing to the formation of a eutectic. It should be remembered, however, that there are cases of unlike crystalline substances that show a higher melting point than either of the two components because of the formation of a new compound. In a number of instances two different compounds melting a few degrees apart may show no depression in melting point when mixed. Thus, naphthalene picrate, m.p. 151°, and thionaphthalene picrate, m.p. 194°, melt when mixed at 149°,[7] and D-dimethyl tartrate, m.p. 48°, and L-dimethyl tartrate, m.p. 43.3°, melt when mixed in equal proportions at 89.4°.[8] It is also well known that two different organic substances, when mixed in different ratios, may form two or more eutectics, which melt considerably below either component, and one or more molecular compounds, which may melt higher than either component. Therefore, the "mixed-melting-point method" should only be used in conjunction with other pertinent data.

From the above considerations it is obvious that failure to observe a lowered capillary melting point in a mixture of two derivatives prepared from a known and an unknown is not a reliable proof of identity unless all other data—solubility tests, functional group tests, and physical constants—are in agreement. With this reservation, the practice of taking mixed melting points, using samples of derivatives from the unknown and the compound tentatively identified as the unknown, may be resorted to whenever two successive crystallizations fail to produce a rise of more than 2° in the melting point of the derivative. The following practice is recommended for beginners.[9]

Prepare one or better two different derivatives of the probable compound. From each derivative save a few milligrams and recrystallize the rest. If the difference between the first and second lots of crystals is not more than 1–2°, check the melting point of the same derivative of the probable compound as listed in the tables. If the difference is not more than 2–3°, prepare the same derivative from a pure sample of the compound

[7] Lock and Notes, *Ber.* **68**, 1200 (1935).

[8] Adriani, *Z. physik. Chem.* **33**, 453 (1900); Gibby and Waters, *J. Chem. Soc.*, (1931) 2151.

[9] In college and university laboratories when "student unknowns" are being identified, this practice is to be modified in accordance with directions from the instructor.

tentatively identified as the unknown, using the same quantities of reagents, procedures, and crystallizations as those used in the preparation of the derivative of the unknown. Mix a few milligrams of each of the two derivatives obtained from the unknown and from the pure sample of the compound tentatively identified as the unknown. If the mixture does not show a variation of more than 1° from the melting point of either component alone, the proof of identity may be considered conclusive. It is recommended that beginners prepare two different derivatives; only after considerable experience has been gained should the conclusive identification be based on the preparation of one derivative and a mixed melting point. A confirmatory method called *mixed fusion*, described in the next section, is considered by many research workers as superior to that of mixed melting points in the proof of whether two derivatives, *A* and *B*, are identical.

An explanation for the above practice may be in order at this point. When semimicro quantities are used, the amount of derivative available after one or two crystallizations is usually 50–100 mg if one begins with 100–200 mg of the unknown; but in some cases it may require a total of four or five crystallizations to obtain a derivative that has the melting point shown in the literature. In other cases, for reasons pointed out above, *the melting point given in the literature will not be obtained, no matter how many recrystallizations are performed.* Therefore, if all other evidence from solubility data, functional-group tests, and physical constants fits a particular probable compound, the procedure outline in the above-recommended practice with reference to derivatives is regarded as sound.

Mixed Fusion Technique for Proof of Identity

When two substances that are not identical (and also not isomorphous) are melted on a glass slide side by side and are allowed to mix, crystals of each pure substance will grow from either side up to the interface where, at the mixing zone, the rate of crystal growth will decrease and finally cease, leaving a thin region of the melt, which is either a *eutectic* or an *addition product* of the two substances, and which may or may not solidify. In most cases, however, the mixing zone melt solidifies. If such a slide is then heated very slowly by passing it over a minute flame, the mixing zone melts first and appears even to the naked eye as a miniature stream.

If the two substances are identical, there will be no zone of mixing, and crystals will grow throughout the sample until the melt has solidified. If such a slide is heated slowly by passing it over a minute flame until the sample melts and is cooled rapidly by placing the slide on a metal surface, the melt will solidify almost instantaneously.

If the two substances are not identical but are isomorphous (this occurs only infrequently), on mixing the two melts will form a solid solution; there will be a gradually increasing rate of growth up to the inter-face, where the rate will decrease and crystal growth will continue through the zone of mixing; then it will increase again as the second substance solidi-fies. If after the mixture solidifies the slide is heated slowly by passing it over a minute flame and then is cooled rapidly by placing it on a metal surface, there will not be rapid solidification as when the two substances are identical.

The apparatus necessary for these observations consists of regular glass slides and cover glasses and a lens or a microscope with a 32-mm or 40-mm objective (an objective of shorter working distance is not suitable because it would be injured by the heat from the preparation). The inexpensive microscope with plastic lenses ($4.00–$7.00) is suitable for this purpose. It can be converted to a polarizing instrument by sealing a small piece of Polaroid film (using Scotch tape) under the opening of the light source and a small round Polaroid film inside the eyepiece. The regular microscope can be converted to a polarizing instrument by inserting a polarizer in the condenser and a cap-type analyzer on the eyepiece[10] or with the Polaroid film as described. In either case the eyepiece can be rotated to obtain various color patterns of the crystals and, though not indispensable, is of aid in mixed fusion.

The steps involved are illustrated in Figure 3.4. About 2 mg of substance A is placed under the cover glass, melted carefully over a micro-flame, and allowed to solidify (if a microburner is not available, use a Bunsen burner and remove the barrel to obtain a minute flame). Substance B is then placed at the end of the cover glass (Figure 3.4, far left) and

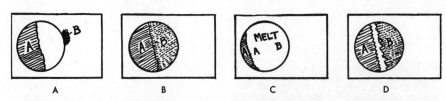

Figure 3.4 Steps in mixing fusion. *Far left:* Substance A melted and allowed to crystallize, then substance B placed near the cover glass. *Left center:* Substance B melted and allowed to run under the cover glass and crystallize. *Right center:* All of B and part of A remelted to form a zone of mixing. *Far right:* Substances A and B recrystallized to zone of mixing. (After McCrone.)

heated so that it melts and runs under the cover glass and comes into con-tact with A (Figure 3.4, left center). The slide is reheated so that all of substance A and some of substance B melt (Figure 3.4, right center). The

---

[10] Bausch and Lomb Optical Co., Rochester, N.Y. Cost about $25.00–$30.00. The cost of the Polaroid film is less than $1.00.

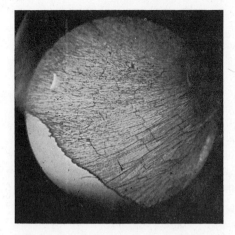

Figure 3.5 Crystallized melt from the fusion of two derivatives (anilides) which have the same melting points, 104–105°. One derivative was prepared from an unknown tentatively identified as isobutyric acid; the other derivative was prepared from a sample of known pure isobutyric acid. The melting point of the anilide of pure isobutyric acid is given in the literature as 106° and 103°. There is no zone of mixing, and the two derivatives are considered identical without determination of mixed melting points. Compare with Figure 3.6.

slide is then examined by means of the lens or under the microscope, using polarized light if possible.

For practice the beginner may use two samples of acetanilide for A and B as an example of two identical substances, and acetanilide and phenacetin as an example when A and B are not identical.

Figure 3.5 shows a photograph of the solidified melt from the fusion of small samples of two compounds. One, melting at 105°, is the anilide of isobutyric acid, and the other, melting at 104°, is the anilide of an unknown tentatively identified as isobutyric acid. Note that there is no zone of mixing (A and B are identical). Figure 3.6 shows the melt from the fusion of the anilide of the unknown tentatively identified as isobutyric acid and the anilide of propionic acid (m.p. 105°). Note that although both compounds melt alone at 105°, there is a eutectic at the zone of mixing (A and B are

Figure 3.6 Crystallized melt from the fusion of two derivatives (anilides) which have the same melting point but are not identical. The crystals to the right are from the anilide of an unknown tentatively identified as isobutyric acid (Figure 3.5) and those to the left from the anilide of propionic acid. Both anilides as prepared and purified gave a melting point of 104–105°. A eutectic forms at the zone of mixing.

not identical). When such a sample is heated slowly by passing it over a minute flame, the eutectic zone melts first, appearing as a small stream. Thus for the experienced worker this proof of nonidentity is simpler and more rapid than with the mixed melting-point method.

Boiling Points[11]

As pointed out in Chapter 1, the boiling point, together with the refractive index, is employed as a criterion of purity of liquid organic substances. Since the boiling point varies with atmospheric pressure, a correction must be applied to the observed value. For small deviations from 760-mm pressure the correction is a fraction of a degree. The rise in boiling points per millimeter increase in atmospheric pressure is approximately one ten-thousandth of the boiling point expressed in absolute degrees. The boiling point of benzene at 760 mm is 80.1°. The correction, therefore, for every millimeter deviation from 760 mm will be $(273 + 80) \times 0.0001 = 0.0353°$. For example, a sample of benzene, when boiled under a pressure of 750 mm, gave 79.7° as the observed boiling point. The correction is $(760 - 750) =$ 10 mm $\times 0.0353° = 0.35°$; the corrected boiling point is $79.7° + 0.35° =$ 80.05°. It should be noted that this rule is only approximate and that it does not apply to pressures that are far removed from 760 mm.

In regions of high altitude and low barometric pressure the correction to be applied is of greater magnitude. However, the use of reference standards obviates the necessity of considering precise determinations of pressure and can be considered as the best method for obtaining reliable values on the boiling points of "unknowns." This method gives an accuracy of 0.5–0.1° for most determinations and may be employed without complex apparatus. The boiling temperature of the liquid under investigation is determined by one of the methods described in this chapter. Immediately afterward a determination is made of the boiling temperatures of the reference standard (see Table 3.2) which is closest in structure and boiling point to the liquid under investigation. The difference between the boiling points of the reference substance measured under standard and under experimental conditions is used to correct the boiling point of the substance under investigation.

Assume, for example, that a compound boils at 84.5°. Under the same conditions the boiling point of benzene, which is the reference standard, is 79.5°; the boiling point of benzene at 760 mm is 80.1°. Hence, the corrected boiling point of the substance under investigation is 84.5° +

[11] For detailed discussion of the boiling temperature of pure liquids, difference between boiling and condensation temperatures, and methods for their measurement, consult *Physical Methods of Organic Chemistry*, Part I, of Weissberger (ed.), *Technique of Organic Chemistry* (3rd ed., N.Y.: Interscience, 1959) pp. 357-91.

TABLE 3.2

Primary Standards for Determinations of Boiling Temperatures

Substance	Boiling point (°C)	Substance	Boiling point (°C)
Ethyl bromide	38.40	Cyclohexanol	161.10
Acetone	56.11	Aniline	184.40
Chloroform	61.27	Methyl benzoate	199.50
Carbon tetrachloride	76.75	Nitrobenzene	210.85
Benzene	80.10	Methyl salicylate	222.95
Water	100.00	p-Nitrotoluene	238.34
Toluene	110.62	Diphenylmethane	264.4
Chlorobenzene	131.84	α-Bromonaphthalene	281.2
Bromobenzene	156.15	Benzophenone	306.10

Source: Cheronis, *Micro Methods*, p. 189.

$0.6° = 85.1°$. For more precise work the boiling temperature of the compound under investigation is determined at a specified pressure at which the boiling temperature of the reference compound is known with an accuracy of 0.01° or less. A Beckmann microthermometer and a manostat are required for such determinations.

Most methods for the microdetermination of the boiling points employ the same apparatus as that used in the determination of melting points. Of the many procedures that have been proposed, one will be described.

Microdetermination of Boiling Points

The apparatus, shown in Figure 3.7, consists of a "boiler" tube, which contains the liquid for boiling-point determination and an inverted capillary, attached to a thermometer. The boiler is constructed of glass tubing 3–4 mm in (outside) diameter, 80–100 mm in length, and sealed at one end. The inverted capillary is made from a melting point capillary (about 120 mm in length) which is sealed about 30 mm from the end by placing it carefully at the outer flame and pushing inward gently from the ends. The open short end is cut carefully about 10 mm from the seal (Figure 3.7B). If the tube is bent in making the seal of the capillary, the tube will not go down into the boiler. If difficulty arises in making the seal, proceed as follows. Seal a capillary tube at one end so that there is no opening remaining. Bring the closed end of the capillary near the flame and seal it to the end of a second capillary. Now cut the capillary melting-point tube the desired distance from the sealed end.

Using a capillary pipet place about 2–3 drops of liquid in the boiler,

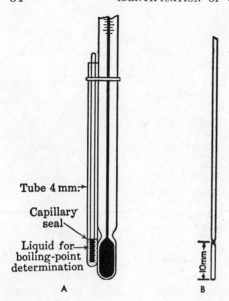

Tube 4 mm.

Capillary seal

Liquid for boiling-point determination

A

10 mm

B

Figure 3.7   A. Boiling-point setup with inverted capillary. B. Construction of inverted capillary.

filling it to a height of 6–8 mm. Insert the capillary, open end down, in the liquid contained in the boiler. Attach the boiler to a thermometer (Figure 3.7A) and immerse in the liquid bath of the melting point apparatus. Heat gradually with a small flame. Bubbles of air trapped in the inverted capillary will force their way out as the temperature rises. When the bubbling becomes rapid, discontinue heating, and allow the temperature to drop 1–2°. As the temperature drops, the liquid recedes in the capillary and completely fills the chamber of the melting point tube. Raise the temperature at the rate of 1° per minute until the vapor of the liquid fills the chamber and begins to issue out as a steady stream of bubbles. Stop heating and record the temperature. When the bubbles stop emerging and the liquid begins to recede in the chamber, note the temperature again. This interval of temperature, which is very small in the case of pure liquids, is the boiling point.

Refractive Indices

The absolute refractive index of a substance represents the ratio of the velocity of light in a vacuum to that in the substance. For the refractive index as commonly determined, air is used as the standard of comparison. The index is denoted by the letter n, with a superscript indicating the temperature of observation and a subscript denoting the wavelength of light used. Thus the refractive index of water $= n_D^{20} = 1.3330$ and $n_D^{50} = 1.3290$. The refractive index decreases as the temperature rises. The variation due to the temperature effect is different with various substances,

but it may be approximated by taking the variation as 0.0004 per degree centigrade. ⁻

The refractive index is one of the most important physical constants of organic compounds and can be determined accurately. As a criterion of the purity of liquids, it is more reliable than the boiling point. The determination of refractive indices is useful for the identification of an unknown organic compound. If the compound is pure, it is very valuable (in conjunction with the boiling point) in excluding other compounds from consideration and often indicates the identity of the unknown. For reviews and more detailed description of the refractometric methods, the reader should consult the literature given in the bibliography at the end of this chapter.

As mentioned in Chapter 1, the refractive index is of value in the examination of the composition of liquids. In most cases examination of the refractive index of several fractions from the distillation of the liquid will indicate whether the analyst is dealing with a mixture or a pure substance. In the fractional distillation of a mixture of two or more liquids, the refractive index may be used to determine the composition of the distillate when the components of the mixture have boiling points that are close together. Finally, the refractive index, together with the boiling point or melting point of the substance, density, and other physical data, are of aid in restricting the possibilities as to the probable nature of the unknown. In many cases, particularly in dealing with isomeric substances having boiling points close together, the refractive index is invaluable in eliminating several of the isomers.

Refractometers and Procedures for Liquids

No matter what instrument or method is used in the determination of refractive indices, it is essential to check the instrument and procedure by means of reference standards as described in the following section.

Several types of refractometers are used for determining the refractive indices of liquids. The Abbe refractometer is widely used because it employs only a few drops of the material and requires but a few minutes for the determination. The Fisher refractometer is particularly adapted to the student's use. The Abbe refractometer, shown in Figure 3.8, consists of: (a) a pair of rotating water-jacketed prisms hinged together; (b) an observing telescope above the prisms for observing the border line of the total reflection that is formed in the prism; and (c) a sector on which the telescope is fastened. The sector is graduated from 1.300 to 1.710 and permits direct reading of the index of refraction; it is adjusted to the sodium D line of the spectrum.

To operate the refractometer, one adjusts the thermometer in place,

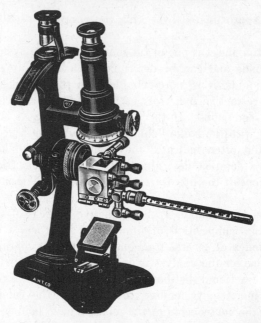

Figure 3.8   Abbe-Spencer refractomer, range nD, 1.300–1.710. Readings can be estimated to the fourth decimal. (Courtesy A. H. Thomas Co.)

and connects the base of the prism enclosure with a reservoir containing water at 20°. The left thumb is placed on the sector, and the right hand is used to open the double prism by pulling down the screwhead that is fastened on the lower prism. The prisms are wiped off carefully with a sheet of lens paper (facial tissue paper is well adapted to the same purpose). If the liquid is of a kind that does not evaporate rapidly, 2 drops placed on the face of the lower prism, which is closed immediately by means of the screwhead and clamped against the upper prism. If the liquid is volatile, the prism is closed, the screw is slackened, and with a pipet dropper a few drops of the liquid are poured into the depression on the side of the prism leading through a narrow channel into the space between the two prisms. The screw is tightened and the prism is rotated by moving the arm at the side of the sector. The mirror is adjusted to obtain maximum illumination, and the cross hairs are brought sharply into focus. As the prism is rotated, the illuminated field is partly darkened by a shadow moving across it. The boundary between the dark and the light field is called the *dividing line*, or *border line*. If the dividing line is not sharp but is hazy with a band of colors, the compensator wheel, which is at the lower part of the observing telescope, is moved until the dividing line becomes sharp and colorless.

The arm of the prism is moved until the dividing line coincides with the intersection of the cross hairs. The refractive index is read through the small eyepiece over the sector. The pointer indicates the last three figures of the refractive index. The first two are read on the left side of the scale a small distance below the pointer. The double prism is opened and cleaned with a piece of lens paper and a few drops of acetone.

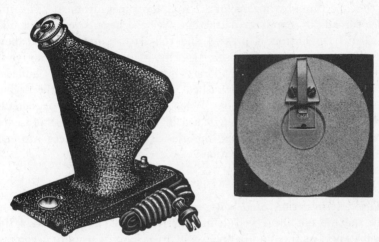

Figure 3.9 Fisher refractomer (*left*) and eyepiece (*right*). (Courtesy Fisher Scientific Co.)

The Fisher refractometer is shown in Figure 3.9. A small glass slide with a beveled edge is fitted on the glass plate of the eyepiece, as shown at the right. A small clamp holds the glass slide down so that a prism-shaped well is formed between the plate of the eyepiece and the glass slide. The instrument has a cord by which it is attached to an electrical outlet. If the push button on the base of the instrument is pressed, and observation is made through the aperture of the eyepiece, an illuminated scale appears. The graduations on the scale are divided at 1.516 by an arrow. This point corresponds to the refractive index of the glass in the eyepiece.

When a very small drop of a liquid is placed in the well formed by the glass slide and glass plate of the eyepiece, the refraction of light passing through the prism of the liquid sample produces a secondary or virtual image of the arrow on the scale. A liquid having a refractive index less than that of the glass employed will cause the light to bend downward, and the secondary image will appear above 1.516 on the scale. Conversely, if the refractive index is higher, the bending is upward, and the secondary image appears below 1.516.

For operation of the Fisher refractometer, the small glass slide is removed from the box and cleaned with lens paper. Likewise, the plate of the eyepiece is wiped clean. The glass slide is placed with the beveled edge downward over the plate so that it just covers the aperture. The plug of the instrument is connected to an electrical outlet. A very small amount (0.01–0.05 ml) of the liquid is added at the edge of the glass slide directly over the aperture, by means of a capillary pipet dropper. The minute droplet spreads by capillary attraction over the beveled edge and fills the well. The push button is pressed and the second arrow either above or below 1.516 on the scale is observed. This reading is the refractive index of the liquid. If the liquid evaporates rapidly, the droplet is added with one hand while the button is pushed with the other, and the secondary image is observed immediately. When the determination is completed, the glass slide is cleaned and replaced in the proper box. If it is desired to make several determinations, the glass slide and eyepiece are cleaned with lens paper between operations.

Density

The determination of density is useful in the identification of compounds that do not form well-defined derivatives. The characterization of such substances as the liquid aliphatic hydrocarbons is usually accomplished through determination of the boiling points, refractive indices, infrared spectra, and densities. The determination of density may also be used as a general index of the relative complexity of the unknown. Compounds that have a density of less than 1.0 usually do not contain more than one functional group, whereas polyfunctional compounds have as a rule a density greater than 1.0. In addition, density and refractive index measurements may be employed to calculate molar refraction and dispersion which are useful in checking the structure of compounds.[12]

The difficulties encountered by beginners in the accurate determination of the densities of organic liquids account for the restricted use of this important physical constant. Compared with the determination of melting points, refractive indices, and boiling points, the measurement of densities is subject to more errors because *simple, rapid,* and *reliable* instruments and techniques have not yet been developed. The macromethods that use 1–5 ml of liquid are reliable if proper pycnometers are used and sufficient time is available for repeated accurate weighings with an analytical balance. The micromethod described in this section gives reliable results if care is exercised. Since the amount of liquid for the determination is 0.02–0.03 ml, small errors in weighing or losses by evaporation will cause large discrep-

---

[12] Cheronis and Entrikin, *op. cit.,* pp. 149–152.

ancies. For determination of the densities of *solids* the student is referred to the literature.[13]

Density measurements are usually expressed as grams of mass per milliliter, $d_4^t = m/V$, g/ml. This is measured by a direct comparison of the weights of equal volumes of the substance at $t°$ and of water at $4°$ $(3.98°)$; specific gravity d_t^t is measured by direct comparison of the weights of equal volumes of the substance and water at $t°$. The temperatures most frequently used are $25/25°$ and $20/20°$. However, the densities and specific gravities of organic compounds recorded in the literature were determined at temperatures ranging from 0 to $40°$. With the Dreisbach[14] tables it is possible to convert rapidly the specific gravity at $25/25°$ to density at any temperature between 0 and $40°$ if the coefficient of cubical expansion, β, is known. A value for specific gravity d_t^t multipled by the density d_4^t of water gives the density d_4^t.

Apparatus and Procedures for 1 ml of Liquid or More

Pycnometers with a capacity of 1–2 ml are commercially available. The pycnometer is cleaned, dried, and then weighed. The bulb is then filled with distilled water to a point above the mark and immersed in a 25-ml beaker containing water at $20°$. After 5–10 minutes the level of water in the pycnometer is adjusted by means of a capillary pipet dropper. The pycnometer is then removed from the beaker, dried rapidly with a small piece of chamois, and weighed. The pycnometer is then emptied, dried, and filled with the liquid under investigation; it is adjusted at $20°$ as before, and weighed. The weight of the sample divided by the weight of water gives the density of the liquid at $20.°$ However, the density of water at $20°$ is not 1.0000, and therefore a correction must be made to express the density with reference to that of water at $4°$ by the following factor:

$$D_4^{20} = \frac{\text{weight of sample}}{\text{weight of water}} \times 0.99823$$

Apparatus and Procedures for Less Than 1 ml of Liquid

Semimicro and micro determinations of density may be made by the beginner through pycnometers that are easily constructed of 3–4-mm tubing or glass capillaries. The pycnometer shown in Figure 3.10, if made of 3-mm glass tubing, has a capacity of 0.2–0.5 ml, and when made from a melting-

[13] Sullivan, *U.S. Bur. Mines, Tech. Paper* **381** (1927); Caley, *Ind. Eng. Chem., Anal. Ed.* **2**, 177 (1930); Blank, *Ind. Eng. Chem., Anal. Ed.* **3**, 9 (1931); Blank and Willard, *J. Chem. Educ.* **10**, 109 (1933); Cheronis, *Micro Methods*, p. 198; Bauer and Lewin in *Physical Methods of Organic Chemistry*, 3rd ed., Part I, pp. 175–88.

[14] Dreisbach and Martin, *Ind. Eng. Chem.* **41**, 2879 (1949).

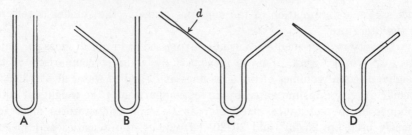

Figure 3.10   Construction of pycnometer.

point capillary, 0.02–0.03 ml. A pycnometer of capacity 0.5 ml is commercially available.

Figure 3.11 shows the Fisher-Davidson gravitometer for routine work. When semimicro quantities are available, this gravitometer, based on

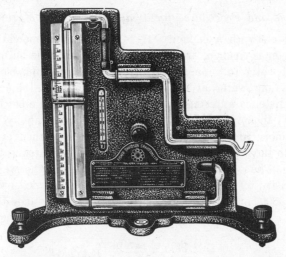

Figure 3.11   Fisher-Davidson gravitometer. (Courtesy Fisher Scientific Co.)

Ciochina's[15] balanced-column method, gives rapid results with an accuracy of about 0.1–0.2 per cent and requires about 0.5 ml of the sample. It gives direct readings, and can be used for the range $D^{20}_4 = 0.6$–2.0. The scale is established by using ethyl benzene as a standard in the L tube of the instrument; the sample is placed in the Z tube and both liquids are drawn up by the pump, which is operated by turning the center knob until the menisci of the liquid in the Z tube rest in the upper and lower arms. The fixed pressure difference between the atmosphere and the connecting lengths of the Z and L tubes varies directly with the density of the liquid

[15] Ciochina, *Z. anal. Chem.* **98**, 416 (1934); **107, 108** (1936).

in the Z tube and is measured on the graduated scale as D_4^{20}. The standard liquid for $D_4^{20} = 0.6$–2.0 is "certified" ethyl benzene supplied with the instrument. For higher densities carbon tetrachloride may be used and the readings multipled by a conversion factor. The Z tube may be replaced with a wider (4-mm) tube for liquids of high viscosity. The amount of sample required for each determination is 0.3–0.7 ml, depending on the substance.

Optical Rotation

The change in the direction of vibration of linearly polarized light during its passage through anisotropic substances is called *optical rotation* and such substances are called *optically active*. The first step in the determination of optical rotation involves the preparation of the solution of the active compound. About 100–500 mg of the substance is accurately weighed and dissolved in 25 ml of the solvent in a volumetric flask. The solvents commonly used are water, methanol or ethanol, chloroform, and a mixture of ethanol and pyridine.

The compound must be purified before the solution is prepared. If the solution is not clear, it should be filtered after being properly diluted. The filtrate is collected in a dry flask and is returned to the funnel until a perfectly clear filtrate passes through the stem. The funnel is placed over another dry flask and the clear filtrate is collected for the determination.

The next step is the filling of the polarimeter tube. The cup is screwed on one end and the tube is held vertically while the solution is poured in until it rises to the top end; the cover glass is placed over the end of the tube in such a manner that no bubbles appear. The cap is screwed on the tube with care. If great pressure is applied to the cover glass, the strain may produce optical activity and a serious error will be introduced in the observation.

The *zero reading* of the instrument is determined by turning the movable prism until the two halves of the field are matched so as to have a uniform illumination. Before the two fields are matched, the eyepiece of the telescope should be checked for focusing; this is accomplished by turning the eyepiece to the right or left until the line that divides the two fields is sharp. The true zero of the instrument may not coincide with the zero of the graduated circle. The main circle is usually divided into degrees and 0.25 division of 1°. The vernier outside is divided into 2.5 divisions, thus enabling a reading of 0.01°. Five readings are made and the values averaged. The polarimeter tube that contains the solution is placed in the opening between the two prisms. The movable prism is turned until uniform illumination is obtained. The reading of the circle is noted and the observation is repeated four or five times. The temperature at the time of the reading is recorded. The values of the readings are averaged; this average rotation, less the zero reading, gives the observed rotation. The specific rotation is calculated from this value, using the formula

$$\text{Specific rotation} = [\alpha]_{D}^{25°} = \frac{\alpha \times 100}{L \times c}$$

where α is the observed rotation of the sample in degrees, L is the length of the tube in decimeters, and c is the concentration of the dissolved active substance in grams per 100 ml of solution.

Molecular Weights

The present discussion deals only with the cryoscopic micromethods. All cryoscopic methods depend on the accurate determination of the melting (or freezing) point of the solvent and of a solution of the sample. The most successful micromethod is based on the use of solvents with high molal depression constants, with which it is possible to obtain a depression of the melting point of 5–20° for 10 per cent solutions of the sample; the measurements are made by the capillary-tube method. This microprocedure was developed by Rast[16] with camphor as the solvent, after Juniaux[17] had noted that, in addition to its excellent solvent properties, camphor exhibited a molecular depression of about 40°.

Although the Rast method is unique in its simplicity, it is subject to limitations and difficulties. The molal freezing point constant of camphor varies between 37 and 40°, depending on the source of the compound. A more serious drawback is the high m.p. (176–180°) of camphor, which limits its use to substances which are stable at these temperatures. Furthermore, it is applicable only to those compounds which are soluble in camphor and which give approximately ideal solutions. Some of these limitations may be surmounted by the use of other solvents which have high molal freezing-point constants. For a solvent to be useful for this purpose, it should dissolve a wide variety of organic compounds; the solution should be clear at the melting point of the mixture; decomposition of the solute should not occur at the melting temperature. Further, the solute should not undergo association in the solvent and the depression of the melting point of the solvent for 5–10 per cent solute concentrations should be large. In Table 3.3 are listed a number of solvents that have been recommended for the Rast microprocedure. Of the substances that are given in the table, few are commercially available for general use.

[16] Rast, *Ber.* **55**, 1051, 3727 (1922).
[17] Juniaux, *Bull. soc. chim. France* **11**, 722, 993 (1912); *Compt. rend.* **154**, 1592, 1692 (1912).

TABLE 3.3
List of Solvents for Rast Microprocedure

Solvent	Melting point (°C)	Molal depression constant (°C)
Cyclohexanol	24.7	42.5
Camphene	49	31
Cyclopentadecanone (Exaltone)	65.6	21.3
Perylene	276	25.7
2,4,6-Trinitrotoluene	82	11.5
Tetrabromethane	—	86.7
Bornyl bromide	—	67.4
Bornylamine	164	40.6
Camphor	176–180	37–40
Camphoquinone	190	45.7
Borneol	202	35.8
Dicyclopentadiene	32	46.2
Dihydro-α-dicyclopentadienone	53	92.0
Tetrahydro-α-dicyclopentadiene	77	35.0
Dihydro-α-dicyclopentadiene-3-ol	53	92.0

Source: Cheronis, *Micro Methods*, p. 201.

The precision of the method depends to a considerable extent on the reproducibility of the solution in the capillary tube. With experience in the preparation of a uniform dispersion of the solute, a precision of about 0.5 per cent is possible. The accuracy of the method depends largely on the nature of the solute. The molecular weight of carboxylic acids cannot be determined with camphor as the solvent; the results are high, probably because of association. Under the most favorable conditions an accuracy of 1–2 per cent is attainable by the experienced worker.

Procedure When Camphor and Other High-Melting Solvents Are Used

The method involves the determination of the melting point of a pure solvent, usually camphor, and the melting point of a solution of the substance under investigation in camphor. The apparatus required includes capillary tubes, which can be readily constructed or obtained from dealers,[18] and a melting-point apparatus of the type described on pages 43–44.

The melting point of the solvent is determined by using the usual method (page 44). If the solvent has not been employed before in the laboratory, the molar depression constant of the particular lot must be made with the

[18] Microware, Inc., Vineland, N.J.; A. H. Thomas Co., Philadelphia, Pa.

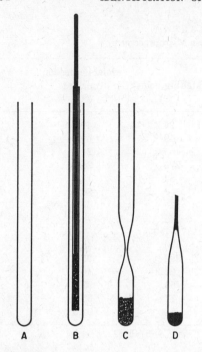

Figure 3.12 Rast method for micro determination of molecular weight. A. Capillary of 50–70 mm length and 3–4 mm diameter. B. A second capillary, containing the material, is inserted into capillary A, and the material is pushed out by means of a small rod. C. Capillary containing solute and solvent is weighed and then sealed. D. The upper part of the capillary is drawn out in the form of a rod, and the sample and solvent are melted and mixed. The tube is now ready for determination of the melting point of the mixture.

same apparatus and thermometer which will be employed in the determination. Substances recommended for standardization are: acetanilide, azobenzene, chloroanthraquinone, sulfonal, and naphthalene.

The solution of the standard or unknown is made in capillary tubes about 50–70 mm in length with an inner diameter of 3–4 mm, constructed of soft glass, unless the sample is affected by traces of alkali; in that event Pyrex glass is used. The capillary, after being thoroughly wiped, is accurately weighed. The sample (0.5–5.0 mg) is introduced by a smaller capillary as described in the caption of Figure 3.12; if the quantity exceeds 1.5 mg, it is advisable to use a capillary with an inner diameter of 4 mm. Sufficient solvent to give a solute concentration of 5–10 per cent is introduced by the same technique. Then, after the capillary has been wiped and reweighed, it is sealed at a point about 20 mm above the level of the solid. The sealing requires some care to minimize the volatilization of the solvent; the filled part of the capillary is wrapped with wet filter paper and then rotated in an inclined position over a microflame at the point where the seal is desired until the glass collapses and forms a solid glass rod. Or the capillary may be sealed by dipping the wrapped part in water and using a fine oxygen flame. Figure 3.12 shows the capillary and the various steps by which the capillary is filled and sealed. The capillary tube is then heated cautiously to melt the solvent and mix it with the solute. Errors will occur unless care is exercised to obtain a homogeneous dispersion, particularly in the case of solvents, such as camphor, which crystallize rapidly below their melting points. The capillary tube is warmed in a bath until the contents melt, and is then withdrawn and the contents are mixed by vigorous shaking; this process is repeated three or four times.

An alternative procedure used in the authors' laboratory consists in rotating the capillary between the thumb and the first two fingers of the right hand over a microflame until the solvent has melted; the left hand is placed over the capillary, which is inclined at an angle of 45°. The contents are subsequently mixed by rotating the tube rapidly between the palms above the microflame at a sufficient height so that the mixture remains liquid. Some experience is needed for this technique since the lower part of the capillary, which contains the solution, is at a temperature above 150°.

The melting points of the pure solvent (camphor cyclohexanol and the like) and of the solution (mixture prepared according to the steps shown in Figure 3.12) are determined several times according to the standard procedure described under melting points in this chapter. Care should be exercised in the observations for the melting point of the solution since the melting point is not sharp; the temperature at which the last crystal disappears should be considered as the melting point of the mixture. For camphor and other solvents melting above 150°, a thermometer graduated in the range 140–230° may be employed.[19] After each determination the thermometer to which the capillary is fastened is removed momentarily from the heating bath and is carefully inverted to mix the contents of the capillary. The melt solidifies rapidly and the determination of the melting point is repeated until the deviation between two consecutive measurements is about 0.2 to 0.5°. The molecular weight is then calculated by the equation

$$M = KS_1/S_2\Delta t$$

where M is the molecular weight of the sample, K the molal depression constant of the solvent multiplied by 1000, S_1 the weight of the sample in milligrams, S_2 the weight of the solvent in milligrams, and Δt the observed difference between the melting points (averaged) of the solvent and the solution.

A modification of the capillary construction for the Rast method which may be found less troublesome by beginners has been published by Pinkus and Barron.[20] A method using cyclohexanol as a solvent has been described recently.[21]

[19] A. H. Thomas Co., Philadelphia, Pa.
[20] Pinkus and Barron, *J. Chem. Educ.* 33, 138 (1956).
[21] Rogozinsky and Cheronis, *Microchem. J.* 5, 595 (1961).

Exercises

1. Discuss the precautions to be observed in the determination of the following constants: (a) melting point; (b) boiling point; (c) refractive index; (d) density.

2. What are the limitations of the cryoscopic method for the determination of molecular weights? What other methods can be employed?

3. Assume that the sample of a solid unknown is more than 25 g. What other method beside the procedure of capillary melting points can be employed for the determination of the melting temperature of the compound which would yield more reliable information?

4. Show how boiling points can be determined without corrections for variation of atmospheric pressure.

5. Give an example for the application of data from physical constants in the tentative identification of an unknown.

6. Indicate how it is possible to check first the precision and then the accuracy of the method employed by a beginner in the determination of melting points which uses an ordinary Thiele tube and an ordinary 0–360° thermometer.

7. Assume that the volume of the liquid unknown is 25 ml. Which method is likely to give more reliable results in the determination of the boiling temperature: (a) distillation of the entire sample; (b) determination of the boiling point by means of a few drops?

8. Indicate all the possible errors that may arise by the use of a partial immersion thermometer and the use of a total immersion thermometer in the measurement of melting points. Outline a method for the precision and accuracy of your method of determining melting points.

9. The boiling point of an unknown liquid was determined by the micromethod as 152–153°. The boiling point of a pure sample of bromobenzene determined with the same procedure and under the same conditions was found to be 154°. What is the corrected boiling point of the unknown substance?

10. Explain the reason for recording the temperature at which bubbles begin to emerge from the capillary as the boiling point in the micromethod for the determination of boiling points.

11. The refractive index on an unknown is $n_D^{20} = 1.3983$. The substances with refractive indices in this range are: n-butyric acid $= 1.3979$; isooctane $= 1.3981$; 2,4-dimethylhexane $= 1.3986$; n-butylamine $= 1.3988$. Give a simple chemical test to determine whether the unknown is an acid, an organic base, or a hydrocarbon.

12. The equation used on page 62 for the calculation of the molecular weight from cryoscopic data may be generally written as $\Delta t = KM$, where Δt is the melting-point depression, M the molality, and K the melting-point depression constant. Consult a text on physical chemistry and derive this equation from considerations of the effect of the addition of a solute to a liquid.

13. Consider the following compounds: n-butane, isobutane, n-butyl alcohol, ethyl ether, sec-butyl alcohol, and t-butyl alcohol. Look up the boiling points in the Tables (page 345 ff.) and then discuss the effect of molecular structure on the boiling points of these compounds.

4

Preliminary Steps in the Examination of an Unknown Substance

The basis of all systematic identification procedures lies in the proof of complete similarity between the physical and chemical properties of the "unknown" substance and the physical and chemical properties of some known compound, the data for which are on record or which may be established by experimentation. Unless the substance being examined is a pure compound, the physical constants will not be reliable enough to use in identification. Furthermore, even small quantities of impurities will often give very misleading results when the substance is subjected to chemical tests for the functional groups that must be identified before the substance can be classified or derivatized. The earlier chapters of this text cover ways to determine the purity of a substance and ways to purify it if it is not essentially pure. These techniques should be acquired by working with known compounds. It is further suggested that known compounds of appropriate chemical classes be used to gain experience and information about the results to be expected from the procedures and chemical reactions involved in the tests in this and the following chapters. This should be done previous to or concurrently with the work with "unknowns."

If the "unknown" is a mixture of compounds and it is desired to identify each compound, Chapter 8 should be consulted for suggested procedures for the separation of a mixture into compounds that may be subsequently identified.

To identify an "unknown," a major objective is to determine the chemical class to which it belongs; for example, to identify it as an alcohol, an amine, a ketone, or as the member of some other class. It is generally not practical to make "spot tests" for each possible class directly on the substance under investigation. Rather, methods are chosen which will detect certain general properties that are characteristic of chemical classes that have in common similar functional groups. Predictions as to the chemical class or classes to which an "unknown" organic compound belongs are largely based on three facts concerning it: (a) the elements present, (b) its solubility behavior in selected solvents, and (c) its acid-base character. Based on the results of these three determinations, it is usually possible to conclude that the "unknown" belongs to one of a very limited number of chemical classes.

Assuming that experience has been gained in the techniques required and in the results to be expected from tests by working with known compounds by the methods suggested, the following steps are recommended as a systematic way to proceed with an "unknown." As experience is gained, the worker may be able to proceed more directly due to his knowledge of the characteristics of many compounds, but the beginner usually saves time by following a systematic procedure carefully and thoughtfully.

1. Determine that the substance is, at least approximately, a pure compound. If necessary, purify it by appropriate methods (see pages 11–38).

2. Determine the elements that are present (see pages 68–76).

3. Determine the solubility in selected solvents (see pages 77–94).

4. Determine the acid-base characteristics (see pages 95–101).

5. Consult the tables on pages 96–97 and make a list of all the classes and subclasses that the "unknown" might belong to, considering the results of determinations 2, 3, and 4 above. Note what functional groups are represented in these classes and subclasses.

6. Perform any of the general observations and tests that have good potential for providing useful information that would relate the "unknown" to one or more of the types of compounds listed in 5 above (see pages 106–12).

7. Based on the information gained thus far, select and perform the tests for specific chemical classes until the unknown is classified (see pages 113–52).

After the "unknown" has been properly assigned to a chemical class, there remains the problem of identifying the specific compound. The directions in Chapters 10 and 11 provide procedures for this purpose.

ANALYSIS FOR ELEMENTS

Although it is assumed for the purposes of this text that the substances being examined are organic compounds, it is perhaps useful to include a simple test for the presence of carbon. With rare exceptions, such as the compounds in which halogen has completely replaced hydrogen, all compounds that contain carbon also contain hydrogen. However, this text only attempts coverage of the most common types of organic compounds. Provision is made for the identification of nitrogen, sulfur, the halogens, and, in certain types of compounds, oxygen. The number of elements that *may* be found in organic compounds is quite large. Such elements as phosphorus, boron, silicon, and several of the metals are increasingly important (see note p. 70). More detailed methods and alternate methods of analysis are presented in a larger work[1] by the authors of this abridged text. Methods of detecting additional elements in organic compounds may be found in the literature.

Detection of Carbon

This test[2] is based on the fact that any organic compound will reduce yellow molybdenum trioxide to "molybdenum blue" (Mo_2O_5) when the two are heated together. Other compounds which are oxidizable by molybdenum trioxide, such as ammonium salts and sulfites, must be proved absent before this test is specific for carbon compounds.

Place a few milligrams of the substance to be tested in the bottom of a 3-inch test tube[3] and add finely powdered molybdenum trioxide to a depth of 6–8 mm. Using a semimicro burner, heat the tube so that the upper portion of the mixture is heated before the bottom of the tube. Continue heating the mixture for 1–2 minutes. The appearance of a blue zone in the area of contact between the sample and the yellow oxide indicates the presence of carbon in the sample.

Fusion with Sodium

To detect nitrogen, sulfur, and the halogens in organic compounds, these elements must first be converted to ions. When an organic compound is fused with an alkali metal, extensive decomposition occurs; nitrogen, in the presence of carbon, is converted to the cyanide ion; sulfur and the halogens are converted to sulfide ions and the halide ions, respectively. The ions thus formed may be identified by conventional methods. Sodium is the most commonly used alkali metal for such fusions. Metallic potassium may be used and for some types of compounds, such as the pyrrole derivatives, it is preferred.[4] For fusion techniques using magnesium, zinc, and calcium oxide, see the larger text by these authors.

Certain comments relative to the use of sodium are warranted at this point. Since sodium reacts violently with water, it is essential that liquids

[1] Cheronis and Entrikin, *Semimicro Qualitative Organic Analysis* (2nd ed., N.Y.: Interscience, 1957).

[2] Feigl and Goldstein, *Mikrochim. Acta* **1956**, 1317.

[3] All glassware is assumed to be Pyrex, Kimex, KG-33, or similar borosilicate glass.

[4] Kainz and Resch, *Mikrochemie* **39**, 75 (1952).

containing more than traces of water not be added to hot sodium. If there is any possibility that water may be present in the "unknown," test for water.

Test for the Presence of Water

Several reagents may be used to detect water. Detection of appreciable amounts of water may be made by using anhydrous copper sulfate (which turns blue when hydrated) or by adding a small crystal of potassium permanganate to the liquid sample. Potassium permanganate is not appreciably soluble in organic compounds, but dissolves to give its characteristic color to water or mixtures that contain appreciable amounts of water.

Tetraisopropyl titanate is a very sensitive reagent for detecting water. The test may be made by adding a drop of the reagent to a few drops of the liquid (or a solution of the compound in anhydrous methanol). Water hydrolyzes the reagent to produce chalky precipitates of the hydrated titanium oxides.

If water is found present, it must be removed by using dehydrating agents before proceeding with a sodium fusion. The test tube in which the fusion is to be carried out must be completely dry. Sodium must be handled with care and should not come in direct contact with the skin. A few compounds react rather explosively when contacted by hot sodium; hence, it is a good practice to wear safety goggles or a face shield when performing a sodium fusion. The fact that very small amounts of the organic compound are being added to the sodium reduces any potential hazard to a minimum.

Note: In addition to fusion with metallic sodium, it is sometimes advisable to fuse the unknown substance with sodium peroxide. This may be done by adding 0.5 g of sodium peroxide to about 50 mg of the compound in a dry 6-inch test tube and fusing the mass. The fused mass should be heated for 3 minutes. After cooling, dissolve the residue in 2 ml of water and carefully neutralize the solution by adding dilute nitric acid. Portions of this solution may be tested for phosphates, sulfates, borates, and silicates by conventional methods. The following method is one scheme for detecting phosphates and sulfates.

Place 0.5 ml of the sodium fusion filtrate in a 4-inch test tube, neutralize the solution by adding $2M$ nitric acid, and add 2–3 drops of a saturated solution of calcium nitrate. A precipitate would indicate either fluorides or phosphates, or both. If a precipitate forms, remove it by filtration or centrifugation and reserve it for further testing. To the filtrate, add 2–3 drops of a saturated solution of barium nitrate. A precipitate would indicate the presence of sulfate ions. If a precipitate forms, remove it. The filtrate may now be tested for the halide ions by the procedures given on pages 72 and 73. In the absence of fluorides, the calcium precipitate may be assumed to be calcium phosphate. Since calcium phosphate may be dissolved by acetic acid, whereas calcium fluoride is not soluble in acetic acid, differentiation between fluorides and phosphates may be made by treating the precipitate with hot 6M acetic acid and then testing the filtrate for phosphates by conventional methods.

If the organic compound is fused with sodium peroxide rather than with sodium, it should be noted that iodine, if present, will be converted to the iodate ion.

Procedure for Fusing with Sodium

Add a *small* drop of the liquid to be tested, or a few milligrams if it is a solid, to a clean, dry, 4-inch test tube. Remove a piece of sodium (about ⅛-inch cube) from the reagent bottle, press it between folds of filter paper to remove the adhering liquid, and drop it in the test tube. Warm the bottom of the tube *slightly* and allow the mixture to stand for 2–3 minutes. Clamp the tube in a vertical position and gradually apply heat to the bottom of the tube until the sodium has melted. Being careful not to hit the sides of the hot tube with the sample, drop a *few milligrams* of the substance being tested directly onto the melted sodium. Heat the bottom of the tube until the glass is red-hot and continue heating it for 2 minutes. Allow the tube to cool to room temperature and add 5 drops of methanol. If the residue is a globular mass, break it up with a clean glass rod to allow contact between the alcohol and any excess sodium metal. If gas bubbles are produced, wait for the reaction to be completed. Then add 2 ml of distilled water. Boil the mixture for a few seconds and filter it. Instead of filtering, the mixture may be centrifuged and the clear centrifugate decanted or removed by means of a pipet. Dilute the filtrate or centrifugate to 4 ml with distilled water. If at this point the solution is dark-colored so as to obscure future tests, it is probable that the amount of sample used was too large or that the fused mixture was not heated to a sufficiently high temperature; the fusion procedure should be repeated.

Detection of Sulfur

Two methods for testing for the presence of sulfide ions are provided.

A. Place 0.2 ml of the alkaline filtrate from the sodium fusion in a 3-inch test tube and add 1 drop of 5 per cent lead acetate solution. Add, dropwise, 6M acetic acid until the solution is acidic. A brown-to-black precipitate or coloration indicates the formation of lead sulfide.

B. Place 0.2 ml of the sodium fusion filtrate in a 3-inch test tube and add 1 drop of a 0.1 per cent solution of sodium nitroprusside. A deep red color indicates the presence of the sulfide ion.

Note: Even sulfates are reduced to sulfides by fusion with sodium. Fusion of the unknown with sodium peroxide converts all forms of sulfur, including sulfides, into sulfate ions. The addition of a barium nitrate solution to a few drops of the neutralized solution from the sodium peroxide fusion will cause the precipitation of barium sulfate if sulfur is present in the original substance.

Detection of Nitrogen

If sulfur is present, a slight modification of the procedures for nitrogen described in the next two paragraphs is desirable. Ferrous sulfide is only slightly soluble; hence, to be sure that adequate ferrous ions are available for both the sulfide and cyanide ions, 20–30 mg of ferrous sulfate should be used instead of 15–20 mg. Furthermore, when the sulfuric acid is added to dissolve the oxides of iron, a residue of ferrous sulfide may remain in the bottom of the tube. This residue will not interfere with the test for nitrogen.

A. In the presence of hot sodium and carbon, nitrogen is converted to the cyanide ion. It is this ion that is detected. Place 0.8 ml of the filtrate in a 4-inch test tube and add 1 drop of 10 per cent sodium hydroxide, 1 drop of 30 per cent potassium fluoride, and 15–20 mg of solid, crystalline ferrous sulfate. Boil the mixture gently for one minute. Add 1 drop of 1 per cent ferric chloride solution and again bring the mixture to boiling. Add $6N$ sulfuric acid dropwise until the oxides of iron have just been dissolved. Allow the tube to stand for 2–3 minutes. A blue color or blue precipitate is a positive test for nitrogen. A green or greenish blue coloration indicates a weak test for nitrogen. Such a solution should be filtered; if a blue color shows on the filter paper, the presence of nitrogen may be assumed.

B. In the copper acetate–benzidine test for cyanide ion[5] about 0.1–0.2 ml of the filtrate or centrifugate is acidified with a drop of 10 per cent acetic acid, and then 1–4 drops of the reagent are added carefully by means of a capillary pipet so that there is no appreciable mixing of the two solutions. If cyanide ion is present, a blue ring develops. The copper acetate–benzidine reagent is prepared from two stock solutions, one containing 150 mg of benzidine in 100 ml of water and 1 ml of acetic acid, the other 286 mg of copper acetate in 100 ml of water. The two solutions are kept separately (in dark bottles) and the reagent is prepared by mixing equal volumes of each just before use.

Detection of Halogens

The sodium fusion filtrate may contain one or more of the following halide ions: fluoride, chloride, bromide, and iodide. Silver fluoride is highly soluble in water. Hence, the presence of the fluoride ion cannot be detected by the addition of silver ion. A separate test for fluorides is given later in this chapter. Since silver cyanide and silver sulfide are both sparingly soluble in water, cyanide and sulfide ions must be removed before adding silver ions to test for the halides. The fusion filtrate must be acidified with

[5] Campbell and Campbell, *J. Chem. Educ.* **27**, 261 (1950); Sieverts and Hermsdorf, *Z. angew. Chem.* **34**, 3 (1921); Feigl, *Spot Tests*, in *Organic Analysis* (6th ed., Amsterdam: Elsevier, 1960), p. 183.

nitric acid before adding silver ions to prevent the precipitation of silver hydroxide or oxide. The presence of one or more of the halide ions, excepting fluoride, may be established by the following method.

Place 0.5 ml of the fusion filtrate (or centrifugate) in a 4-inch test tube. Acidify the solution by adding, dropwise, dilute nitric acid. If neither nitrogen nor sulfur is found present, proceed directly by adding 1 drop of 5 per cent silver nitrate solution. If either nitrogen or sulfur is found present, it must be removed before adding the silver nitrate. To do this, add 1 ml of distilled water to the acidified fusion filtrate and boil the mixture until the volume is reduced to about 0.5 ml. (*Caution: Heat the mixture in a hood.*) Cool the solution and add 3 drops of 5 per cent silver nitrate. A white to yellow precipitate or suspension indicates the presence of chloride, bromide, or iodide ions.

Detection of the Individual Elements: Chlorine, Bromine, and Iodine

Two methods have been found successful for the detection of these halogens even if they are all present in the same substance.

A.[6] Nitric acid will oxidize iodide ions to free iodine at room temperature even when the acid is reasonably dilute. Excess concentrated nitric acid will oxidize bromide ions to bromine at room temperature. Chloride ions are not oxidized under these conditions.

Place 0.5 ml of the sodium fusion filtrate in a 4-inch test tube. Add 0.5 ml of carbon tetrachloride and 3 drops of concentrated nitric acid. Shake and allow to stratify. The presence of iodine is indicated by the carbon tetrachloride layer becoming a violet color. If iodine is present, remove the carbon tetrachloride layer by pipet and add 0.5 ml of carbon tetrachloride to the original test solution. Add 1 drop of concentrated nitric acid and shake the mixture. If the color shows that iodine is still present, remove the carbon tetrachloride layer and add a third portion of carbon tetrachloride. Shake and allow to stratify. Repeat the procedure until the newly added carbon tetrachloride layer remains essentially colorless. Add 2 ml of concentrated nitric acid. Shake the mixture and allow to stratify. The presence of bromine is indicated by a tan or tannish red color in the carbon tetrachloride layer. Repeat the extraction until the carbon tetrachloride layer becomes essentially colorless. Add 3 drops of 5 per cent silver nitrate. The immediate formation of a white precipitate indicates the presence of chloride ions.

B. The following procedure is based on the oxidation of iodide and bromide, but not of chloride, ions by permanganate ions in nitric acid solution.

[6] Hanson, *J. Chem. Educ.*, **38**, 412 (1961).

To 8–10 drops of the solution containing the halide ions in a 3-inch tube, 5 drops of $0.1M$ KMnO$_4$ and 5 drops of $6N$ HNO$_3$ are added. The tube is shaken for 1 to 2 minutes. Add 5–6 drops of carbon disulfide, shake the tube for 2 minutes, and allow the mixture to stratify. About 15–20 mg of oxalic acid is added and the tube shaken.

A reddish brown color in the lower layer indicates that bromine *or* both bromine and iodine are present. If iodine, but not bromine, is present, the color of the carbon disulfide layer will be violet or light purple. If the carbon disulfide layer remains colorless, neither bromine nor iodine is present.

If the color of the carbon disulfide layer is reddish brown, 2 drops of allyl alcohol are added and the mixture shaken. If only bromine is present, the reddish brown color of the carbon disulfide layer will disappear and leave the solution colorless. However, if both bromine and iodine are present, the reddish brown color will change to violet or light purple.

By means of a pipet dropper, remove the aqueous layer (upper layer) from the mixture from which the iodine and bromine, if present, have been extracted and place this solution, which may contain chloride ions, in a 4-inch test tube. Add 2 ml of concentrated nitric acid and 3–4 drops of 5 per cent silver nitrate. A white precipitate is a positive test for chloride ions. A faint white suspension which does not coagulate indicates a probable impurity due either to chlorides in the water or to a chlorine-containing impurity in the original organic compound.

Detection of Fluoride Ion in the Fusion Filtrate

A. Place 15–20 mg of lanthanum chloranilate (2,5-dichloro-3,6-dihydroxy-p-benzoquinone, lanthanum salt) in a small test tube. Add 1–2 ml of water and 5 drops of $6M$ acetic acid to the tube, and then add 5–10 drops of the sodium fusion filtrate. The development of a light violet or pink color within 10 minutes is a positive test for fluorine in the compound that was fused with sodium.

B. In another test[7] the reagent consists of a solution containing titanate ions; addition of hydrogen peroxide yields pertitanate ions, which have a yellow color. If fluoride ions are added to the reagent, the complex ion TiF$_6^{-2}$ is formed and the reagent is decolorized:

$$TiO^{+2} + H_2O_2 + H_2O \rightleftharpoons TiO_4^{-2} + 4H^+$$
$$\text{Yellow}$$

$$TiO^{+2} + 6F^- + 2H^+ \rightleftharpoons TiF_6^{-2} + H_2O$$
$$\text{Colorless}$$

The pertitanate test is performed by placing 0.2–0.5 ml of the filtrate or centrifugate from the fusion mixture in a 3-inch tube and the same amount

[7] Bergman, E., private communication.

of distilled water in second tube. One drop of concentrated sulfuric acid is added to each tube followed by 2–3 drops of 3 per cent hydrogen peroxide solution and 2–3 drops of titanium reagent, in the order given. The titanium reagent may be added one drop at a time until the control tube has a distinct yellow color. The colors in the tubes are compared against a white background. The presence of fluoride ion in the test solution is indicated by decolorization of the pertitanate color.

The titanium reagent is made by dissolving 150 mg of titanium ammonium fluoride and 350 mg of ammonium sulfate in 2.5 ml of sulfuric acid and then diluting with 7.5 ml of sulfuric acid and water to 100 ml.

Detection of Oxygen

Ferrox,[8] ferric hexathiocyanatoferriate, an intensely colored salt, is not soluble in hydrocarbons or their halogen derivatives, but it is soluble in compounds that contain oxygen or sulfur and in most compounds that contain nitrogen.

Prepare the ferric hexathiocyanatoferriate by placing a small crystal of ferric ammonium sulfate and one of potassium thiocyanate in a dry 3-inch test tube. Grind the crystals with a glass stirring rod. Insert the rod with its adhering reagent into a dry 3-inch test tube. Pour 2–4 drops of the liquid to be tested down the rod and stir the mixture. The liquid will dissolve the salt and show red–reddish purple colorations if the compound contains oxygen or sulfur and, generally, if it contains nitrogen. The salt does not dissolve in hydrocarbons or halogenated hydrocarbons; hence, these liquids retain their original color or are only faintly tinted. Solids may be tested by using a saturated solution of the compound in warm benzene or purified carbon tetrachloride. A modification of this test has been suggested [9] which uses an ether solution of the reagent.

Note: This test is most useful if applied to compounds that fall in Solubility Division N so as to distinguish the hydrocarbons that are soluble in concentrated sulfuric acid from the oxygen-containing compounds. Since ethers are hard to detect by functional group tests, the ferrox test is useful in distinguishing most of them from hydrocarbons.

Solids that are not sufficiently soluble in benzene or carbon tetrachloride give negative tests even if they contain oxygen. Some high-molecular-weight compounds, such as diphenyl ether, alkyl naphthyl ethers, and triphenyl carbinol, do not give positive tests. Some esters, such as dimethyl oxalate, do not give the test. Most nitrogen-containing compounds that have been tested give positive test except the alkyl amines which generally do not dissolve the salt.

Ferric chloride may be substituted for the ferric ammonium sulfate in the preparation of the reagent.

[8] Davidson, *Ind. Eng. Chem., Anal. Ed.* **12**, 40 (1940); Davidson and Perlman, *A Guide to Qualitative Organic Analysis* (N.Y.: Brooklyn College, 1952).

[9] Goerdeler and Domgorgen, *Mikrochemie ver. Mikrochim. Acta* **40**, 212 (1953).

Test for Peroxides

It has been suggested that the ferrox test may be used indirectly to test for peroxides in organic liquids.[10] The reagent is an acidified solution of ferrous thiocyanate. Peroxides oxidize the ferrous ions to ferric ions, which then form the red ferrox complex.

The reagent is prepared by dissolving ferrous sulfate and potassium thiocyanate in distilled water containing a little sulfuric acid. The pink solution is rendered colorless just before use by adding a minimum amount of zinc dust. In carrying out the test, it is necessary to use a stoppered tube that is almost filled with the reagent and the organic liquid to be tested so that only a very small volume of air is left in contact with the mixture.

Fill halfway a 3-inch test tube (or use a smaller tube if one is available) with the ferrous thiocyanate reagent and then nearly fill the tube with the organic liquid to be tested. Stopper the tube and shake it. The formation of a red color indicates an oxidizing agent, probably peroxides.

Exercises

1. A solid compound melting at 124° gives a positive Ferrox test and a strong test for the presence of chlorine. When 1 mg of the compound is dissolved in 1 ml of water, the pH of the solution is 1.5. What functional groups are indicated by these tests? Examine the proper table (see the Tables, page 345 ff.) and attempt to find a probable compound that fits these properties.

2. What are the advantages and disadvantages of fusing an unknown with sodium peroxide as compared with fusing it with metallic sodium with regard to the detection of the elements that may be present?

3. If it is known that the organic compound being tested is used as an insecticide, what elements should be considered as being potentially present? Outline a scheme for the analysis of such a substance.

4. Assume that a compound has been fused with sodium and that iodine, among other elements, has been found present. If this compound is now fused with sodium peroxide, what interference problems would be prevented in testing for other elements (consult the solubility tables for inorganic salts)? How could this interference be overcome?

5. Consult the literature for tests for borates. Summarize the tests and state which one would be best suited for use with a fusion filtrate.

[10] O'Brien, *Chem. and Eng. News* **33**, 2008 (1955).

5

Classification by Solubility and by Acid-Base Character

The preceding chapter outlined procedures for the preliminary examination of the purified unknown so as to detect the elements present in the compound. The present chapter contains discussion and procedures for the classification of the unknown on the basis of its solubility in a number of selected solvents, and on its reaction to a few acid-base indicators. Thus, the substance may be tentatively assigned to a large division which contains several classes of organic compounds.

Space limitations of this laboratory text dictate that only a minimum of discussion be included on the fundamentals of solubility behavior and acid-base character of organic compounds. For discussions from the more complete theoretical viewpoint, the reader is referred to the extensive literature[1] on such topics as solvent action, types of bonds and bonding forces, acid-base relationships, inductive effect, resonance, molecular orbitals, electronegativity, dipoles, association systems, van der Waals forces, and the like.

This chapter is divided into two parts. In Part I, substances are classified on the basis of their solubility or lack of solubility in certain solvents. The solvents used for classification are: water, ethyl ether, $1.2N$

---

[1] For example: Audrieth and Kleinberg, *Non-aqueous Solvents* (N.Y.: Wiley, 1953); Bell, *Acids and Bases—Their Quantitative Behavior* (N.Y.: Wiley, 1952); Davidson, *J. Chem. Educ.* 19, 154, 221, 532 (1942), reprinted in *More Acids and Bases* (Easton, Pa.: *J. Chem. Educ.*, 1944); Ferguson, *Electron Structures of Organic Molecules* (N.Y.: Prentice-Hall, 1952); Ingold, *Structure and Mechanism in Organic Chemistry* (Ithaca, N.Y.: Cornell Univ. Press, 1953); Ketelaar, *Chemical Constitution* (Houston: Elsevier, 1953); Luder and Zuffanti, *The Electronic Theory of Acids and Bases* (N.Y.: Wiley, 1946); Wheland, *The Theory of Resonance* (N.Y.: Wiley, 1944).

hydrochloric acid, 2.5N sodium hydroxide, 1.5N sodium bicarbonate, and concentrated sulfuric acid. For our purposes, a compound is arbitrarily considered "soluble" if it dissolves to the extent of 30 mg in 1 ml of solvent. Part II of the chapter provides a method for obtaining more detailed information than the solubility method provides concerning the relative acid-base character of a substance by means of indicators in different solvent environments. Table 5.5 on pages 96–97 lists the more common types of organic compounds classified both on the basis of solubility behavior and the elements that are present. In addition, the classification based on acid-base character is shown in parentheses after the name of the class. Since the properties of a particular molecule are determined by the complete molecular structure and since great variations in structure are possible among the molecules within any one class (amines or the phenols for example) it should be obvious that the schemes proposed in this chapter fail to provide conclusive results. If carefully made tests indicate borderline properties in either the solubility classification or the acid-base character classification, both adjacent classes should be considered.

PART I: CLASSIFICATION BY SOLUBILITY

In general, a solution is not as simple a system as that which is described in elementary courses as a *mixture*. Solubility depends on the intermolecular or interionic forces exhibited by both the solute and the solvent. In order for one substance to become dissolved in another, the bonds in each must be broken and the new bonds formed between the molecules or ions of the solute with the molecules or ions of the solvent. To classify substances by solubility, we make use of three major solvents: water, ethyl ether, and concentrated sulfuric acid. Additionally, aqueous solutions of chemical agents are used to produce chemical changes in the original substance that result in water-soluble products from water-insoluble starting material. Fundamentally, then, the classification depends on the fact that water, ether, and sulfuric acid have characteristically different interactions on solutes, which also have characteristic and differing possibilities for interactions.

Molecular Structure and Interactions

As background for a consideration of the causes of solubility or lack of solubility of a certain solute in a given solvent, it is necessary to recall the characteristics of the kinds of bonds that may be involved. The *ionic bond* results from the transfer of electrons. Among organic compounds, the ionic bond is largely limited to the following classes: metallic or ammonium salts of organic acids; metallic salts of phenols, thiophenols, mercaptans, enols, and alcohols; salts formed by the reaction of amines with acids; and the sulfonic acids. The *covalent bond* results from the sharing of electrons and is the bond of chief importance in organic compounds. The covalent bond may be formed by each of two atoms contributing one electron to form the bonding pair, by one atom sharing a pair of electrons with an electron-deficient atom or ion, or by the mutual contribution by atoms to form both sigma bonds and pi bonds between them. Covalent bonds have characteristic bond lengths, bond strengths, and bond angles. Since atoms of different elements have different affinities for electrons (electronegativity), the electrons of a covalent bond tend not to be equally shared by the bonded atoms; that is, the electron cloud is more dense around one atom. Such a bond is called a *polar* bond. It is sometimes referred to as a covalent bond with partial ionic character. The greater the difference in electronegativity of the bonded atoms, the greater the polarity of the bond.

Most molecules contain many atoms and the various bonds holding all of these atoms together have, in general, varying degrees of polarity. The composite effect of these local *dipoles* within the molecule produces, unless the local polarity is arranged with symmetry, a molecule in which the center of positive charge does not coincide with the center of negative charge. Such a molecule constitutes a dipole or is said to be a polar molecule. The effect of a molecular dipole on molecular properties may be measured and expressed in different terms for different purposes. The *dielectric constant* seems to be the best indication of the effective polarity of a liquid as regards its usefulness as a solvent. For purposes of this chapter, liquids with high dielectric constants will be considered as highly polar liquids and liquids with low dielectric constants will be considered as essentially nonpolar solvents.

Highly polar solvents tend to dissolve ionic or highly polar solutes, but not weakly polar solutes. Conversely, weakly polar solvents tend to dissolve weakly polar solutes, but fail to dissolve ionic or highly polar solutes.

When a substance dissolves, the ions or molecules must become separated from each other and remain dispersed among the solvent molecules. During solution, just as in melting a solid or vaporizing a liquid, energy must be supplied to overcome the interionic or intermolecular forces that have been bonding the ions or molecules of the substance being dissolved. The energy required to break the bonds between the solute particles is provided by the release of energy as new bonds are formed between the solute and the solvent. However, the molecules of the solvent must also be separated before the solute can diffuse among its molecules. If the sum of the energies required to break the original bonds in the solute and in the solvent is greater than the energy that can be released by forming new bonds between the solute and the solvent, the substance will not dissolve. Since molecules that are either nonpolar or only weakly polar are held together by very weak bonds, solutes of these types tend to dissolve in solvents of the same types because no strong bonds must be broken among either the solute molecules or the solvent molecules. For example, benzene dissolves in ether because the energy released by forming benzene–ether bonds is sufficient to break the weak benzene–benzene and ether–ether bonds. Benzene does not dissolve in water because the energy released by forming benzene–water bonds is less than that required to break the benzene–benzene bonds plus the rather strong water–water bonds.

A solute that is ionic or highly polar requires more energy for bond breaking than would be provided by the potential bond-forming interactions with a weakly polar solvent; hence, such substances may only be dissolved in a highly polar solvent. For example, sodium acetate dissolves in water, but not in ether. The ionic bonds of the salt require more energy than can be furnished by the formation of bonds between the separate ions and the weakly polar ether molecules.

A most important application of polarity to solubility is the possibility of forming *hydrogen bonds*. The hydrogen bond results from a single proton sharing two sources of electrons rather than being satisfied by one. Usually the hydrogen atom serves as a bridge between two strongly electronegative atoms, being attached to one atom by a covalent bond and attracted to the other atom by an electrostatic force, the strength of which is less than that of the covalent bond. The energy of formation of the hydrogen bond is 5–10 kcal/mole. For the hydrogen bond to have an important effect on solubility, the hydrogen must be covalently bonded to either oxygen or nitrogen. In such a case, the negative end of the other dipole may be nitrogen, oxygen, or fluorine. Thus, hydrogen bonds may be represented by such combinations as $O—H \cdot \cdot N$, $O—H \cdot \cdot O$, $N—H \cdot \cdot O$, $N—H \cdot \cdot F$, and the like. Intermolecular hydrogen bonding is commonly called *association*, and intramolecular hydrogen bonding is called *chelation*.

To predict the solubility of a solute in a solvent, certain characteristics of each must be considered. Regarding the solute, the following are important: (1) the ionic, polar, or essentially nonpolar character; (2) the molecular weight, particularly with respect to the ratio of the molecular weight to the number of polar groups present in the molecule and the magnitude of such polarity; (3) the potentiality of forming hydrogen bonds; and (4) the structure of the molecule, such as the extent of branching of any alkyl radicals or the location of substituents on aromatic structures. For the solvent, the most important properties are: (1) the type and strength of the bonds involved in maintaining the liquid state; (2) the ability to act as an acceptor or a donor, or both, in hydrogen bond formation; (3) the magnitude of the dielectric constant; and (4) the acid-base character of the solvent. In general, the more nearly the properties of the solute and the solvent correspond, the more readily will they form a solution.

Solubility in Water

Water is a strongly associated liquid since its molecules may act as both donor and acceptor molecules in hydrogen bond formation. It has a high dielectric constant (80) and is highly polar. Also, it may act as either an acid or a base. These properties, often with overlapping effects, determine the types of substances that will dissolve in water. Water dissolves ionic materials because (1) its strong dipole aids in breaking the cation-anion bond, (2) the oxygen end of the water dipole attracts a cation and the hydrogen end of the water dipole attracts an anion, thereby hydrating the ions and dispersing the charges over the larger hydrated ions, (3) the water molecules of the hydrated ions can hydrogen bond to additional water molecules, and (4) the high dielectric constant of the water reduces the attraction between oppositely charged hydrated ions, thereby allowing them to remain separate and dispersed.

Salts of the ammonium or alkali metal cations with organic acids are generally soluble in water (Division S_2) (see page 93 for explanation of the symbols used for solubility divisions). In general, salts of other cations are not sufficiently soluble in water to fall in Division S_2. Many of these water-insoluble salts will dissolve in $1.2N$ hydrochloric acid and hence fall in Solubility Division **B**. Therefore, if in testing an unknown, nitrogen was not found present but the substance falls in Solubility Division **B**, a salt should be suspected and a sample should be investigated by ignition (see page 109).

The water molecule is both an acid and a base; hence, it may ionize an amine by donating a proton to the amine or it may ionize an acid by accepting a proton from the acid. These reactions may be shown as follows:

$$B: + HOH \rightleftharpoons B:H^+ + OH^-$$
$$RCOOH + H_2O \rightleftharpoons RCOO^- + H_3O^+$$

The number of acids and bases that can be ionized in this way by water alone is limited, and it is probable that most acids and bases that dissolve in water do so more by hydrogen bonding than as a result of ionization. However, water-insoluble acids and bases may be converted to soluble ions by reaction solvents discussed in later sections of this chapter.

Nonionic substances do not dissolve in water unless they are capable of forming hydrogen bonds with water molecules. The carbon–hydrogen bond is not sufficiently polar to allow hydrogen bonding between water and a hydrocarbon radical. Polar organic compounds contain hydrocarbon radicals of various sizes and shapes, and these radicals do not have properties that make them water-soluble on their own account. Whether or not a given polar compound will be soluble in water will, therefore, depend on what effect the polar groups present have on solubility relative to the effect of the nonpolar hydrocarbon radical or radicals that are also present in the molecule. Water molecules can hydrogen bond to molecules that have either oxygen or nitrogen as the negative end of a strong dipole, or they can bond to molecules that have hydrogen as the positive end of a strong dipole.

Whether compounds are classified as "soluble" or "insoluble" will depend on the arbitrary limits set. *In our scheme, a substance is to be classed as soluble if 30 mg of it dissolve in 1 ml of the solvent at room temperature.* On this basis, it has been found that one polar group capable of hydrogen bonding in a molecule that does not have more than five carbons in it is usually capable of "dragging" the alkyl radical or radicals into solution in water. Isomeric alkyl radicals have different effects on solubility; "normal" radicals inhibit solubility more than "branched" radicals. For solubility purposes, the phenyl radical has about the same effect as a butyl radical. If two polar groups are present in the same molecule, more carbon atoms may, in general, be present and still allow the substance to dissolve in water; however, the number of carbon atoms per polar group can only be three or four.

While isomerism within the hydrocarbon radical affects solubility, isomerism with respect to the location of the polar group within the molecule has a much greater effect. The solubilities of the isomeric pentanols in water illustrate both of these factors, as is shown in Table 5.1. It will be noted (Table 5.1) that, for primary alcohols, isomerism within the alkyl

TABLE 5.1

Solubility of Pentanols in H_2O

Name	Formula	Solubility[a]	Boiling point
Primary			
1-Pentanol	CCCCC—OH	2.36	138
3-Methyl-1-butanol	$CC(CH_3)CC$—OH	2.85	134
2-Methyl-1-butanol	$CCC(CH_3)C$—OH	3.18	120
2,2-Dimethyl-1-propanol	$CC(CH_3)_2C$—OH	3.74	113
Secondary			
2-Pentanol	CCCC(OH)C	4.86	120
3-Pentanol	CCC(OH)CC	5.61	116
3-Methyl-2-butanol	$CC(CH_3)C(OH)C$	6.07	114
Tertiary			
2-Methyl-2-butanol	$CCC(CH_3)(OH)C$	12.15	102

[a] Expressed in g/100 g of H_2O at 20°.

radical increases the solubility about 50 per cent above the 1-pentanol value, whereas shifting the hydroxyl group from the first to the second carbon in the normal chain more than doubles the solubility, and the 3-pentanol is even more soluble. Where the effects of both types of isomerism are combined, the solubility is still greater. The tertiary alcohol is more than five times as soluble in water as the normal primary alcohol. The inductive effect of the groups attached to the alpha carbon may be the most important factor in the solubility differences mentioned. Palit[2] considers the solubility differences to be due to the effect of the hydroxyl group on the electron displacements in the different alkyl radicals.

The diffusion of solute molecules into a solvent during the process of solution is a phenomenon somewhat resembling the evaporation of a liquid. If the boiling points of the alcohols in Table 5.1 are noted and compared with the solubilities of these alcohols, it will be seen that the lower boiling alcohols of each subclass are more soluble in water. Jordan[3] has published very extensive data on the vapor pressures of organic compounds at various temperatures. Interesting studies may be made regarding the water solubility of members of a given homologous series by noting which members of the series have vapor pressures of at least 5 mm at 20°.

Prediction of probable solubility based on the number of carbon atoms present in the radicals is more reliable when applied to liquids than to solids of equal carbon content, because the solid state involves greater intermolecular forces that must be overcome before solubility may occur. If two solids have the same heats of fusion, the one with the lower melting point will be the more soluble in any given solvent. Similarly, if two solids have the same melting point, the one with the lower heat of fusion will be the more soluble. Table 5.2 shows the relationship of both carbon content and melting point to solubility in water for some dibasic acids. It will be noted that the diprotic acids with an even number of carbon atoms have higher melting points and lower solubilities than the next higher homolog. In the case of cis-trans isomers, the cis form usually has the lower melting point and the higher solubility.

Compounds that are not appreciably soluble in water may react with water to form water-soluble products. Even the low-molecular-weight anhydrides and acid halides are not very soluble in water, but they are hydrolyzed to form water-soluble acids. Such compounds are classed as water-soluble.

Since water dissolves low-molecular-weight acids, bases, and neutral compounds, the acid-base character classification in Part II of this chapter is particularly suitable for compounds that are soluble in water.

[2] Palit, *J. Phys. & Colloid Chem.* **51**, 837 (1947).
[3] Jordan, *Vapor Pressure of Organic Compounds* (N.Y.: Interscience, 1954).

TABLE 5.2

Solubility of Diprotic Acids in H_2O

Name	Formula	Solubility[a]	Melting point
Oxalic	HOOC—COOH	9.5	189
Malonic	HOOC—CH_2—COOH	73.5	135
Succinic	HOOC$(CH_2)_2$COOH	6.8	188
Glutaric	HOOC$(CH_2)_3$COOH	64	97
Adipic	HOOC$(CH_2)_4$COOH	2	153
Pimelic	HOOC$(CH_2)_5$COOH	5	105
Suberic	HOOC$(CH_2)_6$COOH	0.16	144
Azelic	HOOC$(CH_2)_7$COOH	0.24	107
Fumaric (trans)	HOOC$(CH_2)_2$COOH	0.7	300
Maleic (cis)	HOOC$(CH_2)_2$COOH	75	130

[a] Expressed in g/100 g of H_2O at 20°.

Solubility in Ethyl Ether

As a potential solvent, ether differs from water in two important respects, (a) it has a dielectric constant of 4.3 compared to 80 for water, and (b) it is unassociated, that is, its molecules are not hydrogen-bonded to each other. The low dielectric constant indicates that ether can act only as a weakly polar solvent. Ionic compounds, such as salts, are not soluble in ether because the cation-anion attraction is too great to be overcome by the weak dipole of the ether and therefore the ions do not separate from their lattice structure. Furthermore, because of the effectively weak dipole, ether does not have a tendency to solvate ions comparable to the ability of water molecules to hydrate ions. The C—H··O type of hydrogen bond is too weak to allow ether molecules to form association complexes among themselves. For the same reason, ether cannot bond to the molecules of other compounds through the hydrogen atoms of its ethyl radicals. The negative electrostatic charge on the oxygen atom of ether is weak, but it is sufficiently strong to allow hydrogen bonding with molecules that have strong dipoles with hydrogen as the positive end. For example, several crystallizations from hexane fail to remove benzoic acid as an impurity in a benzoate ester, but the addition of 5 per cent ether to the hexane causes the benzoic acid to be held in solution by hydrogen bonding to the ether.

Nonpolar and slightly polar substances will, in general, dissolve in ether because, like ether itself, they are largely unassociated, that is, only very weak bonds have to be broken. Whether or not a polar compound will dissolve in ether depends on the influence of the polar group or groups relative to the influence of the nonpolar radical or radicals that are present.

In general, compounds that have only one polar group per molecule will dissolve in ether unless they are very highly associated or of extreme polarity, such as the sulfonic acids.

Most organic compounds that are not soluble in water are soluble in ether. Hence, solubility in ether is not a useful criterion for classification except for those compounds that are also soluble in water. Some generalizations may be deduced from the foregoing discussions. If a substance is soluble in both water and ether, it most likely is (1) nonionic, (2) contains five or less carbon atoms, (3) has a functional group that is polar and capable of forming hydrogen bonds, and (4) does not contain more than one strongly polar group. Such compounds are classified as Division S_1 compounds in Table 5.5 on pages 96–97. If a compound is soluble in water but not in ether, it may (1) be ionic, or (2) contain two or more polar groups with not more than four carbon atoms per polar group. Such compounds are classified as Division S_2 compounds in the table. As with all generalizations, there are exceptions to the above "rules."

Since neutral, acidic, and basic substances belong in both the S_1 and S_2 Solubility Divisions, the use of indicators to detect these characteristics is strongly recommended as a step in the identification of substances in these solubility divisions.

Solubility in Dilute Hydrochloric Acid

As has been mentioned, the low-molecular-weight amines are soluble in water and hence belong in either Division S_1 or S_2. Amines that are not soluble in water may be dissolved provided they can be converted into the ionic form while in water. The basicity of an amine is due to the presence of an unshared pair of electrons on the nitrogen atom. The strength of an amine as a base will, therefore, be determined by the availability of this electron pair for reaction with an acid. Since amines have one, two, or three alkyl or aryl radicals attached to the nitrogen atom, and since these radicals have varying electron-attracting or electron-repelling tendencies, it is obvious that amines will vary greatly with regard to basicity. Those amines that are basic enough to be converted to ionic salts by a $1.2N$ solution of hydrochloric acid, thus dissolving in water even though the molecular amine was not soluble in water, are designated as Solubility Division **B** compounds in this scheme. Certain types of amines are not basic enough to react with this concentration of acid and thus fail to be detected as amines by this reagent. Generally, if the amine is too weak a base to dissolve in $1.2N$ hydrochloric acid, it will not be basic enough to affect the indicators used in Part II of this chapter. Such compounds are classed in Solubility Division **M** and must be identified by chemical tests.

Effect of Different Radicals on the Basicity of Amines

Alkyl radicals have about the same electronegativity as hydrogen. Therefore alkyl amines have about the same basicity as ammonia. However, alkyl radicals do supply a small induced negative charge to the nitrogen atom, that is, these radicals attract an electron pair a little less (or repel it a little more) than does hydrogen. The alkyl amines are about 25 times as basic as ammonia, as the following K_b values show: ammonia, 1.8×10^{-5}; methylamine, 4.4×10^{-4}; dimethylamine, 5.1×10^{-4}; ethylamine, 4.7×10^{-4}; and diethylamine, 9.5×10^{-4}.

Alkyl amines have K_b values of the order of 10^{-4}, whereas the aryl amines have values of the order of 10^{-10} or less. Aniline is only one-millionth as strong a base as methylamine. As is true of all aromatic compounds, both the free aryl amines and their conjugate cation acids (substituted ammonium ions) are resonating systems. In aryl amines the extra pair of electrons on nitrogen is partly shared with the ring, becoming involved in the pi electron cloud that surrounds the ring, and the basicity is low.

Aryl amines may have one or more substituents on the aryl radical or radicals that are present. The effect of these substituents on the basicity of the aryl amines is of considerable importance. If the substituent group tends to release electrons to the ring (amino, methyl, and methoxy groups are examples), the basicity of the amine is increased. On the other hand, if the substituent group tends to withdraw electrons from the ring (such groups as NH_3^+, $-NO_2$, NO, $-COOH$, $-CN$, or the halogens), the basicity of the amine is decreased. One exception must be mentioned. If the substituent is in the ortho position with respect to the amino nitrogen, the basicity is decreased regardless of whether the group normally tends to release electrons or not. Electron-withdrawing groups reduce the basicity much more than electron-releasing groups do. This "ortho effect" has not been satisfactorily explained. For meta or para substituents, electron release tends to disperse the positive charge on the ion, thus stabilizing it with respect to the free amine form. Electron withdrawal by a substituent tends to intensify the positive charge on the ion, thus destabilizing it with respect to the free amine. Another way of explaining these facts is to say that electron-releasing groups push electrons toward the nitrogen, thereby making the fourth electron pair on nitrogen more readily available for sharing with an acid. Electron withdrawal from the ring by a substituent tends to pull electrons from nitrogen, thereby making the fourth pair less available for reaction.

In addition to electronegativity effects, some of the substituted aryl amines have reduced basicity as a result of association. Intermolecular hydrogen bonding occurs among the meta and para isomers of the nitroanilines and the aminophenols. Chelation occurs in the case of the ortho isomers of these classes.

The presence of an aromatic structure in an amine does not appreciably affect the basicity unless the ring is attached directly to the nitrogen so that the nitrogen electrons may be directly involved in the pi electron system of the ring. For example, benzylamine is very soluble in water and has a K_b of 2×10^{-5}, which is comparable with methylamine rather than with the isomeric toluidides, which are sparingly soluble in water and have K_b values of the order of 10^{-10}. The direct attachment of two aryl radicals to a nitrogen atom almost completely removes the basic character of the nitrogen. For example, the K_b of diphenylamine is 7×10^{-14}. It is not always possible to predict the basicity or the solubility in water of a given amine from the data for a closely related amine. For example, p-phenylenediamine boils 20° lower than m-phenylenediamine and has a K_b that is 14 times greater than the meta isomer; yet, the meta isomer is more than 6 times as soluble in water as the para isomer. In this case, a hint is given as to the reason for the greater solubility of the meta compound. The para isomer melts at 141° while the meta isomer melts at 63°. Obviously, the bonding energy is much greater in the para compound, and this would reduce the solubility in water.

Table 5.3 provides data to illustrate the effect of different radicals on

TABLE 5.3

K_b's of Selected Amine Bases

Amine	$10^{10} K_b$	Amine	$10^{10} K_b$
Aniline	4	Aniline	4
N-Methylaniline	7	o-Chloroaniline	0.05
N,N-Dimethylaniline	11	m-Chloroaniline	0.3
o-Toluidine	3	p-Chloroaniline	1.5
m-Toluidine	5	o-Nitroaniline	0.0004
p-Toluidine	12	m-Nitroaniline	0.03
o-Phenylenediamine	3	p-Nitroaniline	0.001
m-Phenylenediamine	8	Benzidine	0.007
p-Phenylenediamine	110	Diphenylamine	0.0007

the basicity of aryl amines. The K_b values listed are for water as a solvent and are approximate rather than exact values.

In addition to the amines discussed above, certain other classes of water-insoluble compounds may be dissolved by $1.2N$ hydrochloric acid. These include many amphoteric compounds, such as amino acids, amino phenols, amino sulfonamides, and some ketoximes. Certain amides of the type

RCONR$_2$ also dissolve in the acid. In many cases, the amphoteric compounds exist as the dipolar ion in water, and their solubility will increase on either side of the isoelectric point.

Solubility in Dilute Sodium Hydroxide

Water-insoluble compounds that are capable of donating protons may react with bases to form products that are soluble in water. The stronger the acid, the weaker the base that is required to cause a reaction. Three bases are used in this scheme for classification of solubility: water, the bicarbonate ion, and the hydroxyl ion. The action of water as a base was discussed in an earlier section of this chapter. The usefulness of the bicarbonate ion will be covered in a later section, which compares the solubility in sodium bicarbonate with solubility in sodium hydroxide solution. For water-insoluble acidic compounds, aqueous sodium hydroxide may be called the detecting solvent and aqueous sodium bicarbonate a subclassifying solvent. The bicarbonate ion is a much weaker base than the hydroxyl ion and hence only reacts with the more acidic substances. In practice, the water-insoluble compound is tested with $2.5N$ sodium hydroxide, and, if it is soluble, another sample of the compound is tested in $1.5N$ sodium bicarbonate. Substances that dissolve in both reagents are classified in Solubility Division A_1, while substances that are soluble in sodium hydroxide but not in sodium bicarbonate are classified as A_2.

Acids owe their acidic character to the presence in the molecule of —OH, —SH, =NH, or, in special cases, —NH$_2$ groups that are attached to electronegative radicals. The relative electronegativity of alkyl and aryl radicals and the effect of substituents on these radicals was partially discussed in the section of this chapter that deals with substances soluble in dilute hydrochloric acid. Particular attention must be given to the sulfonyl, aroyl, and acyl radicals and their substitution products. The order of electronegativity is:

$$\text{sulfonyl} > \text{aroyl} > \text{acyl} > \text{aryl} > \text{alkyl}$$

All of these radicals are potential components of molecules that are acids. The greater the electronegativity of the radical or radicals, the stronger the compound will be an acid. Classes of acids in which the proton is removed from a hydroxyl group include the sulfonic, sulfinic, and carboxylic acids; phenols, enols, hydroxamic acids, oximes, the *aci* form of primary and secondary nitroparaffins, and several less common classes. Thioacids, thiophenols, and mercaptans represent the classes which contain the proton donating group —SH. Classes from which the proton may be removed from a nitrogen atom by the hydroxyl ion include the sulfonamides and *N*-monoalkyl sulfonamides, the imides of both aliphatic and aromatic acids, and some *N*-monomethyl aromatic amides.

The water-insoluble sulfonic and carboxylic acids, phenols and thio-phenols, and mercaptans (thiols) dissolve in dilute sodium hydroxide because of the production of water-soluble salts of the conventional types. In other cases, the total mechanism is less clear. For example, the acidity of triacylmethanes, 1,3-diketones, beta ketoesters, and the like, may be explained either by the direct ionization of a carbon–hydrogen bond (C—H → C$^-$ + H$^+$) or by the enolization of the molecule before reaction with the hydroxyl ion. Similarly, the acidity of molecules such as nitromethane may be explained by the assumption that the normal CH_3NO_2 isomerizes to CH_2=$N(O)OH$, or it may be due to the resonance stabilization of the ion form after a proton is removed.

The Effect of Substituents on the Strength of Acids

Among the aromatic acids, substituents that release electrons to the ring decrease the acidity and substituents that withdraw electrons from the ring increase the acidity. In general, the same statements hold for the substituted phenols. The ortho-substituted acids have K_a values that are abnormal.

Although substituted alkyl radicals are not common in the amines, they are quite common among the acids. Electron-withdrawing groups substituted on the carbon atom adjacent to the carboxyl group increase the acidity of the acid in the order of their electronegativity. When the electronegative group is moved farther down-chain from the carboxyl group, the effect is very greatly reduced. By comparing the K_a values in Table 5.4, it will be noted that the substituted aliphatic and benzoic acids as well as substituted phenols support the statements. Electron-releasing groups, such as the alkyl radicals, reduce the acidity if they are attached to the carbon adjacent to the carboxyl group.

Comparing the effect of various substituents on the basicity of aryl amines, such as aniline, with the effect of these same groups on the acidity of an aromatic acid, such as benzoic acid, it is logical that the effects should be opposite.

Solubility in Dilute Sodium Bicarbonate

Both sodium hydroxide and sodium bicarbonate solutions are used to detect acidic, water-insoluble substances. The use of the bicarbonate solution only serves to distinguish relatively strong acids from weak acids. Indicators serve the same purpose and should be used particularly if the solubility difference is inconclusive. The most common classes of compounds that are soluble in both of the basic reagents are the carboxylic and sulfonic acids and their anhydrides and acid halides.

Carboxylic acids have K_a's in the range of 10^{-3} to 10^{-5}, while phenols have K_a's in the range of 10^{-9} to 10^{-10} (except for those with two or more

TABLE 5.4[a]
Effect of Substituents on K_a of Aliphatic Acids

Position on the chain	$10^5 K_a$	Number of substituents	$10^5 K_a$
Butyric	1.5	Acetic	1.8
α-Chlorobutyric	140	Chloroacetic	140
β-Chlorobutyric	9	Dichloroacetic	5,000
γ-Chlorobutyric	3	Trichloroacetic	13,000

Different Substituents as X in XCH_2COOH

X-group	$10^5 K_a$	X-group	$10^5 K_a$
H	1.8	I	71
CN	342	CH_3O	30
F	213	C_6H_5	5
Cl	140	C_2H_5	1.5
Br	138	CH_3	1.3

Different Substituents as X in XC_6H_4COOH
($10^5 K_a$)

X-group	Ortho	Meta	Para
H	6.3	6.3	6.3
NO_2	671	32	38
Br	140	15	11
Cl	120	15	11
F	54	14	7
CH_3	12	5	4
CH_3O	8	4	4

Selected Phenols
($10^{10} K_a$)

Phenol	1.1	o-Nitrophenol	600
o-Cresol	0.63	m-Nitrophenol	50
m-Cresol	0.98	p-Nitrophenol	690
p-Cresol	0.67	o-Aminophenol	2
o-Chlorophenol	77	m-Aminophenol	69
m-Chlorophenol	16	Catechol	1
p-Chlorophenol	6.3	Resorcinol	3
o-Bromophenol	41	Hydroquinone	2
m-Bromophenol	14	2,4-Dinitrophenol	(10^{-4})
p-Bromophenol	5.6	2,4,6-Trinitrophenol	(10^{-1})

[a] The data for acids in the above table were adapted from Ferguson, *Electron Structures of Organic Molecules* (N.Y.: Prentice-Hall, 1952); the data for phenols were adapted from Morrison and Boyd, *Organic Chemistry* (Boston: Allyn and Bacon, 1959).

very strong electronegative groups present on the ring). The K_a for the first ionization of H_2CO_3 is about 10^{-7}. The phenols and similarly weak acids are stronger acids than water, but weaker than carbonic acid. Therefore, such substances can react with sodium hydroxide solutions, but not with sodium bicarbonate solutions.

Phenols that have two or three nitro groups are also soluble in the sodium bicarbonate solution as well as in sodium hydroxide. Carboxylic acids that contain more than ten carbon atoms form colloidal dispersions in sodium hydroxide and disperse so slowly in sodium bicarbonate that they are frequently not properly classified as soluble unless the mixture is shaken for several minutes.

Amino-substituted aromatic acids, amino-substituted sulfonic acids, and N,N-diarylamino acids are soluble in sodium bicarbonate if only one amino group is present. See Table 5.5 for other classes that belong to Solubility Divisions A_1 and A_2.

Solubility in Concentrated Sulfuric Acid

If a compound is not soluble in water, dilute hydrochloric acid, or dilute sodium hydroxide, and if it contains nitrogen or sulfur, no further solubility tests are generally useful and it is assigned to the miscellaneous class, Solubility Division **M**. However, if the compound does not contain nitrogen or sulfur, it may be further classified by using concentrated sulfuric acid as a solvent. Three major types of compounds that have failed to dissolve in the previously used solvents do dissolve in concentrated sulfuric acid: (1) oxygen-containing compounds that act as Lewis bases and react with sulfuric acid to form oxonium salts; (2) alkenes, cycloalkenes, and alkynes; and (3) a limited number of easily sulfonated aromatic hydrocarbons. The classes that do not dissolve in sulfuric acid are the alkanes, cycloalkanes, most aromatic hydrocarbons, the halogen derivatives of these hydrocarbons, the diaryl ethers, and perfluoro esters, ethers, aldehydes, and ketones.

Organic compounds that contain oxygen, such as the alcohols, ethers, esters, aldehydes, ketones, and the like, are all weak bases due to the pairs of unshared electrons on the oxygen atom. These bases react with sulfuric acid to form oxonium salts which are soluble in excess sulfuric acid. The oxygen compounds are much weaker bases than the corresponding nitrogen compounds; for example, consider the relative basicity of water and ammonia or ethyl ether and diethylamine. Since these oxonium salts are derived from such weak bases, they react with the stronger base, water. If a solution of such compounds in concentrated sulfuric acid is poured over chipped ice, the weak bases are liberated and separated from the diluted acid.

$$R : \overset{..}{\underset{..}{O}} : R + H_2SO_4 \rightleftharpoons R : \overset{..}{\underset{\overset{..}{H^+}}{O}} : R + HSO_4^-$$

$$R : \overset{..}{\underset{\overset{..}{H^+}}{O}} : R + H : \overset{..}{\underset{..}{O}} : H \rightleftharpoons R : \overset{..}{\underset{..}{O}} : R + H : \overset{..}{\underset{\overset{..}{H^+}}{O}} : H$$

Actively unsaturated hydrocarbons generally react with sulfuric acid by addition to form alkyl sulfuric acids (alkyl hydrogen sulfates), which, through hydrogen bonding, are soluble in the excess acid. In some cases, the hydrocarbons polymerize, and the polymerization product is not soluble in the acid. For purposes of classification such hydrocarbons should be listed as "soluble" in the reagent.

Some aromatic hydrocarbons, particularly the meta isomers of di- or polyalkyl-substituted benzene, are easily sulfonated, and the sulfonic acids thus produced are soluble. The majority of aromatic hydrocarbons are not soluble. Perfluoro-alcohols, esters, aldehydes, ketones, and the like are not soluble in concentrated sulfuric acid and therefore fall in Division **I**.

Determining the Solubility for Classification

As already stated in this text, a compound is considered soluble in a solvent if 30 mg of it dissolve in 1 ml of the solvent at room temperature. In doubtful cases, the mixture should be shaken for 2 minutes before deciding whether the compound has dissolved or not. A narrow, 3-inch test tube is recommended for the solubility determinations. Until enough experience has been gained to make reasonably reliable estimates of the weights of small samples of solids, the test substances should be weighed on a balance that is sensitive to 5 mg. Liquids may, of course, be weighed, or the drops delivered by a pipet may be weighed for several representative liquids until a good estimate is obtained as to the weight of an average liquid drop.

The solvents for the solubility classification are, in the order in which they should be used, *water, ethyl ether,* (benzene may be used instead of ether and about the same classification will result), 1.2*N hydrochloric acid,* 2.5*N sodium hydroxide,* 1.5*N sodium bicarbonate,* and *concentrated sulfuric acid.* All solubilities are to be determined at room temperature. However, it is permissible to warm the mixture *slightly* during the test providing it is cooled to room temperature before making a decision as to solubility. Experience should be gained by determining the solubility classification of several known compounds.

It is *not necessary* to test every compound in *all* of the solvents. Rather, the solvents should be used in the order listed above and the first solubility class found should be accepted. If the compound is soluble in water, it should be tested in ether, but in no other solvent. If the compound is not soluble in water, it need not be tested in ether; rather, proceed to test it in hydrochloric acid. Compounds that are soluble in hydrochloric acid may be amphoteric; hence it is wise also to test them in sodium hydroxide to resolve that possibility. Water-insoluble compounds that are soluble in sodium hydroxide also should be tested in the bicarbonate solution to distinguish between weak and relatively strong acids. If an obvious chemical reaction occurs in any solvent (such as the liberation of carbon dioxide when the substance is placed in sodium bicarbonate solution), the compound should be classed as probably soluble in that reagent even though it fails to dissolve completely.

In making the solubility tests, keep in mind the elements that were found present in the compound. Unless nitrogen is present, there is no need to test the solubility in hydrochloric acid. If the compound contains nitrogen or sulfur and is insoluble in water, hydrochloric acid, and sodium hydroxide, it is arbitrarily placed in a special "miscellaneous" division without further testing.

In some cases, additional information may be obtained by using stronger concentrations of the reagents. For example, the nitroanilines may not be classified as soluble in $1.2N$ acid, but some dissolve in $3N$ acid. The use of the stronger acid may, however, lead to some confusion since compounds such as the N-alkyl acetanilides also dissolve in the stronger acid.

If there is doubt as to whether or not an appreciable amount of a substance has dissolved in the acid or base reagents, the clear solution should be separated from the remaining compound and neutralized to see if the original compound will separate.

The quantity of the compound to be used for solubility determinations may be conserved in several ways when necessary. For determining the solubility in water or ether, a few milligrams of the compound may be added to 2 or 3 drops of the solvent. The substance may be recovered from the ether by evaporating the ether. If the substance is not soluble in water, 5 drops of $6N$ hydrochloric acid may be added to the milliliter of water to test the solubility in the acid. If the compound is insoluble in the acid solution, $6N$ sodium hydroxide may be added, dropwise, until the solution is definitely basic. Such methods are not equivalent to the regular method, but they are adequate for detecting lack of solubility and at least approximate the regular classification solvents for situations where the quantity of the compound to be examined is very limited.

Designations for the Solubility Divisions

Division S_1. This division includes compounds that are soluble in both water and ether.

Division S_2. This division includes compounds that are soluble in water, but insoluble in ether.

Division B. This division includes compounds that are insoluble in water, but dissolve in $1.2N$ hydrochloric acid. They all contain nitrogen.

Not all amines will dissolve in this dilute acid and some are found in **Division M.**

Division A₁. The compounds in this division are soluble in both $2.5N$ sodium hydroxide and $1.5N$ sodium bicarbonate.

Division A₂. This division includes the compounds that are not soluble in water, do dissolve in $2.5N$ sodium hydroxide, but are insoluble in $1.5N$ sodium bicarbonate.

Division M. Compounds that contain nitrogen or sulfur and that have been found insoluble in water, hydrochloric acid, and sodium hydroxide are placed in this miscellaneous division. Halogens may, of course, be present in addition to nitrogen or sulfur. Only the most common classes are listed in Table 5.5.

Division N. Compounds that are soluble in concentrated sulfuric acid and that did not belong in any of the previous divisions belong in this division. Of course, they do not contain nitrogen or sulfur; halogens are not common in compounds that belong in this division.

An outline of the solubility classification procedures appears below.

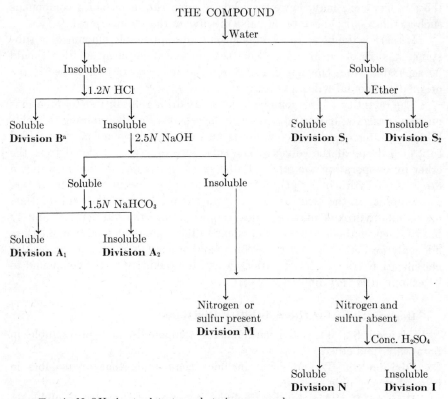

ᵃ Test in NaOH also to detect amphoteric compounds.

Division I. Compounds that are insoluble in all of the classification solvents and that do not contain nitrogen or sulfur are placed in this "insoluble" class.

PART II. CLASSIFICATION BY THE ACID-BASE CHARACTER

The classification of an unknown by its solubility behavior and elemental analysis is often an adequate basis for proceeding directly to functional group tests to identify the chemical class of the unknown. However, indicators may give assistance in classifying the compound according to its relative acidity, neutrality, or basicity, particularly with water-soluble compounds and with compounds that show borderline solubility behavior. Acids and bases among organic compounds were covered in a general way in Part I of this chapter; but for detailed coverage, reference should be made to the literature. The use of indicators for classification purposes as used in this text is based on the work of Davidson.[2]

We shall consider an acid as a proton donor and a base as a proton acceptor. Using generalized formulas, without regard to the charges present, and assuming that HB represents the acid form and B^- the conjugate base of that acid, we may write the equilibrium equation

$$HB \rightleftharpoons H^+ + B^- \tag{1}$$

However, the relative concentrations of HB and B^- at equilibrium in a solution will be determined, in part, by the other acids besides HB and the other bases besides B^- that are present in the solution. Consider, for example, the solvents water and methanol, both of which are capable to different degrees of acting as either acids or bases. If the acid HB is placed separately in water and in methanol, and if these solvents are acting as bases with respect to acid HB, we may write

$$HB + H_2O \rightleftharpoons H_3O^+ + B^- \tag{2}$$

and

$$HB + CH_3OH \rightleftharpoons CH_3OH_2^+ + B^- \tag{3}$$

Since water and methanol are not equally strong bases, it is obvious that the ratio of the acid HB to its conjugate base B^- at equilibrium will not be the same in the two solvents. For equation (1) above, we can write the equation for the value of K_a as

$$K_a = \frac{[H^+][B^-]}{[HB]} \tag{4}$$

This equation is also applicable to equations (2) and (3), but the numerical values of K_a will not be the same in the two solvents. As the numerical values

---

[2] Davidson, *loc. cit.*; Davidson and Perlman, *A Guide to Qualitative Organic Analysis* (N.Y.: Brooklyn College, 1952).

TABLE 5.5

Divisional Solubility and Acid-Base Classifications[a,b]

Division S_2[c] (soluble in water; insoluble in ether)	**Division S_1[d]** (soluble in water and ether)

1. Only C, H, and O present:[b]
  DIBASIC AND POLYBASIC ACIDS (A_i or A_s)
  HYDROXY ACIDS (A_i)
  POLYHYDROXY ALCOHOLS (N)
  POLYHYDROXY PHENOLS (A_w)
  Simple carbohydrates (N)
2. Metals present:
  SALTS OF ACIDS AND PHENOLS
  Miscellaneous metallic compounds (A_m)
3. Nitrogen present:
  AMMONIUM SALTS (B_w)
  AMINE SALTS OF ORGANIC ACIDS (A_m)
  AMINO ACIDS (A_m)
  Amino alcohols (B_w)
  Amides (N)
  Amines (B_i or B_w)
  Semicarbazides (B_w)
  Semicarbazones (N)
  Ureas (N or B_w)
4. Halogen present:
  HALO ACIDS (A_i or A_s)
  Halo alcohols, aldehydes, etc. (N)
  Acyl halides (by hydrolysis) (A_s)
5. Sulfur present:
  SULFONIC ACIDS (A_s)
  Alkyl sulfuric acids (A_s)
6. Nitrogen and halogen present:
  Amine salts of halogen acids (A_w or A_i)
7. Nitrogen and sulfur present:
  AMINOSULFONIC ACIDS (A_i)
  BISULFATES OF WEAK BASES (A_i)
  Cyanosulfonic acids (A_s)
  Nitrosulfonic acids (A_s)
	1. Only C, H, and O present:[b]
	  CARBOXYLIC ACIDS (A_i)
	  ALCOHOLS (N)
	  ALDEHYDES AND KETONES (N)
	  Anhydrides (A_i or A_w)
	  Esters (A_w or N)
	  Ethers (N)
	  Some phenols (A_w)
	  Some glycols (N)
	  Some lactones (A_w or A_i)
	  Some acetals (N)
	2. Nitrogen present:
	  AMIDES (N or A_w)
	  AMINES (B_i or B_w)
	  Nitriles (N)
	  Nitroparaffins (N or A_w)
	  Oximes (A_w)
	  Amino heterocyclics (B_w)
	3. Halogen present:
	  Halogen-substituted compounds of 1
	  above
	4. Sulfur present:
	  Hydroxy heterocyclic sulfur compounds
	    (A_w)
	  Mercapto acids (A_i or A_s)
	  Thio acids (A_i or A_s)
	5. Nitrogen and halogen present:
	  Halogenated amines, amides, and nitriles
	    (B_w or N)
	6. Nitrogen and sulfur present:
	  Amino heterocyclic sulfur compounds
	    (B_w or N)
	
	       Division B[e]
	      (soluble in HCl)
	
	AMINES (B_i or B_w)
	Amino acids (A_m)
	Aryl-substituted hydrazines (B_w)
	N,N-Dialkylamides (B_w)
	Amphoteric compounds
	  (A_iB_w, A_wB_w, A_wB_i)
	Metallic salts

   [a] Nitrogen, halogens, and sulfur are absent unless specified.

   [b] In this table, the more common classes are printed in SMALL CAPITAL letters.

   [c] Moderate-weight compounds with two or more polar groups, except for the sulfonic and sulfinic acids when only one polar group is necessary.

   [d] Generally monofunctional compounds with five carbons or less.

   [e] Amines with sufficiently strong negative substituents as well as diaryl and triaryl amines fall in Division M.

   [f] Generally with ten carbons or less; many form colloidal soap solutions.

   [g] High-molecular-weight acids form colloidal soaps.

   [h] Including N-monoalkylamides.

   [i] Only the most common classes are listed.

   [j] Noncyclic unsaturated hydrocarbons and those unsaturated cyclics that are easily sulfonated, such as di- or polyalkyl-substituted benzenes.

   [k] Char in the acid.

   [l] Including most of the cyclic hydrocarbons and all of the saturated, noncyclic hydrocarbons.

TABLE 5.5 (Continued)

Divisional Solubility and Acid-Base Classifications[a,b]

Division A₁
(soluble in NaOH and NaHCO₃)

1. Only C, H, and O present:
   ACIDS[f] AND ANHYDRIDES (A_i, A_w, A_s)
2. Nitrogen present:
   AMINO ACIDS (A_m)
   NITRO ACIDS (A_i)
   Cyano acids (A_i or A_w)
   Polynitro phenols (A_i)
   Heterocyclic carboxylic acids (A_i)
3. Halogen present:
   HALO ACIDS (A_i or A_s)
   Polyhalo phenols (A_i)
   Acid halides (A_i or A_s)
4. Sulfur present:
   SULFONIC ACIDS (A_s)
   Sulfuric acids (A_s)
   Mercaptans (A_w)
5. Nitrogen and sulfur present:
   Nitro thiophenols (A_i)
   Sulfates of weak bases (A_i or A_s)
   Sulfonamides (A_w)
   Aminosulfonic acids (A_i)
6. Sulfur and halogens present:
   SULFONHALIDES (A_w or A_i)

Division M[i]

1. Nitrogen present:
   ANILIDES AND TOLUIDIDES (N)
   AMIDES (N or A_w)
   NITRO ARYL AMINES (B_w or N)
   NITRO HYDROCARBONS (N)
   Diaryl amines (N)
   Azo, hydrazo, and azoxy compounds (N)
   Dinitro phenylhydrazines (B_w)
   Nitriles (N)
   Aminophenols (A_m)
   Triaryl amines (N)
2. Sulfur present:
   Sulfides (N)
   Sulfones (N)
   Thio esters (N)
   Sulfates (N)
   Bisulfates (A_i)
3. Nitrogen and sulfur present:
   N,N-Dialkyl sulfonamides (N)
   Thiourea derivatives (N or A_w)
   Sulfonamides (A_w)
   Nitrogen-containing sulfonates (N)
   Thiocyanates (N)
4. Nitrogen and halogen present:
   Halogenated amines (N or B_w)
   Halogenated amides (N, B_w)
   Halogenated nitriles (N, B_w)
   Halogenated nitro compounds (N)

Division A₂
(soluble in NaOH; insoluble in NaHCO₃)

1. Only C, H, and O present:
   PHENOLS (A_w or A_i)
   Enols (A_w)
   Some acids[g] and anhydrides (A_i)
2. Nitrogen present:
   AMINO ACIDS (A_m)
   NITRO PHENOLS (A_i)
   Amides[h] (A_w)
   Aminophenols (A_m)
   Cyanophenols (A_i)
   p- and s-Nitroparaffins (A_w)
   Trinitro aromatic hydrocarbons (A_w)
   Oximes (A_w or A_m)
   Ureides (A_w or A_i)
   Imides (A_w)
   Amphoteric compounds (A_m)
3. Halogen present:
   HALO PHENOLS (A_w or A_i)
   Perfluoro secondary and tertiary alcohols (A_w)
4. Sulfur present:
   Mercaptans (thiols) (A_w)
   Thiophenols (A_i)
5. Nitrogen and halogen present:
   Polynitro halogenated aromatic hydrocarbons (A_w)
   Substituted phenols (A_m)
6. Nitrogen and sulfur present:
   Sulfonamides (A_w)
   Aminothiophenols (A_m)
   Aminosulfonic acids (A_i or A_w)
   Thioamides (A_w)
   Aminosulfonamides (A_m)

Division N

ALCOHOLS (N)
ALDEHYDES AND KETONES (N)
ESTERS (A_w or N)
ETHERS (N)
UNSATURATED HYDROCARBONS[j] (N)
Anhydrides (A_w)
Lactones (A_w or N)
Acetals (N)
Polysaccharides[k] (N)

Division I

Hydrocarbons[l] (N)
Halogen derivatives of hydrocarbons (N)
Diaryl ethers (N)
All perfluoro esters, ethers, aldehydes, and ketones (N)

of K_a increase, the acids are said to be increasing in strength. The values of K_a are often given in the form of the negative logarithm of K_a, designated pK_a.

$$pK_a = -\log K_a \qquad (5)$$

It should be noted that K_a and pK_a values vary inversely; a strong acid will have a small pK_a value.

From equation (4)

$$[H^+] = K_a \times \frac{[HB]}{[B^-]} \qquad (6)$$

Since pH $= -\log [H^+]$, we may write

$$pH = pK_a + \log \frac{[B^-]}{[HB]} \qquad (7)$$

From equation (7) we see that if the concentration of an acid and its conjugate base are equal in a solution, the pH of the solution is equal to the pK_a of the acid. Even if the ratio of the concentration of the acid to its conjugate base varies a hundredfold, it would still be true that the pH = $pK_a \pm 1$. By controlling the buffer system in a solution, the pH, which can be determined by indicators, may be approximately equated to the pK_a of the acid.

We have been considering the acidity of acids. It is also important to determine the basicity of bases, expressed as K_b ($-\log K_b = pK_b$). The strength of a base may be determined by the nature of the "proton donor" that is required to convert the base, B$^-$ to the conjugate acid-base system, HB.

Assuming that we wish to consider or measure the basicity of B$^-$ in three separate solvents—water, methanol, and acetic acid, each of which may act as an acid of different acidities—we may write

$$B^- + HOH \rightleftharpoons HB + OH^-, \quad \text{or} \quad K_b = \frac{[HB][OH^-]}{[B^-]} \qquad (8)$$

$$B^- + CH_3OH \rightleftharpoons HB + CH_3O^-, \quad \text{or} \quad K_b = \frac{[HB][CH_3O^-]}{[B^-]} \qquad (9)$$

$$B^- + CH_3COOH \rightleftharpoons HB + CH_3COO^-, \quad \text{or} \quad K_b = \frac{[HB][CH_3COO^-]}{[B^-]} \qquad (10)$$

As for K_a of acids, K_b of bases depends on the acid-base character of the solvent environment in which they are placed. The basic character of a solute will be enhanced in an acidic solvent, and the acidic character of a solute will be enhanced in a basic solvent.

Mixed Indicators in Nonaqueous Solvents

An indicator is a conjugate acid-base system in which the colors of the acid and base are different. The acid form of the indicator may be either a molecule or a cation, thus making the conjugate base an anion or a molecule.

(In rare cases, the acid form may be an anion and the conjugate base a higher valent anion.) The value of the pK_a of an acid differs with different solvents. For example, the pK_a of acetic acid is 4.8 in water, but is 9.7 in methanol. To distinguish clearly weak acids from feeble acids or nonacids, a solvent that is even weaker as an acid than methanol must be used. A mixture of methanol and pyridine is satisfactory for that purpose. Similarly, to classify properly the weak bases, an acid solvent, such as acetic acid, is used. By the use of mixed indicators in appropriate solvents, acids and bases may each be divided into four subclasses: *strong, intermediate, weak,* and *feeble,* designated, respectively, by A_s, A_i, A_w, A_f and B_s, B_i, B_w, B_f. The stronger a compound is as an acid, the weaker will be its conjugate as a base. The converse is, of course, also true. Normally, a compound would be classified, using indicators, on the basis of which conjugate, HB or B^-, is dominant. For example, if benzoic acid is being tested, the fact that the molecular form is an intermediate acid, A_i, provides more useful information than the fact that benzoate ion is a weak base, B_w. Only in the case of the amphoteric substances is it important to detect the strength of both conjugates. Ampholytes are designated by the symbol A_m, and may be assumed to have one of the following combinations: A_iB_w, A_wB_w, or A_wB_i. In case both conjugates are "feeble," that is, HB is an extremely weak acid and its conjugate base, B^-, is an extremely weak base, for all practical purposes the compound is neither an acid nor a base and is called a "neutral," designated by the letter N.

Davidson's Indicator Systems

In Davidson's system of classification, the indicator reagent also serves as the solvent for the compound to be examined. Directions for the preparation of the five indicator reagent-solvents may be found in the Appendix. The reagents should be prepared fresh every few months. They should be stored in brown, tightly stoppered bottles, and dispensed from dropper-bottles to avoid contamination. To avoid using contaminated or deteriorated reagents, the color of the side-shelf reagents should always be noted before they are used for a test. The normal colors are: A-I, light purple; A-II, blue-violet; B-I, light purple; B-II, yellow; and Inverted A-I, yellow. Colored compounds or compounds that give abnormal colors with the reagents should be tested in the same solvent mixtures that are used in preparing the indicator reagents, but without the indicators being present. The colors produced in such cases must be considered in evaluating the color changes that occur with the indicator reagents. If a solid is to be tested, it should be finely powdered before introducing it into the reagent and any solid that does not dissolve should be ground against the side of the tube by means of a clean glass rod.

The outline below shows the sequence to be used in testing with these indicators. Compounds that do not change in either A-I or B-I are neutral.

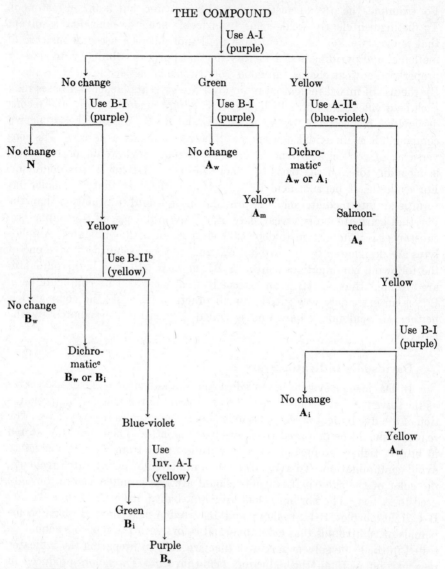

THE COMPOUND

[a] With A-II, acids between A_w and A_i may produce a green color.

[b] With B-II, bases between B_f and B_w give various colors between purple and yellow.

[c] In their dichromatic zones the indicator reagents A-II and B-II appear blue when viewed through a thin section of the solution (detected by tilting the tube), and red when viewed through the full depth. With reagent A-II, the dichromatic character indicates intermediate, uncharged acids or weak cation acids. With reagent B-II, the dichromatic character indicates intermediate, uncharged bases or weak anion bases.

Procedures for Using the Indicators

To test a compound by the use of the classification indicators, place 0.5 ml of the proper indicator reagent in a clean, dry 3-inch test tube and add 15–30 mg of the compound that is being examined. Record the color change, if any, that occurs. The following steps are suggested:

1. Test the compound in reagent A-I and in reagent B-I. Record the results. Deductions: If the color of A-I does not change, the compound is neutral or basic. If the color of B-I does not change, the compound is neutral, amphoteric, or acidic. Therefore, if the compound did not change the color of *either* A-I or B-I, it is a neutral substance, **N**, and no further testing with indicators is required.

2. If there is a color change in A-I, but no change in B-I, proceed to Step 5. If there is a color change in B-I, but no change in A-I, proceed to Step 3.

Basic Compounds

3. Test the compound in reagent B-II and record the results. Deductions: No change in color in B-II indicates a weak base, B_w. If the solution becomes dichromatic (see footnotes to outline, page 100), the substance is a weak or intermediate base, B_w or B_i. If the color is blue-violet, proceed to Step 4.

4. Test the compound in the *inverted* A-I reagent and record the results. Deductions: A green color indicates an intermediate base, B_i, and a purple color indicates a strong base, B_s. The *inverted* A-I reagent is prepared by adding a drop of $2N$ hydrochloric acid to 1 ml of the A-1 reagent so as to *just* change the original purple color to yellow.

Acidic Compounds

5. If the color of A-I changes to green, recall the color produced in B-I. Deductions: If the color in B-I is yellow, the compound is amphoteric, A_m. If the compound did not change the color of B-I, the compound is a weak acid, A_w. If the color of A-I changes to yellow, proceed to Step 6.

6. Test the compound in reagent A-II and record the results. Deductions: If A-II becomes dichromatic (see footnote to outline, page 100), the compound is a weak or intermediate acid, A_w or A_i. If A-II changes to salmon-red, the compound is a strong acid, A_s. If the color of A-II is yellow, recall the color produced in B-I. If the color in B-I was yellow, the compound is amphoteric, A_m, whereas if B-I remained purple, the compound is an intermediate acid, A_i.

On the basis of the colors shown in the classification indicators, the predominant acid-base character of the compound being examined may be deduced and the compound may be classified.

Exercises

1. Predict the proper solubility division and the indicator classification of:

Resorcinol	Allyl acetate	*m*-Nitroaniline
o-Cresol	Propionamide	Citric acid
Acetophenone	Acetonitrile	Butylamine
Dextrose	Salicylaldehyde	Dibutyl ether
Aminobenzoic acid	Ethylene glycol	
Nitrobenzene	Cyclopentanone	

2. Predict the solubility classification of the following and state the reason for the classification of:

2-Aminoethanol	*N*-Ethylbenzenesulfonamide	Diphenyl ether
o-Nitroaniline	Carbon disulfide	Benzylamine
2-Nitropropane	Butyl alcohol	
2-Nitro-2-methylpropane	Butyl mercaptan	

3. Explain the reasons for the following facts:
   a. Oxamide is only slightly soluble in water.
   b. Adipic (hexanoic) acid is very much more soluble in water than glutaric (pentanoic) acid.
   c. *p*-Phenylenediamine is less soluble in ether than the *o*- or the *m*-isomer.
   d. 2-Pentanol is approximately twice as soluble in water as 1-pentanol.
   e. 2-Methyl-2-butanol is approximately twice as soluble as 3-methyl-2-butanol in water.
   f. *m*-Xylene is more soluble than toluene in sulfuric acid.
   g. Dioxane is more soluble than ethyl ether in water.
   h. Benzamide is a very weak acid whereas *N,N*-diethylbenzamide is a weak base.
   i. The dissociation constants of phenol and 1,3-pentandione are of the same order of magnitude but phenol dissolves more readily in aqueous sodium hydroxide.

4. In each of the following, list two compounds that would probably be "borderline" cases as regards solubility in the solubility divisions indicated:
   a. Acids: S_1 and A_1
   b. Ketones: S_1 and **N**
   c. Amines: S_1 and **B**
   d. Esters: S_1 and **N**
   e. Amides: S_1 and **M**

5. Give the names and structural formulas for ten compounds that are amphoteric and would classify in Solubility Divisions **B** and A_2.

6. Give the names and structural formulas for ten compounds that might represent "borderline" cases between A_1 and A_2.

7. Give the names and structural formulas for ten amines that would not likely be soluble in 1.2N HCl.

8. Explain why each of the following is a true statement:
   a. Oxamide is insoluble in water; *N,N'*-dimethyloxamide is soluble in water but not in ether; tetramethyloxamide is soluble in both water and ether.
   b. Aminoacetic acid is less soluble than acetic acid in water.
   c. Propanthiol is less soluble than propanol in water.

d. The addition of certain alcohols to concentrated sulfuric acid causes a precipitate to form.

9. By the use of the classification solvents, show how the following mixtures could be separated and each component recovered in relatively pure form:
   a. Sucrose and hydroquinone.
   b. Picric acid and phenol.
   c. Butyl ether and butyl chloride.
   d. Chloroacetic acid and p-chlorophenol.

10. By using the solubility solvents in multiple steps, show how the following mixtures might be effectively separated:
    a. Ethylene glycol, p-cresol, and benzoic acid.
    b. p-Toluidine, 3-pentanone, p-toluic acid, and nitroethane.

11. Arrange the following compounds in the order of increasing acidity: Specify the solubility division and the indicator classification of each.

a. Acetonitrile	b. Propylamine	c. Salicylaldehyde
d. Benzanilide	e. Benzenesulfonic acid	f. m-Nitrophenol
g. Chloroacetic acid	h. Benzoic acid	

12. Arrange the following in the order of increasing basicity. Specify the solubility division and indicator classification of each.

a. N-Methylbenzenesulfonamide	b. Aniline	c. Propyl ether
d. o-Chloroaniline	e. Diphenylamine	f. p-Nitroaniline
g. Benzylamine	h. Phenylhydrazine	

13. Explain the following facts by a detailed discussion of molecular structures:
    a. Although they are all of the same order of magnitude, the order of increase in strength as bases is as follows: p-toluidine, m-toluidine, o-toluidine.
    b. The nitro group as a substituent in phenols increases their strength as acids but the nitro group as a substituent in aryl amines decreases their strength as bases.
    c. As the X group in XCH_2COOH, the following groups have increasing effect on the strength of the acid: CH_3, H, C_6H_5, CH_3O, I, Br, Cl, F, CN.
    d. Compared to butanoic acid, 4-chlorobutanoic is twice as strong an acid; the 3-chloro isomer is six times as strong; and the 2-chloro isomer is more than ninety times as strong as the unsubstituted acid.
    e. The K_a for the nitro-substituted benzoic acids are: para, 3.75×10^{-4}; meta, 3.21×10^{-4}; and ortho, $6.7; \times 10^{-3}$.

14. Give the names and structual formulas for three compounds that would fall in each of the classes as determined by indicators. In what solubility division would each of these compounds fall?

15. Would it be (a) highly desirable, (b) probably useful, or (c) a waste of time to classify a substance by the indicator method if its solubility had already been determined to be in Division S_2, S_1, A_2, A_1, B, M, N, or I, respectively?

16. Explain the fact that the pK_a of acetic acid is 4.8 in water but is 9.7 in methanol.

17. Write structural ionic formulas for the conjugate acids of the following bases: aniline, p-nitroaniline, and butylamine. Which of these acids would be the strongest?

18. Considering your answers to exercise 13c, 13d, and 13e above, which would be the strongest conjugate base in each group of acids?

19. The primary dissociation constant for carbonic acid is 4×10^{-7}. Other things being equal, what is the maximum acidity that a compound may have and fail to dissolve in sodium bicarbonate solution?

20. Explain the effect of each of the following properties of a compound as they affect its properties as a solvent; dipole moment, dielectric constant, ability for forming hydrogen bonds.

21. Explain why chloroform is a better solvent than carbon tetrachloride for certain types of compounds. Name four compounds that probably would be soluble in chloroform but sparingly soluble in carbon tetrachloride.

22. Comment on the probable cause of the greater solubility of many compounds in methanol than in either water or ether.

23. What types of water insoluble amines should dissolve in an acetic acid–sodium acetate buffer solution with a pH of 5?

24. In making solubility determinations for classification purposes, why is it undesirable to heat the solutions even though they are cooled before the solubility is noted?

25. What would be the advantages and the disadvantages of using 85% phosphoric acid instead of concentrated sulfuric acid as a classification solvent? What subclassifications are possible by the use of both of these acids separately?

26. Explain why benzylamine is a much stronger base than aniline, but weaker than butylamine.

6

Tests for the Classification
of an Unknown

This chapter consists of two parts: Part I contains introductory material and preliminary tests, and Part II consists of specific tests for certain functional groups that characterize the various classes of organic compounds.

PART I: AN INVENTORY AND A FORWARD LOOK

In order that further work may be taken in proper sequence, an inventory should now be made of all of the facts that have been learned about the compound under examination. These facts should be recorded in logical order, and, as further examination of the substance progresses, a complete record should be kept of the results of each test made, together with any deductions that may be logically drawn from these results. A test or observation that eliminates a class is significant.

It is logical to suspect a class that is commonly used in laboratories and industry before considering an uncommon one. When one is considering the probable classes of a compound on the basis of the elements present, the most common combinations should be considered first.

Up to this point, the compound, purified if necessary, has been examined to establish the elements present, the solubility behavior, and the apparent acidity, basicity, or neutrality of the substance. On the basis of these facts one may restrict the substance to a relatively few possible classes by examining the tables on pages 96 and 97. In some cases, the information available at this point may definitely suggest a certain classification of compounds. In some cases it is wise to proceed directly to tests for that specific class as given in Part II of this chapter. However, in many cases it is advantageous to use one or more of the following preliminary tests which are not specific for any one class but which serve as guides for a logical selection for tests for a specific class.

The following examples illustrate how the information acquired previously may serve as a basis for further steps. Nitrogen, sulfur, the halogens, and metals may be assumed absent unless they are specified as being present.

1. Assume that the compound failed to give positive tests for oxygen and that it was a neutral compound in Solubility Division **N**. Table 5.5 lists only hydrocarbons as meeting these specifications. In this case it would be worthwhile to use the preliminary tests P-2 and P-5 to detect general characteristics of the hydrocarbons before performing test 6.13 which is specific for hydrocarbons.

2. Assume a neutral compound in Division S_2. The ignition test, P-3, would serve as a useful guide as to whether 6.3 or 6.8 should be tried first. Since most polyhydroxy alcohols are liquids and carbohydrates are solids, these facts would also suggest which test should be performed first. Obviously, if test 6.8 gave positive results for carbohydrates, the tests for other classes should ordinarily be omitted.

3. Assume a neutral compound in Solubility Division S_1. In this case several organic classes have to be considered: alcohols, aldehydes, acetals, esters, ethers, glycols, and ketones. Since only the low-molecular-weight members of these homologous series would be soluble in water as well as in ether, most of the potential compounds would have rather characteristic odors, and this fact would probably be the best guide to the order in which the class tests should be made. Test P-6 might be useful in further characterizing the compound after its class has been tentatively established by tests 6.3, 6.9, 6.10, or 6.11. Esters and ethers do not produce iodoform (P-6), but only certain alcohols, aldehydes, and ketones give positive results (see note to test P-6).

4. Assume an S_2 compound for which the acid-base classification was indefinite with the probability of the compound being amphoteric, A_m. The compound contained nitrogen. Since the tests for amino acids can be made quickly, test 6.1C should be performed first to determine if the compound is an amino acid or to rule out that class. If this test fails, it would be wise to consider salts by tests P-3 and P-4.

P-1　Gross Observations

Note the color and odor of the substance and try to recall what class of compounds has similar properties. Many phenols and aryl amines develop color in storage due to oxidation. Colors due to impurities usually diminish or disappear when the substance is purified. Low-molecular-weight acids, many phenols, amines, alcohols, esters, hydrocarbons, and carbonyl compounds have characteristic odors. Regularly, students of organic chemistry should carefully note the color, odor, and other observable characteristics of each compound that they use and try to relate its properties to the chemical class represented.

The great majority of carbon compounds are colorless. Therefore, if the pure compound is colored, certain groups may be suspected, and the selection of further tests to be applied would take those facts into account. Although the exact cause of color in molecules is still somewhat uncertain, several authors have logically summarized the present theories on the subject. Among compounds containing carbon, hydrogen, oxygen, halogens, or sulfur, there are a few colored substances, mainly quinones, aromatic ketones that have unsaturated side-chains, and a very few alkanediones.

If nitrogen is present, with or without the additional presence of halogens or sulfur, the most common colored compounds are substituted anilines, toluidines, polycyclic amines, or hydrazines; nitro-, nitroso-, or amino-phenols; polynitro- or polyaminohydrocarbons; nitro- or aminoquinones; azo or diazo compounds; picrates; hydrazones or osazones. Aside from the azo compounds, which are generally orange or red, compounds that contain only chromophoric groups, such as $-N{=}N-$, $={C}{=}S$, $={C}{=}O$, $-N{=}O$, $-NO_2$, and the o- or p-quinoid structures are usually yellow. If, however, the compound also contains auxochromic groups, such as $-NH_2$, $-NHR$, $-OH$ or $-SH$, the color is deepened and intensified. Some of the halogenated nitro-hydrocarbons are colored. The color tends to deepen as the halogen is changed from chlorine to bromine and from bromine to iodine.

Unsubstituted anilides are colorless. Many compounds tend to darken on exposure to light or air, and hence a purified sample should always be used for color estimation.

P-2   Tests for Aromatic Structure

It is frequently an advantage to know as soon as possible if the unknown has aromatic structural characteristics. The following two tests, although not completely reliable or specific, are very useful.

A. The Chloroform–Aluminum Chloride Test

Compounds that have an aromatic structure usually react with chloroform in the presence of aluminum chloride to produce colored products.

Place 100 mg of aluminum chloride in a dry 4-inch test tube and heat it in a strong flame to sublime aluminum chloride up onto the sides of the tube. Allow the tube to cool. Prepare a solution of 10–20 mg of the compound in 5 to 8 drops of chloroform and run this solution down the side of the test tube containing the sublimed aluminum chloride. Note any color produced by contact of the solution with the salt.

Note: Stock samples of aluminum chloride generally have absorbed and reacted with water vapor. Freshly sublimed salt is a much more efficient catalyst for this reaction. If this test is carefully performed it is useful but not completely reliable for detecting aromatic type structures. It is most useful for distinguishing aromatic hydrocarbons or their chlorine compounds from nonaromatic hydrocarbons and their chlorine compounds. Many nonaromatic compounds that contain bromine produce yellow colors, and many nonaromatic compounds that contain iodine produce violet colorations. As a rule, nonaromatic compounds fail to produce a color on the aluminum chloride, whereas monocyclic aromatic compounds give rise to a yellow-orange or red color; bicyclic aromatics give blue or purple, and more complex aromatics produce green colorations on the salt.

B. The Formaldehyde–Sulfuric Acid Test[1]

The formaldehyde–sulfuric acid test is useful in distinguishing aromatic compounds from nonaromatic compounds for those substances that fall in Solubility Division **I**, that is, compounds that are insoluble in concentrated sulfuric acid.

Prepare a solution of about 30 mg of the compound to be tested in 1 ml of a nonaromatic solvent (hexane, cyclohexane, or carbon tetrachloride). Add 1–2 drops of this solution to 1 ml of the reagent, which is prepared at the time of use by adding 1 drop of formalin (37–40 per cent formaldehyde) to 1 ml of concentrated sulfuric acid and shaking the solution slightly. Note the color of the surface layer of the reagent after the test solution has been added and the color of the reagent after the tube has been shaken.

Note: With time, colors for most of the compounds that give positive tests change to various shades of brown or black. It is advisable to run a blank test on the solvent before it is used, since it may contain aromatic contaminants. A few milligrams of trioxane may be substituted for the formalin in preparing the reagent. Typical colors produced in this test are:[2] benzene, toluene, and *n*-butyl benzene give red; *sec*-butyl benzene gives pink; *tert*-butyl benzene and mesitylene produce orange; diphenyl and triphenylbenzene give blue or greenish blue; naphthalene and phenanthrene produce blue-green to green; aryl halides produce pink to purple colors; naphthyl ethers give purple colors. Alkanes, cycloalkanes, and their halogen derivatives produce either no color or a pale yellow. Often, but not always, precipitates form.

The formaldehyde–sulfuric acid reagent will also produce colors with many other classes of aromatic compounds besides the classes that are insoluble in concentrated sulfuric acid, but in those classes the color production is not consistent enough to be useful in classification work. Unsaturated, noncyclic hydrocarbons react with the reagent and usually produce brown precipitates.

The mechanism of color production appears to include polymerization involving carbonium ions.

[1] Morris, Stiles, and Lane, *Ind. Eng. Chem., Anal. Ed.* **18,** 294 (1946); LeRosen, Moravek, and Carlton, *Anal. Chem.* **24,** 1335 (1952); Silverman and Bradshaw, *Anal. Chem.* **27,** 96 (1955); Rosen, *Anal. Chem.* **27,** 111 (1955).

[2] LeRosen, Moravek, and Carlton, *loc cit.*

P-3 The Ignition Test

Much may be learned about a substance by an ignition test. Place 1–2 drops, or a few milligrams of a solid, on the inverted lid of a crucible or in a shallow evaporating dish. Heat the vessel with a small flame. From time to time, contact the substance directly with the flame so that it may ignite before it volatilizes. If gases are given off by the heated substance before it ignites, a pH paper that has been moistened with distilled water should be held in the fumes to detect any acidic or basic properties. If the substance ignites, note the odor and the general appearance of the flame. If the substance carbonizes, apply a hot flame to the vessel until the carbon is all oxidized. The residue, if any, is most likely the oxide or carbonate of some metal, usually white but sometimes colored. Such a residue should be tested for metals by the semimicro methods for cation analysis as published in texts on that subject.

Most aromatic compounds, particularly if they do not contain oxygen, burn with a smoky flame. Low-molecular-weight nonaromatic compounds burn with a nonsmoky flame which tends to be bluish if oxygen is present in the organic compound. Halogen compounds are difficult to ignite, but burn with a smoky flame when ignited. Sugars and proteins tend to carbonize and produce characteristic odors.

P-4 Tests for Salts

Commonly encountered organic salts may be divided into three classes: ammonium salts of organic acids, metallic salts of acids or phenols, and salts of amines with acids. Such compounds will generally be suspected as a result of the ignition test (test P-3), the elemental analysis, and the solubility determinations.

A. Ammonium Salts

1. Place 25–50 mg of the suspected salt in a 4-inch test tube and add 0.5 ml of 10 per cent sodium hydroxide. Place a small filter paper over the top of the tube and crush it down around the tube. Add 2 drops of 10 per cent copper sulfate on the filter paper covering the mouth of the test tube. Heat the tube with a small flame to boil the mixture. Ammonium salts liberate ammonia gas which will react with the copper ions on the paper to develop a blue color. This test is more reliable than the "sniff test" or the litmus paper test for ammonia arising from the alkaline hydrolysis of an ammonium salt.

2. A very sensitive test for ammonia is based on the oxidation of ammonia to nitrous acid which reacts with phenol to produce nitrosophenol. The

equilibrium isomer of nitrosophenol is the monoxime of quinone. It may be assumed that the color produced in the test is the result of a reaction between phenol and nitrosophenol.

Mix 1 ml of a 4 per cent solution of phenol in water with 1 ml of a 5 per cent aqueous solution of sodium hypochlorite in a 4-inch test tube. Add a few milligrams of a solid that is to be tested or a few drops of a solution of the compound in water. Warm the mixture. A blue color develops if ammonia or ammonium ions are present (or substances that yield ammonia or ammonium ions under the conditions of the experiment).

Note: This is a very sensitive test. It has been found that many amides also give a blue color when used in this test. This reaction may be accounted for either by assuming that some ammonium salts are present in the amide as impurities or on the basis of some hydrolysis of the amide in the warm alkaline solution. The test has not been investigated sufficiently to establish certainty of what additional interferences may exist in this method of detecting ammonia and ammonium salts. Nessler's reagent, as used in inorganic chemistry, may also be used to detect ammonia.

B. Metallic Salts

The presence of metals is indicated by a white or light-colored residue on ignition of a sample (test P-3). The identity of the metal can be determined by treating this oxide or carbonate by the conventional methods of inorganic analysis.

C. Salts of Amines

Salts formed by reacting amines with acids are generally ionic and soluble in water. Most commonly, the anion present is the chloride or sulfate ion. Less commonly, the anion will be phosphate, acetate, or some other ion. Detection methods are given in Chapter 4 for the halide, sulfate, and phosphate ions.

Note: Salts of amines that are not sufficiently soluble in water for direct testing may be decomposed by treatment with 5 per cent sodium hydroxide. The amine may be extracted by ether, the aqueous layer acidified with acetic acid and then tested for the anions.

The detection of the acetate ion in amine acetates is much more involved. One method is as follows: Place the salt in a test tube and acidify it with phosphoric acid. Distill the evolved acetic acid into 0.5 ml of water in a 4-inch test tube. Neutralize the distillate with sodium bicarbonate and evaporate it to dryness. Add a few milligrams of dry sodium formate to the dry sodium acetate in the test tube and cover the mouth of the test tube with a filter paper to which has been added 1 drop of a 20 per cent solution of morpholine and 1 drop of a 5 per cent solution of sodium nitroprusside. Heat the salt in the test tube with a small flame. The development of a blue color on the test paper is a positive test for acetaldehyde which is formed by the reaction of sodium formate with sodium acetate.

P-5 Tests for Active Unsaturation

Both test A and test B should be performed and the results compared, as shown in Table 6.1, before reaching general conclusions regarding the nature

TABLE 6.1

Comparison of Results of Permanganate Ion and Bromine Tests

		Bromine	
Types of structure	Permanganate	Addition	Substitution
Most alkenes and alkynes	Positive	Positive	
$Ar_2C{=}CAr_2$; many $ArC{=}CAr$[a]	Positive	Negative	
Phenols; aryl amines	Positive		Positive
Ketones[b]	Negative		Positive
Many aldehydes[c]	Positive		Positive
Primary and secondary alcohols[d]	Positive		Negative
Mercaptans; sulfides	Positive		Negative
Thiophenols	Positive		Positive

[a] Alkenes with electronegative groups attached to each of the unsaturated carbon atoms react slowly to very slowly with bromine under the conditions of the test (e.g., stilbene and cinnamic acid).

[b] Particularly the methyl ketones.

[c] Formaldehyde, formates, and benzaldehyde do not react appreciably with bromine.

[d] Secondary alcohols react more readily than primary alcohols. High-molecular-weight alcohols react so slowly that the test may appear negative.

of the substance being examined.

A. Bromine (in CCl₄) Test

Bromine will form addition compounds with almost all actively unsaturated compounds. It will also react to substitute bromine for hydrogen and liberate hydrogen bromide with compounds that are easily brominated. Typical equations are

$$\begin{array}{cc} H\ H & H\ H \\ HC{=}C{-}CH_2OH + Br_2 \rightarrow HC{-}C{-}CH_2OH \\ & Br\ Br \end{array}$$

Dissolve 50–100 mg of the compound to be tested in 1–2 ml of carbon tetrachloride and add, dropwise, a 2 per cent solution of bromine in carbon tetrachloride. If more than 2 drops of the bromine solution are required to cause the bromine color to remain for at least 1 minute, a reaction by either addition or substitution is indicated.

Note: Most compounds that possess alkene or alkyne structures add bromine quite rapidly, with the exception of molecules that have electronegative groups attached to both of the carbon atoms that are involved in the unsaturation. A compound that readily reacts with bromine without the evolution of hydrogen bromide may be presumed to be unsaturated, but this conclusion should be confirmed by the permanganate test (Test B).

Discharge of the color of more than a drop or two of the bromine solution, accompanied by the liberation of hydrogen bromide gas, indicates a phenol, amine, enol, aldehyde, ketone, or some other compound containing an active methylene group. Amines do not evolve hydrogen bromide after the first substitution by bromine because the amine reacts with the first hydrogen bromide produced to form a salt. Not all members of these classes react with bromine under the specified conditions. The fact that hydrogen bromide is being evolved may usually be determined by blowing the breath across the top of the reaction tube and noting the fog that is produced.

B. Permanganate Ion Test

The results of the permanganate ion test, when compared with the results of the bromine test, may be used to detect active unsaturated linkages or to indicate certain other type structures. The test is based on the fact that easily oxidized compounds will reduce the permanganate ion, thus causing the disappearance of the purple color and the appearance of the brown-colored hydrated oxides of manganese. Applied to hydrocarbons, the test is positive for the alkenes and alkynes (Baeyer test).

$$3R_2C{=}CR_2 + 2MnO_4^- + 4H_2O \rightarrow 3R_2C{-}CR_2 + 2MnO_2 + 2OH^-$$

$$\underset{H}{\overset{|}{O}} \quad \underset{H}{\overset{|}{O}}$$

Dissolve 25–30 mg of the compound in 2 ml of water or acetone (free of alcohols) in a 4-inch test tube. Add a 1 per cent aqueous solution of potassium permanganate drop by drop, with vigorous shaking of the tube. If more than 1 drop of the permanganate is reduced, an unsaturated hydrocarbon or some other easily oxidized compound is indicated.

Note: Recall that bromine reacts by addition with many unsaturated compounds and by substitution with many others. The permanganate ion, in the presence of water, oxidizes alkene and alkyne types of bonds to glycols which are cleaved and oxidized to the corresponding acids by more vigorous treatment with permanganate. However, the permanganate ion will also oxidize several other types of compounds. Table 6.1 summarizes the results of the bromine test and the permanganate test on different types of structures.

P-6   The Iodoform Formation Test

Compounds that contain the CH_3CO group or that may be readily oxidized to contain this group may be converted to iodoform by an alkaline solution of sodium hypoiodite.[3]

[3] Fuson and Tullock, *J. Am. Chem. Soc.* **56,** 1638 (1934); Fuson and Bull, *Chem. Revs.* **15,** 275 (1934); McElvain, Bright, and Johnson, *J. Am. Chem. Soc.* **63,** 1558 (1941); Stodola, *Ind. Eng. Chem., Anal. Ed.* **15,** 72 (1943).

Dissolve 100 mg of the compound being tested in 1 ml of water (use dioxane if the compound is insoluble in water). Add 3 ml of 10 per cent sodium hydroxide solution and then add dropwise a 10 per cent solution of iodine in a 20 per cent solution of potassium iodide in water, until a slight excess of iodine exists in the solution. Place the tube in a beaker of 60° water. Add more iodine until the iodine color persists for 2 minutes and then add drops of 10 per cent sodium hydroxide solution until the brown iodine color just disappears. Remove the tube from the warm water and add 10 ml of water. Iodoform precipitates as a yellow solid, which melts at 120°.

Note: Acetaldehyde is the only alkanal that yields iodoform. Acrolein, furfural, and aldehydes of similar structure also form iodoform. Ethanol is the only primary alcohol that gives a positive test. Tertiary alcohols give negative results. Secondary alcohols (2-alkanols) that can be oxidized to methyl ketones by sodium hypoiodite yield iodoform as do the methyl ketones (2-alkanones). Compounds that contain the CH_3CO grouping but which, on hydrolysis, produce acetic acid do not give the iodoform test (e.g., acetoacetic acid and its esters, acetanilide, and the like). If dioxane is used as a solvent, it should be previously tested for impurities that might give a positive iodoform test.

Exercises

1. During ignition, what classes of compounds would be expected to evolve acidic vapors? basic vapors?

2. If a black residue results from an ignition, how can one determine whether this residue is unoxidized carbon or a metallic oxide?

3. A sample of flour is submitted for examination with the allegation that it contains arsenic oxide. How would you proceed to prove or disprove the presence of arsenic?

4. Consult the literature, and outline a method by which it could be established that an amine salt under examination is (a) an acetate, or (b) an oxalate.

5. Why does an amine not give a positive reaction in test P-4A2?

6. What is the probable chemistry involved in the chloroform–aluminum chloride test for aromatic structure? Why is chloroform used in this test rather than some alkyl chloride?

7. If a compound only slowly reacts to form iodoform, the sodium hypoiodite may be consumed by a competitive reaction that produces two inorganic compounds. What are they?

PART II: TESTS FOR SPECIAL CLASSES

The specific classes of organic compounds are generally recognized by the detection of the main functional group that is present in the molecule. For example, detection of the carbonyl group indicates an aldehyde or a ketone; the presence of nitrogen with basic properties indicates an amine; detection of hydroxyl groups, depending on the properties exhibited, indicates either alcohols or phenols. In the case of polyfunctional compounds, two or more functional groups may be detected during the course of the investigation, and a decision must be made as to which functional group dominates the properties of the compound so that the compound may be found in the proper table that gives its physical constants and those of suitable derivatives. For example, a cyano-substituted acid should be classified as an acid rather than as a nitrile, and an aminophenol should be sought in the table of amines since it would be derivatized as an amine rather than as a phenol.

In every case, an attempt should be made to determine the major functional group first. For instance, a compound of Solubility Division A_2 that proves to be weakly acidic and contains nitrogen should be tested for phenols before an attempt is made to determine the nature of the nitrogen substituent. Probably, if it proves to be a phenol, the compound can be identified by its physical constant and the physical constants of its derivatives as a phenol without the necessity of testing specifically for the nature of the nitrogen-containing group. Similarly, since acids and phenols with halogen substituents are much more common than acid halides, a compound that falls in Solubility Division A_1 and contains halogen should be tested for acids and phenols before acid halides.

When testing for functional groups in organic compounds, it must be recognized that the group is not an individual identity such as an ion would be in inorganic analysis, but rather it is a component of the molecule, the total structure of which affects the properties of all functional groups at least to some extent. In some cases, this influence may be so great that the functional group for which the particular test is being made may fail to give a positive reaction. For this reason, it is advisable to use more than one method of identifying the groups whenever possible.

Students should perform the tests given in this chapter using known compounds of the appropriate classes to gain experience in the procedures and to observe the reactions which are characteristic of each class. Also, it is often worthwhile when working with an unknown to simultaneously test a known compound that is assumed to be similar to the unknown.

Methods for the preparation of special reagents used in the tests described in this chapter will be found in the Appendix.

6.1 Acids and Anhydrides

The probability than an unknown is an acid would be detected by its reaction with the indicators in the acid-base classification and, if it is not soluble in water, by its classification in Solubility Divisions A_1 or A_2. However, not all acidic substances are classed as *acids* for purposes of identification. Furthermore, some true acids are so very weak or dissolve in alkaline solutions so very slowly that they may escape detection by the solubility or indicator methods. A method of detecting even very weak acids is, therefore, given below to be used if there is doubt.

A. Iodate–Iodide Test for Acids[4]

The iodate-iodide test will detect acids even if they are so weak that their reaction with indicators is indecisive.

About 5 mg of the substance to be tested (or a saturated solution of it in 2 drops of neutral alcohol) is placed in a 3-inch test tube. Add 2 drops of a 2 per cent solution of potassium iodide and 2 drops of a 4 per cent solution of potassium iodate. Stopper the test tube and hold it in boiling water for 1 minute. Cool the tube and add 1–4 drops of a freshly prepared 0.1 per cent solution of starch. If the substance is an acid, a blue color appears.

$$5I^- + IO_3^- + 6H^+ \rightarrow 3H_2O + 3I_2$$

Note: In some cases it is necessary to add more starch solution to cause the characteristic blue iodine–starch complex to appear.

Solid acids may also be detected by grinding together a few milligrams of the acid with a few milligrams of the dry potassium iodide and potassium iodate. The development of a brown color due to free iodine is a positive test. In case of doubt, add 5 drops of water and 2–4 drops of starch solution.

B. Carboxylic Acids

Three classes of derivatives of carboxylic acids may be detected by conversion to hydroxamic acids which may then be reacted with ferric ions to produce the highly colored ferric hydroxamates. Acids may be converted into one of these classes—acid chlorides, anhydrides, or esters—and thus detected indirectly. Obviously, it is necessary to establish that the unknown substance is not an anhydride (test 6.1E), an acid halide (test 6.2B), or an ester (test 6.10) before proceeding to test for acids. Furthermore, since ferric ions produce colored products with other classes, particularly with phenols, the unknown should be tested with a solution of ferric chloride (test 6.19A) before making any test for a hydroxamic acid. Procedures for converting acids to derivatives suitable for further testing follow.

1. *Conversion to acid chlorides.* Place 100 mg (or 3 drops) of the compound in a 6-inch test tube and add 8 drops of thionyl chloride. Insert a microcondenser and gently reflux the material for 10 minutes.

$$RCOOH + SOCl_2 \rightarrow RCOCl + SO_2 + HCl$$

Test for the acid halide by test 6.2B.

Note: Thionyl chloride may convert dibasic acids to the anhydride instead of the acid chloride (particularly, in case of $HOOC(CH_2)_nCOOH$, where n is 2 or 3, or to the cyclic ketone where n is 4 or 5). Anhydrides respond to the same hydroxamate test

[4] Feigl, *Spot Tests in Organic Analysis* (6th ed., Princeton and N.Y.: Elsevier, 1960), p. 117.

as do acid halides. α-Hydroxyacids react with thionyl chloride to produce formic acid and an aldehyde or ketone.

Phosphorus pentachloride may be used instead of thionyl chloride for the conversion of acids to acid chlorides.

2. *Conversion to esters.* If the original substance is an acid that has been converted to an acid chloride by procedure (1) above, it may be converted to an ester by the following method.

Add 0.5 ml of an alcohol (e.g., butanol) to the tube in which the acid halide has been prepared and gently reflux the mixture for 2 minutes. If a precipitate exists, add more alcohol dropwise and continue heating until the precipitate dissolves. Cool the tube and add 1 ml of water to hydrolyze any excess thionyl chloride. Remove the water-insoluble layer by means of a pipet and test it for esters by test 6.10.

Note: Another method for converting an acid to an ester for detection by the hydroxamate test is as follows: Dissolve or suspend 30–40 mg of the acid in 0.5 ml of ethylene glycol. Add 1 drop of concentrated sulfuric acid and gently reflux the mixture for 5 minutes. Cool, and apply the ester test.

3. Acids may be identified by determining the neutralization equivalent; see the Appendix.

C. Amino Acids

Ninhydrin[5] (triketohydrindene hydrate) reacts with α-amino acids and hydrolysis products of proteins to produce violet, blue, or purple products. The final colored products are not single compounds, but are condensation products, some of which depend on the amino acids originally present; but one product that has been observed in all cases is the violet bis-1, 3-diketoindyl.[6]

Dissolve 1–5 mg of the substance to be tested in 1–3 ml of water in a 4-inch test tube. Add 1 ml of a 0.1 per cent solution of ninhydrin in water. Heat the contents of the tube to boiling for 1–2 minutes. The production of a violet, blue, or purple color is a positive test.

Note: Proteins, peptones, and peptides give positive tests with ninhydrin after hydrolysis, so that at least some free amino acids are present. Proline and hydroxyproline do not give the characteristic blue colored products, but produce yellow-red colors instead. Feigl[7] states that β-amino acids and primary and secondary alkyl amines give positive tests with ninhydrin in some cases.

[5] Runemann, *J. Chem. Soc.* **97**, 1438 and 2025 (1910); **99**, 792 and 1486 (1911).

[6] Moubacher and Ibrahim, *J. Chem. Soc.* 702 (1949); Schoenberg and Moubacher, *Chem. Rev.* **50**, 272 (1952).

[7] Feigl, *op. cit.*, p. 293.

D. Sulfonic Acids

Sulfonic acids are usually detected by the elemental analysis coupled with their solubility behavior and decided acidity. If further testing is desired, the sulfonic acid may be converted to the acid chloride by using thionyl chloride and then treating with concentrated ammonium hydroxide to prepare the sulfonamide which can be detected by test 6.5D. Feigl [8] suggests detecting sulfonic acids by converting them to ferric acet-hydroxamate.

E. Acid Anhydrides

Most anhydrides can be converted to hydroxamic acids by the method given in test 6.2B, since both anhydrides and acid halides may be converted to hydroxamic acids by this procedure. If an anhydride is suspected even though the hydroxamate test is negative, heat 50–75 mg of the compound with 3 drops of butanol for 3–5 minutes and test for an ester (test 6.10).

Note: Containers of "anhydrides" that have been frequently opened and thus exposed to moist air will often be found to contain the acid rather than the anhydride. Samples of anhydrides frequently fail to give the expected physical constants because of partial or complete hydrolysis.

Exercises

1. For what kind of acids would thionyl chloride not be a good reagent for converting the acid to an acid chloride?
2. If phosphorous pentachloride is used, how would you separate the acid chloride from the phosphorous oxychloride?
3. Would acidic substances such as phenols give a positive test for acids by this method?
4. What product would be formed if an amino acid were treated with nitrous acid?
5. What are the problems involved in trying to convert an amino acid into an ester?
6. What pretreatment of an amino acid must be made before it is practical to determine its neutralization equivalent?
7. How could an alkyl sulfuric acid be distinguished from an alkane sulfonic acid?
8. Will acids as well as anhydrides convert to esters by the method given in Section E?

6.2 Acid Halides

A. Amide Formation

Acid halides may be reacted with an amine such as benzylamine or aniline to form substituted amides that are only slightly soluble in cold water.

Add 2 drops of the compound to 3 drops of benzylamine in a 4-inch test

[8] Feigl, *op. cit.*, p. 265.

tube. After the reaction has subsided, add 2 ml of cold water and shake the tube vigorously. The white benzylamide may be recrystallized from water.

Note: Benzylamine reacts more readily with acid halides, particularly high-molecular-weight aroyl halides or sulfonyl halides, than does aniline. Acyl, aroyl, and sulfonyl halides all react with aniline. If it is suspected that the acid halide is a sulfonyl halide, 5 drops of pyridine should be added to the test tube and the mixture gently heated for 3 minutes before the water is added. Sulfonyl halides react with pyridine to form a sulfonpyridinium cation which is a good acylating agent. Acid anhydrides will not react appreciably to form anilides under the conditions specified. Acids may react with the aniline to form salts but not anilides.

Exercises

1. If acetic acid, acetic anhydride, and acetyl chloride were separately reacted with aniline by the above method, what would the products of each reaction be?

2. What would be the appearance of each mixture after treatment with water?

3. If it is suspected that unreacted acid halides or aniline are present in the aqueous mixture, how could their presence be detected?

4. Would it ever be desirable to adapt the Schotten-Baumann reaction to this experiment?

5. Compare the solubilities of aniline acetate and acetanilide in water and in cold 5 per cent sodium hydroxide solution. Do these facts suggest a method for distinguishing salts of aniline from anilides?

6. Under what conditions would the analysis for the elements not distinguish acid halides from acid anhydrides?

B. The Ferric Hydroxamate Test[9]

Hydroxamic acids react with ferric chloride in acidic solutions to form soluble ferric hydroxamates that are highly colored. The most common color for these salts is bluish red (magenta) but a deep red color is often noted, particularly if the concentration is high. Many classes of organic compounds may be converted into hydroxamic acids and hence may be tested for either directly or indirectly by the ferric hydroxamate test.

It must be recalled that ferric chloride reacts directly with several classes of organic compounds, particularly with phenols, to produce colored products, some of which have colors very similar to the colors produced with hydroxamic acids. Hence, *it is advisable to test the compound with ferric chloride alone before applying any hydroxamate test.* This may be done as follows:

Dissolve 30 mg of the compound in 1 ml of ethanol in a 4-inch test tube and add 1 ml of $1N$ hydrochloric acid, followed by 1 drop of 10 per cent

[9] Buckles and Thelen, *Anal. Chem.* **22,** 676 (1950); Cheronis, *Micro and Semimicro Methods,* Vol. VI of Weissberger (ed.), *Technique of Organic Chemistry* (2nd ed., N.Y.: Interscience, 1957), pp. 460–462; Davidson, *J. Chem. Ed.* **17,** 81 (1940); Feigl, Anger, and Frehden, *Mikrochemie* **15,** 18 (1934); Feigl, *op. cit.,* p. 249.

ferric chloride. If the mixture has any color except yellow, it is very doubtful that reliable conclusions could be drawn from performing a hydroxamate test.

Procedure for acid halides and anhydrides. Add 30–40 mg of the compound being examined to 0.5 ml of a $1N$ solution of hydroxylamine hydrochloride in alcohol. Add 2 drops of $6M$ hydrochloric acid to the mixture; warm it slightly for 2 minutes and then boil it for a few seconds. Cool the solution and add 1 drop of 10 per cent ferric chloride. The formation of a reddish blue or bluish red color is a positive test. If the color is more red than blue, adjust the pH of the solution to a pH of 2–3 by adding dropwise $2N$ hydrochloric acid. The color should shift towards purple (see Note). Typical reactions are:

$$RCOX + NH_2OH \rightarrow RCO(NHOH) + HX$$
$$(RCO)_2O + NH_2OH \rightarrow RCO(NHOH) + RCOOH$$
$$RCO(NHOH) + Fe^{+++} \rightarrow Fe(RCONHO)^{++} + H^+$$

Note: It has been determined [10] that the color of the soluble complex formed between ferric ions and the hydroxamate ions varies with the pH of the solution. The probable predominating species are: $Fe(RCONHO)^{++}$ in strongly acidic solution, $Fe(RCONHO)^+_2$ in weakly acidic solution, and $Fe(RCONHO)_3$ in neutral and weakly alkaline solution.

6.3 Alcohols

A. Hydroxamate Test

Alcohols may be converted to esters by reaction with acetyl chloride. Tertiary alcohols are partly converted to alkyl chlorides when they are treated with acetyl chloride, owing to the reaction of the liberated hydrogen chloride on another molecule of the alcohol. This result can be avoided by introducing dimethylaniline into the mixture to react preferentially with the hydrogen chloride. The dimethylaniline may be omitted if the alcohol is known not to be a tertiary alcohol.

Mix 0.1 ml of acetyl chloride with 0.1 ml of dimethylaniline and add 0.2 ml of the alcohol. Shake the mixture frequently for 5 minutes. Add about 1 g of ice; or add, dropwise and with shaking, 1 ml of cold water to decompose the remaining acetyl chloride. Pour the mixture into a small test tube so that the stratified layers may be easily distinguished. Remove 2–3 drops of the upper layer and test for esters by test 6.10.

B. Xanthate Test

It has been shown[11] that the alcohols may be qualitatively detected and quantitatively estimated by reacting the potassium alkoxides with carbon

[10] Aksnes, *Acta Chem. Scand.* **11**, 710 (1957).
[11] Whitmore and Lieber, *Ind. Eng. Chem., Anal. Ed.* **7**, 127 (1935).

disulfide to form the potassium alkyl xanthates. The test is also satisfactory for the cellosolves (monoalkyl ethers of glycol), but the carbitols (monoalkyl ethers of diethyleneglycol) yield heavy red oils rather than the customary light yellow precipitates. The xanthates of tertiary alcohols are rather easily hydrolyzed, but enough precipitate is formed to give a positive test.

$$ROH + KOH \rightarrow ROK + H_2O$$

$$ROK + CS_2 \rightarrow ROCSSK$$

Add 1 pellet of solid potassium hydroxide to 0.5 ml of the alcohol *in a dry test tube* and heat until the KOH dissolves (very volatile alcohols require a reflux condenser). Cool the tube and add 1 ml of ether. Add, dropwise, carbon disulfide until a pale yellow precipitate forms. If a precipitate does not form by the time 0.5 ml of carbon disulfide has been added, the test should be considered negative.

Note: Ketones that can enolize effectively give a positive xanthate test.

C. Tests to Distinguish Primary, Secondary, and Tertiary Alcohols

1. Kruse, Grist, and McCoy[12] recommend the use of *N*-bromosuccinimide for distinguishing the three subclasses of alcohols.

Dissolve 50–75 mg of the alcohol in 1–2 ml of a 0.01 per cent (by weight) solution of bromine in carbon tetrachloride in a 4-inch test tube. Add 20–30 mg of *N*-bromosuccinimide and place the test tube in a water bath at 78–80°. At this temperature, the solution will boil gently. Any color change from the initial pale yellow will occur within 13 minutes (usually within 5 minutes). Primary alcohols give a permanent orange color. Secondary alcohols produce a transitory orange color which fades (usually very rapidly), leaving the solution colorless after continued boiling. After the tube is cooled, an orange precipitate is often apparent in the case of a secondary alcohol. Tertiary alcohols generally do not produce color changes with the reagent.

Note: Allyl, benzyl, and *tert*-amyl alcohols all give false results in that all three give the test for secondary alcohols. Primary alcohols up to octadecanol react regularly. The monoalkyl ethers of ethylene glycol give the test for primary alcohols. The cyclohexanols react regularly (secondary alcohols).

Although the chemistry involved in this test has not been established, Kruse, Grist and McCoy, basing their deductions on other studies,[13] suggest that the primary and secondary alcoholics react to form hypobromites which split out hydrogen bromide to yield aldehydes and ketones. The hydrogen bromide reacts with unchanged *N*-bromosuccinimide to yield succinimide and bromine which is probably the cause of the color. The fading of the color in the test for secondary alcohols would then be explained by the greater reaction of the hydrogen bromide with secondary alcohols, thus allowing less

[12] Kruse, Grist, and McCoy, *Anal. Chem.* **26**, 1319 (1954).

[13] Feiser and Rajagopalan, *J. Am. Chem. Soc.* **71**, 3935, 3938 (1949); Barakat and Mousa, *J. Pharm. and Pharmacol.* **4**, 115, (1952).

bromine formation, or to the greater reaction of bromine with the ketones than with the aldehydes.

2. A distinction may be made among the three subclasses of water-soluble alcohols by using the Lucas[14] reagent (see Appendix).

Add 3–4 drops of the alcohol to 2 ml of the reagent in a 3-inch test tube. Shake the tube vigorously and then allow the mixture to stand at room temperature. A reaction is indicated by the clouding of the solution due to the formation of an insoluble alkyl chloride. Tertiary alcohols (also allyl, benzyl, and cinnamyl alcohols) react immediately; secondary alcohols react within 2–3 minutes; primary alcohols require a much longer time.

3. Bordwell and Wellman[14a] have recently developed a rapid test for distinguishing tertiary alcohols from others.

Dissolve 15–30 mg of the alcohol in 1 ml of alcohol-free acetone and add 1 drop of the reagent (see Note). Shake the mixture. Primary and secondary alcohols react within 10 seconds to give an opaque, blue-green suspension. Tertiary alcohols do not react with the reagent.

Note: The reagent is prepared by dissolving 1 g of chromic oxide in 1 ml of concentrated sulfuric acid and then diluting with 3 ml of water. The orange color of the reagent completely disappears if the alcohol is primary or secondary. Impurities in a tertiary alcohol may cause a small amount of precipitate in the orange-colored solution. Easily oxidized substances, such as aldehydes, phenols, enols, and the like, react with this reagent. It would appear that this reagent might be useful in distinguishing aldehydes from ketones.

D. Additional Tests

Among the many other tests reported in the literature, the following are representative: (1) Feigl[15] recommends the use of vanadium oxinate; (2) Ritter[16] distinguishes the subclasses by using potassium permanganate and acetic acid; (3) Duke and Whitman[17] oxidize the primary alcohols to aldehydes and the secondary alcohols to ketones for detection; and (4) Duke and Smith[18] use the ceric nitrate reagent.

These tests are described in the larger work by the present authors.[18a]

Exercises

1. By using 1-butanol, 2-butanol, and 2-methyl-2-propanol as representative alcohols with each of the three methods given for distinguishing primary, secondary, and tertiary alcohols, determine the relative sensitivity and reliability of the three methods.

[14] Lucas, *J. Am. Chem. Soc.* **52**, 802 (1930).
[14a] Bordwell and Wellman, *J. Chem. Educ.* **39**, 308 (1962).
[15] Feigl, *op. cit.*, p. 184.
[16] Ritter, *J. Chem. Ed.* **30**, 395 (1953).
[17] Duke and Whitman, *Anal. Chem.* **20**, 490 (1948).
[18] Duke and Smith, *Ind. Eng. Chem., Anal. Ed.* **12**, 201 (1940).
[18a] Cheronis and Entrikin, *Semimicro Qualitative Organic Analysis* (N.Y.: Interscience, 1957), pp. 225, 246, and 252.

2. In terms of the mechanisms of the reactions, explain the differences in reactivity of primary, secondary, and tertiary alcohols of similar carbon content.

3. Suggest reasons why allyl alcohol and benzyl alcohol react similarly in test C, and why each gives the reaction for a secondary alcohol instead of a primary alcohol.

4. Explain why phenols do not give positive reactions to these tests for alcohols. Would you expect vinyl alcohol to give these tests?

5. Suggest a reason why 1-chloro-2-propanol reacts like a primary alcohol in the Lucas test (6.3C2).

6.4 Alkyl and Aryl Halides

The halogens may occur as substituents in all classes of organic compounds. However, with the exception of several classes of perfluoro compounds, only the halogenated hydrocarbons fall in Solubility Division **I**. The following tests are useful in distinguishing alkyl halides from aryl halides with the limitations noted. Shriner, Fuson, and Curtin[19] give a fairly extensive discussion of the reactivity of silver ions and of iodide ions in acetone on various types of halogen compounds.

A. Alcoholic Silver Nitrate

1. Add 30–40 mg of the compound to 0.5 ml of a saturated solution of silver nitrate in ethanol. Do not heat the solution. Note any precipitate that forms within 2 minutes.

Alkyl bromides and iodides and tertiary alkyl chlorides precipitate silver halide by this test. Alicyclic bromides and iodides, allyl halides, and 1,2-dibromoalkanes also give a positive test at room temperature. Aryl halides and primary and secondary chlorides do not form precipitates by this method.

2. If no precipitate forms within 2 minutes at room temperature, heat the solution from (1) above and boil it for 30 seconds. Silver chloride will precipitate from primary and secondary alkyl halides. Aryl halides, vinyl halides, and compounds such as chloroform do not cause precipitation.

B. Hydrolysis

The alkyl halides hydrolyze much more readily than the aryl halides. In the following test, all of the alkyl halides (except fluorides) will cause the precipitation of the silver halide, but the aryl halides produce no more than a light cloudiness.

Mix 100 mg of the halogen compound with 5 ml of 5 per cent alcoholic solution of potassium hydroxide and reflux the mixture for 5 minutes. Cool the mixture and add 10 ml of distilled water. Acidify the solution with dilute nitric acid. Unless the solution is clear, filter it. Add 2 drops of 5 per cent silver nitrate solution.

[19] Shriner, Fuson, and Curtin, *Identification of Organic Compounds* (4th ed., N.Y.: Wiley, 1956), pp. 136–147.

C. *Formaldehyde–Sulfuric Acid Test*

According to a limited study,[20] the following test produces pink, red, or bluish red colors with aryl halides, whereas alkyl halides produce yellow, amber, or brown colors.

Add 1 drop of the compound to be tested to 1 ml of hexane or carbon tetrachloride. Add 1–2 drops of this solution to 1 ml of a reagent which is prepared at time of use by adding 1 drop of formalin (37 per cent formaldehyde) to 1 ml of concentrated sulfuric acid. Shake the tube gently and note the color.

Exercises

1. Considering the bond energies involved, explain the order of reactivity of the alkyl halides containing the same alkyl group but different halogens with silver nitrate.

2. Explain the probable reasons for the known order of reactivity of primary, secondary, and tertiary butyl chlorides.

3. What type of mechanism (S_{n2}, S_{n1}, etc.) seems most probable for the reactions of alkyl halides with alcoholic silver nitrate? Does the hydrolysis of alkyl halides in alcoholic potassium hydroxide involve the same mechanism?

4. Why are the aryl halides less reactive than the alkyl halides?

5. Chlorobenzene does not react appreciably with hot, alcoholic silver nitrate, but 2,4-dinitrochlorobenzene causes a precipitation of silver chloride under the same conditions; explain.

6. The addition of one mole of bromine to 1-phenyl-1,3-butadiene yields only 1-phenyl-3,4-dibromo-1-butene. Explain the reaction. What products would you expect as a result of permanganate oxidation of the original compound (specify the conditions).

7. Tell exactly what you would do (and see) to distinguish: (a) ethylene bromohydrin from ethylene bromide; (b) bromocyclohexane from bromobenzene; (c) benzyl chloride from p-chlorotoluene.

6.5 Unsubstituted Amides

Procedures for the detection of amides that are not N-substituted will be given in this section. Methods for detecting N-substituted amides will be given in 6.6.

A. *Liberation of Ammonia*

Amides may be hydrolyzed by treatment with alkali to form the salt of the acid and liberate ammonia, a compound that may be easily detected. Ammonium salts also may liberate ammonia by the same treatment; however, ammonium salts do not give the hydroxamate test (test B below). Ammonia may be liberated from unsubstituted amides by (a) boiling a mixture of 50 mg of the substance in 2 ml of 20 per cent sodium hydroxide for 1 minute, or (b) by dry fusion of a mixture of 50 mg of the compound with 200–300 mg of pulverized sodium hydroxide. In either case, the vaporized

[20] Berry, *Proc. Louisiana Acad. Sci.* **18**, 92 (1955).

ammonia may be detected by test P-4A, page 109. N-Substituted amides in which the alkyl groups have only a few carbon atoms will also give the blue coloration to copper sulfate by this method, since low-weight amines as well as ammonia form blue complexes with copper ions (test P-4A1). They do not give positive results by the phenol–hypochlorite test (test P-4A2). See tests 6.7E and 6.7G to establish the presence of amines. To perform these tests, the vapors from the hydroylsis reaction must be collected in aqueous solution; see test 6.6A.

B. Hydroxamate Test

Amides that are not N-substituted may be detected by the hydroxamate test,[21] provided it is known that the substance does not belong to some other class that will also yield a hydroxamic acid under the same conditions.

Add 30 mg of the substance to 2 ml of $1N$ hydroxylamine hydrochloride in propylene glycol in a 4-inch test tube. Boil for 2 minutes, cool, and add 0.5–1 ml of 5 per cent ferric chloride solution. A red-to-violet color is a positive test.

C. Distinguishing Aliphatic Amides from Aromatic Amides

Most aromatic amides are converted directly to the hydroxamic acid by hydrogen peroxide, whereas the aliphatic amides fail to form the hydroxamic acid in this way. Most aliphatic amides are converted to hydroxamic acid by hydroxylamine in aqueous or ethanolic solution; aromatic amides react much less readily.

$$ArCONH_2 + H_2O_2 \rightarrow ArCONHOH + H_2O$$
$$RCONH_2 + H_2NOH \cdot HCl \rightarrow RCONHOH + NH_4Cl$$

1. *Aliphatic amides.* Add 50 mg of the amide to 1 ml of $1N$ hydroxylamine hydrochloride in ethanol and boil the mixture for 3 minutes. Cool the tube and add 1–2 drops of 5 per cent ferric chloride. A bluish red color is a positive test.

2. *Aromatic amides.* Suspend 50 mg of the amide in 2–3 ml of water. Stopper the tube and shake it vigorously for a few seconds. Add 4–5 drops of 6 per cent hydrogen peroxide and heat the mixture to near boiling. If the amide does not completely dissolve add a few drops more of hydrogen peroxide. Cool the solution and add 1 drop of 5 per cent ferric chloride. If a bluish red color does not develop within 1 minute, warm the tube gently but do not boil the solution. In most cases the reaction eventually goes past the hydroxamate stage and a brown color develops, gradually settling out as a brown precipitate.

[21] Soloway and Lipschitz, *Anal. Chem.* **24**, 898 (1952).

It has been observed that the addition of a few milliliters of 10 per cent sodium hydroxide to the final mixtures obtained in this test produces a clear, deep, reddish brown-colored solution.

Note: Any time ferric chloride is to be used as one of the reagents in a test, the reader must remember that ferric chloride reacts directly with several classes of compounds to give colored products (see test 6.19A). For example, salicylamide is a phenol as well as an amide; hence, the hydroxamate test on this compound is not significant because of the interfering color produced by the interaction of ferric chloride with the phenolic group.

D. Sulfonamides

To prepare the test paper for this test, mix equal volumes of a 1 per cent solution of N,N-dimethyl-α-naphthylamine in methanol and a 1 per cent aqueous solution of sodium nitrite. Dip a piece of filter paper in the mixture and allow it to dry in the dark.

In making this test[22] for sulfonamides, place 1–2 drops of a solution or a suspension of the compound in water on the test paper and touch the spot with a drop of 0.2–0.5 per cent hydrochloric acid. A red or dark rose color develops quickly if a sulfonamide is present.

Note: This is an extremely sensitive test for sulfonamides. It may be used when the concentration of the sulfonamide is of the order of 3 micrograms in the solution being tested. The test may be made on blood by treating the blood with an equal volume of 10 per cent trichloroacetic acid before applying it to the test paper.

Amides other than sulfonamides do not give this test but an orange-red ring of color may appear around the spot where the compound was placed on the paper.

E. Urea, Thiourea, and Substituted Ureas

When urea, N-substituted urea, or N,N-symmetrically substituted urea is heated with an excess of phenylhydrazine at about 200°, diphenylcarbazide is formed and ammonia or an amine is liberated. For example

$$H_2NCONHR + 2H_2NNHC_6H_5 \rightarrow (C_6H_5NHNH)_2CO + NH_3 + RNH_2$$

Thiourea gives an analogous reaction and forms diphenylthiocarbazide. The diphenylcarbazide reacts with nickel ions to form violet inner complex salts that are soluble in chloroform.

Place 10–20 mg of the compound in a 3-inch test tube and add 2–3 drops of phenylhydrazine. Heat the tube in an oil bath at about 195° for 5 minutes. After cooling the tube, add 6 drops of concentrated ammonium hydroxide and 6 drops of a 10 per cent solution of nickel sulfate. Shake the tube vigorously and allow it to stand for 3 minutes. Extract the mixture with 10 drops of

[22] Hackmann, *Deut. med. Wechschr.* **72,** 71 (1947); *C.A.* **41,** 4824 (1947).

chloroform. A red-violet or violet color in the chloroform layer constitutes a positive test.

Note: At the temperature used, part of the urea will be converted to biuret, but this also reacts with phenylhydrazine to produce diphenylcarbazide. Many urethanes also give a positive test with this procedure. The test is not as sensitive with thiourea as with urea. The presence of thiourea may be detected by heating the dry sample of the original compound to about 200° and noting the evolution of hydrogen sulfide.

Exercises

1. How does nitrous acid react with amides?
2. Explain the interferences and complications which make the use of nitrous acid generally unsatisfactory for the detection of amides.
3. How can amides be converted to nitriles?

6.6 Substituted Amides

Substituted amides vary greatly in their compositions and in the methods by which they may be detected. In fact, no specific test for all types of amides is on record. Most methods for detecting N-substituted amides depend on first hydrolyzing the amide and then, by one method or another, identifying the amine that is liberated. Two major problems present themselves. First, many amides, particularly those with nitro and halogen substituents, are resistant to hydrolysis. Second, the recovery of the amine, produced by hydrolysis, is sometimes difficult. The following tests cover the more common types of substituted amides. For the more difficult types, reference should be made to larger works.

A. N-*Alkyl-Substituted Amines*

Reference has been made in test 6.5A to the fact that low-weight alkyl amines as well as ammonia can be detected by the formation of blue complexes with copper ions. To differentiate ammonia from primary or secondary amines, proceed as follows:

Place 200–300 mg of the amide in a 6-inch test tube and add 5 ml of 10 per cent sodium hydroxide. Insert a microcondenser into the tube and gently reflux the mixture for 15 minutes. After cooling the test tube, remove the condenser and arrange to distill from the test tube into another 6-inch test tube which contains 5 ml of water to which 2 drops of concentrated hydrochloric acid have been added. Have the delivery tube extend to within 3–4 mm of the surface of the acidified water. *Slowly* distill the hydrolyzed mixture until about 0.5 ml has passed into the receiving tube. Neutralize the distillate by adding, dropwise, 1.5N sodium bicarbonate solution. Divide the distillate into two equal portions and apply test 6.7E for primary alkyl amines and test 6.7G for secondary alkyl amines. Tertiary amines would not be produced by hydrolyzing an amide.

B. Anilides

The most common amides of this type are anilides; they may be detected by the following two methods with the limitations noted.

1. Anilides that do not have substituents on the ring produce a rose color when treated at room temperature with sulfuric acid and potassium dichromate.

Add 100 mg of the compound to 3 ml of concentrated sulfuric acid. Stopper the tube and shake it vigorously. Add 50 mg of finely powdered potassium dichromate. A bluish pink color is a positive test.

2. A modification of the aniline acetate test for carbohydrates (compare test 6.8A2) is quite satisfactory for determining anilides, both simple anilides and anilides with certain substituents (see Note below).

Mix 100 mg of the compound with one pellet of crushed sodium hydroxide. Add the mixture to a dry 4-inch test tube and wrap a piece of filter paper over the mouth of the tube. Moisten the paper with 1–2 drops of 4M acetic acid. Heat the tube and fuse the mixture. Aniline and certain substituted anilines will vaporize and react with the acetic acid on the paper. Place 50 mg of a sugar in a dry 4-inch test tube and transfer the filter paper that had covered the fusion mixture over to the tube which contains the sugar. Heat the sugar until it chars. Decomposition of the sugar produces furfural, which reacts with the aniline acetate to give a pink color on the paper.

Note: The procedure produces positive results even when halogens, methyl, or hydroxy groups are present on the aniline ring. Obviously, only those anilines which are at least moderately volatile will give this test.

C. Other Substituted Amides

Proof that the compound is an amide involves hydrolysis and identification of the amine and the acid thus produced. Since hydrolysis of these amides is difficult, it is desirable to detect the products by derivatizing them in the same operation, thus detecting the classes and determining the individual compounds produced at the same time. The sections of this book relating to the preparation of derivatives of amides, amines, and acids should be consulted before selecting a procedure for hydrolyzing the amide.

Hydrolysis of the simpler substituted amides is effected by boiling the compound for a few hours with either 6N hydrochloric acid or with 20 per cent sodium hydroxide. Alkaline hydrolysis is usually faster than acid hydrolysis. If alkaline hydrolysis is used, the freed amine may be recovered by filtration (if it is a solid), by steam distillation, or by extraction by ether. The aqueous solution of the sodium salt of the acid should be reserved for recovery of the acid. If acid hydrolysis is used, the organic acid may be separated by filtration, distillation, or extraction. The acid solution of the amine salt may be made alkaline and the amine recovered as above.

Other alkaline-hydrolyzing media include (a) a saturated solution of potassium hydroxide in ethanol, (b) a 5 per cent solution of sodium ethoxide in ethanol, and (c) a 5–20 per cent solution of potassium hydroxide in propylene glycol or glycerol. Amides that contain halogens or nitro groups in the amine moiety are quite resistant to hydrolysis. In such cases, 100 per cent phosphoric acid, made by mixing the 85 per cent acid with phosphorus pentoxide, is recommended. N-Alkyl-substituted sulfonamides are very resistant to hydrolysis. It is reported that they require 10 to 40 hours of refluxing with dilute hydrochloric acid. A mixture of 48 per cent hydrobromic acid and phenol has been recommended [23] as the best medium for hydrolyzing the substituted sulfonamides.

Exercises

1. What mechanism is probably involved in the acid hydrolysis of amides? Is the same mechanism involved in hydrolysis in alkaline solution?

2. In hydrolyzing the less complex amides, the rate of hydrolysis is usually higher in alkaline than in acidic solution. Why?

3. Can you offer any probable explanation for the fact that p-nitro acetanilide is very resistant to hydrolysis whereas acetanilide can be hydrolyzed normally?

4. What reason might be offered for the fact that N-alkyl-substituted sulfonamides are very resistant to hydrolysis as compared to unsubstituted sulfonamides?

5. If p-nitro-N,N-dimethylaniline fails to be classified as an amine and is subjected to the alkaline hydrolysis test for amides, dimethylamine will be recovered from the hydrolyzate. What will the other product be?

6.7 Amines

Simple amines are easily identified by solubility, basic reaction, and specific tests. Many of the substituted aromatic amines, even if they are primary amines, fail to dissolve in dilute acids. Since no specific test works perfectly for all amines, it is wise to test any nitrogen-containing compound that is not easily classified by at least two of the tests for amines.

It is reported[24] that a buffer solution of pH 5.5, made by adding 24 g of glacial acetic acid to 164 g of anhydrous sodium acetate and diluting to 1 liter, may be used as a solvent to differentiate between alkyl amines which are soluble and aryl amines which fail to dissolve. It is probable that this method is not completely dependable, but it has proved useful as a general guide.

A study has been made of the dependability of a large number of tests that have been recommended for the detection of amines. Forty representa-

[23] Snyder and Heckert, *J. Am. Chem. Soc.* **74,** 2006 (1952); Snyder and Geller, *J. Am. Chem. Soc.* **74,** 4864 (1952).

[24] Petrarca, *J. Org. Chem.* **24,** 1171 (1959).

tive amines were selected and subjected to many recently proposed tests as well as to the older tests.[25] The following tests seemed to be the most dependable. Note that the first four tests apply to more than one subclass of amines, whereas the later tests apply to primary, secondary, and tertiary amines.

A. N-Halo Succinimide Test[26]

N-Bromosuccinimide and N-iodosuccinimide have been shown to be good detecting and differentiating reagents for primary, secondary, and tertiary amines. The N-bromosuccinimide may also be used to differentiate primary, secondary, and tertiary alcohols (test 6.3C). Confirmatory tests should be run for amines and alcohols when using this reagent.

Add 100 mg of the compound to 1 ml of carbon tetrachloride in a 4-inch test tube. Add 30 mg of N-iodosuccinimide and a few mg of benzoyl peroxide. Wash the sides of the test tube with 1 ml of carbon tetrachloride. Place the test tube in a water bath at 80° and keep the tube in the bath at that temperature for 10 minutes. Note the formation of a brown color and note whether the color remains for the full 10 minutes or fades to a yellow-tan or to no color within a few minutes in the water bath. If a brown color develops and remains for the full time, dissolve 30 mg of the original compound being tested in 1 ml of carbon tetrachloride in a 4-inch test tube and add 30 mg of N-bromosuccinimide. Look for the immediate appearance of an orange precipitate. Primary and tertiary amines react with N-iodosuccinimide to give brown colors that do not fade during heating. Secondary amines produce brown colors that do fade to yellow-tan or colorless within a few minutes in the bath. Tertiary amines, but not primary amines, produce an orange precipitate when contacted with N-bromosuccinimide.

Note: The original reference cited covers only the application of these reagents to alkyl amines. However, experience has shown that the tests are also useful for many aromatic amines. The chemistry involved in this test is not well established, but suggestions regarding the probable mechanisms are given in the original article. The test for tertiary amines is particularly useful as a confirmatory test where other tests are used for detecting amines.

B. Hinsberg Test

1. p-Toluenesulfonchloride reacts with primary and secondary amines to form N-substituted sulfonamides. The derivatives from primary amines are soluble in dilute sodium hydroxide, whereas the derivatives from secondary amines are not soluble in dilute sodium hydroxide unless the original compound was amphoteric, such as a N-monoalkyl-substituted

[25] Brown and Wolt, *Proc. Louisiana Acad. Sci.* **18**, 95 (1955).
[26] Kruse, Grist, and McCoy, *loc cit.* (ref. 12).

amino acid or aminophenol.

Note: Benzenesulfonchloride may be substituted for the *p*-toluenesulfonchloride, but is less efficient.

2. In this test, it is important that the measurements be made accurately. Add 2 drops of a liquid amine to 2 ml of pyridine; then add 0.8 ml of *freshly prepared* 2 per cent aqueous sodium hydroxide. Shake the mixture thoroughly. Add 1 drop of benzene sulfonchloride and again shake the mixture. A yellow color indicates a primary amine; an orange color indicates a secondary amine; a deep red or purple color, a tertiary amine. The test appears reliable for alkyl amines and for many aryl amines that have only one ring.

C. Diazonium Salt Test

Most aryl amines and phenols (see test 6.19B) undergo coupling reactions with diazonium salts to form colored azo compounds. Since amines and phenols fall in different solubility divisions and give other characteristic tests, these two classes need not be confused by the fact that they both undergo azo compound formation. Alkyl amines do react with diazonium salts, but the color (usually yellow) is much less intense than with aryl amines. While several diazonium salts may be used (see Notes below), experience in the author's laboratories indicates that *p*-nitrobenzenediazonium tetrafluoroborate[27] is the most satisfactory reagent that has been tried.

Place 25 mg of the compound in a 3-inch test tube and add, dropwise, while slightly warming the tube, $1.2N$ hydrochloric acid until the substance dissolves. Cool the tube and add, dropwise, 10 per cent ammonium hydroxide until the solution *just* begins to become cloudy. Coupling with amines is optimum in a slightly acidic solution. Add 3–5 drops of a freshly prepared 1 per cent aqueous solution of *p*-nitrobenzenediazonium tetrafluoroborate reagent. A decidedly yellow, orange, or red coloration or precipitate is a positive test.

Note: A precipitate will form in a 1 per cent aqueous solution of the reagent if the solution is allowed to stand for a few days. Experience has shown, however, that the precipitates that form may be removed by filtration and still leave a reagent of sufficient strength to give positive tests.

The reagent gives positive tests with virtually all aryl amines and phenols that will undergo coupling reactions with diazonium compounds. Of course, compounds that have both of the ortho positions and the para positions substituted do not react. Compounds such as *p*-aminobenzoic acid and *p*-aminoacetophenone produce orange-red precipitates; anilides do not give the test. Alkyl amines generally give very weak or negative tests. However, this test is not considered reliable as a method of distinguishing alkyl amines from aryl amines.

[27] LeRosen, Monahan, Rivet, Smith, and Suter, *Anal. Chem.* **22**, 809 (1950).

Amines couple with diazonium salts most effectively at pH 3.5 to 7. Since the aryl amines are not appreciably soluble in water, the addition of acid is necessary to put them in solution. The solution should not be too acidic—hence the addition of ammonium hydroxide or sodium acetate to reduce the acidity.

D. Lignin Test

Webster[28] has proposed a very simple test that has proved quite reliable for primary and secondary amines—especially so if they are aryl amines. The test depends on the action of lignin in newsprint paper, but the chemistry of the reaction is not known.

Dissolve 10–20 mg of the compound in a few drops of ethanol and moisten a small area of newsprint paper with the solution. Place 2 drops of $6N$ hydrochloric acid on the moistened spot. The immediate development of a yellow or orange color is a positive test for a primary or secondary aryl amine. If the test is negative, repeat it, using a hot solution of the amine in ethanol and hot hydrochloric acid. Primary and secondary alkyl and alicyclic amines do not give the yellow or orange colors at room temperature but do give them when hot solutions are used. Tertiary amines, aliphatic amino acids, and amides do not give the test. Negatively substituted aryl amines that are too weakly basic to show definite basic properties will give this test (even aminosulfonic acids in most cases).

E. Tests for Primary Amines: Alkyl Amines

In addition to the above tests which are positive for primary amines and for other classes, the following tests are generally specific for primary alkyl amines.

1. Rimini test. Place 5 ml of a very dilute solution of the amine (or 1 or 2 drops of the amine in 5 ml of water) in a 4-inch test tube and add 1 ml of acetone. Add 1 drop of a 1 per cent solution of nitroprusside. The development of a definite violet-red color within 2 minutes is a positive test.

Note: The amine must be fairly soluble in water to give satisfactory results. This test for primary alkyl amines is positive, even in the presence of secondary alkyl amines. The acetone used must be free of acetaldehyde.

2. 2,4-Dinitrofluorobenzene test.[29] Moisten a piece of filter paper with a saturated solution of 2,4-dinitrofluorobenzene in ethanol and add a drop of a solution of the amine in water. An intense yellow color is a positive test for primary alkyl amines. Other amines may produce orange, red, or brown colors. Ammonia does not give this test.

[28] Webster, *Proc. S. Dakota Acad. Sci.* **24**, 85 (1944).
[29] Smith and Jones, *A Scheme of Qualitative Organic Analysis* (London: Blackie and Son, 1948), p. 110.

F. Tests for Primary Amines: Aryl Amines

1. *Diazotization test.* Primary aryl amines are converted to diazonium salts by nitrous acid. At low temperatures, these salts are stable and will couple in the alpha position with the sodium salt of β-naphthol to form a red coloration or precipitate.

Prepare about 100 ml of a mixture of crushed ice, salt, and water to be used as a chilling bath. In one 4-inch test tube, mix 30–50 mg of the amine with 1 ml of water and 4 drops of concentrated sulfuric acid. In a second tube, place 1 ml of 10 per cent sodium nitrite. In a third tube, dissolve 100 mg of β-naphthol in 2 ml of 10 per cent sodium hydroxide. Chill all three solutions in the ice bath. After the solutions are thoroughly chilled, add the sodium nitrite solution, dropwise and with shaking, to the acidified amine solution. Now add, dropwise, the sodium naphthoxide solution. A red color or precipitate indicates a primary aryl amine.

Note: Some electronegatively substituted amines, such as 2,4-dinitroaniline, do not diazotize by this method. *o*-Diamines do not diazotize, but form dark-colored azoimides. *m*-Diamines diazotize, but react with undiazotized amine to form brown dyes. *p*-Diamines diazotize and couple with the naphthol in the normal way.

It must be remembered that nitrous acid also reacts with secondary amines with the formation of *N*-nitroso products. These products are usually yellow or red oils that are insoluble in water. Nitrous acid reacts with *N,N*-dialkyl amines to form *p*-nitroso compounds. It also reacts with phenols and some other classes of compounds, but none of these reactions yield diazonium salts that will couple with naphthol.

2. *Isocyanide test.* Most primary amines will react with chloroform and potassium hydroxide to form the nauseatingly odored isocyanides. The test is very delicate and the reaction will be given by small concentrations of primary amines if they are present in other amines as impurities. *This test is more useful with aromatic amines than with others.*

$$C_6H_5NH_2 + CHCl_3 + 3KOH \rightarrow C_6H_5NC + 3KCl + 3HOH$$

Mix 50 mg of the primary arylamine with 2 drops of chloroform and 1 ml of 2*N* potassium hydroxide in methanol. Warm the mixture slightly and note the odor.

G. Tests for Secondary Amines

In addition to the *N*-halo succinimide and Hinsberg tests already discussed, the following tests are particularly useful for the detection of secondary amines.

1. *Nickel–dithiocarbamate test.*[30] The original article by Duke gives separate tests for primary and secondary amines; the test for secondary amines is particularly useful. The reagent is prepared by dissolving 0.5 g of

[30] Duke, *Ind. Eng. Chem., Anal Ed.* **17**, 196 (1945).

nickel chloride hexahydrate in 100 ml of water and adding enough carbon disulfide to saturate the solution and leave a small globule of carbon disulfide.

Add 30–50 mg of the amine to 5 ml of water in a 4-inch test tube. If the amine does not dissolve, add a drop of hydrochloric acid and shake the solution. In another 4-inch test tube, place 1 ml of the nickel chloride–carbon disulfide reagent and add 0.5–1 ml of concentrated ammonium hydroxide, followed by 0.5-1 ml of the amine solution. Secondary alkyl amines usually give a greenish yellow precipitate, and secondary aryl amines give a white or tan precipitate.

2. *Simon test for alkyl amines.* The following test is selective for secondary alkyl amines.

Add 1 ml of a freshly prepared 5 per cent solution of acetaldehyde to 5 ml of a dilute aqueous solution of the amine (for example, 1–2 drops of amine in 5 ml of water). Add 1–2 drops of 10 per cent sodium nitroprusside and 2 drops 1.5N NaHCO$_3$. A blue color will develop within 3 minutes. On standing, the solution may change color to green and finally to yellow. Primary amines produce a purple color in this test.

H. Tests for Tertiary Amines

Tertiary amines may be detected by the N-bromosuccinimide reaction (test 6.7A) and by the modified Hinsberg test (test 6.7B2).

A reagent that seems very selective for tertiary amines of both the aliphatic and aromatic types has been proposed by Feigl[31] and Ohkuma.[32] The reagent consists of a solution prepared by heating 2 g of citric acid in a 100 ml of acetic anhydride.

Add 2–3 drops of the amine or its solution in alcohol to 3 drops of the reagent in a small test tube. Place the test tube in a bath of boiling water or heat briefly over a small flame. The development of a red to purple color within 1 to 2 minutes is a positive test.

Exercises

1. Why are the alkyl amines slightly stronger bases than ammonia?

2. Place the following amines in the order of their decreasing basicity and explain the reasons for the order: aniline, butylamine, diphenylamine, p-nitrotoluidine, and o-nitrotoluidine.

3. Suggest a practical way to distinguish between the following: (a) ethylamine and diethylamine; (b) benzylamine and aniline; (c) N-ethylaniline and N,N-diethylaniline; (d) N-methylaniline and o-toluidine.

4. Explain why trimethylamine is a weaker base than dimethylamine.

5. Why are the diazonium salts of arylamines more stable than those of alkyl amines?

[31] Feigl, *op. cit.*, p. 281.
[32] S. Ohkuma, *J. Pharm. Soc. Japan* **75**, 1124 (1955).

6. In coupling a diazotized reagent with an amine, a weakly acidic solution is used, whereas in coupling with a phenol a weakly basic solution is desirable. Explain the reason for this difference.

7. On the basis of the results you have obtained in using known amines with the various tests provided in section 6.7, make an outline of the order in which you would apply tests to determine the characteristics of an unknown amine.

6.8 Carbohydrates

A. General Tests for Carbohydrates

1. *Anthrone test.* Anthrone reacts with virtually all types of carbohydrates to produce a green color.[33] It is relatively specific for carbohydrates.

Place 1 ml of water containing 1–3 mg of the compound in a 4-inch test tube. While holding the tube at an angle, pour 2 ml of a 0.2 per cent solution of anthrone in 95 per cent sulfuric acid down the side of the tube. Shake the tube very gently. If a green zone does not appear in 30 seconds, shake the mixture and warm it slightly. Carbohydrates produce a green color which changes to blue-green.

Note: The solution of anthrone in sulfuric acid should be prepared fresh every few days. The final mixture must be at least 50 per cent with respect to sulfuric acid to hold the anthrone in solution. Mono-, di-, and polysaccharides and their acetates, dextrins, dextrans, gums, glucosides, and starches give a positive test with anthrone. Furfural produces a transitory green color which rapidly changes to brown. Aldehydes, alcohols, and proteins often give a red color with the reagent. Easily dehydrated organic compounds may produce a tan or brown color because of the sulfuric acid; however, in the very small quantities used in this test such coloration is not confusing.

This reagent has been found very useful in the quantitative determination of extremely small amounts of carbohydrates.[34]

2. p-*Toluidine acetate test.* Carbohydrates are decomposed by heat or by hot phosphoric acid to yield furfural or furfural derivatives that produce a red color with p-toluidine acetate.

Place 5–10 mg of the compound in a microcrucible (or evaporate a few drops of a solution of the compound to dryness in a crucible). Cover the crucible with a filter paper that has been moistened with 2 drops of a 10 per cent solution of p-toluidine in 10 per cent acetic acid. Heat the bottom of the crucible with a microburner for 1 minute.

If a pink to red color does not form on the paper, repeat the experiment, adding a drop of syrupy phosphoric acid to the compound before heating it.

[33] Dreywood, *Ind. Eng. Chem., Anal. Ed.* **18**, 499 (1946); Sattler and Zerban, *Science* **108**, 207 (1948); Sattler and Zerban, *J. Am. Chem. Soc.* **72**, 3814 (1950); Koehler, *Anal. Chem.* **24**, 1576 (1952).

[34] Morris, *Science* **107**, 254 (1948); Viles and Silverman, *Anal. Chem.* **21**, 951 (1949); Lowald and McCormack, *Anal. Chem.* **21**, 383 (1949); **24**, 1576 (1952).

Note: Furfural and furfural derivatives may be detected by this test by heating an aqueous solution of the aldehyde and allowing the vapors to contact the aniline treated paper. Tested dry, carbohydrates, including agar, starch, many gums, and alkyl- and acyl-substituted cellulose, give a positive test, especially if the phosphoric acid is used.

Aniline[35] may be substituted for the *p*-toluidine in this test but it has been found that the *p*-toluidine gives a deeper pink to red color and that the color persists much longer on the paper.

B. *Test for Water-Soluble Carbohydrates*

1. *Molisch test.*[36]   Disolve 20 mg of the compound in 1 ml of water and add 2 drops of a 5 per cent solution of α-naphthol in methanol. Place 1 ml of concentrated sulfuric acid in a 4-inch test tube and, while holding the tube at an angle, slowly introduce the solution to be tested by means of a pipet so that it stratifies on top of the acid. The development of a violet-purple color at the interface is a positive test.

Note: Pentoses and their disaccharides are decomposed by concentrated sulfuric acid to form furfural. Hexoses and their disaccharides analogously produce hydroxy-methylfurfurals. These furfurals produce colored condensation products with α-naphthol. Some more complex carbohydrates give faintly positive tests.

2. *Resorcinol test.*[37]   Add 20 mg of the compound to 1 ml of a 0.1 per cent solution of resorcinol in water. Stratify this mixture on top of 2 ml of concentrated sulfuric acid. An orange to red zone at the interface is a positive test.

3. *Aminoguanidine test.*   It is reported [38] that aminoguanidine and a very dilute solution of potassium dichromate in sulfuric acid may be used to distinguish pentoses, carbohydrates that contain ketohexose units, and carbohydrates that contain only aldohexose units.

C. *Distinguishing Monosaccharides from Disaccharides*

The monosaccharides are more easily oxidized than the disaccharides. Barfoed's solution will oxidize monosaccharides within 2 minutes, but it will not oxidize disaccharides unless heated for several minutes. Benedict's solution will oxidize all of the common sugars except sucrose. Methods of preparation of the reagents are given in the Appendix.

Place 2 ml of Barfoed's reagent or Benedict's reagent in a 4-inch test tube and add 10–20 mg of the carbohydrate (or 1 ml of a dilute solution of it in water) to the reagent. Place the tube in a bath of boiling water for 3 minutes. Remove the tube from the bath and allow it to cool. A yellow-

[35] Feigl, *op. cit.*, p. 426.
[36] Molisch, *Monatsh.* 7, 198 (1886).
[37] Morgenstern, *Centr. Zuckerind* 50, 226 (1942); *C.A.* 38, 530 (1944).
[38] Tauber, *Anal. Chem.* 25, 826 (1953).

orange or orange-red precipitate is a positive test. It should be remembered that a yellow suspension in a blue solution appears green.

Note: These reagents are not specific tests for sugars since they also oxidize other easily oxidized compounds, particularly α-hydroxyaldehydes and ketones and α-keto-aldehydes. Their use in this test is to distinguish monosaccharides from disaccharides and both of these types form more complex carbohydrates.

D. *Ketoses*

The Seliwanoff test for ketoses is based on the conversion of the ketose to hydroxy methylfurfural and its subsequent condensation with resorcinol to form colored complexes.

Mix 1 ml of Seliwanoff's reagent with 1 ml of about a 5 per cent solution of the sugar in water. Heat the mixture to boiling. A red color develops within 2 minutes if the sugar is a ketose. Long standing, or prolonged heating, will develop the color with aldoses. See the Appendix for the preparation of the reagent.

E. *Pentoses*

Tollen's test for pentoses is based on the reaction of the pentose with hydrochloric acid to form furfural, which is then condensed with phloroglucinol to yield red complexes. Other sugars may produce yellow, orange, or brown colors.

Dissolve about 10 mg of the sugar in 5 ml of $6N$ hydrochloric acid and add about 10 mg of phloroglucinol. Boil the mixture for 1 minute. A red coloration indicates a pentose.

F. *Osazone Formation*

Methods for forming osazones and discussion of them are given in Chapter 10, pages 281–85.

Exercises

1. Which disaccharide is not fermentable by yeast?
2. How can hot nitric acid be used to differentiate galactose from glucose?
3. Why is fructose (a ketose) easily oxidized?
4. How could you detect glucose as an impurity in maltose?
5. Why is sucrose not a reducing sugar?

6.9 Carbonyl Compounds

A. *Aldehydes and Ketones*

Most aldehydes and ketones react alike with all reagents that condense with the carbonyl group. Of the several useful carbonyl reagents, 2,4-dinitrophenylhydrazine has proven to be very effective in detecting carbonyl compounds. However, it is possible for erroneous deductions to be drawn from the formation of a precipitate when using this reagent, particularly if it has been used for a preliminary "spot test" without having properly classified the unknown by solubility. Like many dinitro compounds, 2,4-dinitrophenylhydrazine can form slightly soluble adducts with phenols and with many hydrocarbons, aryl halides, and ethers. Hence, if there is any doubt about the validity of the test as given, repeat the test substituting p-nitrophenylhydrazine for the 2,4-dinitrophenylhydrazine. Although this reagent is not as selective a carbonyl reagent as the 2,4-dinitrophenylhydrazine, it does not form adducts with other classes.

$$R_2CO + H_2NNHC_6H_3(NO_2)_2 \rightarrow R_2C{=}NNHC_6H_3(NO_2)_2 + H_2O$$

$$\underset{H}{RCO} + H_2NNHC_6H_3(NO_2)_2 \rightarrow \underset{H}{RC}{=}NNHC_6H_3(NO_2)_2 + H_2O$$

Place 5 ml of a saturated solution of 2,4-dinitrophenylhydrazine in $2N$ hydrochloric acid in a 6-inch test tube. Add a solution of 30–40 mg of the compound in 0.5 ml of methanol. Stopper the test tube and shake it vigorously. If a precipitate does not form, heat the mixture to boiling for 30 seconds and shake it again. A precipitate, generally yellow or orange, is a positive test. A few drops of water may be added to aid precipitation.

Note: Aqueous solutions of aldehydes or ketones may also be tested by this method. If the carbonyl compound is only slightly soluble in water, it may be necessary to add more alcohol than the 0.5 ml recommended.

In the 2,4-dinitrophenylhydrazine test, an excess of the carbonyl compound is to be avoided since the phenylhydrazone is more soluble in the carbonyl compound.

B. Aldehydes

The following tests are given by aldehydes but not by ketones, except where noted.

1. *Benzenesulfonhydroxamic acid test.* Dissolve a few milligrams of benzene sulfonhydroxamic acid in 0.5 ml of methanol in a 4-inch test tube and add 30 mg of the compound to be tested. Now add 0.5 ml of $2N$ potassium hydroxide in methanol. Heat the mixture just to boiling. Cool the tube, acidify the mixture with dilute hydrochloric acid, and add 1 drop of 10 per cent ferric chloride. A bluish red color (ferric hydroxamate) is a positive test.

Note: The exact chemistry of this reaction is not known, but one explanation is shown by the following equations:

$$C_6H_5SO_2NHOH + 2KOH \rightarrow C_6H_5SO_2K + KNO + 2H_2O$$

$$KNO + HCl + RCHO \rightarrow RCONHOH + KCl$$

$$3RCONHOH + FeCl_3 \rightarrow (RCONHO)_3Fe + 3HCl$$

It is reported that some nitro- and hydroxyaromatic aldehydes do not give this test. Benzyl ketones do give a positive test.

2. *Schiff's test.* p-Rosaniline hydrochloride will react with sulfurous acid to form a leuco-sulfonic acid, which will react with more sulfurous acid to form the colorless bis-N-sulfinic acid. This acid will react with 2 moles of an aldehyde to give an addition complex that is unstable and loses 1 mole of sulfurous acid to produce a wine-purple quinonoid-type dye. Add 3 drops of an aldehyde to 2 ml of colorless Schiff's reagent (see the Appendix for preparation). Do not warm the mixture. A wine-purple coloration will develop within 10 minutes. A few ketones give faint colorations with this test. There are compounds other than aldehydes that will give light-pink colorations, but these colors lack the blue cast characteristic of aldehydes.

3. *Methone test.* The compound 5,5-dimethylcyclohexane-1,-3-dione is often called methone, dimethone, or dimethol and is generally listed in the chemical catalogs as dimethyldihydroresorcinol. It is recommended [39] as a reagent for aldeyhdes and does not give the test with ketones. A milky suspension forms immediately when the reagent is added to very small amounts of aldehydes.

Add 50 mg of the aldehyde to 1 ml of water and then add 3 drops of a 5 per cent solution of methone in ethanol. Shake the mixture. The formation of a milky suspension within 2 minutes is a positive test for aldehydes.

4. *Oxidation tests.*   *(a) Benedict's reagent.* It has long been believed that Benedict's reagent (test 6.8C) would oxidize all aliphatic aldehydes but not aromatic aldehydes or any of the ketones except the α-hydroxyketones. Daniels, Rush, and Bauer [40] have recently claimed that the precipitate formed by the reaction of Benedict's reagent on common aliphatic aldehydes is not cuprous oxide and that, therefore, this reagent really gives a negative test for these compounds. It is their contention that the reagent only gives a positive test of the α-hydroxyaldehydes, α-ketoaldehydes, and the α-hydroxyketones. Regardless of the composition of the precipitate, it is true that aliphatic aldehydes in general do produce a yellow- to orange-colored precipitate or suspension when heated with Benedict's reagent. A yellow suspension in a blue solution appears green. Other classes of compounds are also oxidizable by Benedict's reagent, and the test is mentioned here only as a means of helping to distinguish between aliphatic aldehydes and aromatic aldehydes and ketones. Tollen's reagent, discussed below, gives a positive test for both aliphatic and aromatic aldehydes but a negative test for ketones, except those substituted with hydroxyl, alkoxy, or dialkylamino groups on the α-carbon.

(b) Tollen's reagent. The silver ions in a solution containing silver–ammonia complex ions are reduced to metallic silver by most aldehydes,

[39] Weinberger, *Ind. Eng. Chem., Anal. Ed.* **3**, 365 (1931).
[40] Daniels, Rush, and Bauer, *J. Chem. Educ.* **37**, 205 (1960).

readily oxidized sugars, polyhydroxyphenols, aminophenols, hydroxylamines, and other reducing agents.[41]

Add 30–50 mg of the compound to be tested to 2 ml of freshly prepared reagent. Shake the tube and allow it to stand for 10 minutes. If no reaction occurs in this time, place the tube in a beaker of water at about 35° for 5 minutes. A precipitate of silver is a positive test.

The reagent is prepared as follows: Add 2 drops of 5 per cent sodium hydroxide to 1 ml of 5 per cent aqueous silver nitrate. Shake the tube and add 2N ammonium hydroxide dropwise and with shaking until the precipitated silver hydroxide just dissolves.

Note: The test tube to be used in this test must be clean. It is best to clean the tube by boiling it in a 10 per cent solution of sodium hydroxide and then discarding this solution. The silver produced in this test will generally precipitate as a "mirror" on the glass tube if it is thoroughly clean but the formation of black metallic silver in the mixture is also a positive test.

Caution: *Silver fulminate, which is very explosive when dry, may be present in the residues from the use of Tollen's solution.* Hence, as soon as the test is completed the contents of the tube should be poured down the sink and washed through the trap with water. Also, rinse out the test tube with dilute nitric acid.

5. *Detection of aliphatic aldehydes with 2-hydrazinobenzothiazole.* A spot test for detecting aliphatic aldehydes has been suggested by Sawicki and Hauser.[42] The following is a slight modification of one of the methods suggested by these authors.

Place 1 drop or 30 mg of the aldeyhde on a spot plate. Add 2–3 drops of dimethylformamide. Add a few milligrams of 2-hydrazinobenzothiazole to the mixture and allow to stand for 1–2 minutes. Add 1 drop of 1 per cent aqueous potassium ferricyanide solution. Let the mixture stand for 2–3 minutes and then add 2–3 drops of 20 per cent potassium hydroxide solution. A deep blue color develops within 5 minutes if aliphatic aldehydes are present.

Note: Higher-weight aldehydes require more than 5 minues (up to 30 minutes) for the development of the blue color. The aromatic aldehydes do not develop the blue color, but those which have been tested give an orange or brown-green coloration. Ketones failed to produce a color. A mechanism for the chemistry involved in this test is suggested in the reference cited, in which the authors give quantitative data and directions for additional methods for using this reagent.

Drucker and Rosen[43] distinguish ketones from saturated aliphatic aldehydes by oxidation with peroxytrifluoroacetic acid, which produces esters or lactones only with ketones. The esters or lactones are detected by the ferric hydroxamate test (see test 6.10).

[41] Morgan and Mickelwait, *J. Soc. Chem. Ind. (London)* **21**, 1375 (1902).
[42] Sawicki and Hauser, *Anal. Chem.* **32**, 1434 (1960).
[43] Drucker and Rosen, *Anal. Chem.* **33**, 273 (1961).

Exercises

1. How does the silver ion compare as an oxidizing agent with the cupric ion?

2. How can Tollen's solution and Benedict's solution be used in the investigation of aldehydes to distinguish aliphatic aldehydes from aromatic aldehydes?

3. What are the advantages and disadvantages of phenylhydrazine as compared to 2,4-dinitrophenylhydrazine as a reagent for carbonyl compounds?

4. Name three other reagents that might be used to detect carbonyl compounds.

5. Nucleophilic reagents attack the carbon of the carbonyl group. The rate of reaction is largely dependent on the degree of positive character (electron deficiency) of this carbon. Thus, groups that increase the electron deficiency on this carbon increase the rate of reaction, and groups which decrease the electron deficiency decrease the reaction rate between the carbonyl group and the nucleophilic reagent. Arrange the following compounds in order of increasing reactivity toward a nucleophilic reagent: 2-propanone, methyl, vinyl ketone, propionaldehyde, bromoacetaldehyde, and trichloroacetaldehyde.

6.10 Esters

The best test for esters is the ferric hydroxamate test. Unlike the acid anhydrides and acid halides (compare with test 6.2B), esters react with hydroxylamine to form hydroxamic acids only when the reaction is carried out in alkaline solution.

Place 0.5 ml of 1N hydroxylamine hydrochloride in methanol in a 4-inch test tube and add 30 mg of the compound to be tested. Now add, dropwise, a 2N solution of potassium hydroxide in methanol until the mixture is alkaline to litmus and then add 4 drops more of the potassium hydroxide solution. Heat the mixture just to boiling, then cool it, and add, dropwise and with shaking, 2N hydrochloric acid until the pH of the mixture is approximately 3 (use Hydrion paper or a similar indicator). Add 1 drop of 10 per cent ferric chloride and note the color. A bluish red or reddish blue color is a positive test.

Note: A white precipitate of potassium chloride often forms during the performance of this test. This salt does not interfere with the test, but, if desired, it may be dissolved by adding a few drops of water.

A few esters, particularly the esters of carbonic, carbamic, and sulfonic acids, do not give the test.

In general, the ferric chloride test for phenols (test 6.19A) does not interfere with the ferric hydroxamate test for esters and the like because, in the acid solution employed in this test, the ferric phenolate ion is largely converted to the phenol.

Methods are given in the Appendix for determining the saponification equivalent, saponification number, and the iodine number. These values are frequently useful in identifying esters.

Exercises

1. Discuss the mechanism involved in the hydrolysis of esters.

2. The identification of an ester usually requires hydrolysis as a preliminary step. Discuss the relative advantages and disadvantages of hydrolyzing esters in alkaline and acidic media.

3. Define saponification equivalent. What are the limitations related to the use of the saponification equivalent in determining the identity of an ester?

4. Define iodine number. For what types of commercial esters does the determination of the iodine number provide a very useful value?

6.11 Ethers

Ethers are quite unreactive to all chemical reagents. Since they are the least easily detected class in Solubility Division N, tests for all of the other classes should be made first. The tests given below are not specific for ethers and the results may be misinterpreted unless it has been established that other classes of similar solubility are not present.

A. Esterification[43]

A great many of the ethers may be hydrolyzed and converted into acetate esters by heating a mixture of the ether, acetic acid, and concentrated sulfuric acid. The reaction is not complete, but sufficient ester is formed to give the hydroxamate test for esters (test 6.10). The presence of unchanged ether does not interfere with this test.

$$2CH_3COOH + R_2O + (H_2SO_4) \rightarrow H_2O + 2CH_3COOR$$

Mix 0.5 ml of the ether with 2 ml of glacial acid and 0.5 ml of concentrated sulfuric acid. Reflux the mixture for 5 minutes and distill 1 drop. Test this drop for esters by test 6.10 (be sure that enough potassium hydroxide solution is used to make the mixture alkaline). If the drop of distillate does not give a positive test for esters, cool the mixture that was refluxed and add 5 ml of ice water to it. If a separate liquid phase separates, test it for esters. It is sometimes advisable to extract the mixture with 0.5 ml of benzene and test the benzene extract for esters.

B. Additional Tests

Owing to their lack of activity, ethers may be confused with hydrocarbons. Alkyl ethers may be distinguished from hydrocarbons by the Ferrox test (page 75).

Aryl ethers may be distinguished from alkyl ethers by the formaldehyde–sulfuric acid reagent (P–2B).

Alkyl ethers are generally soluble in concentrated hydrochloric acid, whereas aryl ethers and alkyl aryl ethers are not soluble.

Exercises

1. Why are ethers more soluble in concentrated hydrochloric acid or sulfuric acid than they are in water?

[43] Davidson and Perlman, *A Guide to Qualitative Organic Analysis* (Brooklyn, N.Y.: Brooklyn College, 1952).

2. Why are alkyl ethers more soluble than aryl ethers in concentrated hydrochloric acid?

3. It has been suggested that dilute solutions of iodine in various solvents change color from violet to yellow-tan depending on the basicity of the solvent. Predict the colors of solutions of iodine in the following solvents: ethyl ether, benzene, ethanol, and hexane. Check your predictions by performing the experiment.

6.12 Hydrazines

Hydrazines may be detected by condensing them with some aldehyde or ketone.

Suspend 25–50 mg of the compound in 1 ml of water and add enough acetic acid to dissolve the hydrazine. Add a few drops of a 5 per cent solution of acetaldehyde. The formation of a precipitate strongly indicates a hydrazine.

Note: In case the hydrazine does not readily dissolve in acetic acid, use dilute hydrochloric acid instead. Hydrazines liberate ammonia when heated with 10 per cent sodium hydroxide (test P-4A). They are readily oxidized by Tollen's (test 6.8E) and Benedict's (test 6.8C) reagents.

Exercises

1. Would semicarbazide be considered a hydrazine? Would it give a positive reaction in this test?

2. Write equations for the reactions of five different hydrazines with five different carbonyl compounds.

3. Would the formation of a precipitate by reacting an unknown with a carbonyl compound specifically identify the unknown as a hydrazine? Why?

6.13 Hydrocarbons

Hydrocarbons are detected more by elimination than by direct proof. Knowledge of the solubility behavior of the substance is a definite aid, as is information regarding the elements that are present. Hydrocarbons that are soluble in concentrated sulfuric acid are either actively unsaturated or easily sulfonated compounds. The Ferrox test (page 75) will usually distinguish all hydrocarbons (negative reaction) from other classes (positive reaction) that fall in Solubility Division N. Confirmatory tests may be made on compounds that do not give positive reactions for functional groups (esters, alcohols, aldehydes, ketones, and the like) of this solubility division. Such tests would include the tests for active unsaturation (tests P–5) and the formaldehyde–sulfuric acid test (test P–2B).

A recently proposed test for olefins[44] employs the Friedel-Crafts acetylation to produce a ketone which is then detected by addition of 2,4-dinitrophenylhydrazine, followed by the addition of a base to produce a deep red color.

[44] Sharefkin and Sulzberg, *Anal. Chem.* **32**, 993 (1960).

Acetylene and monosubstituted acetylenes react with Nessler's solution to precipitate mercuric acetylides. The mercuric acetylides are not as explosive as the silver and copper acetylides. Hydrocarbons that are not soluble in concentrated sulfuric acid include the alkanes, cycloalkanes, and most of the aromatic compounds. These Solubility Division I compounds may be distinguished from the alkyl and aryl halides by elemental analysis. The formaldehyde–sulfuric acid reagent will distinguish the alkanes from the aromatic hydrocarbons. Test P–2A is particularly useful for distinguishing aromatic hydrocarbons from cycloalkanes and alkanes. Ethers of aryl radicals frequently are not soluble in concentrated sulfuric acid. Unfortunately, many of these ethers do not give positive results with the formaldehyde–sulfuric acid reagent; hence, no direct way is known to distinguish them from aromatic hydrocarbons, except by physical methods such as infrared spectroscopy (see Chapter 7).

Exercise

1. What instrumental methods might be used to differentiate ethers from hydrocarbons?

6.14 Nitrates and Nitrites

A. Diphenylamine Test

Both nitrites and nitrates act as oxidizing agents on diphenylamine in sulfuric acid. Since the test is very sensitive, enough nitrate radicals are liberated from organic nitrates by sulfuric acid to give the test. Diphenylamine is first oxidized to diphenylbenzidine, which is further oxidized to the quinonoid form.

$$2(C_6H_5)_2NH \rightarrow (C_6H_5NHC_6H_4)_2 \rightarrow C_6H_5N{=}C_6H_4{=}C_6H_4{=}NC_6H_5$$

Add 100 mg of the compound to be tested to 3 ml of a reagent made by dissolving 200 mg of diphenylamine in 100 ml of concentrated sulfuric acid. A blue color is a positive test.

B. Alkyl Nitrites

Care should be taken in handling the alkyl nitrites as they have a pronounced action on the heart. They may be detected by the fact that they will react with 2-phenylindole to precipitate 3-isonitroso-2-phenylindole.

Dissolve 100 mg of 2-phenylindole in boiling ethanol and add 100 mg of the nitrite. On cooling, the 3-isonitroso-2-phenylindole will precipitate.

It may be recrystallized from amyl acetate as yellow needles and has a melting point of 280°.

Note: Nitrates and nitrites oxidize ferrous hydroxide to ferric hydroxide by test 6.16A.

6.15 Nitriles

A nitrogen-containing compound that does not give positive results with the hydroxamate tests for esters (for example, nitro-substituted esters) or amides (including anilides and the like) may be tentatively identified as a nitrile if it reacts positively to the following test.[45] Confirmation may be made by hydrolysis.

Add 30 mg of the compound to 2 ml of $1N$ hydroxylamine hydrochloride in propylene glycol. Then add 1 ml of $1N$ potassium hydroxide in propylene glycol and boil the mixture for 2 minutes. Cool the test tube and add 0.5–1 ml of 5 per cent ferric chloride. A red to violet color is a positive test.

Note: Owing to the high boiling point of propylene glycol, these conditions are the most rigorous used to convert a compound into a hydroxamic acid. Hence, it must be established that the substance does not belong to any more readily converted class before this test is significant. Anilides and similar substituted amides, like the nitriles, are not converted to hydroxamic acids by the less drastic methods; hence, they may be confused with nitriles by this test and must be distinguished from them by other tests (see test 6.6C).

6.16 Nitro Compounds

Nitro groups are frequently present as substituents in several classes of aromatic compounds, such as acids, aldehydes, ketones, amines, azo-compounds, and ethers, as well as hydrocarbons. Less commonly, nitro groups are found in nonaromatic acids, alcohols, and the like. In these cases, except for the nitro hydrocarbons and possibly the nitro ethers, the class to which the substance belongs would generally be established and the compound identified by solubility data, acid-base character, functional group tests for the groups that are more chemically active than the nitro group and derivatization of this more active group. The presence of the nitro group or groups would, therefore, generally not have to be proved directly, but would be indicated by the physical constants of the compounds and of its derivatives. Nitro compounds can be reduced and the resulting amine identified. See pages 325–26 for methods of reducing nitro compounds. The presence of one or more nitro groups in a compound may be detected by proving the absence of other classes of compounds that also act as oxidizing agents on ferrous hydroxide. Nitro groups may be detected by infrared spectroscopy.

[45] Soloway and Lipschitz, *loc. cit.* (ref. 21).

A. Ferrous Hydroxide Test

It has been shown[46] that organic compounds that are oxidizing agents will oxidize ferrous hydroxide to ferric hydroxide with a change of color from blue to red-brown. The most common organic compounds that are oxidizing agents are the nitro compounds; less common classes are the nitroso compounds, quinones, hydroxylamines, nitrates, and nitrites.

$$C_6H_5NO_2 + 4H_2O + 6Fe(OH)_2 \rightarrow C_6H_5NH_2 + 6Fe(OH)_3$$

In a 3-inch test tube, mix about 20 mg of the compound with 1.5 ml of freshly prepared 5 per cent solution of ferrous ammonium sulfate. Add 1 drop of $3N$ sulfuric acid and 1 ml of $2N$ potassium hydroxide in methanol. Stopper the tube quickly and shake it. A positive test is indicated by the precipitate turning red-brown within 1 minute.

Note: The use of a small tube is required so that very little air will come in contact with the ferrous hydroxide.

B. Dinitro and Trinitro Hydrocarbons

These nitro derivatives of benzene and its homologs may usually be classified by the following test.[47]

Add 50 mg of the compound to 5 ml of acetone in a test tube and then add, while shaking the tube, 2 ml of 5 per cent sodium hydroxide. Mono-nitro compounds do not produce colors, but colors develop quickly for dinitro compounds (purplish blue) and trinitrocompounds (deep red).

Note: The chemistry involved in this test has not been proven. Two dinitrobenzenes give irregular colors: 1,2-dinitrobenzene (no marked color formation) and 1,4-dinitrobenzene (greenish yellow). The presence of amino, alkylamino, acylamino, hydroxy, or acylated hydroxy groups on the benzene nucleus interfere with the test. Most dinitro and trinitro phenols tested gave yellow, yellow-orange, or greenish yellow colors. 2,4-Dinitroaniline gives a red color that is confusing with the trinitro hydrocarbons.

It is stated[48] that *m*-dinitro compounds may be detected, even in the presence of other dinitro compounds, by heating a few milligrams of the compound with a drop of 10 per cent aqueous potassium cyanide solution in a microcrucible. A red or violet color appears on heating and this color is not affected by the addition of a few drops of $2N$ hydrochloric acid.

C. Nitroparaffins

Nitrous acid test. The test for the nitrocompounds (test 6.16) is given by some but not all of the nitroparaffins. For example, nitromethane and

[46] Hearon and Gustavson, *Ind. Eng. Chem., Anal. Ed.* **9**, 352 (1937).

[47] Bost and Nicholson, *Ind. Eng. Chem., Anal. Ed.* **7**, 190 (1935); English, *Anal. Chem.* **20**, 745 (1948).

[48] Feigl, *op. cit.*, p. 174.

2-nitropropane give positive results, but nitroethane and 1-nitropropane fail to oxidize the ferrous hydroxide under the conditions of this test. By vigorous reduction the nitroparaffins may be converted to primary amines.

The action of nitrous acid on the nitroparaffins in alkaline solution may be used to distinguish primary, secondary, and tertiary nitroparaffins. Under the conditions given in the test below, primary nitroparaffins produce a reddish amber color, secondary nitroparaffins a sky-blue color, and tertiary nitroparaffins do not produce any color.

Nitrous acid reacts with a primary nitroparaffin to yield a nitrolic acid. The salts of the nitrolic acids are red in solution (these salts are explosive when dry). Nitrous acid reacts with a secondary nitroparaffin to yield a pseudonitrole. These compounds are blue in solution. Nitrous acid fails to react with tertiary nitroparaffins.

Add 5 drops of the nitroparaffin to 2 ml of 10 per cent sodium hydroxide. Allow the mixture to stand for 3 minutes. Add 1 ml of 10 per cent sodium nitrite solution and then add dropwise 10 per cent sulfuric acid, but do not add enough acid to neutralize the mixture completely.

Note: Ferric chloride was suggested as a reagent for nitroparaffins by Scott and Treon.[49] Jones and Riddick[50] recommend resorcinol in 66 per cent sulfuric acid as a colorimetric reagent.

D. Nitrophenols

Nitrophenols give a rather intense yellow or yellow-orange color immediately when they are dissolved in alkaline solutions. See test 6.19 for additional tests.

Add 30 mg of the compound to 2 ml of a 10 per cent aqueous solution of sodium hydroxide. A yellow color is given by most *p*-nitrophenols, whereas *o*-nitrophenols usually give an orange color.

Note: Many compounds besides nitrophenols produce some color in sodium hydroxide, but the intensity and rapidity of formation of the yellow color in this test is quite distinctive for nitrophenols. Nitrosophenols tend to produce a yellow-green color. Compounds like *o*-nitrophenol, that do not give the color test with ferric chloride, will give a good positive test by this method.

6.17 Nitroso Compounds

A. The following procedure gives positive tests for most common nitroso compounds.

Dissolve 30–40 mg of the compound in 2 ml of concentrated sulfuric acid and add 50 mg of phenol. Shake the tube and warm it slightly. The development of a blue or green color, which changes to red when water is added dropwise to the mixture, constitutes a positive test.

[49] Scott and Treon, *Ind. Eng. Chem., Anal. Ed.* **12,** 189 (1940).
[50] Jones and Riddick, *Anal. Chem.* **24,** 1533 (1952).

Note: This is the same as the Liebermann test for phenols by using nitrous acid, except that the organic nitroso compound is used instead of sodium nitrite. Nitroso compounds may be detected by test 6.16A.

It is claimed that *C*-nitroso compounds liberate iodine immediately when added to an acidified solution of potassium iodide, whereas *N*-nitroso compounds do not. *C*-Nitroso compounds are usually yellow-green in color. They dissolve in ether to form colorless solutions which turn blue on warming.

B. It has been reported[51] recently that all true nitroso compounds (but not isonitroso compounds) give positive results by the following procedure.

Prepare a mixture of 7 ml concentrated sulfuric acid in 3 ml of water. After cooling the acid, add 10 mg of *N,N'*-diphenylbenzidine. Add 0.5 ml of this solution to a test tube containing a few milligrams of the substance to be tested. True nitroso compounds develop a blue color immediately or after brief warming in a boiling water bath.

Exercises

1. What other groups besides the nitro group may act as oxidizing agents?
2. What intermediate compounds are, or may be, formed during the reduction of a nitrotoluene to the toluidine?
3. What are the major uses of the nitroparaffins?

6.18 Oximes, Hydrazones, and Semicarbazones

All three of the classes of oximes, hydrazones, and semicarbazones may be hydrolyzed by concentrated hydrochloric acid and thus converted into the hydrochloride salts of hydroxylamine, the hydrazine, and semicarbazide, respectively. These classes may be detected and identified by reactions used to prepare derivatives of carbonyl compounds.

6.19 Phenols

Phenols are to be suspected when a compound falls in Solubility Division A_2 and proves to be a weak or intermediate acid in the acid-base classification. Generally, more than one test for phenols must be made before a conclusion can be drawn, since the nature and location of substituent groups on a phenol markedly affect the reactions that it will undergo.

A. Ferric Chloride Test

Most phenols, enols, hydroxamic acids, many hydroxy acids, some oximes, and enolizable compounds in which the enolic structure is present to the extent of at least 5 per cent of the compound react with ferric chloride to produce colored complexes. The colors produced by a large number of the common phenols have been published by Wesp and Brode[52] and by Soloway

---

[51] Anger, *Mikrochim. Acta* **1960**, 58.
[52] Wesp and Brode, *J. Am. Chem. Soc.* **56**, 1037 (1934).

and Wilen.[53] The colors vary somewhat depending on the solvent used, the concentration of the reactants, and the elapse of time between the reaction and observation.

Dissolve 30–50 mg of the compound in 1–2 ml of water, or a mixture of water and alcohol, and add up to 3 drops of a 2.5 per cent aqueous solution of ferric chloride. Note any change in color or the formation of a precipitate.

Note: Most phenols produce red, blue, purple, or green colorations with ferric chloride. Soloway and Wilen have shown that the use of an anhydrous solvent (chloroform) and a weak base (pyridine) causes the test to be much more sensitive and to allow the detection of a large number of phenols that give negative results when tested in water. The function of the pyridine is apparently that of a proton acceptor, thus increasing the concentration of the phenolate ion. The composition of the colored complexes has not been established. Most nitrophenols, hydroquinone, guiacol, *m*- and *p*-hydroxybenzoic acids and their esters, and 2,6-ditertbutyl-*p*-cresol give negative tests.

Aliphatic hydroxy acids produce distinctly yellow solutions with ferric chloride. Many aromatic acids yield tan precipitates (gallic acid gives a black precipitate). Enols usually produce tan, red, or red-violet colorations. Oximes, if they give positive tests usually give red colors, as do the sulfinic acids. Hydroxypyridines and hydroxyquinolines give red, blue, or green colors.

B. Coupling with a Diazonium Salt

As was pointed out in test 6.7C, phenols as well as amines couple with diazonium salts to form colored azo compounds. The reaction apparently involves the diazonium ion, an electrophilic agent, substituting in the para (preferably) or ortho positions on the ring. Since the oxide ion is more of an electron-releasing group than the hydroxyl group, the phenoxide ions are more readily reacted with the diazonium ion than are the phenol molecules. The test should be carried out in slightly basic solution.

Dissolve 25 mg of the compound in a few drops of 2 per cent sodium hydroxide. If necessary, warm the mixture to form the solution, but cool it before proceeding to add 3–4 drops of a 1 per cent solution of *p*-nitrobenzenediazonium tetrafluoroborate. A decided coloration or a precipitate that is red, orange, yellow-green, or blue is a positive test. Read the note at the end of test 6.7C.

C. Indicator Formation

Most phenols condense with phthalic anhydride to form indicators that have blue, purple, red, or green colors in alkaline solution.

Place about 200 mg of anhydrous zinc chloride in a 4-inch test tube and heat it to be sure that it is anhydrous. Add 300 mg of phthalic anhydride and 50 mg of the compound to be tested. Heat the mixture sufficiently to fuse it and then cool the tube. Add 1 ml of 2 per cent sodium hydroxide and stir the mixture with a strong glass rod to break up the fused

[53] Soloway and Wilen, *Anal. Chem.* **24,** 979 (1952).

mass. Add more sodium hydroxide until the mixture is alkaline. Note the color.

Note: It is important not to add too much excess sodium hydroxide, since many indicators lose their characteristic colors in excess alkali. The fact that the compound is an indicator may be established by successively making the mixture alkaline and acidic.

D. Nitrous Acid Test

This test is given by many phenols which yield *p*-nitroso derivatives which, in turn, react with excess phenol to form indophenols. The indophenols are acid-base indicators.

$$C_6H_5OH \xrightarrow{\text{HONO}} ONC_6H_4OH \rightleftarrows O{=}C_6H_4{=}NOH$$

$$O{=}C_6H_4NOH + C_6H_5OH \xrightarrow{H_2SO_4} O{=}C_6H_4{=}NC_6H_4{-}OH$$

$$(HOC_6H_4N{=}C_6H_4{=}OH^+)SO_4^-H \xleftarrow{H_2SO_4} O{=}C_6H_4{=}NC_6H_4OH \xrightarrow{NaOH}$$

$$(O{=}C_6H_4{=}NC_6H_4O)^-Na^+$$

Add about 50 mg of the compound to 1 ml of concentrated sulfuric acid in a 4-inch test tube and then add about 20 mg of sodium nitrite. Shake the tube and warm it slightly. A positive test is indicated by a green, blue, or purple color. Cautiously pour the mixture into 5 ml of water. The color will generally change to red or blue-red. Make the solution alkaline by adding 20 per cent sodium hydroxide. The color in alkaline solution is generally blue or green.

Note: Nitrophenols and para-substituted phenols do not give this test. Of the dihydroxybenzenes, only resorcinol gives a positive test. It is reported [54] that —CHO, —COOH, and —COCH₃ groups on the ring also prevent the reaction.

E. Millon's Test

Millon's test is for monohydroxy phenols that have at least one ortho position open. It is also given by tyrosine, tyrosine-containing proteins, phenolic acids, and other compounds that have one phenolic group with an ortho position open. Add 50 mg of the phenol to 1 ml of Millon's reagent. Place the tube in a beaker of water and heat it to boiling. A red color will develop. The chemistry involved in this color formation is not clear.

F. Aminoantipyrine Test [55]

4-Aminoantipyrine reacts with many phenols in the presence of ferricyanide ions and in alkaline solution to produce quinoid-type compounds.

[54] Turney, *J. Org. Chem.* **22**, 1692 (1957).

[55] Emerson, *J. Org. Chem.* **8**, 417 (1943); Gottlieb and Marsh, *Ind. Eng. Chem., Anal. Ed.* **18**, 16 (1946); Martin, *Anal. Chem.* **21**, 1419 (1949).

The ferricyanide ions act as oxidizing agents in the reaction. The amine group of the aminoantipyrine is converted, in the case of phenol, to the —N=C_6H_4=O group.

Add 30 mg of the compound to 10–20 ml of water and then add 0.3 ml of a 2 per cent aqueous solution of 4-aminoantipyrine. Add 1 ml of 2N ammonium hydroxide and shake the mixture. Test the solution with phenolphthalein paper and, if it is not basic, add more ammonium hydroxide dropwise until it is basic. Now add 1 ml of a 2 per cent aqueous solution of potassium ferricyanide. Many phenols produce a raspberry-red color, but some give a green color.

Note: It is important that the mixture be basic before the addition of the ferri-cyanide solution, since other classes of compounds produce colorations in neutral or acidic solutions.

This test gives weak or negative results with phenols that have the para position sub-stituted by any of the following radicals: alkyl, aryl, nitro, nitrose, benzoyl, or aldehydic groups. The following groups do not prevent the test even if they are present in the para position: hydroxyl, halogen, carboxyl, sulfonic acid, and methoxy.

Exercises

1. Why do the acetoacetic esters and dibenzoylmethane give positive reactions with ferric chloride?

2. Using resonance structures, show why amines should couple with diazonium reagents effectively in acidic solutions whereas phenols couple better in weakly basic solutions.

3. Millon's reagent is sometimes recommended as a specific test for proteins. Why is this not a valid method of detecting all proteins?

6.20 Sulfides, Disulfides, and Sulfones

These three classes of compounds all undergo decomposition during fusion with sodium hydroxide. The sulfides and disulfides produce sodium sulfide, while the sulfones produce sodium sulfite. After acidification of the fused mass, hydrogen sulfide or sulfur dioxide may be detected by its odor or by conventional chemical methods.

A. *Sulfides*

The chemical tests for sulfides and thiols (mercaptans) are sufficiently similar as to be confusing. However, since these two classes fall in different solubility divisions and have quite different odors, they are not usually confused. If sulfides are used instead of thiols in test 6.21C, the first pre-cipitate is a very light yellow, rather than a golden yellow. The addition of free sulfur gives an orange color, but unless the compound is hydrolyzed it does not turn black. If alkyl sulfides are substituted for thiols in test 6.21D, the color is red rather than bluish red and tends to become yellow.

B. Disulfides

Disulfides may be reduced easily to the corresponding thiols which may be detected by test 6.21A or other test for thiols.

Add 30 mg of the compound to 1 ml of $1N$ hydroxylamine hydrochloride in methanol. Add a few mg of zinc dust and shake the mixture for 1 minute. Decant the liquid after the excess zinc has settled and test the solution for thiols.

6.21 Thiols (Mercaptans and Thiophenols)

The low-molecular-weight mercaptans and thiophenols are only slightly soluble in water but dissolve in sodium hydroxide to form salts. Both have penetrating, objectionable odors. The thiophenols are not very common. They may be detected by the test for aromatic structures and by their ease of nitration or bromination. The tests given below are designed particularly for mercaptans.

A. Isatin Test

Mercaptans give a green color of unknown cause in this test. Alkyl sulfides and hydrogen sulfide do not interfere, since they do not give such colorations.

Add 3 drops of a dilute solution of the mercaptan in ethanol to 2 ml of a 1 per cent solution of isatin in concentrated sulfuric acid.

B. Lead Mercaptides

Mercaptans react with lead or mercuric salts of weak acids to form lead mercaptides or mercuric mercaptides. Lead acetate or mercuric cyanide are generally used for these tests.

$$2HOH + 2RSH + Pb^{+2} \rightarrow Pb(SR)_2 + 2H_3O^+$$

Add 2 drops of the mercaptan to 5 ml of a saturated solution of lead acetate in ethanol. Lead mercaptide (yellow) precipitates.

C. Lead Sulfide

The petroleum industry makes extensive use of a so-called *doctor test* for *sour* distillates, that is, those containing mercaptans. With mercaptans, the reagent (sodium plumbite) first forms the yellow lead mercaptides, which are converted by sulfur to the black lead sulfide and the alkyl disulfides. The chemical changes in this test are variable, but one pair of equations may be written as follows:

$$Pb(OH)_2 + 2RSH \rightarrow Pb(SR)_2 + HOH$$
$$Pb(SR)_2 + S \rightarrow PbS + RSSR$$

Add 1 drop of a mercaptan to 2 ml of the sodium plumbite solution and shake the mixture vigorously. A yellow precipitate forms. Now add about 50 mg of finely powdered sulfur. The color may first change to orange, but will become black within a few minutes.

D. Nitroprusside

Mercaptans, in a slightly alkaline solution of sodium nitroprusside, give about the same deep wine color as is given by hydrogen sulfide. The exact composition of the color complex is not known, but it is believed to involve a union of the sulfur with the nitroso group of the nitroprusside. Alkyl sulfides also react with sodium nitroprusside, but the color is more red than blue. Thiophenols also give this test if ammonium hydroxide is substituted for the sodium hydroxide.

Add 1 drop of the mercaptan to 2 ml of a 1 per cent solution of sodium nitroprusside and then add 3 drops of 10 per cent sodium hydroxide. A deep wine color forms. The color changes to yellow if the solution is acidified with hydrochloric acid. Aryl sulfides do not give this test.

Problems

The following problems are designed to serve one or more of the following purposes:

1. A review of the classifications that may be made on the basis of elemental analysis, acid-base reactions, and solubility, together with the types of compounds that come under the various subclasses thus made.

2. A review of the tests for functional groups and general molecular structures as provided in Chapter 6.

3. The acquisition of experience in making logical deductions as to the nature of the substance from the facts at hand.

Assignment One

In the investigation of unknown compounds the following types of behavior are observed frequently. Indicate in each instance the deductions that may be made as to the nature of the compound.

1. A Division N compound, when heated with a solution of sodium hydroxide, reacted and was converted to a Division S_2 compound.

2. A Division I compound, when treated with chloroform and anhydrous aluminun chloride, gave a purple coloration.

3. A Division I compound gave a precipitate with alcoholic silver nitrate solution.

4. A Division S_1 compound gave a positive test with Tollen's reagent.

5. A compound containing only carbon, hydrogen, and oxygen reacted with acetyl chloride but not with 2,4-dinitrophenylhydrazine. Treatment with periodic acid converted it into a compound that reacted with 2,4-dinitrophenylhydrazine but not with acetyl chloride.

6. An alcohol gave a positive iodoform test and a negative test with Lucas reagent.

7. A Division S_2 compound contained nitrogen and sulfur. Addition of barium chloride to an aqueous solution produced a precipitate insoluble in acids. Addition of alkali to an aqueous solution caused the separation of a Division **B** compound.

Assignment Two

Select a classification test that may be used to distinguish between the following pairs of compounds. State what you would see when the test is applied to each of the pair of compounds.

1. $(p)HOC_6H_4COC_2H_5$ and $C_6H_5COOC_2H_5$.
2. $(p)C_2H_5C_6H_4OH$ and $(o)C_2H_5C_6H_4OH$.
3. $C_6H_5CONH_2$ and $C_6H_5CH_2NH_2$.
4. $C_6H_{13}NH_2$ and $C_6H_5NH_2$.
5. $C_6H_5N(C_2H_5)_2$ and $(C_2H_5)_3N$.
6. $C_6H_5NO_2$ and $C_6H_4(NO_2)_2(1,3)$.
7. $C_6H_5NO_2$ and $C_6H_5NHCOCH_3$.

Assignment Three

What class or classes of compounds would give the following reactions (nitrogen, sulfur, and halogens are absent unless specified)?

1. Solubility Division S_2; amphoteric; contains nitrogen; positive test with ninhydrin.

2. Solubility Division S_2; contains nitrogen and sulfur; intermediate acid; can be diazotized.

3. Solubility Division S_1; neutral; contains nitrogen; does not react with acetyl chloride; fusion with dry sodium hydroxide liberates a gas that gives a blue color when passed into an aqueous mixture of acetaldehyde and sodium nitroprusside; acidification of the fusion residue with phosphoric acid followed by distillation yields an intermediate acid.

4. Solubility Division S_1; weak acid; reacts with hydroxylamine and ferric chloride to give a bluish red color; treated with 10 per cent sodium hydroxide and then tested with hydroxylamine and ferric chloride, it gives a negative test.

5. Solubility Division A_2; weak acid; gives a red color with ferric chloride; gives a yellow precipitate with 2,4-dintrophenylhydrazine.

6. Solubility Division **N**; neutral; negative test with Ferrox.

Assignment Four

What observations or chemical tests would you make to distinguish the following pairs of compounds?

1. Chlorobenzene and benzyl chloride.
2. Toluene and heptane.
3. Phenol and cyclohexanol.
4. Benzoic acid and salicylic acid.
5. Benzene and cyclohexane.
6. Methyl acetate and ethyl formate.
7. Sucrose and maltose.
8. Nitrobenzene and bromobenzene.
9. Toluene and benzoic acid.
10. Aniline hydrochloride and p-chloroaniline.

Assignment Five

Without using their physical constants, what would be the simplest way to distinguish the following?

1. Butanol and propyl ether.
2. Butyl chloride and butanol.
3. Pentanol and pentane.
4. Hexane and hexene.
5. Hexyl ether and toluene.
6. Naphthalene and 1-naphthol.
7. *p*-Cresol and benzoic acid.
8. 3-Hexanone and hexanal.
9. Butyl iodide and iodobenzene.
10. Ethanol and ethyl mercaptan.

Assignment Six

Suggest two ways, not including physical methods, to distinguish between the following pairs of compounds.

1. Propionic acid and propionyl chloride.
2. *tert*-Butyl chloride and chlorobenzene.
3. *tert*-Butyl alcohol and isobutyl alcohol.
4. *N,N*-Diethylaniline and *N*-ethylaniline.
5. Propanol and propanthiol.
6. Benzamide and *N*-methylbenzamide.
7. Aniline and benzamide.
8. Butanal and butanone.

Assignment Seven

Assuming that an unlabeled bottle contained one or the other of the following pairs of chemicals, describe a practical chemical method for proving the identity of the unknown without using elemental analysis, physical constants, solubility, or acid-base classification (tell what you would do and what you would see).

1. Butyl ether and butyl acetate.
2. Allyl chloride and propyl chloride.
3. Heptanal and benzaldehyde.
4. Aniline and benzylamine.
5. Dextrose and levulose.
6. Benzamide and ammonium benzoate.
7. Aniline and *N*-ethylaniline.
8. 3-Pentanol and butanol.

Assignment Eight

Assuming that a compound is one of the following chemicals in a group of unlabeled bottles, tell what chemical tests you would perform and what results you would observe in the process of putting the correct label on each bottle: hexanol, hexanamide, *p*-chlorophenol, 2-naphthol, ammonium oxalate, and butanal.

Assignment Nine

Assuming that a large collection of unlabeled bottles containing organic compounds is given to you to test each compound and to separate out all compounds that belong to each of the following classes, outline the series of tests that you would follow: alkyl halides, aryl halides, aldehydes, ketones, alcohols, esters, amines, amides, phenols, and acids.

Assignment Ten

Assuming that the compounds in assignment nine above have been properly grouped into the classes specified, what further tests could profitably be made on the contents of these unlabeled bottles before attempting to identify the specific compounds?

Assignment Eleven

What is the saponification equivalent of propyl acetate? Write the structural formulas for eight other esters that would have the same saponification equivalent as propyl acetate. How could these nine esters be distinguished?

Assignment Twelve

By using only chemical tests, how could the following pairs of substances be distinguished? Note that their physical properties, elemental analysis, solubility, and acid-base character are not to be used, nor are the physical constants of derivatives.
1. Benzoic acid and salicylic acid.
2. 2-Propanol and 1-propanol.
3. N-Ethyl-2-naphthylamine and 2-naphthylamine.
4. Hexanal and benzaldehyde.
5. Ethyl butyrate and methyl palmitate.
6. Hexane and benzene.
7. Butyl chloride and ethyl bromide.
8. A pentose and a hexose.
9. Maltose and sucrose.
10. Vinyl ether and 2-pentene.

Assignment Thirteen

Identify compounds A, B, C, D, and E in the following problem. After compound A has been identified, select two compounds that would be isomeric with it ($C_8H_{10}O$), and show the products that each of them would yield if subjected to the same treatment as given in this problem.

An aromatic compound, A, has a molecular formula of $C_8H_{10}O$. By mild oxidation a compound, B, is produced, which has a formula of C_8H_8O and reacts with 2,4-dinitrophenylhydrazine. When compound B is refluxed with alkaline potassium permanganate, a compound, C, is recovered by acidification, which has the formula $C_7H_6O_2$. Compound A reacts with acetyl chloride to form compound D, which has the formula $C_{10}H_{12}O_2$. Compound A reacts with sodium hypoiodite to yield iodoform and a salt, E, that converts to compound C on acidification.

Assignment Fourteen

Identify compounds A, B, C, D, and E in the following problem.

A compound, A, has a formula of $C_5H_{12}O$. It reacts with a saturated solution of anhydrous zinc chloride in concentrated hydrochloric to form (in 7 minutes) an insoluble liquid, B, with a formula of $C_5H_{11}Cl$. This compound does not precipitate alcoholic silver nitrate, but does form a precipitate with a solution of potassium iodide in acetone. Compound B reacts with alcoholic potassium hydroxide to form compound C, which has a formula of C_5H_{10} and reacts with bromine in carbon tetrachloride to form compound D with a formula of $C_5H_{10}Br_2$. Compound A does not yield iodoform when treated with sodium hypoiodite, but it is oxidized by hot aqueous permanganate solution to form compound E with a formula $C_5H_{10}O$. Compound E reacts with 2,4-dinitrophenylhydrazine but not with Schiff's reagent.

After compound A has been identified, select three isomers of it (same chemical class) and show what differences in the reactions would occur in each as compared to compound A.

Assignment Fifteen

Deduce structures A, B, C, D, and E from the following facts.

Compound A, $C_{10}H_{13}Br$, was boiled with alcoholic sodium hydroxide solution to give B, $C_{10}H_{12}$. When B was ozonized, formaldehyde and compound C, $C_9H_{10}O$, were isolated. Substance C was reduced with hydrogen over platinum to give D, $C_9H_{12}O$. When D was heated with a sulfuric acid at 160°, substance E, C_9H_{10}, was obtained, which, when treated with potassium permanganate, gave terephthalic acid. When E was ozonized, however, the products were formaldehyde and p-tolualdehyde.

Write the formula for an isomer of A, in which the bromine is in a different position, and indicate the products that would be obtained from this isomer if it were subjected to the same reactions as suggested for compound A above.

Assignment Sixteen

The colorless compound A, $C_9H_{10}O_2$, was soluble in aqueous sodium hydroxide solution, but not in sodium bicarbonate solution. Compound A readily reacted with bromine water, hydroxylamine, and semicarbazide, but not with Tollen's or Benedict's reagent. Reduction with lithium aluminum hydride afforded B, $C_9H_{12}O_2$. Both A and B gave positive haloform tests. Clemensen reduction (zinc amalgam, hydrochloric acid) of A yielded substance C, $C_9H_{12}O$. When C was reacted with sodium hydroxide, $C_9H_{11}ONa$ was isolated, which was boiled with methyl iodide to give compound D, $C_{10}H_{14}O$. Permanganate oxidation of D produced p-methoxybenzoic acid as the sole product.

Write structures for A, B, C, and D as deduced from the above data.

Assignment Seventeen

1. A compound, $C_{14}H_{11}OCl$, will react with sodium benzoate to produce an ester, and it will also react with phenylhydrazine to produce a phenylhydrazone. What is a probable structure of this reagent?

2. A compound A, $C_{13}H_9NO_3Cl$, on hydrolysis by aqueous sodium hydroxide, produces a water-soluble product, B, and a water-insoluble product, C. Compound C is extracted from the hydrolysis mixture by ether, and is found to have a formula of C_6H_6NCl. Upon acidification of the alkaline solution (after compound C has been removed by extraction), a white solid separates which may be converted to compound B by treatment with sodium bicarbonate. Compound B has a formula of $C_7H_5NO_4$. To what chemical classes would compounds A, B, and C belong? Write structural formulas for nine possible formulas for compound A.

Assignment Eighteen

Based on the probable reaction mechanisms involved, explain:

1. The relative reaction rates of primary, secondary, and tertiary alcohols with Lucas' reagent.

2. The fact that esters, acid halides, and acid anhydrides can all be converted to hydroxamic acids by hydroxylamine, whereas acids do not react with hydroxylamine to produce hydroxamic acids.

3. The fact that aryl halides are less reactive than alkyl halides with silver nitrate.

7

Physical Methods for the Determination of Functional Groups in Organic Compounds

In Chapter 1 (page 4) a summary was given of the physical methods which either are used extensively or have restricted applications in the identification of organic substances. Of the methods listed in Table 1.2, *paper chromatography* and *infrared spectroscopy* are the most widely used, and will be briefly discussed in this chapter. Extensive bibliographical references are given at the end of the chapter for further discussion of these and other physical methods.

CHROMATOGRAPHIC PROCEDURES

Introduction

The term *chromatography* is applied to fractionation procedures based on the differential migration of the components of a mixture when adsorbed on paper or other stationary phases, and brought under the influence of a solvent. For example, if paper is employed as the adsorbent of the mixture and a solvent is passed through the paper, it causes the components of the mixture to move at different rates through the adsorbent, thereby effecting resolution; this procedure is called *paper chromatography*. If the immobile adsorbent phase is a solid such as alumina, magnesia, silica, or the like (contained in a glass column), and the solvent is passed through to effect resolution, the procedure is called *column chromatography*. If the stationary phase is a solid covered with a nonvolatile partition liquid and the sample of the mixture is passed by means of a "carrier gas," the procedure is called *gas chromatography.* Of these, the simplest and most widely applied to identification work is paper chromatography. The main reasons for the wide application of this procedure are: (*a*) its relative simplicity, (*b*) the inexpensiveness of the equipment required, and (*c*) the ease of application of the method to the separation and identification of microgram quantities of organic compounds, which is difficult or impossible to accomplish by other procedures.

For a discussion of column chromatography, and its application to the separation and identification of organic compounds, the student is referred to the more extensive work of the authors. The application of gas chromatography is primarily to fractionation of mixtures, and as yet it has not been applied extensively to qualitative identification. References at the end of the chapter provide a guide to recent developments of this method.

Principle of Chromatographic Fractionation

In chromatographic separations we are dealing with two phases: (*a*) the *immobile phase* of the water molecules bound into the cellulose network of the paper, and (*b*) the *mobile phase* of the solvent (or mixture of solvents) which is called the *developer*. The mixture of organic compounds which is applied as a *spot* on the paper may be considered as the solute that is to be distributed between the two phases. Distribution of the solute between "bound" water (on the paper) and the mobile phase (developing solvent) results in movement of the solute through the paper. However, the developer moves more rapidly than the solute; the ratio of the movement of the solute to rate of movement of the developing solvent is called the R_f *value.* In paper chromatography, if the developing solvent and the solute are started at the same time on the paper, the ratio can be expressed in terms of the distances moved by each:

$$R_f = \frac{\text{Distance solute moved}}{\text{Distance solvent moved}}$$

To illustrate R_f calculation refer to Figure 7.1 and assume that an amino acid is placed at position A. After the strip of paper has been developed for 1.5 hours with a mixture of 70 per cent *tert*-butyl alcohol and 30 per cent water, the solvent front has reached D. This point is quickly marked with a pencil so that it will not be lost on evaporation of the solvent. After drying and spraying with a suitable color reagent, the amino acid is now found at position B. If the distance CD is 14 cm, and the distance AB is 9.5 cm, then the R_f value of this compound from *tert*-butyl alcohol–water in the percentages given above is

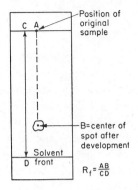

Figure 7.1 Illustration of determination of the R_f value of an organic substance on a paper chromatogram.

$$R_f = \frac{AB}{CD} = \frac{9.5}{14} = 0.67$$

The organic compound has traveled along the paper at two-thirds the rate of the developing solvent. It should be noted that the R_f value of a compound (under specified conditions) is a physical constant and serves together with other data as a means of *identification*.

The factors that affect the movement of the solute through the immobile phase are many and varied. In paper chromatography the partition coefficient, the ratio of the concentration of solute in the immobile phase to that in the mobile phase, has been shown to be quantitatively related to the R_f value by the expression

$$R_f = \frac{A_1}{A_1 + \alpha A_s}$$

where α = partition coefficient, A_1 = cross-sectional area of the mobile phase, and A_s = cross-sectional area of the immobile phase. This equation was arrived at from investigation in partition chromatography with columns from which the values of A_1 and A_s are more readily obtained than from paper strips or sheets. The limited number of partition coefficients available, particularly for the unusual solvent systems commonly encountered in paper chromatography, render impractical the use of this equation in a preliminary calculation of the R_f value. Assuming that adsorption between the solute and cellulose is negligible, one need only consider the solubility of the solute in the two liquid phases. If the solute is highly soluble in the organic phase used as the developing solvent, it will move rapidly and will exhibit a high R_f value. On the other hand, should the solute show a low solubility in organic solvents and a high water solubility, it will move slowly, if at all, and will have a low R_f value. Thus it is seen that in paper chromatography factors affecting the solubility of the solute in the two liquid phases account for its behavior in moving through the immobile water phase.

General Procedure for Paper Chromatography

Strips of filter paper are cut from sheets or may be obtained in rolls of ¾ or 1 inch width from most chemical houses. Care should be exercised in handling these papers since grease from the hands and rough edges produced in cutting strips from sheets can interrupt the movement of the developing solvent up the strip and give rise to erratic R_f values. A few microliters (μl)[1] of a solution of the solute mixture are applied at a marked point near one end of the paper strip. This end of the strip is then dipped into the developing solvent, and the development is allowed to proceed until the solvent has moved a predetermined distance. The solvent may be allowed to ascend or to descend the paper strip. In the former case, the process is somewhat slower; but conditions of equilibrium with respect to the distribution of the solute between the two phases are more nearly attained, and the isolated zones of the components of the mixture are usually more compact. If the mixture is simple and good separation is indicated, the descending method may be preferred because of the time it saves. These two methods are shown diagrammatically in Figures 7.2 and 7.3. The arrangement shown in Figure 7.2 may be employed for two-dimensional chromatography.

After development is complete, the distance that the solvent has moved

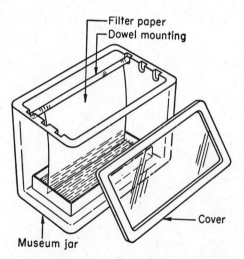

Figure 7.2 Apparatus for ascending paper chromatography.

[1] A microliter ($= 1 \times 10^{-6}$ liter or 0.001 ml) is often designated by the symbol μl or in earlier literature by the Greek letter λ (lambda), also pronounced microliter. Similarly, a microgram ($= 1 \times 10^{-6}$ g or 0.001 mg) is often designated by μg or in earlier literature by the Greek letter γ (gamma).

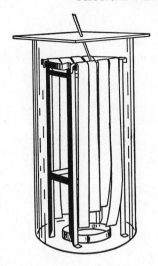

Figure 7.3 Apparatus for descending
paper chromatography.

is noted and the paper strip is hung to dry. If the components of the mixture in the spot are colorless, the paper is then sprayed or streaked with a
color-forming reagent, or in some cases it is dipped in a solution of the reagent. This process is called *color development* or *location of spots* by color
formation. With the appearance of color in the various zones the distance
each solute or component has moved can be measured from the point of
application of the mixture to the center or edge of the respective zones, and
thus the R_f values are obtained. It should be noted that in a number of cases
the spots are located by ultraviolet light, which produces fluorescence.

Often the need arises for further development of a chromatogram when
the solvent has reached the limit of the paper strip. The investigator may
take advantage of further development without the trouble of transferring
partially separated components to other strips. This technique is *two-
dimensional*, or two-directional, chromatography. The method is to apply
the solute mixture near one corner of a sheet of chromatographic paper, perhaps an inch and a half from each edge. One of these edges is dipped into
the solvent and the process of development takes place as before. After the
solvent has moved a reasonable distance the paper is removed and dried.
The original solvent may be replaced by a second solvent known to have
particularly good developer characteristics for those components of the
solute mixture that, up to this point, have not become well separated. The
edge of the dried paper along which development took place with the original
solvent is dipped into the new solvent, and development with the new solvent
proceeds. When the solvent front has reached the prescribed limit, the
paper is removed, dried, and sprayed with reagent to reveal the location of
the components of the mixture.

Application of the Sample on Paper

The amount of material initially applied to the paper depends on a number of factors. First, enough of the sample must be used to produce a distinct zone that will be found when the chromatogram is subjected to color development. This amount will depend, obviously, on the sensitivity of the detection method. If an insufficient amount of a mixture is applied, a component of low percentage may be missed, even though other fractions show up clearly. If too much material is applied, the spots appear too large and may obscure other components whose R_f values lie close to those of the original substance; resolution will be incomplete and overlapping of the spots will hinder identification. This problem occurs not only when too much material is applied but also when the R_f values of the components are close to each other. It has been suggested that R_f values must differ by at least 10 per cent for perfect resolution in one dimension, no matter how little material is used. If the compound is fairly insoluble in the solvent, the application of too much will cause an elongation of the spot. However, the appearance of "tails" on chromatograms does not necessarily mean that too much material is being used, for certain substances normally run in this manner. Since the actual amount to be used depends on all of these factors, it is difficult to suggest any set amount as being optimum. In one-dimensional analysis, 10–20 micrograms is sufficient, and for the test tube technique an initial application of 5–10 micrograms is satisfactory. For two-dimensional chromatograms, 20–30 micrograms of substance is required. Generally, spots of less than 0.25 μg are difficult to detect.

The application of the sample should be restricted to a very small area. The material is best applied with a micropipet. A droplet of about 3–5 μl at the tip of the pipet is touched to the marked spot on the paper, which is spread a few inches above a hot plate. After the solution on the paper has evaporated, another droplet is added on the same spot until the required quantity has been delivered. If the sample is added in large drops, the solution spreads over too large an area and a diffuse spot will result in the developed chromatogram.

The sample may be dissolved in any solvent provided it can be conveniently evaporated after application. However, since all substances which are to be chromatographed must be soluble to some extent in water, this solvent is preferred. In some cases in which the compound has a low solubility in water, it is necessary to convert it to a more soluble form by addition of an acid or a base and, after application on the paper, to reconvert it to the original compound. Thus cystine, tyrosine, histidine, arginine, and other amino acids are applied to the paper as solutions of the hydrochlorides and then neutralized by holding the paper over ammonia for a few minutes.

Selection of Solvent Systems for Development of Paper Chromatograms

The selection of proper solvents is very important in paper chromatography. For partition on the paper, the solvent should be mixed with water to some degree to provide water for absorption by the paper at the same time that the solvent passes over it. However, too much water is undesirable, and indiscriminate saturation of organic solvents may well lead to poor chromatograms. As a rule, the solvent should not contain more than 10–20 per cent of water by weight. On the other hand, there are cases of solvents completely miscible with water in which the proportion of water may be higher; isopropanol mixtures containing as much as 40–50 per cent water have been employed. Solvents with low vapor pressures are unsatisfactory because it is difficult to remove them completely from the paper by heating and traces may interfere with color development or allow the dissolved substances to diffuse over a large area, yielding a poor chromatogram. Collidine cannot be used, for example, with the iodoplatinate indicator for methionine, even if the chromatogram is heated at 120°C. for an hour, since traces that remain on the paper will decolorize the indicator. Solvents with high vapor pressures must be used with caution as they are sensitive to temperature fluctuations and are likely to distill off or condense on the paper, thus causing phase irregularities if the temperature is not carefully controlled. The solvent need not be a single substance. In the chromatography of various sugars, for example, two good solvents are butanol-acetic acid and a mixture of ethyl acetate and pyridine saturated with water. Table 7.1 lists some of the solvents and reagents used for a few types of organic compounds. The literature should be consulted for further information.

Elution of Spots

It is possible to elute the spots from a paper chromatogram so as to obtain the relatively pure components of a fractionated mixture. For this purpose 100 micrograms or more of the mixture is placed on a strip and developed in a solvent mixture by the proper procedure. A guide strip is run alongside which is treated with the indicator after the evaporation of the solvent in order to locate the areas of the various components while the main strip is not treated with the indicator solution. The guide strip is placed alongside the main strip from which the solvent has been evaporated and the areas of the various spots are located, marked with light pencil, and cut out. The small strips with individual components are placed between small glass plates, the tip of one end of the strip hanging in a solvent and the other tilting into a small beaker. The solvent rises by capillary

TABLE 7.1

Solvents and Reagents Commonly Used for Paper Chromatography
of a Few Organic Compounds[a]

Compound	Solvent system	Reagents for locating spots
Organic acids	Mixtures of formic acid with: ethanol, 2-propanol, 1-butanol, and other alcohols or ketones	Bromcresol green or bromphenol blue solution in ethanol
2,4-Dinitrophenyl-hydrazides of acids	Buffer pH 11.6 saturated with EtCOMe	Compounds colored; no indicator required
Hydroxy and keto acids	Toluene–acetic acid; 1-butanol–propionic acid	Ammonia and Nessler's solution; o-phenylenediamine with ultraviolet light
2,4-Dinitrophenyl derivatives of keto acids and all carbonyl compounds	1-Butanol–ammonia; 1-Butanol–acetic acid–water	KOH solution (10 per cent in methanol)
Hydroxyamate derivatives of organic acids	Phenol–isobutyric acid	Ferric chloride
Amino acids	1-Butanol–acetic acid–water; phenol–water; phenol–ammonia; 2,4-lutidine–water	Ninhydrin solution (0.1–0.2 per cent in ethanol)
Amino acids (containing S)	tert-Butanol–methylethyl ketone–water–diethylamine	Potassium iodoplatinate solution (aqueous)
2,4-Dinitrophenyl derivatives of amino acids	1-Butanol–acetic acid; collidine; phenol	Compounds colored; no indicator required
3,5-Dinitrobenzoates of alcohols	Methanol–acetone; methanol–hexane; 2-propanol; pyridine	KOH solution (5 per cent)
Amines	1-Butanol–acetic acid	Ninhydrin solution; iodine solution
2,4-Dinitrophenylhydrazones of carbonyl compounds	Ether–hexane; acetone–hexane	KOH solution (10 per cent)
Phenol derivatives; methylolphenols	1-Butanol–ammonia	p-Nitrobenzenediazonium fluoborate
Phenylazobenzene sulfonates	2-Butanol–aqueous solution Na_2CO_3	Compound colored
Sugars	1-Butanol–acetic acid; ethyl acetate–pyridine–water; 1-butanol–collidine; isopropanol–water	p-Aminodimethylaniline (tin salt); silver nitrate solution; 3,5-dinitrosalicylic acid solution; 0.3 per cent solutions of p-aminohippuric acid in ethanol

[a] For more detailed information see Appendix.

attraction, passes through the strip, and drips into the beaker. After 2–3 hours the solvent in the beaker is evaporated to obtain the relatively pure components eluted from an individual spot. This procedure requires considerable practice before good results are obtained. The recovered compound may be used for further chemical identification or estimation by chemical reaction or used for infrared or other spectra.

Test Tube Technique for Paper Chromatography

For beginners the most convenient method to practice is the test tube technique. Figure 7.4 shows a regular 6-inch Pyrex test tube which is used for single spots; the short Pyrex tube (25 × 150 mm) is more suitable when two spots are placed on the same strip (if the solvent travels rapidly the regular 8 inch tube [25 × 200] is employed). When the number of strips required is small, they may be easily cut from a pattern made from a yellow manila folder. The pattern for the 6 inch tube is 135 mm in length, 15 mm in width at the top, and 10 mm at the bottom. The pattern for the wider tube is 135 or 165 mm in length, 25 mm in width at the top, and 15 mm at the bottom. The strips are cut from three or four discs of Whatman No. 1 filter paper, 15 cm in diameter, folded in half. The pattern is placed on the paper, the outlines are marked lightly with a pencil, and the paper is cut with scissors along the inside of the pencil mark, so that the strip becomes approximately as wide as the pattern. The upper outside strips are discarded (since they have been handled); the inner ones are handled only at the extreme edges of the broad ends. The strips are now pierced at the center, about 4–5 mm from the top, to facilitate suspending during the drying operations. A light pencil line is drawn about 8 mm from the narrow

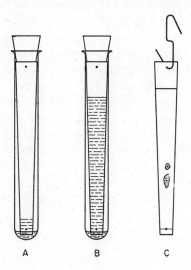

Figure 7.4 Test tube technique for ascending paper chromatography. A. Test tube with paper strip a short time after being placed in the developing solvent. B. Same strip after development and before removal from tube. C. Same strip after solvent has been evaporated and the spots located by application of the indicator solution.

A   B   C

end on which the sample is to be placed. For the preparation of a larger number of strips, sheets of Whatman No. 1 (45.7 × 56.3 cm) are employed, and the strips are cut with a paper cutter.

The sample is applied with a micropipet[2] at the center of the pencil line as described in a preceding section and allowed to dry. A test tube is provided with a well-fitting cork so that, when the tube is stoppered, the cork fails to touch the paper. About 0.4–0.5 ml of the appropriate solvent saturated with water is placed in the bottom of the tube, being released through a capillary pipet in such manner that the sides of the tube above the surface of the liquid are perfectly dry. The tube is placed on the rack, and the paper is inserted lightly so that the narrow end dips into the liquid, but the surface of the solvent is well below the pencil line. The tube is then corked and allowed to stand (Figure 7.4) until the solvent ascends near to the top of the strip or about 5 mm from the broad end. This depends on the type of solvent, but usually takes about 1.5–3 hours. The tube is opened, the paper removed with a forceps, and the solvent front marked with a light pencil mark. The strip is suspended by a clip as shown in Figure 7.4, allowed to dry, and then sprayed with an indicator as described in the following section.

Location of Zones on Paper Chromatograms

After the chromatogram has been developed and the solvent evaporated, a solution is applied that will react with the chromatographed compound to produce a color so that its position on the paper may be determined. A solution of the indicator (Table 7.1) is usually sprayed over the surface of the paper with an atomizer, although sometimes—for example, with chromatograms of the 3,5-dinitrobenzoates of alcohols—brushing on the reagent gives better results. In many cases dipping the strip in the reagent solution is simpler. The choice of indicator is wholly dependent on the nature of the substance being chromatographed. In the detection of amino acids by chromatography, ninhydrin (triketohydrindene hydrate) is almost exclusively used. When a 0.1–0.2 per cent solution of ninhydrin in ethanol or acetone is sprayed on a chromatogram, colored spots appear within a few hours. This reaction may be accelerated by carefully heating the chromatogram, after it has been sprayed, at about 90–100° for 10 minutes. Colored compounds (dyes, nitro compounds, and the like) do not require the application of indicator, although many times dilute solutions of alkalies and complexing agents are sprayed in order to intensify the spots. Compounds that fluoresce under ultraviolet light are often located by passing the chromatogram in front of a small ultraviolet lamp.

---

[2] See the Appendix for making pipets with long capillaries.

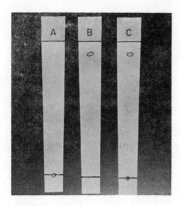

Figure 7.5 Three paper chromatograms obtained by the test tube technique. A. 2,4-Dinitrophenyldrazine hydrochloride: spot did not move, hence $R_f = 0.0$. B. Pure butanone–2,4-dinitrophenylhydrazone: $R_f = 0.91$. C. A droplet (10µl) from a reaction mixture of 0.7 mg of an unknown suspected of being butanone and 1 mg of 2,4-dinitrophenylhydrazine hydrochloride. The R_f value of C is the same as B. The chromatograms were developed for 1 hour with a 95:5 hexane-ether mixture saturated with water. After evaporation of the solvent, the strips were dipped in 5 per cent methanolic potassium hydroxide. The strips were dried and the spots outlined with pencil and photographed.

Since fading occurs, the chromatogram is examined immediately after the color has been developed and the spots are outlined in pencil. Figure 7.5 is an actual photograph of three strips of paper chromatograms made by the test tube ascending technique. The spots after development were outlined lightly with a pencil and then photographed. Approximate quantitative information may be obtained by measurement of the area of the spot and comparison with the area of a standard run made at the same time. For quantitative determinations of the colored spots by use of photoelectric colorimeters (densitometers), the literature at the end of the chapter should be consulted.

Use of Paper Chromatography in Identification Work

Figure 7.5 illustrates the application of paper chromatography to the identification of organic compounds. In strip C of Figure 7.5, a droplet (10 µl) of a solution from a reaction mixture of 0.7 mg of an unknown suspected to be butanone ($CH_3COCH_2CH_3$) and 1 mg of 2,4-dinitrophenylhydrazine hydrochloride were placed on the penciled mark at the lower end of the strip. By referring to page 259, it will be found that butanone and the carbonyl compounds in general react with 2,4-dinitrophenylhydrazine hydrochloride to give derivatives known as 2,4-dinitrophenylhydrazones. The strip was placed in an 8-inch tube containing about 0.5 ml of a mixture (95:5) of hexane–ether saturated with water and allowed to develop for about 60 minutes when the solvent reached the upper part of the strip marked by a line near the letter C. The strip was removed from the tube and the solvent front was marked; after evaporation of the solvent, the strip was sprayed with 5% solution of sodium hydroxide in methanol. A red spot caused by the 2,4-dinitrophenylhydrazone of butanone appears at the top, and is outlined by a pencil mark, as shown in Figure 7.5C. In a separate tube another strip was placed (B in Figure 7.5), on the lower part of which was placed a droplet of a solution of known pure butanone-2,4-dinitrophenylhydrazone, and developed and sprayed in the same manner. It will be noted that the R_f value of the reagent (Strip A: 2,4-dinitrophenyl-hydrazone hydrochloride) is zero; in other words, the reagent with this particular solvent system does not migrate. From this relatively simple experiment, it is concluded that the derivatives used in strips B and C are identical. Examples for practice in the laboratory on the use of paper chromatography for the identification of organic compounds will be found on pages 233–34 and 264.

Mixed R_f Values

It will be recalled that mixed melting points of the same derivative prepared from an unknown and a pure sample of a known compound may be employed as the final step in the identification of an organic compound (see page 47). In the same manner a mixed R_f value may be employed when the derivative cannot be isolated in sufficient quantity for a melting-point determination, but can be chromatographed. In the above example, a fourth strip could be prepared by placing at the point of origin 5 μl of the solution used in strip B (pure known) and then 5 μl of the solution used in strip C (unknown). The application is made in such a manner that both solutions are placed on the same spot, and the area of the spot is about the same as in B and C. The strip with the mixed derivatives is developed and sprayed in the same manner as described for the other strips. After the chromatogram is developed, dried, and sprayed, there will be only one well-defined spot, if the two derivatives are identical. If the derivatives are not identical, and their R_f values differ by 10 per cent or more, there will be two spots above the point of origin. For example, even acetone–2,4-dinitro-phenylhydrazone $(R_f = 0.85)$, and butanone–2,4-dinitrophenylhydrazone $(R_f = 0.91)$, which differ by only 6 per cent in R_f values, will give a spot which is not well defined but which will have tails and will be blurred. Thus it is also possible under certain conditions to use simple chromatographic procedures to determine the purity of a compound, since a pure substance on development with an appropriate solvent and use of a suitable indicator will give only one well-defined spot.

INFRARED SPECTROSCOPY

Introduction

The present section aims to give a brief introduction to the application of infrared spectra in the detection of the functional groups that are most commonly encountered in organic compounds. The use of infrared spec-trometers, both in industrial and academic laboratories, is steadily increasing. One of the most important reasons for this increase is the availability of rugged instruments at a cost which is not prohibitive. For example, for student use it is possible to obtain an instrument at the cost of about $4,000.00, while a few years ago the lowest-priced infrared spectrometer was upwards of $10,000. The student is referred to the literature cited at the end of the chapter for a more extensive discussion of infrared spectroscopy as applied to the identification of functional groups and structural studies.

Infrared Radiation and Units of Expression

Visible light is defined as that portion of the electromagnetic spectrum which has a wavelength from 4.1×10^{-5} cm to 7.5×10^{-5} cm. This corresponds to frequencies of about 4.3×10^{14} to 7.3×10^{14} vibrations per second. If wavelength is expressed by λ and the absolute frequency in cycles per second by ν and the speed of light by c (186,000 miles/sec or 3×10^{10} cm/sec), the expression for the relation of wavelength and frequency is $\lambda = c$, $\lambda = c/\nu$, and $\nu = c/\lambda$. The unit of wavelength is usually the *micron* ($\mu = 10^{-4}$ cm). The frequency is expressed in fresnels (f), the number of complete vibrations per second, or in *wave numbers* (ν'), which is the reciprocal of the wavelength in centimeters (cm^{-1}) or waves per centimeter. It has been the practice in designating the infrared region and infrared spectra to use both wavelengths expressed in microns, μ, and wave numbers in cm^{-1}. The longest wavelength to which the eye is sensitive is in the region of about 0.7 μ. The wavelength from about 0.7 to 350 μ, sometimes called heat rays, comprise the infrared region. For the infrared spectra of organic substances, the important wavelengths are about 2.5–50 μ. However, the common range of infrared measurement extends from 2.5–15.5 μ or from 4000–650 cm^{-1}. The region between 2.5μ and the visible spectrum 0.7 μ is called the near infrared region and that beyond 50 μ the far infrared region. At the present time, the most common method of expressing infrared spectra is to use both wavelength units (μ) and wave number units (cm^{-1}). Figures 7.8 to 7.11 illustrate the method employed.

Origin of Infrared Absorption Spectra

When light strikes a substance, a portion of the incident light is reflected, another is transmitted, and a portion is absorbed by the molecules of the substance. As is well known, when energy is absorbed by the molecules of a substance, it is transformed into complex motions of the molecule as a whole, the atoms which compose the molecule and the electrons which form the various bonds between atoms. If the energy has sufficient magnitude (high frequency and low wavelength) the motion imparted to the electrons may be sufficient to cause complete displacement or ionization. In the case of infrared light, the magnitude of the energy is small, and hence there is very little electronic displacement within the molecule. However, the absorption of the infrared radiation displaces the atoms from their normal positions and causes them to oscillate. A rough analogy of the vibrations caused when infrared light is absorbed may be obtained by considering a simple molecule such as methane in space. The carbon atom may be pictured as a steel ball at the center of the tetrahedron and the four hydrogen atoms as small aluminum balls attached by means of metal springs to the central steel ball. Let it be assumed that this mechanical model in space is activated by striking it a blow; the model will undergo a series of motions which will consist of *bending* and *stretching* of the springs and the motions of the metal balls. These motions properly observed and examined can be resolved into a pattern having definite frequencies whose values are related to the mass of the various metal balls, to the type of springs by which the balls are connected together and to the arrangement of the balls in space.

With the above analogy in mind, the interaction between infrared radiation and the molecules of an organic substance can be considered. As infrared light strikes a certain number of molecules, there will be radiation frequencies which correspond to the molecular frequencies, and under such conditions there is absorption of these energy frequencies in the same manner as a vibrating tuning fork brought near another fork of the same frequency causes it to vibrate. The molecules which absorb energy undergo complex vibrations which involve stretching and bending of the bonds as well as changes in the position of the charges within the molecule, that is, momentary changes in the dipole moments of the molecules. The vibrational energy in turn is changed to translatory energy and by collisions with neighboring molecules the over-all kinetic energy or temperature of the molecular system rises under the effect of infrared heating.

Measurement of Infrared Absorption Spectra

In order to measure the particular wavelengths which a homogeneous system of organic molecules will absorb, it is necessary to employ a beam of infrared light and allow it to fall on a sample of the molecular species under investigation and then measure which frequencies or wavelengths were absorbed by determining from the transmitted light which frequencies are missing.

A simplified schematic representation of an infrared spectrometer is shown in Figure 7.6. A source of radiation A emits infrared light, which is focused on a mirror B and passes through the sample C and then through the entrance slit D, the width of which can be adjusted from 0.01 mm to 1 mm. The light passing through the slit is focused into parallel light by mirror E and then passes through an optical prism P (usually made of highly polished monocrystal of sodium chloride) which is transparent to infrared and re-

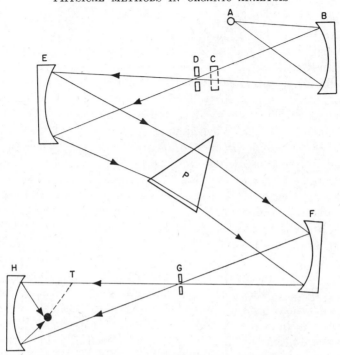

Figure 7.6 Schematic representation of the optics in a simple infrared spectrometer. (After Pimental, *J. Chem. Educ.* **37,** 651 (1960))

fracts each frequency through a definite angle. A narrow range of frequencies of the refracted beam is focused by mirror F into the exit slit G and by means of the ellipsoidal mirror H on the detector T. By slow rotation of mirror F it is possible to alter continuously the frequency which passes through the exit slit G, and thus the detector response gives a spectrum as a function of the angle of rotation of F. The response is measured by a system of sensitive fast thermocouples as percentage of light transmitted. By means of an electronic balancing circuit, the per cent transmittance is recorded as a function of the wavelength. By proper rotation of mirror F, the sample can be scanned with successive infrared frequencies, and from the transmission data the regions in which the infrared radiation is weakened or absorbed can be plotted.[3] These values are assumed to correspond to

[3] The optics of commercial spectrophotometers are more complex since the radiation from the source is split into two beams; one passes through the sample, the other serves as a reference beam and is compared with the sample beam in order to determine the absorbed radiation. The student will find a discussion of the optics of the instrument in the operating instructions furnished by the manufacturer (*Operating Instructions of Model 137B Infracord Spectrophotometer,* Bulletin 990-9177, Perkin-Elmer Corp., Norwalk Conn., pp. 15-27; *Operating Directions of Beckman IR5 Infrared Spectrophotometer,* Beckman Instruments Co., Fullerton, Cal., pp. 1-3).

mechanical vibration frequencies of the molecules composing the sample. Figures 7.8 to 7.11 show typical infrared spectra.

The usual type of cells of 0.1 mm thickness have volumes in the order of 0.2–0.5 ml; cells of 0.025 mm thickness require a sample of about 0.1 ml. Micro cells for smaller volumes are also available. For liquids which are not viscous the most accurate method is the use of cells, being careful to dry the liquid as to eliminate water, because water first absorbs extensively through much of the infrared region, and secondly it corrodes the sodium chloride (or potassium bromide or cesium bromide), which is used as "windows" in the construction of the cells. If the liquid is viscous, two sodium chloride plates or "windows" are used. A droplet of the viscous liquid is placed on one plate, and the second plate is placed on top and rubbed back and forth so as to spread a uniform film. The excess fluid is wiped off, and the window is placed in the sample area of the instrument.

There are three general methods in which to prepare a solid sample for scanning. The first is to use a solvent. With the ordinary cells, 0.5–5 mg of the solid is sufficient; with a micro cell the quantity of the solute needed can be reduced to 50–100 μg. However, the solvent must be selected with consideration to its own infrared spectrum. Hence the application of this method is limited to a few "transparent" solvents such as carbon tetrachloride and carbon disulfide and others such as chloroform, bromoform, methylene chloride, tetrachloroethylene, acetonitrile, and cyclohexane. The solvents must be highly purified or the commercially available "Spectro" grades used. It is advisable to employ the "mull" dispersion first.

The second method is to prepare a fine dispersion or "mull" of the solid in mineral oil (Nujol) or hexachlorobutadiene. A few drops of the mulling liquid is placed in an agate mortar and a few milligrams of the solid is added and ground into a fine paste. A small amount of the paste is spread on one plate; the second plate is placed on top and rubbed gently back and forth so as to obtain uniform distribution of the solid and avoid "coalescence," which will result in poor resolution. Since Nujol absorbs in the 6.5–7.5 μ region and is less strong at 3.3 μ, another mull is made with hexachlorobutadiene or Halocarbon oil.[4]

The plates or cells should never be cleaned with water, but with toluene ("Spectro" grade), and then dried by placing in a desiccator and evacuating. The plates are polished frequently by rubbing lightly with fine aluminum oxide on filter paper or fine cloth.

The third method is to disperse the sample in specially purified potassium bromide and press the mixture into a pellet or "window" which is then inserted directly in the sample area of the instrument. About 1 part of the sample and 200 parts of potassium bromide are ground together, usually

[4] Halocarbon Oil Co., Hackensack, N.J.

by means of a mechanical vibrator (Wig-L-Bug)[5] so as to obtain thorough grinding and uniform dispersion. The mixture is inserted in a special die and then subjected to a pressure of several tons under vacuum to produce a disk which is often transparent, but at times is hazy. Pellets which are not highly transparent are satisfactory for most qualitative work. For small amounts of water-soluble samples it is possible to disperse the solution in potassium bromide and then freeze the resulting solution and evaporate it under vacuum so as to disperse or "lyopholize" the sample. This technique has been used with samples of a few micrograms.

Each of the above three methods for obtaining infrared spectra of solids has drawbacks, discussed in detail in the instruction manual furnished with the instrument and in the more extensive treatises cited at the end of the chapter.

For the detection of functional groups in an unknown, the sample is run for 20 to 30 minutes at median values of operating variables in order to obtain a survey of the characteristic spectra which are present. The runs are made in the range of 2.5–15 μ or 4000–650 cm^{-1}, which is the region most suitable to the standard chloride prism of the student-type instruments. With a supplementary cesium bromide prism, the range can be extended to 40 μ or 250 cm^{-1}.

Characteristic Infrared Absorption Bands of Functional Groups

If the infrared spectra of several aliphatic aldehydes and ketones are examined, there will be found a common absorption band in the region of wavelength 6 μ or wave number about 1740–1600 cm^{-1}. This band is ascribed to the vibrations of the carbonyl function, C=O, and is to be found in about the same region in aromatic aldehydes and ketones, as in carboxylic acids, amides, and esters which have the following functions, respectively

$$\underset{\text{—C—OH,}}{\overset{\overset{\textstyle O}{\|}}{}} \qquad \underset{\text{—C—NH}_2,}{\overset{\overset{\textstyle O}{\|}}{}} \qquad \underset{\text{—C—OR}}{\overset{\overset{\textstyle O}{\|}}{}}$$

and in general in all compounds which contain the carbonyl group in some part of their molecues.

By means of painstaking statistical studies of infrared spectra of pure organic compounds, it has been possible to arrive at reliable "fingerprints" of the characteristic infrared absorption bands exhibited by the various functional groups and atomic linkages such as: —CO, —COOH, —OH, —NH$_2$, —NH, —CN, —CH$_2$, —CH$_3$, —C$_6$H$_5$, and the like. Thus it has been possible by a considerable amount of study and correlation of the

[5] Crescent Dental Manufacturing Co.

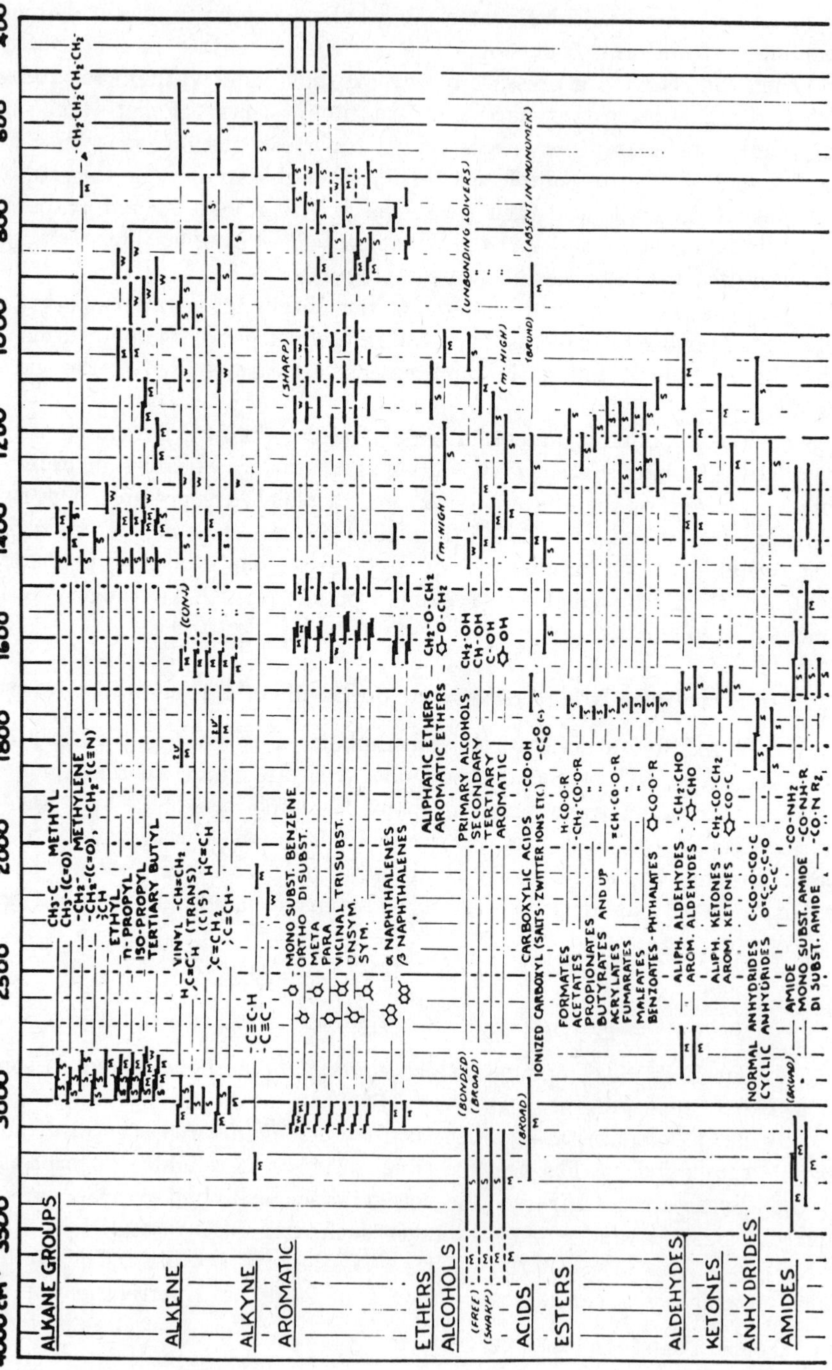

4000 cm⁻¹ 3500 3000 2500 2000 1800 1600 1400 1200 1000 800 600 400

ALKANE GROUPS

- CH₃-C METHYL
- CH₃-(C=O)
- -CH₂- METHYLENE
- -CH₃-(C=O), -CH₂-(C≡N)
- -CH₃
- ETHYL
- -CH
- n-PROPYL
- ISO-PROPYL
- TERTIARY BUTYL

ALKENE

- VINYL -CH=CH₂
- H₂C=CH (TRANS)
-)C=CH (CIS)
-)C=CH₂
-)C=CH-

ALKYNE

- -C≡C-H
- -C≡C-

AROMATIC

- MONO SUBST. BENZENE
- ORTHO DISUBST.
- META
- PARA
- VICINAL TRISUBST.
- UNSYM.
- SYM.
- α NAPHTHALENES
- β NAPHTHALENES

ETHERS

- ALIPHATIC ETHERS
- AROMATIC ETHERS

ALCOHOLS

- CH₂-OH
- CH-OH
- C-OH
- -O-OH
- (FREE)
- (SHARP)
- PRIMARY ALCOHOLS
- SECONDARY
- TERTIARY
- AROMATIC
- (UNBONDING LOWERS)

ACIDS

- CARBOXYLIC ACIDS -CO-OH
- IONIZED CARBOXYL (SALTS-ZWITTER IONS ETC)
- -C=O
- (BONDED)
- (BROAD)

ESTERS

- FORMATES H·CO-O-R
- ACETATES -CH₂-CO-O-R
- PROPIONATES
- BUTYRATES AND UP
- ACRYLATES
- FUMARATES
- MALEATES
- BENZOATES-PHTHALATES
- R·CH-CO-O-R
- Ø-CO-O-R

ALDEHYDES

- ALIPH. ALDEHYDES -CH₂-CHO
- AROM. ALDEHYDES -Ø-CHO

KETONES

- ALIPH. KETONES -CH₂-CO-CH₂
- AROM. KETONES -Ø-CO-C

ANHYDRIDES

- NORMAL ANHYDRIDES C-CO-O-CO-C
- CYCLIC ANHYDRIDES Ø·C-CO-CO

AMIDES

- AMIDE -CO-NH₂
- MONO SUBST. AMIDE -CO-NH-R
- DI SUBST. AMIDE -CO-N R₂
- (BROAD)

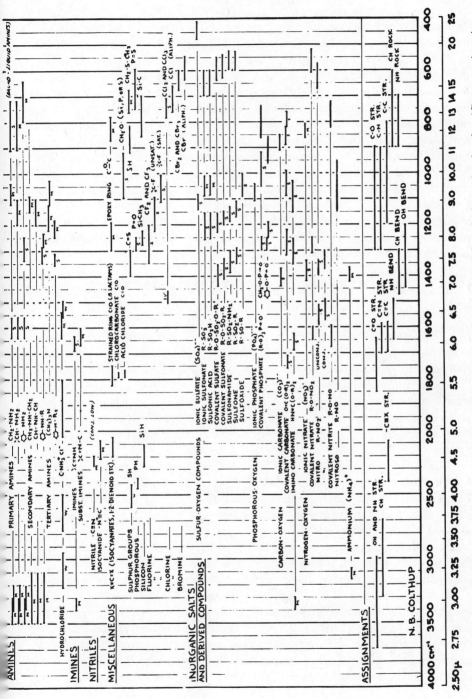

Figure 7.7 Spectra-structure correlations. Probable position of the characteristic infrared absorption bands. (Courtesy American Cyanamid Co.)

experimental data to compile a reference table of the characteristic absorption bands of the various functional groups and atomic linkages encountered in organic compounds. Such a reference table compiled by Colthup is shown in Figure 7.7. It will be noted that the symbols under the absorption bands in Figure 7.7—*s*, *m*, and *w*—indicate the intensity of the band as *strong*, *medium*, and *weak*, respectively. For routine or exploratory work the information furnished from this excellent compilation is very useful for obtaining hints and suggestions first as to *what groups are not likely to be present* and second as to *what groups and atomic linkages may be present*. For complete identification, however, it is necessary to acquire experience in the interpretation of the infrared spectra as outlined in the next section. Once a tentative assumption has been made as to the nature of the unknown, confirmation is obtained by the preparation of derivative or matching the infrared spectrum of the unknown with one which is found in the catalogues or atlases of infrared spectra of pure organic compounds (see references at end of chapter).

INTERPRETATION OF INFRARED SPECTRA

Introduction

The interpretation of the infrared spectra of unknown compounds depends on the position and intensity of the absorption bands. The interpretation aims to correlate the absorption data in the spectra with the presence of certain functional groups and their environment. However, the characteristic absorption frequencies of functional groups are subject to considerable alteration by the rest of the molecular structure or environment, so that some experience is required to interpret the infrared spectrum of an unknown.

An introduction to the problems of interpretation is obtained by examination and brief discussion of the spectra shown in Figures 7.8 and 7.9. The absorption maxima for *n*-hexane (7–8A) is at the regions 3.5 μ (2950 cm^{-1}), 6.8 μ (1470 cm^{-1}), and 7.25 μ (1375 cm^{-1}). The first is ascribed to the carbon–hydrogen stretching vibrations, the second to the —CH$_2$ deformations, and the last to the —CH$_3$ group. These absorption maxima, though present in aliphatic hydrocarbons and in general in aliphatic compounds with carbon–hydrogen linkages, are influenced by the presence of other groups as shown in Figures 7.8B and 7.8C; such groups contain essentially the *n*-hexane structure modified by the introduction of a hydroxyl group in the hexanol (7.8B) and a hydroxyl plus a carbonyl in the hexanoic (caproic) acid (7.8C). The introduction of the hydroxyl group has produced a distinct intense absorption maximum at 2.8 μ (3660 cm^{-1}) and a number of other bands between 8.5–11 μ (1180–910 cm^{-1}). The introduction of the

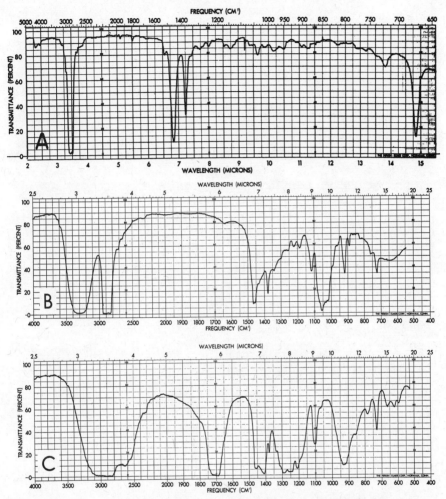

Figure 7.8 A. n-Hexane. B. 1-Hexanol. C. Hexanoic (n-caproic) acid. Phrase, liquid; thickness, 0.025 mm; prism, NaCl; gain, 4.5; scan time, 12 min; suppression, 10; scale, 2.5 cm/μ. (Courtesy Perkin-Elmer Corp.)

carboxyl group gives rise not only to the bands from the hydroxyl group but also to those due to the carbonyl vibrations at 5.7–6.2 μ (1740–1650 cm^{-1}) region. If comparison is made of the region between 3–4 μ (3330–2500 cm^{-1}) of the three spectra of the related structures in Figures 7.8A, B, C, the effect of the alteration in the absorption maxima by other groups in the molecule will be realized.

A further illustration of the problems of interpretation is afforded by examination of the spectra of cyclohexane, cyclohexene, and benzene. If

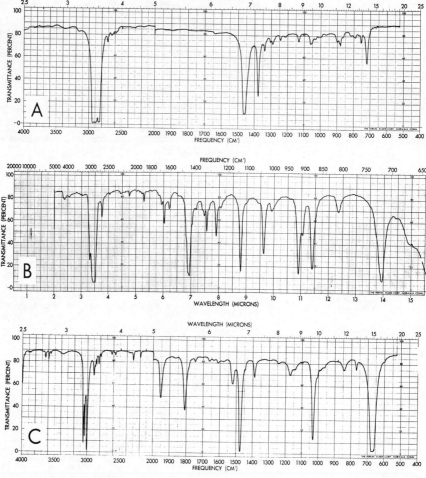

Figure 7.9. A. Cyclohexane. B. Cyclohexene. C. Benzene. Phase, liquid; thickness, 0.025 mm; prism, NaCl; gain, 4.5; scan time, 12 min; suppression, 10; scale, 2.5 cm/μ. (Courtesy Perkin-Elmer Corp.)

the spectrum of cyclohexane (7.9A) is compared to that of hexane (7.8A), it will be noted that they are similar in the higher frequencies, but that cyclohexane shows two bands in the lower frequencies which are also present in the spectra of cyclohexene (7.9B). The presence of the double bond in cyclohexene is indicated by the absorption at 6.1 μ (1650 cm^{-1}). The absorption maxima of benzene are in the 3.25 μ (3030 cm^{-1}) region due to the =C—H stretching vibrations and at 5–7 μ (2000–1430) due to the C=C vibrations. Generally the presence of an aromatic structure is recognized by the presence of these bands, which are not affected appreciably by substituents that cause intense absorption at the lower frequencies $(8\text{--}9\ \mu = 1250\text{--}1100\ \text{cm}^{-1})$.

The following discussion is a summary of a few generalizations which will serve as a guide; however, the student is referred to more extensive discussions of the interpretation of the infrared data given in the literature cited at the end of the chapter. In order to avoid confusion the absorption bands will be designated only by wave numbers.

Alkanes, Alkenes, Alkynes, and Aromatic Structures

Alkanes

A large number of aliphatic hydrocarbons containing the —CH_3 and —CH_2— groups exhibit strong absorption bands due to C—H stretching vibrations at the region $3600\text{--}2850\ \text{cm}^{-1}$; there is also absorption at the regions $1460\ \text{cm}^{-1}$ and $1375\ \text{cm}^{-1}$ due to deformation vibrations of the —CH_2— and CH_3—C groups. Since the last two lie close together, they may not be readily differentiated, but the intensity has been used together with other data to calculate the relative proportion of —CH_2— and —CH_3 groups. Other C—H stretching frequencies are: $\equiv$C—H, $3330\text{--}3230\ \text{cm}^{-1}$; $R_2C{=}CH_2$; $3100\text{--}3070\ \text{cm}^{-1}$; and aromatic hydrogen; $3070\text{--}3040\ \text{cm}^{-1}$. These stretching frequencies are affected by: (a) adjacent carbon–carbon double bonds, (b) adjacent carbonyl groups, (c) strain in the ring system, and (d) replacement of the hydrogen ($_1^1$H) by its isotopes. As the chain of the hydrocarbon increases, a wide variety of frequencies appear; of these a strong absorption band near $720\ \text{cm}^{-1}$ is of value for —$(CH_2)_n$, where n = 4 or more. Other skeletal vibrations of value are $(CH_3)_3$—C—R, $1250\text{--}1230\ \text{cm}^{-1}$, and $(CH_3)_2$—C—R, $1175\text{--}1180\ \text{cm}^{-1}$.

Cycloalkanes show characteristic frequencies; cyclopropane, $3020\text{--}2983\ \text{cm}^{-1}$, $1020\text{--}100\ \text{cm}^{-1}$, and also a strong band at $866\ \text{cm}^{-1}$; cyclobutane, $3010\text{--}2982\ \text{cm}^{-1}$ and $1000\text{--}960\ \text{cm}^{-1}$; cyclopentane; $2930\text{--}2910\ \text{cm}^{-1}$ and $980\text{--}930\ \text{cm}^{-1}$; cyclohexane, $2920\text{--}2910\ \text{cm}^{-1}$ and $1055\text{--}890\ \text{cm}^{-1}$.

Alkenes and Alkynes

Alkenes and in general compounds which contain an isolated double bond show a relatively weak absorption at $1680\text{--}1620\ \text{cm}^{-1}$ and mainly at $1660\text{--}1640\ \text{cm}^{-1}$. The intensity of the C$=$C stretching vibrations varies considerably with the substituents around the double bond, so that there may be little absorption in this region. In simple conjugated aliphatic dienes and trienes the number of absorption bands increases in number and remains in the region $1650\text{--}1600\ \text{cm}^{-1}$. Conjugation with an aromatic ring shifts the absorption to the lower part of this region with resulting increase in intensity. The same occurs by conjugation of a double bond to one or more carbonyl groups, as in the unsaturated ketones and diketones.

A characteristic absorption band due to the out-of-plane deformation of the attached hydrogen atoms, =C—H, occurs in the region of 1000–800 cm^{-1}. The exact position often permits determination of the configuration. For example, in the olefin RHC=CHR the trans absorbs at 970–960 cm^{-1}, while in the cis isomers this band is missing and often a peak is found in the 720–690 cm^{-1} region.

Monosubstituted acetylenes of the type RC≡CH exhibit a characteristic band of moderate intensity at 2140–2100 cm^{-1} due to C≡C stretching and a band at 3300 cm^{-1} due to C—H stretching. In the type R'C≡CR the latter band is absent, while the former is of variable intensity; in the symmetrically substituted RC≡CR the band is too weak to be detected. As with the double bond, conjugation and aromatic substitution cause shifts and intensification of the absorption.

Aromatic Structures

Aromatic compounds show the C—H stretching band at the region of 3000–3050 cm^{-1} and the C=C skeletal bands, which are medium to weak, at 1600–1475 cm^{-1}; three bands are often found at 1620–1600 cm^{-1}, 1600–1580 cm^{-1}, and 1520–1475 cm^{-1}. The monosubstituted and o-disubstituted compounds show characteristic out-of-plane C—H absorption at 770–735 cm^{-1}. The monosubstituted usually are differentiated by the presence of a relatively intense band at 710–690 cm^{-1}, which is either absent or extremely weak in the o-disubstituted compounds. m-Disubstituted aromatic compounds show a strong band at 810–750 cm^{-1} and a second band of medium intensity at 723–680 cm^{-1}. The p-disubstituted compounds show a strong band at the region of 855–800 cm^{-1}.

Carbon–Halogen Functions

Fluorine gives the strongest absorption bands of halogens, in the region of 1400–1000 cm^{-1}. In monofluorinated compounds the absorption band is in the 1100–1000 cm^{-1} region, while in polyfluorinated compounds several intense bands occur in the 1400–1000 cm^{-1} region. The introduction of fluorine also causes pronounced shifts in the adjacent C—H vibrations.

The C—Cl band appears in the 800–600 cm^{-1} region in the monochloro, and mainly near the range of 700–750 cm^{-1}; in the polychloro compounds a medium secondary band occurs at 1510–1470 cm^{-1}. Several chlorine atoms attached to the same carbon atom give higher C—Cl frequencies.

The C—Br absorption occurs in the 600–500 cm^{-1} region, and in the C—I linkages near the 500 cm^{-1} region.

Carbon–Oxygen Functions—Carbonyl Compounds

The absorption bands due to the C=O stretching vibrations occur in the region of 1740–1700 cm^{-1} for saturated aliphatic aldehydes and ketones.

Aldehydes

In the simple aliphatic aldehydes the C=O absorption band occurs in the 1740–1720 cm^{-1} region; in unsaturated aldehydes this value is lowered to 1705–1660 cm^{-1}, and in aromatic structures to 1715–1695 cm^{-1}. The vibration of the hydrogen atom attached to the carbonyl group is different from the normal C—H stretching (of the methylene groups), and characteristic bands occur in the 2820–2700 cm^{-1} region. For many aliphatic aldehydes the bands are near 2600–2720 cm^{-1} and 2820 cm^{-1}; for many aromatic aldehydes, at about 2730 cm^{-1} and 2820 cm^{-1}.

Ketones

In the simple aliphatic ketones the C=O absorption occurs at 1725–1706 cm^{-1} (mainly at 1710 cm^{-1}) if there is no interference from neighboring groups. This frequency is lowered to 1690–1660 cm^{-1} by unsaturated ketones. The same effect is observed in aryl and diaryl ketones in which the band occurs at the 1700–1680 cm^{-1} and 1670–1660 cm^{-1} regions, respectively. In the α-diketones there is only a small frequency rise; however, in the β-diketones there is a strong, broad band in the range of 1640–1540 cm^{-1} ascribed to enolization and formation of resonance hybrids.

Cyclic ketones show absorption bands similar to open-chain ketones except that ring strain produces shifts, as can be seen by the following values: cyclobutanone, 1775 cm^{-1}; cyclopentanone, 1740 cm^{-1}; cyclohexanone, 1710 cm^{-1}; and cycloheptanone, 1700 cm^{-1}. The effect of large rings is very small. However, there is pronounced influence from steric effects, bulky substituents, and inductive effects from α-halogen atoms. Generally steric strain due to bulky substituents lowers the frequency, while the α-halogens produce a rise. For example, di-t-butylketone absorbs at 1686 cm^{-1}, and 2,4-dibromo-3-ketones near 1760 cm^{-1}. However, the effect of the halogen is also related to its orientation with respect to the carbonyl group. Other absorption bands found in alkyl ketones are in the region of 1325–1215 cm^{-1}; in the aryl ketones absorption occurs in the 1225–1075 cm^{-1} region.

Quinones

Quinones show one strong band and one weak band in the 1690–1660 cm^{-1} region. If there are two carbonyls in one ring, as in p-benzoquinone, the bands are located at 1669 cm^{-1} and 1656 cm^{-1}; if there are two car-bonyl groups in two cyclic structures, the bands are in the region of 1655–1635 cm^{-1}.

Carbon–Oxygen Functions—Hydroxy Compounds: Alcohols

The absorption bands due to the hydroxyl group are found in the 3700–3200 cm^{-1} region due to oxygen–hydrogen stretching and in the 1400–1050 cm^{-1} region due to C—O stretching and OH deformations. However, these lower bands are of limited value, since in many cases the stretching vibra-tions overlap with those of other functional groups. Therefore the most useful region for the detection of the hydroxyl group is the 3700–3200 cm^{-1}. However, the absorption at this region is complicated, since the hydroxyl group being highly polar may exhibit hydrogen bonding with other mole-cules. The free hydroxyl group found in some tertiary alcohols in which hydrogen bonding is restricted exhibits strong absorption at 3650–3590 cm^{-1}; if bonding occurs between two or more molecules or with molecules of the solvent (as it occurs in most cases), a broad band occurs at 3450–3200 cm^{-1}, usually near 3400 cm^{-1}. Change of solvent and dilution tend to diminish the intensity and to broaden or shift the band toward the region of the free hydroxyl group, since the extent of intermolecular bonding is decreased. When hydrogen bonding occurs between two groups within the same mole-cule (as in the enolic β-diketones) or when chelation occurs (as in the ortho-nitrophenols), a strong band appears in the region 3200–2500 cm^{-1} which is not affected by dilution.

The carbon–oxygen stretching bands occur in the 1300 cm^{-1} region and in the 1100–1000 cm^{-1} region. The former is subject to considerable shifts by structural effects. The latter can often be used to differentiate between primary (range 1075–1010 cm^{-1}), secondary (range 1119–1105 cm^{-1}), and tertiary alcohols (range near 1140 cm^{-1}).

Phenols

The characteristic frequencies of the oxygen–hydrogen stretching are slightly lowered by the aromatic ring. Values of 3616–3592 cm^{-1} for the OH stretching have been obtained in many substituted phenol and catechols. The effects of hydrogen bonding on the o-substituted phenol and the chela-tion of the o-nitrophenols have been mentioned. The OH deformations and C—O stretching absorptions for phenols are near the 1400 cm^{-1} and 1200 cm^{-1} regions.

Ethers

The characteristic frequencies are due to the carbon–oxygen vibration which for alkyl ethers (—CH$_2$—O—CH$_2$—) occurs in the region of 1150–

1060 cm^{-1}, for aryl ethers at 1270–1230 cm^{-1}, and for cyclic ethers at 1250–1070 cm^{-1}, depending on the nature of the ring. Infrared evidence alone is inconclusive for the presence of ether linkages.

Carboxylic Acids

The characteristic frequencies of carboxylic acids would be expected to be those due to the hydroxyl group (and those of the carbonyl group which already have been discussed). However, the tendency to dimerization by hydrogen bonding between the hydroxyl of one and the carbonyl groups of another molecule is extensive, so that position of the O—H and C=O stretching frequencies is altered. In the solid or liquid state carboxylic acids give broad OH bands at 3000–2500 cm^{-1} with the main peak near 3000 cm^{-1}. A weaker band of 2700–2500 cm^{-1} appears in strongly hydrogen-bonded compounds. The carbonyl frequency in unsubstituted saturated monobasic aliphatic acids occurs in the 1725–1705 cm^{-1} region. Substituents such as α-halogens tend to effect a rise (1740 cm^{-1}), and aryl groups or conjugation tend to lower (1700 cm^{-1}) the carbonyl frequency. Other absorption bands in carboxylic acids may be found in the regions 1450–1200 cm^{-1} and 950–900 cm^{-1}, ascribed to C—O stretching and OH deformation vibrations. Long-chain fatty acids exhibit a series of bands in the region of 1350–1180 cm^{-1} which can be correlated with the carbon chain. Finally it should be noted that in the salts of carboxylic acids the carbonyl frequency is missing and is replaced by two bands in the 1610–1550 cm^{-1} and 1400–1300 cm^{-1} regions. This is ascribed to the formation of the resonating carboxylate ion due to ionization.

Acid Anhydrides

Anhydrides show two carbonyl absorption bands, one at about 1800–1825 cm^{-1} and the second at 1760–1750 cm^{-1}. The two bands are approximately 60 cm^{-1} apart. In cyclic anhydrides there is a shift to higher frequencies.

Acid Halides

The carbonyl frequency due to the proximity of the halogen atom rises to the 1800 cm^{-1} region. The observed values in acid halides studies are at 1815–1770 cm^{-1} with the aryl and conjugated halides in the lower region.

Esters

There are two general absorption bands in esters. One is due to the C=O and the other to the C—O linkage in R—CO—OR'. The first falls in the 1750–1735 cm^{-1} region and the second in the 1300–1100 cm^{-1} region. The latter absorption bands also occur in acids, alcohols, and ethers and are not extensively used except in certain individual classes where considerable work has been done. The carbonyl absorption band in normal saturated esters occurs near 1740 cm^{-1}, except in the formates (1724–1722 cm^{-1}). The shift due to environmental effects (conjugation, α-halogen effect, cyclic structures, etc.) follows the same trends as discussed under ketones. The same is true of enols of β-keto esters as indicated under β-diketones, namely, a strong band at 1650 cm^{-1}.

Lactones

Six-membered lactones (δ-lactones) show the same ester carbonyl absorption as open-chain esters; the five-membered compounds (γ-lactones) absorb in the 1800–1740 cm^{-1} region; the four-membered (β-lactones) absorb in the 1840–1820 cm^{-1} range.

Carbon–Nitrogen Functions

Amines

There are three types of vibrations which give rise to absorption bands in amines: those due to NH stretching, those from NH deformation, and the ones from C—N stretching.

Primary amines show two absorption bands in the NH stretching region 3500–3300 cm^{-1}—one near 3500 cm^{-1} and the other near 3400 cm^{-1}. Hydrogen bonding tends to shift to lower frequencies. As previously mentioned, there is an overlapping between NH- and OH-bonded hydrogen vibration in this region, but with dilution it is often possible to make a differentiation since the intermolecular hydrogen bonding is minimized and the OH absorption, if present, shifts to the free hydroxyl region (3600 cm^{-1}). The absorption due to NH deformations occurs in the 1650–1590 cm^{-1} region and is of medium intensity, while the C—N absorption band occurs in the 1340–1250 cm^{-1} region with aromatic primary amines. In aliphatic amines this band is weak and occurs in the 1220–1020 cm^{-1} region.

Secondary amines show only one NH stretching band at 3500–3200 cm^{-1}; further the NH deformation frequency is very weak, and the C—N absorption band for secondary aromatic amines occurs in the 1350–1280 cm^{-1} region. Tertiary amines show only a C—N absorption band in the 1360–1310 cm^{-1} region, which is useful in some cases.

Heterocycles

Pyridines and quinolines show ring vibrations similar to benzene, but differ in the hydrogen deformations which shift to lower frequencies. These occur in the 1200 cm^{-1}, 1100–1000 cm^{-1}, and 900–700 cm^{-1} regions. The C—H stretching vibration bands occur near the 3020 cm^{-1} region; the C=C and C=CN bands occur in the same general region as in benzene—at 1660–1590 cm^{-1} and near 1500 cm^{-1}.

Amides

All amides show the carbonyl absorption band at 1680–1650 cm^{-1} in the solid form and near 1700 cm^{-1} in chloroform solution. Primary amides show two N—H bands near 3500 cm^{-1} and 3400 cm^{-1}, while the secondary amide shows a single bond in the 3400–3100 cm^{-1} region. The position of this single bond in secondary amides depends on the extent of hydrogen bonding and whether the amide exists in the cis or trans form. Both primary and secondary amides show a second band very close to the main carbonyl absorption in the region of 1600–1500 cm^{-1}; in primary amides it usually occurs near the 1650–1620 cm^{-1}, and in secondary amides near 1550 cm^{-1}. This band is called the amide II band, while that due to the carbonyl is referred to as amide I.

Nitriles, Isocyanates, Oximes, Imines, Azines, and Diazo Compounds

Nitriles show the C≡N absorptions band in the 2240 cm^{-1} region; conjugation with a double bond and direct linkage of the cyano group to an aromatic nucleus tend to lower this frequency slightly. The intensity is diminished and even quenched by the introduction of functional groups containing oxygen.

Isocyanates absorb strongly in the 2275–2240 cm^{-1} region with the majority showing this intense absorption near 2270 cm^{-1}. Although cyanides, as indicated above, absorb near the same region, the intensity is not as great as that of the N=C=O linkage. The absorption of oximes, imines, azines, and compounds containing the C=N linkage is in the 1690–1630 cm^{-1} region. The —N=N— linkage has an absorption frequency near 1600 cm^{-1}, but its intensity is too weak to be useful. However, diazomethane absorbs at 2101 cm^{-1}, and diazonium salts absorb in the region 2290–2240 cm^{-1}.

Nitro and Nitroso Compounds

The nitro, —NO$_2$ group exhibits two very strong absorption bands in the 1650–1500 cm^{-1} and 1350–1250 cm^{-1} regions, corresponding to the asymmetric and symmetric stretching of the group. The influence of the radical R, in R—NO$_2$, is shown by the nitroparaffins, which have the two absorptions in the ranges of 1567–1550 cm^{-1} and 1379–1368 cm^{-1}, and by the aromatic nitro compounds, in which the bands occur near 1530 cm^{-1} and 1360 cm^{-1}.

The isomeric nitrites R—O—N—O show two absorption bands near

1650 cm^{-1} and near 1610 cm^{-1}. Nitroso aliphatic compounds (tertiary) absorb near 1550 cm^{-1}, and aromatic near 1500 cm^{-1}, while the N—nitroso (nitrosoamines) absorb near the 1400 cm^{-1} region.

Sulfur Functions

The sulfur function linkages have not been as extensively investigated as the oxygen and nitrogen functions. The following absorption ranges have been reported: thiols or mercaptans, 2600–2550 cm^{-1}; thioureas, 1400–1200 cm^{-1}; sulfoxides, near 1050 cm^{-1}; sulfones, two bands, one near 1150 cm^{-1} and another at 1350 cm^{-1}; sulfonamides, two bands, one at 1180–1140 cm^{-1} and the second at 1350–1300 cm^{-1}.

APPLICATION OF INFRARED SPECTRA
TO STUDENTS' UNKNOWNS

The following two examples illustrate one method of the use of infrared spectra for identification of students' unknowns. After the student has obtained individual instruction in the operation and care of the instrument and the preparation of samples, and has also become familiar with the spectra-correlation chart (Figure 7.7) and is able to interpret the spectra of a few simple compounds which illustrate the common functional groups, he is given two unknown compounds, one solid and one liquid. The student is advised, before running the spectrum, to determine the elements present, the boiling point, presence of traces of water (if the unknown is liquid), the melting point (if the sample is a solid). Based on the interpretation of the spectrum, one or two derivatives are then prepared to confirm the identification. After the report is filed with the instructor, the spectrum submitted by the student is returned for comparison with the known spectrum of the pure compound which is identical with the unknown. The reports, as shown by two different students, are reproduced here as they were submitted and before correction by the instructor. In both cases the final conclusions were correct.

Student's Report on Unknown A

The unknown compound was a solid of faintly green color and sweet odor, with a melting point of 49–$52°$ before crystallization and 51–$52°$ after crystallization. Test for elements showed nitrogen present, but not halogens or sulfur. The spectrum was obtained by using a Nujol mull with a hexachlorobutadiene mull substituted at wavelengths where Nujol absorbs (3.2–3.5 and 6.7–7.5 μ). The spectrum (Figure 7.10) was interpreted as follows: (a) absorption at 1350 cm^{-1} indicates symmetrical C—NO$_2$ stretching and at 1530 cm^{-1} asymmetrical C—NO$_2$ stretching, while the band at 740 and

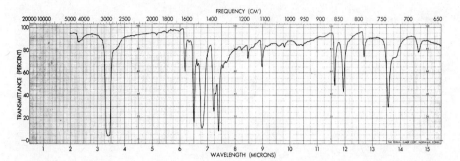

Figure 7.10 Spectrum of unknown identified as p-nitrotoluene. Phase, null in Nujol; prism, NaCl; gain, 6; speed, 4; scan time, 15 min.

850 cm^{-1} indicates RNO$_2$ deformation and skeletal vibration; (b) the absorption at 1475, 1500, and 1600 cm^{-1} shows aromatic character; (c) the absorption at 2960 and 2860 cm^{-1} indicates CH$_3$ stretching, at 1370 cm^{-1} symmetrical C—CH$_3$ deformation, at 1460 cm^{-1} asymmetric C—CH$_3$ deformation; (d) the absorption at 1225, 1175, 1110, 1090, 1045, and 1020 cm^{-1} indicates either 1:2 or 1:4 or 1:2:4 substitution; (e) the absorption at 835 and 860 cm^{-1} indicates out-of-plane deformation vibration of two adjacent ring hydrogen atoms, and hence the substitution is 1:4 or para.

From these data, it was tentatively assumed that the unknown was p-nitrotoluene. The identification was completed by the preparation of 2,4-dinitrotoluene from the unknown, which gave a melting point (after crystallization) of 69°; the literature value for this derivative is listed as 70°.

Student's Report on Unknown B

The unknown was a liquid with an ammoniacal odor; the boiling point was 133–134°. Test for elements was positive for nitrogen and negative for halogens and sulfur. Test for water was negative, but liquid was dried over CaH$_2$ before placing the sample in the cell. The spectrum (Figure 7.11) was interpreted as follows: (a) two absorptions bands in the 3500 cm^{-1} region indicate a primary amine (—NH stretching absorptions), and a medium band at 1590 cm^{-1} is characteristic of primary amines (—NH deformation); (b) a small band at 1360 cm^{-1} may be due to C—N vibrations, but this is questionable; (c) the strong band at 2900 cm^{-1} is due to —CH$_2$ stretching, while the weak one at 1340 cm^{-1} is due to —CH— deformation; the 1450 cm^{-1} strong absorption may indicate CH$_2$ deformation; (d) the absorption at 970 and 1040 cm^{-1} indicates a cyclohexyl structure, although the two bands are reported in the literature as tentative. However, these two bands are of some relation to this structure. From this it was concluded

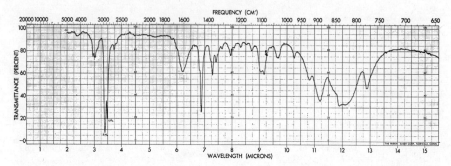

Figure 7.11 Spectrum of unknown identified as cyclohexylamine. Phase, liquid; prism, NaCl; gain, 6; speed, 4; scan time, 15 min.

that the compound may be cyclohexylamine, since the boiling point of cyclohexylamine (134°) is close to that of the unknown (133–134°). Two derivatives were prepared and purified: benzenesulfonamide, m.p. 103–104° (literature value, 104°); and benzamide, m.p. 148° (literature value, 149°). Mixed fusion and mixed melting points of each derivative prepared from the unknown with the same type of derivative prepared from pure cyclohexylamine showed that the two derivatives were identical.

References

Paper and Column Chromatography

Block, R. J., E. L. Darrum, and G. Zweig, *Paper Chromatography and Paper Electrophoresis* (N.Y.: Academic Press, 1955).

Block, R. J., R. de Strange, and G. Zweig, *Paper Chromatography* (N.Y.: Academie Press, 1952).

Brimley, R. C., and F. C. Barrett, *Practical Chromatography* (Londdon: Chapman and Hall, 1953).

Cassidy, H. G., *Adsorption and Chromatography*, Vol. X of Weissberger (ed.) *Technique of Organic Chemistry* (2nd ed., N.Y.: Interscience, 1957).

Cheronis, N.D., and J. B. Entrikin, *Semimicro Qualitative Organic Analysis* (N.Y.: Interscience, 1957), pp. 77–103.

Cramer, F., *Paper Chromatography* (2nd ed., N.Y.: Macmillan, 1954).

Glick, D., *Methods of Biochemical Analysis*, Vol. I (N.Y.: Interscience, 1954).

Lederer, M., *Introduction to Paper Electrophoresis and Related Methods* (N.Y. and Houston: Elsevier, 1955).

Lederer, M., and E. Lederer, *Chromatography: A Review of Principles and Applications* (N.Y. and Houston: Elsevier, 1953).

Smith, I., *Chromatographic and Electrophoretic Techniques:* Vol. I, *Chromatography;* Vol. II, *Zone Electrophoresis* (2nd ed., N.Y.: Interscience, 1960). An excellent up-to-date treatise on organic compounds encountered in biological systems.

Underwood and Rockland, *Anal. Chem.* **26**, 1553 (1954). Small-scale filter paper chromatography. An excellent article for the beginner.

Williams, T. I., *Introduction to Chromatography* (London: Blackie and Son, 1948). A good introduction for a beginner to column chromatography.

Gas-Liquid Chromatography

Keulemans, A. I. M., *Gas Chromatography* (N.Y.: Reinhold, 1957), pp. 1–17. General theory.

Kolthoff, I. M., *et al.*, eds. *Organic Analysis*, Vol. IV (N.Y.: Interscience, 1960), pp. 91–227. Gas chromatography as applied to organic analysis.

Phillips Co., *Gas Chromatography* (N.Y.: Academic Press, 1956). Specific organic separations.

Reviews

Bendich, A., in *Methods in Enzymology*, III (N.Y.: Academic Press, 1957), p. 715.

Lederer, M., ed., *Chromatographic Reviews*, Vol. I (N.Y. and Houston: Elsevier, 1959).

Strain, *Anal. Chem.* 1950, 1952, 1954, 1956, 1958, and 1960. Annual reviews.

Direct print photographic paper for rapid analysis of paper chromatograms containing ultra violet absorbing materials: Ganis, *Anal Chem.* 30, 2068 (1958).

Methods for determining the area of spots in paper chromatography: Knypl and Antoszewski, *J. Chem. Ed.* 37, 49 (1960).

Method of elution of spots from paper chromatograms: Chargaff and Vischer, *J. Biol. Chem.* 176, 703 (1948); 176, 715 (1948); 186, 37 (1950).

Permanent record of chromatograms: Markham and Smith, *Biochem. J.* 45, 294 (1949); 46, 509 (1950); 49, 401 (1951).

Progress in fractionation procedures: Mistretta, *Microchim, J.* 3, 305 (1959); 4, 305 (1960).

Dal Nogare, *Anal. Chem.* 32, 19 (1960). Gas-liquid chromatography.

Hardy and Pollard, *J. Chromatog.* 2, 1 (1959). Gas-liquid chromatography.

Infrared Spectroscopy

Anderson, D. H., N. D. Woodall, and W. West, "Infrared Spectroscopy," in Weissberger (ed.) *Technique of Organic Chemistry*, Vol. I, Part III (3rd ed., N.Y.: Interscience, 1960), pp. 1959–2019.

Barnes, Gore, Stafford, and Williams, *Anal. Chem.* 20, 402 (1948).

Bellamy, L., *The Infrared Spectra of Complex Molecules* (2nd ed., N.Y.: Wiley, 1958). An excellent text for the student.

Brown *et al.*, *Infrared Spectroscopy* (Chicago: Chicago Society for Paint Technology, 1959). The use of IR spectroscopy as an analytical tool in the field of paints and coatings. A good introduction for the application of infrared to polymer technology.

Conley, *J. Chem. Educ.* 35, 453 (1958).

Crendon and Loveel, *Anal. Chem.* 32, 300 (1960).

Crocket and Haendler, *Anal. Chem.* 61, 626 (1959).

Gilkey, *Anal. Chem.* 30, 1931 (1950).

Gore, R. C., and E. S. Waight, "Infrared Light Absorption," in Braude and Nachod (eds.) *Determining of Organic Structures by Physical Methods* (N.Y.: Academic Press, 1955), pp. 195–230.

Gore, R. C., *Anal. Chem.* 22, 7 (1950); 23, 7 (1951); 24, 8 (1952); 26, 11 (1954); 28, 575 (1956); 30, 570 (1958); 32, 238 (1960).

Jones, R. N., and C. Sandorfy, "The Applications of Infrared and Raman Spectrometry to the Elucidation of Molecular Structure," in Weissberger (ed.) *Technique of Organic Chemistry*, Vol. IX (N.Y.: Interscience, 1956), pp. 187–581.

Mason, W. B., *Infrared Microsampling in Bio-Medical Investigations* (Norwalk,

Conn.: Perkin-Elmer Corp., 1959). Reprint on request. "Infrared Microspectrophotometry," in Cheronis (ed.), *Submicrogram Experimentation* (N.Y.: Interscience, 1961), pp. 293–309.

Miller, F., "Applications of Infrared Spectra to Organic Chemistry," in Gilman (ed.) *Organic Chemistry,* Vol. III (N.Y.: Wiley, 1953), pp. 123–57.

Pimental, *J. Chem. Educ.* **37,** 651 (1960).

Randall, H. M., N. Fuson, R. G. Fowler, and J. R. Dangl, *Infrared Determination of Organic Structures* (N.Y.: Van Nostrand, 1949).

White *et al., Anal. Chem.* **30,** 1694 (1958).

Infrared Spectra for the Detection of Functional Groups

Alkanes: Pitzen and Kilpatrick, *Chem. Rev.* **39,** 435 (1946); Sheppard and Simson, *Quart. Revs. (London)* **7,** 19 (1953); Saier and Coggeshall, *Anal. Chem.* **20,** 212 (1948); Jones, McKay, and Sinclair, *J. Am. Chem. Soc.* **74,** 2575 (1952); Brini, *Bull. soc. chim. France* **1955,** 996.

Alkenes: Sheppard and Simson, Quart. Revs. *(London)* **6,** 1 (1952); O'Connor, *J. Am. Oil Chemists' Soc.* **33,** 1 (1956); Kitson, *Anal. Chem.* **25,** 1470 (1953); Jones and Herling, *J. Org. Chem.* **19,** 1252 (1954); Silas *et al., Anal. Chem.* **31,** 529 (1959), (polybutadienes) *Anal. Chem.* **31,** 529 (1959); Goddu *et al.* (unsaturation by near infrared), *Anal. Chem.* **29,** 1790 (1959).

Aromatic compounds: Young, DuVall, and Wright, *Anal. Chem.* **23,** 709 (1951); Ferguson and Levant, *Anal. Chem.* **23,** 1510 (1951); Corio and Dailey, *J. Am. Chem. Soc.* **78,** 3043 (1956); Durie *et al.* (Polynuclear), *Austr. J. Chem.* **10,** 431 (1957).

Carbonyl compounds: Jones and Herling, *J. Org. Chem.* **19,** 1252 (1954); Calthup, *J. Opt. Soc. Amer.* **40,** 379 (1950); Pozefsky and Coggeshall, *Anal. Chem.* **23,** 1611 (1951); Pinchas, *Anal. Chem.* **27,** 2 (1955), **29,** 334 (1957); Jones *et al., J. Am. Chem. Soc.* **74,** 80, 2822, 2828, (1952); **70,** 2024 (1948); **72,** 956 (1950); Rasmussen, Tunnicliff, and Brattain, *J. Am. Chem. Soc.* **71,** 1068 (1949); Davison and Christ (semicarbazones), *J. Chem. Soc.* **1955,** 3989; Brini (long-chain ketones) *Bull. soc. chim. France* **1957,** 516. Durie *et al.* (polynuclear quinones) *Austr. J. Chem.* **10,** 431 (1957); Cetina and Mateos (cycloalkanones), *J. Org. Chem.* **25,** 704 (1960); Wiley *et al.* (dimethylhydrazones), *J. Org. Chem.* **22,** 204 (1957); Jones and Hancock (2,4-DNPH's), *J. Org. Chem.* **25,** 226 (1960); Powers, Harper, and Tai (aromatic aldehydes), *Anal. Chem.* **32,** 1287 (1960).

Alcohols and phenols: Zeiss and Tsutsui, *J. Am. Chem. Soc.* **75,** 897 (1953); Stuart and Sutherland, *J. Chem. Phys.* **24,** 559 (1956); Widom, Philippe, and Hobbs, *J. Am. Chem. Soc.* **79,** 1383 (1957); Hadzi and Sheppard, *Trans. Faraday Soc.* **50,** 911 (1954); Kuhn, *J. Am. Chem. Soc.* **74,** 2492 (1952); Coggeshall and Saier, *J. Am. Chem. Soc.* **73,** 5414 (1951); Rassmussen, Tunnicliff, and Brattain, *J. Am. Chem. Soc.* **71,** 1068 (1949); Brini, *Bull. soc. chim. France* **1957** (long-chain alcohols); Goddu *et al.* (phenols by near infrared), *Anal. Chem.* **30,** 2009, 3013 (1958).

Carboxylic acids: Shreve *et al., Anal. Chem.* **22,** 1498 (1950); Flett, *J. Chem. Soc.* **1951,** 962; Gore *et al., Anal. Chem.* **21,** 382 (1949); Jones, McKay, and Sinclair, *J. Am. Chem. Soc.* **74,** 2575 (1952); Cardwell, Dunitz, and Orgel, *J. Chem. Soc.* **1953,** 3740; Meikjeljohn *et al., Anal. Chem.* **29,** 329 (1957); Susi (stearic acid derivatives), *Anal. Chem.* **31,** 910 (1959).

Esters: Jones *et al., J. Am. Chem. Soc.* **71,** 241 (1949); **73,** 3215 (1951); **74,** 80 (1952); Searles *et al., J. Am. Chem. Soc.* **75,** 71 (1953); Jones and Herling, *J. Org. Chem.* **19,** 1252 (1954); Rasmussen and Brattain, *J. Am. Chem. Soc.* **71,** 1073 (1949). Abramovitch (malonates, oxalates), *Can. J. Chem.* **36,** 151 (1958), **37,** 361, 1146 (1959).

Amines: Bellamy and Williams, *Spectrochim. Acta* **9,** 341 (1957); Richtering, *Z. Phys. Chem.,* **9,** 393 (1956), Russell and Thompson, *J. Chem. Soc.* **1955,** 483; Colthup,

J. Opt. Soc. Amer. **40**, 397 (1950); Mason, *J. Chem. Soc.* **1958**, 3619; Hill and Meakins, *J. Chem. Soc.* **1958**, 760; Whetsel *et al., Anal. Chem.* **30**, 1594 (1958).

Heterocycles: Finnegan, Henry, and Olsen, *J. Am. Chem. Soc.* **77**, 4420 (1955); Tallent and Siewers, *Anal. Chem.* **28**, 953 (1956).

Amides: Badger and Rubaeava, *Proc. Nat. Acad. Sci.* **40**, 12 (1954); Richards and Thompson, *J. Chem. Soc.* **1947**, 1248; Beer, Kessler, and Sutherland, *J. Chem. Phys.* **29**, 1097 (1958).

Nitriles: Kitson and Griffith, *Anal. Chem.* **24**, 334 (1952); Wiley and Wakefield, *J. Org. Chem.* **25**, 546 (1960); Cotton and Lingales, *J. Am. Chem. Soc.* **83**, 351 (1961).

Isocyanates: Davison, *J. Chem. Soc.* **1953**, 3712.

Diazonium salts: Whetsel, Hawkins, and Johnson, *J. Am. Chem. Soc.* **78**, 3360 (1956).

Steroids: Tapley and Vitello, *App. Spectroscopy* **9**, 69 (1955); Smakula *et al., J. Am. Chem. Soc.* **81**, 1708 (1959).

Narcotics: Manning. *J. Bull. Narcotics.* U.N. Dept. Social Affairs, **7**, 85 (1955).

Ureids: Segal *et al., J. Am. Chem. Soc.* **82**, 2807 (1960).

Carbohydrates: Resnik *et al., Anal. Chem.* **29**, 1874 (1959).

Disaccharides: White *et al., Anal. Chem.* **30**, 506 (1958).

Oxirane compounds: Bromstein, *Anal. Chem.* **30**, 544 (1958).

Catalog and Atlases on Infrared Spectra

National Research Council, Committee on Infrared Spectra, Washington, 25, D.C.; A.S.T.M., 1916 Race St., Philadelphia, 3, Pa.; American Petroleum Institute, Research Projects, Carnegie Institute, Pittsburgh, 13, Pa.; Sadtler Research Laboratories, 1517 Vine Street, Philadelphia, 2, Pa. (catalogue contains 19,000 spectra).

8

Separation of Mixtures

The purification of impure organic compounds, which in reality involves the separation of mixtures, was discussed in the fractionation procedures covered in Chapter 2 to which the student is referred. The present chapter deals with the systematic procedures employed in the fractionation of a mixture of unknown composition, so that its various components may be separated in sufficiently pure condition to allow identification of each component.

General Principles

The final separation of a mixture is based on differences in the physical properties of the components at the time of separation. The two physical properties that are most useful are vapor pressure (volatility) and solubility. Sufficient differences may exist among the components of a mixture with regard to one or the other of these properties to make possible the separation of the mixture directly by some form of distillation or extraction by inert solvents, or by fractional crystallization. However, *it is usually necessary to produce the requisite differences in physical properties of the components of the mixture by the use of one or more chemical agents* that change the chemical nature of one or more of the components, thus producing the necessary changes in physical properties (volatility or solubility). Chapter 5, which discusses the relationship of molecular structure and polarity to volatility and solubility, should be thoroughly reviewed at this time. The relationship of polarity and of other properties to selective absorbability on solids should also be reviewed; the relationships are discussed in the section of Chapter 7 that deals with chromatography.

The more important methods for separating mixtures are:

1. Extraction by solvents without chemical change.
2. Extraction by solvents that produce chemical change.

3. Fractional crystallization.
4. Fractional distillation.
   a. Direct distillation at atmospheric pressure.
   b. Distillation under reduced pressure.
   c. Steam-distillation.
   d. Sublimation.
5. Chromatographic techniques and ion exchangers.
6. Various combinations of the above five methods.

Preliminary Tests for a General Mixture

If the mixture consists of more than one liquid phase, or of a solid phase in a liquid, these phases should be separated, and treated individually. In such mixtures, it is probable that the same compounds will exist in more than one of the phases. It is important that the preliminary tests be run on representative samples of the mixture, so that all the components will be tested. A complete record should be kept of all the tests made, including the results of the tests and the deductions that are made. It is wise, also, to keep a record of classes that are eliminated as possibilities because of the results of the tests. All samples that are to be set aside for later examination or use should be *adequately labeled*.

The following tests should be made, together with any others that the nature or behavior of the mixture may suggest:

Composition

The tests suggested in Chapter 4 are recommended. Care should be taken, in making the analysis for the elements, to insure that all components of the mixture are present in the sample taken for the fusion with sodium (test for water; dehydrate before fusing with sodium). In the preparation of the scheme for separating the mixture, it is extremely important to know what elements are present.

Solubility

Solubility determinations should be made on well-mixed samples, utilizing all of the solubility classification solvents, as discussed in Chapter 5. It should be recalled that one or more of the components of the mixture may dissolve in any one solvent, and also that the same compound may partially dissolve in more than one solvent. The solubility in ether should be determined even if the material is not soluble in water. Other solvents that may be used to advantage include methanol, ethanol, carbon tetrachloride, and chloroform. In cases where it is difficult to determine whether or not the solvent has dissolved appreciable amounts of the mixture, the solvent should be separated from the residue and distilled. Exceptions to this technique would be solutions in sodium hydroxide or sulfuric acid. An alkaline extract should be tested by acidifying it and extracting with ether. In the case of concentrated sulfuric acid as a solvent, some classes of compounds that dissolve in it may be recovered by pouring the acid onto an excess of cracked ice.

In evaluating the data from the solubility tests, it is essential not to lose sight of the elements that were found present in the mixture.

Acid-Base Character

The mixture should be examined with the indicators used in Chapter 5. The information thus obtained is helpful, but not completely reliable as regards possible components of the mixture. The use of the indicator solutions on various fractions as separated is also recommended.

Distillation

The student should consult Chapter 2 for procedures for fractional distillation of small quantities.

Any evidence of thermal decomposition of any of the compounds during the distillation should be noted, and, if decomposition is evident, the distillation should be abandoned. In other cases, it should be observed whether a solid residue remains after distillation is complete; if such a residue exists, it should be steam-distilled. The various fractions obtained by distillation should be examined. The distillation under reduced pressure should be considered if normal distillation is unsatisfactory.

Solid Mixtures

In the case of solid mixtures that did not appear to be separable by cold solvents, hot solvents should be used in an attempt to separate the mixture by fractional crystallization. Steam distillation may often be used to advantage on solids. Examination of the mixture by means of a hand lens or microscope often determines the number of components.

Chemical Tests

Selected tests from Chapter 6 should be applied to the original mixture, or to fractions that have been separated from it by the preliminary testing methods. Every ascertainable fact about the presence or absence of the various chemical classes in the mixture will aid in devising a scheme of separation that will have maximum effectiveness with the minimum number of operations.

A brief discussion of ways to separate some binary mixtures will illustrate the application of general principles to methods of separation. The products as separated by the procedures given would require, in most cases, further purification before derivatives could be prepared satisfactorily.

Example 8.1: Salicylic Acid and Malonic Acid

Although both salicylic acid and malonic acid are polar and both are soluble in water, malonic acid is not appreciably soluble in ether because of its two carboxyl groups (highly polar). Therefore, the salicylic acid may be extracted from the malonic acid by ether.

Example 8.2: p-Nitrobenzoic Acid and Methyl-p-Nitrobenzoate

Dissolve the mixture in hot ethanol and allow the solution to cool. The acid will crystallize out of the solution and leave the ester in solution. As precipitated, the acid will be relatively pure, but enough of the acid will remain in the alcoholic solution of the ester to contaminate it. After the ester is recovered by evaporating the alcohol, the residue should be washed with 2 per cent sodium carbonate to remove the acid.

Example 8.3: Aniline and Nitrobenzene

The boiling points of aniline and nitrobenzene are too close together to make fractional distillation a practical method of separating these compounds in pure enough condition to get reliable physical constants for each component. Neither compound is soluble in water, but both are soluble in ether. The addition of dilute hydrochloric acid to the mixture will not affect the nitrobenzene, but it will convert the aniline into a highly polar salt (anilinium chloride, which is soluble in water but not in ether). The nitrobenzene may then be separated from the aqueous layer, washed with dilute hydrochloric acid to remove any remaining aniline, washed with water and very dilute sodium hydroxide to remove traces of the hydrochloric acid, dried with an anhydrous salt, and finally distilled. The aniline may be recovered from the solution of its hydrochloride salt by making the solution distinctly alkaline, extracting with ether, and separating the ether-aniline mixture by fractional distillation.

Note: Better separation is accomplished if the mixture is dissolved in three times its volume of ether before adding the hydrochloric acid to extract the aniline.

Example 8.4: Benzene and Cyclohexylamine

Benzene and cyclohexylamine could be separated by the same method as used in the preceding example. However, since the boiling points of these two compounds are more than 50° apart, it would be possible to separate them by direct distillation using a good fractionating column.

Example 8.5: Benzaldehyde and Benzoic Acid

A dilute sodium bicarbonate solution will convert benzoic acid into a salt and the benzaldehyde may then be separated by extraction with ether or by steam distillation. In the case of steam distillation the acid could be recovered by making the aqueous solution acidic with sulfuric or phosphoric acid and then extracting with ether. The mixture cannot be separated by steam distillation directly because both compounds will steam-distill.

Example 8.6: o-Nitrophenol and p-Nitrophenol

The solubilities of *o*- and *p*-nitrophenol are too similar to allow these isomers to be effectively separated by extraction or fractional crystallization. Moreover, any chemical agent that is added to react with one of them will react with both. However, since *o*-nitrophenol is chelated (therefore exists as a monomer) whereas the *p*-nitrophenol is intermolecularly associated, the ortho isomer can be steam-distilled while the para isomer cannot be. Both of these phenols are sufficiently soluble in water that the loss of material is considerable if one merely separates the phenols from the two aqueous mixtures (distillate and residue) by filtration. This loss may be largely avoided by making each of the aqueous filtrates alkaline with sodium carbonate and then evaporating the solutions nearly to dryness, acidifying with 50 per cent sulfuric acid, and extracting the recovered phenols with benzene. The phenols may then be separated from the benzene by distillation.

For mixtures that contain more than two components, a separation scheme generally involves the use of two or more methods. The following examples illustrate ways that certain mixtures have been separated into reasonably pure products. In each case, further purification of the individual components was found to be necessary before reliable physical constants were obtained and before pure derivatives could be prepared.

Example 8.7: Adipic Acid, p-Nitrobenzoic Acid,
Methyl-p-Nitrobenzoate, and Anthraquinone

The mixture was shaken with warm ethanol and allowed to cool. The undissolved material was removed and the extraction with alcohol repeated. The undissolved residue was treated with 10 per cent aqueous sodium hydroxide. Impure anthraquinone remained as a solid residue. The alkaline solution was acidified and impure *p*-nitrobenzoic acid was recovered by filtration. The alcoholic solution from the original extraction was distilled to remove the alcohol. The residue was extracted by ether and impure adipic acid remained undissolved. The methyl-*p*-nitrobenzoate was recovered by evaporating the ether.

Example 8.8: Butyl Sebacate, Aniline, Isopropylamine, and Acetone

Preliminary experience with distilling a small sample of the mixture indicated the presence of low-boiling components and the presence of high-boiling components that darkened when heated to high temperatures. Furthermore, it was noted that the fumes from the low-boiling fraction were basic. Arrangements were made for fractional distillation of the original mixture under reduced pressure in order to avoid thermal decomposition and to receive the low-boiling distillate in 3N hydrochloric acid which was surrounded by an ice-bath. Distillation was interrupted when it appeared that the low-boiling fraction had been removed from the mixture. The distillate was placed in a small distillation tube and evaporated to dryness, with the vapors being condensed in a cold receiver. Since testing a few drops of this distillate indicated the presence of a carbonyl compound, the 2,4-dinitrophenylhydrazine reagent was added to the distillate and the acetone 2,4-dinitrophenylhydrazone recovered. The isopropylamine was identified from the residual hydrochloride salt.

The high-boiling residue from the original distillation was dissolved in ether and extracted with 10 per cent hydrochloric acid. The insoluble butyl sebacate was recovered and purified. The aniline was recovered by making the acidic solution basic and extracting the amine by ether.

Example 8.9: Heptyl Alcohol, o-Toluidine, and p-Chlorotoluene

The general scheme suggested on page 210 was followed. The heptyl alcohol and *p*-chlorotoluene had to be separated by careful fractional distillation.

Example 8.10: Benzyl Benzoate, 4-Methyl-2-Pentanone, and sec-Butylamine

The general scheme suggested on page 204 was followed. The ketone and ester were separated by steam distillation.

Summary

All pertinent factors should be considered before these various methods are applied to the separation of mixtures. For emphasis, some of the more important factors are listed below:

1. Methods that are quite satisfactory when large quantities of the mixture are available may not be suitable for semimicro quantities.

2. Safety factors should be considered. In particular, in working with

ether (or benzene), extreme care must be used to avoid igniting it by nearby open flames or exposed, highly heated wires (cone-type electric heaters).

3. Some types of compounds, such as low-molecular-weight acid chlorides and anhydrides, are hydrolyzed by water. It is particularly important to remember this fact when carrying out steam distillations.

4. Several classes of compounds are hydrolyzed in hot alkaline solutions or hot acidic solutions. If steam distillation is carried out while a mixture that contains such compounds is definitely alkaline or acidic, hydrolysis may occur, at least to a sufficient extent to complicate the separation of the mixture and the later identification of the original components.

5. Any method that involves the introduction of considerable quantities of water into the mixture is to be avoided if possible, especially if it is believed that a water-soluble component is present, because it may prove very difficult to recover the compound from the water (for example, sugars, glycols, amino acids, sulfonic acids, low-molecular-weight alcohols, carbonyls, and amines, and, in general, compounds that form azeotropic mixtures with water).

6. Attempts to use fractional distillation with high-boiling mixtures (except under reduced pressure) may cause: (a) thermal decomposition of some components, (b) a chemical reaction between components that would not appreciably react at normal temperatures, or (c) oxidation of one or more of the components.

7. Separation of liquids by fractional distillation is limited by the fact that a great number of liquid mixtures form azeotropic solutions.

8. Extraction of a substance from water by ether or from ether by water is not as complete as would be predicted from the solubilities of the substance in each pure solvent. This is due to the fact that ether and water are moderately soluble in each other and the two phases are a saturated solution of ether in water and a saturated solution of water in ether, respectively.

9. The separation of a basic or acidic compound from a neutral compound is usually accomplished by dissolving the mixture in ether or benzene and then extracting the acidic or basic compound from the solution by aqueous solution of a base or an acid. The extraction will be incomplete unless the immiscible layers are very thoroughly mixed and unless the extraction is repeated several times. One way to produce good mixing of the two layers is to draw up portions of the mixture in a pipet and expel the liquid forcibly into the remainder of the mixture. This procedure should be continued for 2–3 minutes.

A General Procedure for the Separation of Mixtures

The following general scheme may be applied to the separation of a mixture into several fractions, many of which correspond to the usual solubility classes of Chapter 5. The scheme is offered with the belief that it will serve as a general guide, and not with the expectation that it is applicable to all problems. This scheme should be modified in accordance with the observed facts for any one mixture. A later section of this chapter takes up the problems of separating mixtures that are present within single fractions as separated by this scheme.

In separating mixtures that are known to contain only two components, the following general scheme may be simplified. At the first point where it appears that the components have been separated from one another, the general scheme should be abandoned and steps should be taken to purify each component.

Note: Students should make up and separate by this scheme (or by one of the alternate methods) a mixture of known composition containing about 0.5–2 g each of four or five compounds. Select compounds that will separate in different fractions. The time consumed in separating such a *known* will be more than repaid in the time saved in handling *unknowns* later. Further practice with mixtures containing 0.1–1 g of each component is desirable.

If suggestions as to the components to include in these practice mixtures are desired, the mixtures used in Examples 8.7 to 8.10 above may be used. Other mixtures for the regular scheme might be: (1) 1 ml each of acetone, aniline, chlorobenzene, and *o*-cresol; or, (2) 0.5 g each of benzoic acid, 2-naphthol, oxalic acid, and naphthalene. A satisfactory mixture for Alternate Method I would be 1 ml each of ethylene glycol, 2-propanol, butylamine, and propionic acid, all dissolved in 10 ml of water. One suggested mixture for Alternate Method II would be 1 ml each of ethyl acetate, *N,N*-dimethyl-aniline, carbon tetrachloride, and oleic acid.

The procedures outlined in the following eight steps are summarized in the flow sheet on page 204. *The Roman numerals in the flow sheet refer to the numbers of the steps.*

If the mixture is a solid, or if the preliminary distillation test showed that there was no distillate below 100°, Step I should be omitted. If the preliminary distillation showed the presence of a low-boiling amine, the distillate in Step I should be absorbed in 3N HCl.

Step I. Place 3–10 ml of the liquid mixture in a 25-ml distilling flask or distilling tube. Using a well-cooled receiver, distill the mixture to remove all the components that distill below 100°. The distillate may contain low-molecular-weight members of practically all the nonaromatic classes of compounds. Generally speaking, the molecules will contain 5 or less carbon atoms. A few saturated cyclic hydrocarbons, a few heterocyclic compounds, and benzene also boil below 100°. Most of these volatile compounds are soluble in both water and ether but the hydrocarbons and their halogen derivatives are not soluble in water. By noting the boiling range of the distillate, a good estimate may be made as to whether or not the distillate is a mixture. Compounds that boil below 100° when pure may not distill below 100° from mixtures.

If the distillate is a mixture, chemical separation and solvent extraction will be possible in a few cases but very careful fractional distillation will be required for most of these mixtures. Test the distillate for the elements and make appropriate classification tests.

It should be noted that chemical reactions may occur between components of the original mixture during this period of heating, even if the compounds did not react in the cold mixture. For example, if a mixture of aniline hydrochloride, sodium benzoate, and ethanol is heated to distill the ethanol, ethyl benzoate is formed in good yield. Examination of a sample of the original mixture in comparison with the final results of the analysis will usually detect any such change in composition of the mixture during its separation.

Test the residue from the distillation for water. If it is present, it must be removed before proceeding to the next step.

Step II. The residue from Step I should be shaken with ether, using 5 ml of ether for each gram of the mixture. Allow the ether to remain in contact with the mixture for 3 minutes (shake occasionally). Treat any undissolved residue by Step III and save the ether solution for Step V.

Step III. Warm the ether-insoluble residue to drive off the ether. Add 5 ml of water for each gram of residue and shake the mixture vigorously. Remove the aqueous solution. Again extract the residue with water, using 10 ml of water for each gram of residue. The water will remove Solubility Division S_2 compounds (see page 96). The two aqueous solutions may be combined, or they may be examined separately. Owing to marked differences in solubility among various components in this group, it is entirely possible that the two aqueous solutions represent a fair separation of Division S_2 compounds. Examine the aqueous solution by evaporating the water out of a 5-ml sample. If the residue is extremely small, the Division S_2 compounds are not represented in the mixture. If a residue exists after evaporating the water, test other samples of the aqueous solution for acidity, for carbohydrates, and for other likely types of compounds of the Division.

If carbohydrates are present, the water may be removed by vacuum distillation or by azeotropic distillation. The compound or compounds introduced for such purpose should be soluble in ether so that any of these liquids remaining after all the water has been distilled may be removed by ether extraction. Water-insoluble acids that are present in the original mixture as their soluble salts may be separated from the aqueous solution by

making the solutions acidic with mineral acids and then distilling or extracting with ether. The salts of amines may be decomposed by sodium hydroxide and the amines removed by distillation or ether extraction.

Step IV. The ether-insoluble, water-insoluble residue from Step III
should be shaken with a volume of cool methanol equal to 5 times the weight
of the residue. The alcoholic solution may be separated from any insoluble
residue by filtration or decantation. The alcohol should then be distilled.
Thus, an alcohol-soluble and an alcohol-insoluble fraction may be obtained.
Examine these fractions for homogeneity. If either fraction appears to be a
mixture, extract such a mixture with 10 per cent hydrochloric acid and with
10 per cent sodium hydroxide in an attempt to separate the components. If
these extractions fail, fractional crystallization from various solvents should
be tried.

Unfortunately, exact and complete data are lacking on the solubilities
of most organic compounds in various solvents, including the common solvents. The attempt to use solvents in the separation of mixtures is further
complicated by the fact that in many cases isomers of the same compound
do not have similar solubilities. However, incomplete lists of some of the
types of compounds that may be expected in the two fractions resulting from
the methanol extraction are given below.

Some compounds insoluble in alcohol, ether, and water:

Many dinitro derivatives of the aromatic hydrocarbons and their
amino, hydroxy, and acid derivatives.

Many trinitro compounds of the above types.

Several dihalo derivatives of anthracene.

Several amino-substituted sulfonic acids; a few amides and imides.

Benzyl and benzoyl ureas; several derivatives of anthraquinone.

Some compounds soluble in alcohol, but insoluble in ether and in water:

Some dibromo- and dinitrobenzoic acids and a few other aromatic acids.

Several polyhydroxy- and polyaminoquinones and quinolines.

A few aminophenols; a few amides and anilides; a very few amines.

Step V. Pour the ether solution from Step II into a distilling flask or
distilling tube the capacity of which is twice the volume of the ether and
distill the ether from a steam bath. Cool the residue and then extract it
twice with water, using 3 ml of water per gram of residue for the first extraction and 7 ml of water per gram of residue for the second extraction. These
aqueous solutions will contain the Solubility Division S_1 compounds (see
page 96). Examine these two aqueous solutions separately, since many

compounds of Division S_1 are highly soluble in water, whereas others are only moderately soluble.

Since many compounds that are slightly soluble in water do not belong to Division S_1, the aqueous extract may be given some color or odor by such compounds. Extract the aqueous component with 5 ml of ether and discard the ether. To determine whether or not the water has removed a Division S_1 fraction, test the solution with litmus. Also distill a 5-ml portion, noting the boiling range, the properties of the distillate, and the residue. If it is concluded that one or more components have been removed by the water, saturate the aqueous solution with potassium carbonate. Any acids originally present will be converted to salts and most of the other compounds will separate from the salt solution. Shake the solution with half its volume of ether. Separate the ether layer and distill the ether. The residue will be the Division S_1 compounds, with the exception of the acids. When acids have been detected in the aqueous extract by the litmus test, the potassium carbonate solution should be neutralized with dilute sulfuric acid to the yellow end point of bromothymol blue and the solution extracted with half its volume of ether. It is best to remove drops of the solution and add them to bromothymol blue test paper, rather than add the indicator to the solution. The ether extraction will remove most of the phenols or amides that were present in the aqueous solution. The aqueous solution should now be made definitely acidic with dilute sulfuric acid and distilled to remove the volatile acids. If some acid is precipitated in the water when it is acidified, it may be removed by ether extraction.

Step VI. The residue that was insoluble in water at the beginning of Step V should be dissolved in ether. (If nitrogen was absent in the residue, omit the remainder of this step and proceed to the next one.) Place the ether solution in a separatory funnel and shake it thoroughly with one-fourth its volume of $1N$ hydrochloric acid. Separate the two layers. Again extract the ether with one-half its volume of 10 per cent hydrochloric acid. The acid will remove Solubility Division **B** compounds. The two acidic solutions should be examined separately on the chance that some separation of the amine components may have been accomplished. Make the solutions slightly alkaline with $1N$ sodium hydroxide. Extract them twice with ether and combine the ether solutions. Dry the ether solution with anhydrous sodium carbonate and distill the ether to obtain the Division **B** fraction.

It should be recalled that many amines are not extractable by dilute acids and will be found in the Division **M** fraction.

Step VII. Shake the ether solution remaining after the Division **B** compounds have been removed in Step VI with half its volume of $2N$ sodium hydroxide solution. Extract the ether again with half as much $2N$ sodium hydroxide solution. Combine the two alkaline extracts and warm the mix-

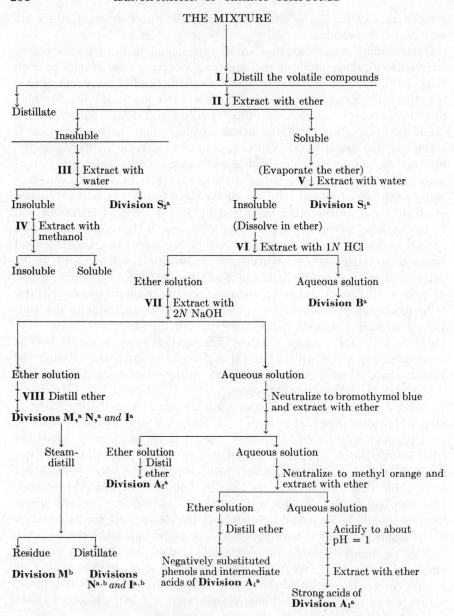

THE MIXTURE

I ↓ Distill the volatile compounds

II ↓ Extract with ether

Distillate

Insoluble

Soluble

III ↓ Extract with water

(Evaporate the ether)
V ↓ Extract with water

Insoluble   Division S_2^a

Insoluble   Division S_1^a

IV ↓ Extract with methanol

(Dissolve in ether)

VI ↓ Extract with 1N HCl

Insoluble   Soluble

Ether solution   Aqueous solution

VII ↓ Extract with 2N NaOH

Division B^a

Ether solution   Aqueous solution

↓ VIII Distill ether

Divisions M^a, N^a and I^a

↓ Neutralize to bromothymol blue and extract with ether

Steam-distill   Ether solution ↓ Distil ether   Aqueous solution

Division A_2^a

↓ Neutralize to methyl orange and extract with ether

Ether solution   Aqueous solution

↓ Distill ether   ↓ Acidify to about pH = 1

Residue   Distillate

Negatively substituted phenols and intermediate acids of Division A_1^a ↓ Extract with ether

Division M^b Divisions $N^{a,b}$ and $I^{a,b}$

Strong acids of Division A_1^a

a Refer to the Divisional Solubility Classifications for the classes and subclasses of compounds that may be present (pages 96–97).

b Division M compounds are absent unless nitrogen or sulfur was found present during analysis. This separation by steam distillation is not always effective since some compounds of Division M will steam-distill, and some compounds of Divisions N and I are not effectively distilled by steam.

ture to drive off the dissolved ether. Neutralize the alkaline solution to the yellow end point of bromothymol blue by adding dilute hydrochloric acid dropwise, while vigorously stirring the solution. To test the solution for the proper pH, remove a drop of the mixture from time to time and apply it to a strip of bromothymol blue test paper (if the indicator is added directly to the solution, it will be extracted by ether and cause confusing colorations). Now extract the aqueous solution twice with ether to remove the Solubility Division A_2 compounds. Dry the ether with anhydrous sodium sulfate. Decant the ether and distill it, leaving the Division A_2 compounds as the residue. Not all the phenols will be extracted at the pH used and the later acid fraction should be tested for phenols.

The aqueous solution from which most of the phenols and other weakly acidic compounds have been extracted should be further acidified to the end point of methyl orange. Ether extraction at this pH will remove most of the negatively substituted phenols and the intermediate acids. This fraction represents the less acidic members of Division **A** compounds.

Concentrate the aqueous solution by evaporation to about half its original volume. Cool the solution and acidify it to the red end point of thymol blue. Extract the solution with ether to remove the most acidic compounds of Division A_1.

Step VIII. The ether solution from which the acids have been removed should be washed twice with 5-ml portions of water to remove any remaining sodium hydroxide. Dry the ether with an anhydrous salt and decant and distill it. The residue will contain compounds that are in Solubility Divisions **M, N,** and **I.** Suggestions for further separation of this mixture are given on pages 207–08.

Suggestions for Separating Intraclass Mixtures

Assuming that the mixture was treated by the scheme suggested, it has been separated into a maximum of ten fractions. It is improbable, however, that any one mixture will contain compounds that would separate in all ten of these fractions. Because of overlapping solubilities, it is entirely possible that some of the compounds have been partially separated in two or more fractions. This fact should be kept in mind when the individual fractions are purified, and tests are being performed on them.

In connection with attempts at purification of the individual fractions, it may be discovered that the fraction represents a mixture of two or more compounds, not counting the impurities due to imperfect separation. No simple set of directions can be given for the separation of such intraclass mixtures. The usual methods of distillation, fractional extraction, and fractional precipitation are often useful. Hot solvents, the less commonly used solvents, and mixed solvents should also be tried. Occasionally, resort may have to be made to chemical reactions that will make separation possible. Benzene may be separated from cyclohexane by nitrating or sulfonating the benzene. A mixture of an ester and an ether that cannot be fractionated may be separated by saponifying the ester.

Mixtures of Division S₂ Compounds

Aqueous mixtures of Division S_2 compounds should be tested for carbohydrates, amine salts, metallic salts, and ammonium salts. If carbohydrates are absent, such mixtures may be distilled to remove the water, but if carbohydrates are present, it is best to distill under reduced pressure. If amine salts are present, make the mixture alkaline and then distill. The salts of acids could, of course, be converted to free acids by adding a mineral acid. The acids thus liberated may or may not be extractable by ether, or be capable of being steam-distilled. In general, molecules having two or more polar groups cannot be steam-distilled.

Hot alcohol is a convenient solvent for separating mixtures of Division S_2 compounds after the water has been removed from the mixture. Sugars do not dissolve in the hot alcohol. Most of the carboxylic acids will dissolve in hot alcohol but will crystallize out on cooling. Many of the other compounds of this class remain in solution in the alcohol and may be recovered by distilling the alcohol.

The hydrogen atom in chloroform is an *acceptor* in hydrogen-bonding. Hence, compounds having functional groups that act as *donors* will dissolve in chloroform, even if they do not dissolve in carbon tetrachloride. Chloroform will extract some types of compounds from nonaqueous mixtures of this class.

Mixtures of Division S₁ Compounds

If the aqueous mixture of Division S_1 compounds is either acidic or basic, neutralize the solution. Steam distillation will separate the volatile components from the salts, the polyhydroxy phenols, and other nonvolatile compounds. The nonvolatile residue may often be separated by fractional crystallization from hot water. Ether and chloroform are good solvents for extracting the residue after the water has been removed.

The volatile compounds, which would be present in the distillate, will include the alcohols, esters, aldehydes, and ketones. If a test for aldehydes and ketones is positive, these classes may be separated from the alcohols and esters by conversion to the sodium bisulfite complexes or to the phenylhydrazones. The alcohols and esters may often be separated by fractional distillation. Another method is to "salt out" the alcohols and esters by saturating the solution with potassium carbonate, separating the alcohol–ester fraction by means of a separatory funnel or pipette, and then adding a few grams of calcium chloride to the alcohol–ester fraction. After a few minutes, add just enough water to dissolve the salt. The alcohol will remain in solution with the calcium chloride, whereas the ester will separate. To recover the alcohol, saturate the salt solution with sodium sulfate and extract with ether.

Mixtures of Division B Compounds

Many, but not all, of the amines of this class are volatile with steam. Hence, steam distillation is sometimes helpful in separating such mixtures. Fractional crystallization and, less often, fractional distillation may be used. Of course, benzenesulfonyl chloride or p-toluenesulfonyl chloride will react with the primary and secondary amines, but not with the tertiary amines. It is occasionally advisable to treat the amine mixture with one of these reagents and then extract the tertiary amine with 10 per cent hydrochloric acid. The derivatives of the primary and secondary amines may be separated by solubility differences. If it is necessary to recover the original amines, the derivatives may be hydrolyzed by prolonged refluxing with dilute hydrochloric acid.

Aromatic amines may be separated from many impurities by converting them into picrates in alcohol solution. The amines may be regenerated from the picrates by treatment with ammonia.

Mixtures of Divisions M, N, and I Compounds

The scheme of separating mixtures proposed in this chapter places, in one residual group, all of the compounds that are soluble in ether but insoluble in water and were not extracted by hydrochloric acid or sodium hydroxide. There is, therefore, considerable probability that this residue will be a mixture of two or more compounds. If neither nitrogen nor sulfur is present in this residue, the Division M compounds are absent. In general, the Division N and Division I compounds are volatile with steam, whereas only a few Division M compounds are volatile with steam. Hence, steam distillation will usually separate the Division M compounds from the other two classes.

Although sulfuric acid and phosphoric acid are not usually satisfactory for the separation of mixtures, the use of these acids is recommended for small samples to help in determining what types of compounds are present. Division I compounds are insoluble in concentrated sulfuric acid. Of the compounds that dissolve in sulfuric acid, only the lower-molecular-weight ones will dissolve in 85 per cent phosphoric acid. It should be recalled that the members of these classes were removed by water extraction if they did not contain more than 4–5 carbon atoms per molecule. The phosphoric acid will dissolve members of these classes if they do not contain more than **8–9** carbon atoms per molecule.

Mixtures of Division **M** compounds may be best separated by fractional extraction or fractional crystallization. Mixed solvents are frequently useful. Several of the more common members of this class may be extracted by hot water, from which they will separate when the solution is cooled. Carbon tetrachloride will dissolve many of the compounds of Division **M,** but it fails to dissolve many of the dinitro and polynitro compounds, anilides, amides, sulfones, and other compounds of similar structure. Chloroform will dissolve most of those compounds that are insoluble in carbon tetrachloride, especially if they contain active *donor* groups for hydrogen bonding. Chloroform is not a good solvent for the sulfonamides. Methanol is useful in fractionating the mixture that is insoluble in carbon tetrachloride; it dissolves the anilides and amides, but not the nitro compounds or sulfones. As previously mentioned, a few of the Division **M** compounds can be steam-distilled.

The mixtures of Division **N** and Division **I** compounds may frequently be fractionally distilled, either at atmospheric pressure or under vacuum. If aldehydes or ketones are present, the mixture may be dissolved in ether and extracted with a saturated solution of sodium bisulfite. For solid mixtures of these divisions, fractional crystallization from hot solvents is often the best method. Aromatic hydrocarbons may usually be separated from nonaromatic hydrocarbons by chromatographic methods using a silica gel column and eluting with alcohol.

Alternate Methods for the Separation of Mixtures

It cannot be emphasized too strongly that no one schematic procedure for the separation of mixtures is equally applicable to all types of mixtures. Two of the many possible schemes are outlined below. Details are omitted since in most cases the separation of the compound or compounds from each fraction obtained by these methods may be accomplished by regular methods or by methods suggested in the general procedure and discussed in more detail in the preceding sections.

Solubility classification may be related to *probable* volatility with steam as follows. Most Solubility Division S_1 compounds are volatile with steam, whereas Division S_2 compounds are not; most Division A_1 and A_2 compounds are not volatile, but there are several exceptions; many Division **B** compounds are volatile; some Division **M** compounds are volatile; most of the Division **N** and **I** compounds are volatile with steam. When steam distillation is used, the matter of possible hydrolysis of compounds must not be overlooked.

The classification of a given mixture as water-soluble or water-insoluble presents difficulties. Some of the components of the mixture may be very soluble in water and the others relatively insoluble. In such a case, the mixture could be separated by water into two mixtures, one of which would be treated as a water-soluble mixture and the other as a water-insoluble mixture. More often, however, various components of the mixture partially dissolve in water but fail to be completely dissolved in the quantity of water that may be used. In that case, the aqueous solution may be considered as a water-soluble unknown mixture and the undissolved material approached as a water-insoluble unknown. Note that the presence of water-soluble organic solvents in the mixture may take into solution in the aqueous phase many compounds that would not dissolve in water alone. The fact that a given mixture completely dissolves in water does not necessarily prove that there are no water-insoluble substances present.

A schematic outline of one way to separate a water-soluble mixture is given on page 210. Solubility Division M compounds are usually not very soluble in aqueous mixtures; however, if such compounds should be present they are generally not volatile with steam and hence would be left in the residue with Division S_2 compounds. Division I compounds would not be present.

The outline on page 211 represents one way to separate a mixture that is not soluble in water. Benzene could be substituted for the ether as a solvent in this procedure.

Separation of Mixtures by Chromatography

Chromatography is a very rapidly expanding field in analytical chemistry. The applications of the various chromatographic techniques to the separation of organic mixtures are almost limitless. An introduction to the use of these methods is given in Chapter 7. It is urged that all students in qualitative organic analysis become acquainted with these effective methods by study and experimentation. Even if the unknown is to be finally separated by other methods, it is often practical to try column chromatography first to estimate how many components are present in the mixture that is being investigated. The number of zones that is indicated by the use of the streak-reagents will at least indicate the *minimum* number of components in the mixture. At the same time, the chemical nature of the components will be indicated by noting which reagents react with the components on the column.

An introduction to the use of column chromatography is given in the authors' larger work.[1] In recent years, gas-liquid chromatography has been applied to the separation of a great variety of mixtures, and the literature

[1] Cheronis and Entrikin, *Semimicro Qualitative Organic Analysis* (2nd ed., N.Y.: Interscience, 1957), pp. 99–103.

ALTERNATE METHOD I

MIXTURES OF WATER-SOLUBLE COMPOUNDS

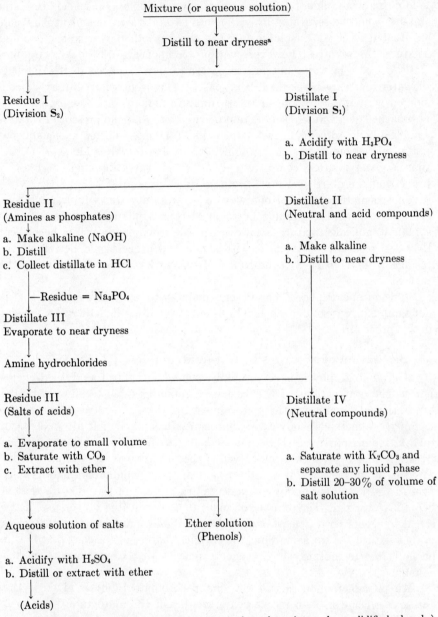

Mixture (or aqueous solution)

Distill to near dryness[a]

Residue I
(Division S_2)

Distillate I
(Division S_1)

a. Acidify with H_3PO_4
b. Distill to near dryness

Residue II
(Amines as phosphates)

a. Make alkaline (NaOH)
b. Distill
c. Collect distillate in HCl

—Residue = Na_3PO_4

Distillate III
Evaporate to near dryness

Amine hydrochlorides

Distillate II
(Neutral and acid compounds)

a. Make alkaline
b. Distill to near dryness

Residue III
(Salts of acids)

a. Evaporate to small volume
b. Saturate with CO_2
c. Extract with ether

Distillate IV
(Neutral compounds)

a. Saturate with K_2CO_3 and
   separate any liquid phase
b. Distill 20–30% of volume of
   salt solution

Aqueous solution of salts

a. Acidify with H_2SO_4
b. Distill or extract with ether

(Acids)

Ether solution
(Phenols)

[a] Interrupt the distillation at 130–140° (unless the mixture has solidified already), change receivers, and introduce steam into the distilling flask to see if any components can be distilled by steam.

[b] Division S_1 contains neutral, acidic, and basic compounds.

ALTERNATE METHOD II

MIXTURES OF WATER-INSOLUBLE COMPOUNDS

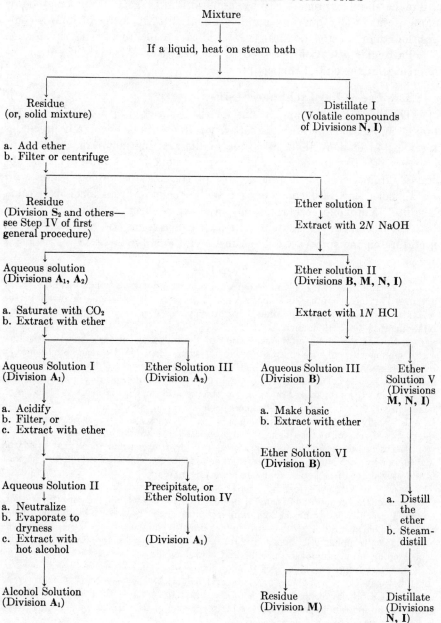

Mixture

↓

If a liquid, heat on steam bath

Residue
(or, solid mixture)

a. Add ether
b. Filter or centrifuge

Distillate I
(Volatile compounds
of Divisions **N, I**)

Residue
(Division **S₂** and others—
see Step IV of first
general procedure)

Ether solution I

Extract with 2N NaOH

Aqueous solution
(Divisions **A₁, A₂**)

a. Saturate with CO₂
b. Extract with ether

Ether solution II
(Divisions **B, M, N, I**)

Extract with 1N HCl

Aqueous Solution I
(Division **A₁**)

a. Acidify
b. Filter, or
c. Extract with ether

Ether Solution III
(Division **A₂**)

Aqueous Solution III
(Division **B**)

a. Make basic
b. Extract with ether

Ether
Solution V
(Divisions
M, N, I)

Ether Solution VI
(Division **B**)

Aqueous Solution II

a. Neutralize
b. Evaporate to
dryness
c. Extract with
hot alcohol

Precipitate, or
Ether Solution IV

(Division **A₁**)

a. Distill
the
ether
b. Steam-
distill

Alcohol Solution
(Division **A₁**)

Residue
(Division **M**)

Distillate
(Divisions
N, I)

on the subject is very extensive. Samples of 1 ml or lesss of a gaseous mixture, or 0.1 ml or less of a liquid may be effectively separated by an appropriate gas-liquid chromatography apparatus. In general, the method is more useful for the quantitative determination of mixtures of known composition than for the qualitative identification of "unknown" substances.

Particular attention is called to the article by Metcalfe listed in the references at the end of this chapter.

The Use of Ion Exchange Resins

The use of ion exchange resins for the separation of mixtures has only limited application, because the capacity of the resin bed generally precludes the removal of more than 30–50 mg of the reacting material. Such resins are very effective in the removal of trace amounts of acidic or basic substances from neutral substances. By pretreating the proper resins with certain ions, carbonyl compounds may be removed from alcohols (resin-bisulfite), and monosaccharides may be separated from other sugars (resin-borate). Several experiments using ion exchange resins and reierences to the literature on the subject are given in the authors' larger text.[2]

Exercises

1. What is the Hinsberg method for separating primary, secondary, and tertiary amines? Outline the procedure and write equations for all reactions involved in the separation of a mixture containing 1-naphthylamine, N-methyl-1-naphthylamine and N,N-dimethyl-1-naphthylamine.

2. Outline a scheme for the separation of the following mixtures and suggest how each component could be purified after its separation: (a) benzene, ethylbenzene, butylamine, and benzamide; (b) methanol, p-bromophenol, 3,5-dinitrobenzoic acid, and amyl alcohol; (c) ethanol, isobutraldehyde, tert-amyl alcohol, p-chlorotoluene, and benzoic acid; (d) propanol, 2-butanone, propionic acid, and lactic acid; (e) methanol, nitrobenzene, 2,4-dinitroaniline, p-bromobenzoic acid, and cyclohexanol.

3. List five amphoteric compounds and predict where each would be separated in the proposed general method or in the appropriate alternate method.

4. Discuss the compounds most likely to be hydrolyzed and those that might be hydrolyzed by each step in Alternate Method I. What modification would you suggest in the procedure if it is known that esters are present?

5. Outline a scheme for the separation of the following mixtures, and give the equations for the preparation of satisfactory derivatives for each component: (a) acetone, aniline hydrochloride, oxalic acid, 2-butanol, p-nitrobenzoic acid, α-naphthol, dinitrobenzene, and benzyl alcohol; (b) o-nitroaniline, dimethylaniline, nitrobenzene, benzylcyanide, diphenyl, diphenylsulphone, glycerol, and benzene; (c) chloroacetic acid, acetic acid, acetone, ethyl propionate, phenol, and benzaldehyde.

6. Select from the volumes of *Organic Syntheses* a method of synthesis of each of the following: (a) a halophenol; (b) an aryl halide; (c) an ester; (d) a heterocyclic amine; (e) a nitrohydrocarbon; (f) a dibasic acid; (g) a thiol. Outline the method of synthesis and indicate by an outline how the separation of the various compounds present in the reaction mixture is accomplished.

[2] Cheronis and Entrikin, *op. cit.*, pp. 103–09.

7. Outline a scheme for the separation of each of the following mixtures and tell how each component could be purified for derivatization after its separation: (a) acetone, decanoic acid, 2-heptanone, and toluene; (b) acetic acid, lauric acid, 2-hexanone, and 3-hexanone (consider the difference in reactivity of the two ketones with sodium bisulfite); (c) tributylamine, heptane, methanol, and aniline.

8. Suggest a method for separating the following mixtures: (a) aniline, N-methylaniline, N,N-dimethylaniline, benzene, and nitrobenzene; (b) maltose, resorcinol, salicylic acid, and m-dinitrobenzene; (c) p-cresol, p-toluic acid, p-nitrotoluene, and 3-pentanore.

9. Practical problems:

a. If a 4 oz. sample of used lubricating oil were submitted to you with the request that you determine if sugar had been added to the oil in an attempt to sabotage the engine, how would you proceed to determine the presence of sugar in the oil?

b. How would you proceed to detect arsenic oxide in a sample of hamburger meat?

c. How would you proceed to detect strychnine in flour?

d. If, when you were preparing a solution of salicylic acid in alcohol taken from an iron drum, the solution had a light pink color, what explanation would you offer for the pink color?

10. For the following mixtures of known composition, certain methods of separation have been proposed. Criticize the suggested methods from the viewpoint of good separation and possible recovery of the components in pure enough form to be derivatized successfully. Can you suggest a better procedure?

a. Mixture: Butanol and castor oil.
Suggested method of separation: Direct distillation.

b. Mixture: Isobutyl alcohol and toluene.
Suggested method: Shake the mixture with cold, concentrated sulfuric acid; separate the toluene layer; pour the acid layer over an equal weight of chipped ice, reflux the diluted acid mixture for 15 minutes, and fractionally distill to recover the alcohol.

c. Mixture: Cyclohexanone and benzyl propionate.
Suggested method of separation: Shake with an acidified solution of 2,4-dinitrophenylhydrazine.

d. Mixture: Methanol, toluene, benzoic acid, and 2-hexanone.
Suggested method of separation: (1) extract with water to remove the methanol, (2) extract with sodium hydroxide solution to remove the benzoic acid, (3) extract with sodium bisulfite solution to remove the 2-hexanone; (4) leave the toluene as residue.

e. Mixture: Pentanol, pentanal, heptylamine, trimethylamine.
Suggested method of separation: (1) extract the trimethylamine with water, (2) extract the heptylamine with dilute hydrochloric acid, (3) extract the pentanal with sodium bisulfite, (4) leave the pentanol as a residue.

References

The following is a selected list of articles dealing with various aspects of the separation of mixtures. Others may be found by consulting *Chemical Abstracts*.

Batt and Alber, *Ind. and Eng. Chem., Anal. Ed.* **13**, 127–32 (1941). Comparative study of procedures of microextraction.

Berg and Parker, *Anal. Chem.* **20**, 456–57 (1948). Determination of aromatics and olefins by acid solubility test.

Cassidy, *J. Chem. Educ.* **23**, 427–32 (1946). A general discussion.

Chu *et al., Ind. and Eng. Chem.* **46**, 754–61 (1954). Vapor-liquid equilibrium *m*- and *p*-xylenes in different solvents.

Dubrissy and Aoditti, *Bull. Soc. Chim.* **51**, 1199–1202 (1932); *C.A.* **27**, 648 (1933). Capillary chemical phenomenon.

Dunn and Drell, *J. Chem. Educ.* **28**, 480–83 (1951). Qualitative identification of amino acids.

Hüyser and Schogfsma, *C.A.* **36**, 6268 (1942). Separation of mixtures of chemically related substances.

Kirchner and Haagen-Smit, *Ind. and Eng. Chem., Anal. Ed.* **18**, 31–32 (1946). Separation of acids.

Metcalfe, *Anal. Chem.* **33**, 1559 (1961). The role of separations in organic analysis (includes 71 references to the recent literature).

Mitchell, J. (ed.), *Organic Analysis,* Vols. I and II (N.Y.: Interscience, 1953 and 1954).

Pagel and McLafferty, *Anal. Chem.* **20**, 272 (1948). Use of tributyl phosphate for extracting organic acids.

Randall *et al., Ind. and Eng. Chem.* **30**, 1063, 1188, 1311 (1938); **31**, 227, 908, 1181, 1295 (1939). Separation processes.

Seaman *et al., J. Am. Chem. Soc.* **67**, 1571–78 (1945). Separation and determination of alkaryl amines.

Sunier and Rosenblum, *Ind. and Eng. Chem., Anal. Ed.* **2**, 109–13 (1930). Separating constant-boiling mixtures.

Swann, *Anal. Chem.* **21**, 1448–1453 (1949). Determination of dibasic acids in alkyd resins.

Tenny and Sturgis, *Anal. Chem.* **26**, 946–953 (1954). Separation of hydrocarbons.

Underwood and Rockland, *Anal. Chem.* **26**, 1553 (1954). Qualitative identification of amino acids.

Whitmore and Wood, *Mikrochimie* **28**, 1–13 (1939) (in English); *C.A.* **34**, 2285 (1940). Microchemical separation of some toxiologically important alkaloids.

Wilson, Anderson, and Donohoe, *Anal. Chem.* **23**, 1032 (1951). Precipitate tertiary amines.

Azeotropic Data, pub. as No. 6 of *Advances in Chemistry Series,* by Amer. Chem. Soc., Washington, 1952.

9

Coordination of Data and Tentative
Identification of the Unknown

The objective of this chapter is to discuss the steps by which the experimental data that have been collected are correlated and a tentative assumption is made as to the nature of the unknown. Thus far in a systematic analysis small samples of the unknown were subjected to: (*a*) preliminary examination; (*b*) determination of the melting point if the unknown is a solid or of the boiling point and perhaps the refractive index if it is a liquid; (*c*) determination of the elements present; (*d*) solubility tests and reaction to indicators leading to the classification of the unknown as probably belonging to one of a restricted number of classes of organic compounds; (*e*) specific class tests by means of which the unknown was restricted to fewer and fewer classes and finally to a single class. At this point all of these data are organized and compared with the information about known compounds listed in the literature. For example, assume that the unknown is a solid melting at 187 or 188° that is classified in Solubility Division A_1 and acid-base class A_i; elemental analysis does not show halogens or nitrogen present. The next step is to consult Table 5B on page 356 which lists the solid acids, and find what acid or acids melt at or very near 187 or 188°. Inspection of the table shows six acids that melt at either 187 or 188°. It is to be noted, however, that three of the listed acids contain nitrogen and one contains iodine. Hence, if the analysis for elements was effectively done, these four acids may be eliminated from consideration, leaving succinic and camphoric acids as possibilities. However, the unknown was found not to be soluble in water; this finding eliminates succinic acid as a probability, since an acid with two polar groups and only four carbon atoms would be soluble in water. Tentatively, the unknown may be assumed to be camphoric acid.

Proof of the identity must await the preparation of one or more derivatives as described in Chapter 10.

Correlation and Interpretation of Data and Literature Search

The brief discussion above indicates that the tentative identification of the unknown compound involves correlation and interpretation of the experimental data that have been obtained together with a literature search to discover a compound whose properties resemble those of the unknown. Extensive tables are provided in later sections of this book, in which the major classes of organic compounds are arranged alphabetically. Separate tables are provided for liquid and solid members of each class. Within each table, the compounds are listed in the order of their boiling points or melting points.

The table of appropriate chemical compounds, indicated by the solubility, indicator, and specific class tests, is consulted and a list is made of all the compounds in that class for which the melting point or the boiling point is within 3° of the value found for the unknown. At this point it can be assumed that the unknown is one of the compounds listed provided its boiling point or its melting point has been accurately determined and also provided that it has been properly classified. It is, of course, possible that the unknown is not one of the compounds in the first list of possibilities, but for the present we must consider that unlikely and proceed with the next step. This consists in eliminating some of the possibilities by considering the elements present or by performing certain tests. For example, if the unknown is an alcohol boiling at 127–128° and analysis shows that it contains chlorine, only two of the seven alcohols boiling between 124.5 and 132° need be considered, since the other five do not contain chlorine. If the unknown is an acid boiling at 139–140°, it might be either acrylic acid or propionic acid. Since acrylic acid is unsaturated and propionic acid is saturated, the results of the test for active unsaturation would indicate the identity of the unknown.

In most cases once a tentative list of possibilities has been made and the number has been restricted to one, two, or three compounds, a literature search is made as to the properties of the compounds listed as possibilities. In addition to the present text it may be necessary to consult reference works available in most libraries in order to find the properties of the probable compounds.

Use of the Library

As should be apparent to the student, this text is neither an all-inclusive treatise on tests and derivatives for all organic compounds nor a compendium of methods for every possible situation. For example, the many important spectographic and microscopic methods of identification are not included in this text. Of necessity, the methods proposed in this text for the detection of compounds are those that apply quite generally to members of a given class. Also, the procedures for the preparation of derivatives and the tabular data on derivatives are those that have general application. It is often true that a single member of a chemical class or a very few members of that class may be detected by reactions that do not apply to the class in general. Therefore, it is necessary to consult more complete listings of organic compounds, derivatives, specific tests, and other identification data. The sections on organic chemistry in Soule's *Library Guide for the Chemist* (McGraw-Hill), and Crane, Patterson, and Marrs' *A Guide to the Literature of Chemistry* (Wiley) discuss the major source materials in a chemical library and how to use them. The Indexes of *Chemical Abstracts* are most useful for locating tests for the detection of specific compounds and for locating data on their derivatives. Attention is called to the fact that many references to the chemical literature are given at the end of several chapters of this text.

Exercises

The exercises given in this section are designed for:
1. A review of the properties of compounds as revealed by tests.
2. Practice in deductive and inductive reasoning.
3. Experience in the use of the tables of compounds and their derivatives.

Assignment One

What derivatives could be prepared or tests made which would distinguish the following compounds:
1. *m*-Ethyl toluene and propyl benzene.
2. Butanol and 3-pentanol.
3. 3-Pentanol and 2-pentanol.
4. Butanol and 2-pentanol.
5. *m*-Chlorobenzoic acid and 2,5-dichlorobenzoic acid.
6. *m*-Chlorobenzoic acid and *m*-bromobenzoic acid.
7. Ethyl benzoate and benzyl acetate.

Assignment Two

Write structural formulas for six compounds that have $C_3H_6O_2$ as the molecular formula. Select chemical tests that could be used to distinguish each compound from all the others.

Assignment Three

The following isomeric compounds have boiling points close together. Select a reaction that would produce a derivative from each isomer which would distinguish it from the others. Write balance equations for the reactions involved: (*a*) the three cresols, (*b*) the three ethyl toluenes and propyl benzene, (*c*) the three chlorotoluenes, (*d*) the methyl butanols, and (*e*) the three *N*-methyl toluidines.

Assignment Four

Tentatively identify the following compounds:

1. A compound of Solubility Division **I** boils at 158°. On vigorous oxidation a compound is formed which melts at 122° and is in Solubility Division A_1.

2. A compound, $C_9H_{12}O$, does not decolorize bromine in carbon tetrachloride, but is oxidized by cold potassium permanganate to yield a product that reacts with *p*-nitrophenylhydrazine; vigorous oxidation by hot alkaline permanganate followed by acidification of the mixture produces benzoic acid; the original substance produces iodoform when treated with iodine in alkaline solution.

3. A Solubility Division **N** compound contains bromine and boils at 218°. It is not affected by bromine in carbon tetrachloride, hot alkaline permanganate, or hot alcoholic silver nitrate. When the compound is vigorously treated with hot, concentrated hydrobromic acid, *o*-bromophenol is recovered as one product.

4. An acid melts at 154–156° and contains halogen; the *p*-nitrobenzyl ester melts at 104–106°.

5. A compound of Solubility Division **B** had a boiling point of 201–203°. It could be diazotized and coupled with 2-naphthol. The original compound reacted readily with bromine water to give a precipitate, and on analysis this precipitate proved to have three atoms of bromine per molecule.

6. A solid melted at 53° and was in Solubility Division **M**. After prolonged refluxing with dilute sodium hydroxide, a liquid layer separated on top of the mixture. This compound, *A*, was separated by ether extraction and purified. Compound *A* boiled at 206–207° and reacted with acetic anhydride to yield a product that melted at 83°. The alkaline solution from which *A* had been extracted was concentrated by evaporation and acidified. An acid was recovered that melted at 121–122°.

7. An unknown was soluble in water, but not in ether. It contained both nitrogen and chlorine. The aqueous solution was acidic to litmus and gave a precipitate with aqueous silver nitrate. Treatment of the aqueous solution with 20 per cent sodium hydroxide produced an emulsion, which was extracted by ether to yield a liquid that boiled at 183–184° and formed a benzene sulfonamide melting at 112°.

Assignment Five

Identify the following compounds (nitrogen, sulfur, and halogens are absent unless specified):

1. A colorless, pleasant-odored liquid falls in Solubility Division **I**. A faint yellow color develops on the aluminum chloride in the test for cyclic structure. The boiling range of the compound is 109–111°. Nitration of the compound gives yellow crystals, which melt at 70°. The aroyl benzoic acid derivative melts at 136–137°.

2. A colorless liquid falls in Solubility Division **N**. The test for cyclic structure is negative. The compound fails to react with 2,4-dinitrophenylhydrazine and gives a negative test for active unsaturation. The original compound does not give the hydroxamate test for esters, but when the compound is treated with acetyl chloride, the product does give an ester test. The boiling range of the compound is 131–132°. It reacts with α-naphthyl isocyanate to form a derivative that melts at 68°.

3. A colorless liquid gives negative tests for active unsaturation and cyclic structure. It falls in Solubility Division S_1. It gives a weakly positive test with Schiff's reagent and gives a positive test with the iodoform reaction. The boiling point is 55–56°. The phenylhydrazone of the compound melts at 41–42°; the *p*-nitrophenylhydrazone melts at 148–149°.

4. A colorless liquid gives a faint yellow color to aluminum chloride. The solubility

determinations seem to indicate a borderline case, with Divisions N and 1 as possibilities. The compound fails to react with phenylhydrazine or acetyl chloride. It gives negative tests in the xanthate and hydroxamic acid reactions. Nitration with fuming nitric acid gives a yellow solid, which gives a test for a dinitro compound by the acetone–sodium hydroxide test. The original liquid boils at 79–80°, and the nitro derivative melts at 89°. The aroyl benzoic acid derivative melts at 127°.

5. A light yellow liquid gives a positive test for nitrogen. It falls in Solubility Division M. The test for cyclic structure is positive. The compound does not darken appreciably when treated with the acetone–sodium hydroxide reagent. It oxidizes ferrous hydroxide. The boiling point of the original liquid is 208–209°, and its nitration product melts at 89–90°.

6. A pinkish tan solid gives negative tests for active unsaturation. It gives a purple color when tested for cyclic structure with aluminum chloride. It falls in Solubility Division A_2. The compound gives a red color with the ceric nitrate reagent and a purple coloration with ferric chloride. It also gives a purple color when treated with a mixture of sulfuric acid and sodium nitrite. The compound melts at 93–94° and gives a derivative with α-naphthyl isocyanate, which melts at 150–152°; its phenylurethan melts at 178°.

7. A slightly yellow liquid gives a positive test for halogen. Further tests show that the halogen is bromine. The compound falls in Solubility Division I. Only a small precipitate is obtained when the compound is shaken with alcoholic silver nitrate. The boiling point of the compound is 156–157°. When the compound is refluxed with metallic sodium, a white solid is formed that melts at 69°. This solid does not contain halogen and has a borderline solubility, being in Division N or Division I. The nitration product of this solid melts at 233°. Nitration of the original compound yields a product that melts at 75°.

Assignment Six

Identify the following compounds (nitrogen, sulfur, and halogens are absent unless specified):

1. A colorless solid falls in Solubility Division A_1. The test for cyclic structure is positive. It gives a wine coloration with ferric chloride. The melting point of the compound is 156–157°. When it is treated with thionyl chloride and then refluxed with aniline, the derivative melts at 133–134°. The anilide of the original compound melts at 135°.

2. A colorless solid gives positive tests for nitrogen and halogen. The cyclic test is positive and the compound falls in Solubility Division B. The precipitate is small when the compound is shaken with alcoholic silver nitrate. The melting point is 70°. The compound reacts with acetyl chloride and gives a derivative that melts at 178–179° The original compound reacts with benzenesulfonyl chloride to give a product that is soluble in sodium hydroxide solution and that melts at 122°.

3. A light tan liquid boils at 193–194°. It contains nitrogen. It gives a positive cyclic test. It seems to partly dissolve in dilute acid but is placed in Solubility Division M. The compound reacts with nitrous acid solution to yield a yellow oil. It does not oxidize ferrous hydroxide. It reacts with benzenesulfonyl chloride to form a solid that does not dissolve in aqueous sodium hydroxide. This solid melts at 79°.

4. A colorless liquid boils at 101–102°. The Solubility Division is S_1. The aqueous solution is neutral. The compound gives a positive test for alcohols by the xanthate test, but when the original compound is treated with acetyl chloride, the product that forms does not give a test for esters. This product falls in Solubility Division I and gives a test for chlorine. However, when the original compound is mixed with dimethylaniline and then treated with acetyl chloride, an ester is formd. The 3,5-dinitro-

benzoate of the original compound melts at 115–116°. The α-naphthylurethan melts at 72°.

5. A colorless liquid falls in Solubility Division S_1 and has a boiling point of 96–97°. It readily decolorizes a bromine solution. Its aqueous solution is neutral. It does not react with phenylhydrazine or ferric chloride. It gives a yellow precipitate when treated with solid sodium hydroxide and carbon disulfide. It fails to give a positive test in the hydroxamic acid reaction before treatment with acetyl chloride, but does give a purple color after treatment with the acyl halide. The original compound reacts with 3,5-dinitrobenzoyl chloride to give a derivative that melts at 48°. The phenylurethan melts at 70°.

6. A colorless liquid that boils at 74° is soluble in both water and ether. The compound reduces an alkaline solution of cupric ions and gives a light red coloration with Schiff's reagent. The compound reacts with 2,4-dinitrophenylhydrazine to give a reddish orange solid that melts at 122°. The semicarbazone melts at 106°.

7. A white solid contains halogens and falls in Solubility Division I. Nitration of the compound gives a light yellow solid that melts at 84°. The original compound melts at 89°.

Assignment Seven

Identify the following compounds (nitrogen, sulfur, and halogens are absent unless specified):

1. A colorless compound melts at 48–49°. It is soluble in sulfuric acid, but not in the other solvents. The compound does not react in the xanthate test or in the hydroxamic acid test. It gives a test for cyclic structure, but does not contain active unsaturation. When the compound is fused with sodium hydroxide, a liquid distills that has a boiling point of 80°. The fused mass is dissolved in water and acidified and at that point a solid precipitates that is found to be in Solubility Division A_1 and to have a melting point of 121°. The original compound forms a phenylhydrazone that melts at 137° and a semicarbazone that melts at 167°.

2. A yellow solid contains nitrogen. It dissolves in sodium bicarbonate with difficulty, but is placed in Division A_1. It gives a purple color with ferric chloride and a yellow-orange color with the acetone-sodium hydroxide reagent. It oxidizes ferrous hydroxide readily and decolorizes a bromine solution. The bromination product melts at 118°. The nitration product of the original compound melts at 121–122°. The original compound melts at 114°.

3. A gray-green solid melts over the range 110–120°. The material is recrystallized from an alcohol–water mixture to yield colorless crystals that melt at 123–125°. Nitrogen and sulfur are found present. The compound is insoluble in water and acid, but reacts with 10 per cent sodium hydroxide solution to form colorless crystals that melt at 54° and contain nitrogen but no sulfur. No organic sulfur compounds can be found in the alkaline solution, but it is noted that, when the solution is acidified with hydrochloric acid and barium chloride is added, a white precipitate forms. When the crystals that form when the original compound is treated with sodium hydroxide are refluxed with acetyl chloride, a solid forms that melts at 101°.

4. A yellow solid contains nitrogen and chlorine. It gives a positive test for cyclic structure and falls in Solubility Division M. It gives a moderate precipitate when treated with an alcoholic solution of silver nitrate and readily forms silver chloride after treatment with an alcoholic solution of potassium hydroxide. The compound oxidizes ferrous hydroxide, but fails to change the color of the acetone–sodium hydroxide reagent. The compound melts over the ranges 45–47°. The compound is refluxed for 3 hours with an alkaline solution of potassium permanganate. The manganese dioxide is

filtered off and the filtrate is acidified. A white precipitate forms that contains nitrogen, but not halogen. This precipitate falls in Solubility Division A_1 and melts at 140–141°. This derivative forms an amide that melts at 143°.

5. A white solid melts at 114° and contains nitrogen. It falls in Division **M**. It does not oxidize ferrous hydroxide. When the compound is strongly heated with soda lime, a liquid distills. This liquid is found to contain nitrogen and to fall in Solubility Division **B**. The reaction of this liquid with acetyl chloride gives the original starting material.

6. A white crystalline solid contains nitrogen. The melting point is 114° and the Solubility Division S_2. The aqueous solution is neutral to litmus. When the solid is heated, it distills at 222°. Water is evolved during this distillation. This distillate reacts with nitrous acid to form a water-soluble acid that boils at 118°.

7. A liquid contains sulfur and falls in Solubility Division **M**. The compound is oxidized to a white solid that melts at 128–129°. When the original compound is heated with a mixture of concentrated nitric and sulfuric acids, it yields a solid that melts at 201°.

Assignment Eight

Identify the following compounds. Write equations for all the chemical reactions involved.

1. A solid of Solubility Division A_1 is found to contain nitrogen, sulfur, and halogens. The melting point is 80°. When an alkaline solution of the compound is acidified, a solid is recovered that contains nitrogen and sulfur, but no halogens. This compound melts at 109–111°. The original compound reacts with aniline to give a compound which melts at 171° and is soluble in 10 per cent sodium hydroxide. The amide melts at 180°.

2. A colorless liquid boils at 83–84°. It adds bromine rather readily and reacts with concentrated sulfuric acid (darkens). The original compound is refluxed with an alkaline solution of potassium permanganate. During this treatment, the compound goes into solution. Acidification of the aqueous solution causes a white precipitate to form. This new compound is found to fall in Solubility Division A_1 and to melt at 152–154°. A saturated solution of this oxidation product is found to be definitely acidic. The solid acid is distilled with solid barium hydroxide to yield an oily liquid that distills between 125° and 130°. This liquid falls in Solubility Division **N** and forms a semicarbazone that melts at 210°.

3. Solubility Division A_1; intermediate acid; melting point, 133–135°; reacts with phosphorous pentachloride to give a compound (*B*) that melts at 36° and gives a positive ferric hydroxamate test; both the original compound and its derivative (*B*) react with aniline to form derivatives that melt at 151–153°.

4. A liquid that boils at 197–198° appears to have borderline solubility (S_2 and S_1); it gives a positive xanthate test and reacts with acetic anhydride to give a product that is insoluble in water, but gives a positive ferric hydroxamate test. The original compound reacts with α-naphthylisocyanate to give a product that melts at 174–175°; it also reacts with 3,5-dinitrobenzoyl chloride to give a product that melts at 168–169°.

5. Solubility Division **M**; nitrogen and chlorine present; melting point 82–83°; gives positive results with test 6.16A and negative results with tests 6.4A and 6.17. Refluxed for 1 hour with 10 per cent NaOH, the compound dissolves, and, on acidification, a compound separates that contains nitrogen, but no chlorine, and melts at 114°; the original compound reacts with zinc and hydrochloric acid and dissolves in the acid solution; when this acid solution is made alkaline and extracted with ether, a compound is recovered that melts at 70–72° and reacts with benzenesulfonchloride to yield an alkali soluble derivative which melts at 121–122°.

Assignment Nine

Identify the following compounds. Write equations, where possible, for all indicated reactions.

1. Solubility Division S_1; weak acid; melting point 169°; has a neutralization equivalent of 55; oxidizes to yield a yellow solid with a melting point of 116°; does not give a color with ferric chloride and decolorizes cold aqueous potassium permanganate; reacts with acetic anhydride to give a solid that melts at 123–124°, and has a saponification equivalent of 97.

2. Orange solid, melting point, 106–107°; Solubility Division **M**; contains nitrogen; gives a blue color with sodium hydroxide and acetone; negative results on tests for amides; after hydrolysis by refluxing with concentrated hydrochloric acid, a product is recovered which does not contain nitrogen, is in Solubility Division **N**, and boils at 103–104°; this product oxidizes to an acid which has a neutralization equivalent of 102.

3. Solubility Division A_2; boiling point 120°; contains nitrogen; reacts with nitrous acid to give a blue color without the evolution of nitrogen; oxidizes ferrous hydroxide; when treated with tin and hydrochloric acid, the compound goes into solution; when this solution is made alkaline and warmed, a compound distills that contains nitrogen and boils at 33–34°.

4. Solubility Division A_1; melting point, 132°; intermediate acid; gives ferric hydroxamate test; forms an anilide that melts at 170° and an imide whose melting point is 235°.

5. Compound *A* is a white solid, melting point 114°; contains nitrogen; Solubility Division A_2; gives violet-red color with ferric chloride; oxidizes ferrous hydroxide, but does not give a color with acetone and sodium hydroxide; after refluxing with zinc and hydrochloric acid and the reaction mixture made slightly alkaline, a white solid *B* separates; compound *B* is soluble in dilute hydrochloric acid and reacts with acetyl chloride to form a white solid.

Assignment Ten

For each class of compounds for which derivatives are given in the tables, select representative compounds of the class and write structural equations for the formation of each derivative listed for that class. Name all of the compounds used and produced.

References

The works listed below are recommended for beginners. A more complete listing may be found in the larger text by the authors.

Comprehensive Treatises

Beilstein, *Handbuch der organischen Chemie,* 4th ed., F. Richter (ed.), (Berlin: Springer, 1918——). More than 70 volumes, including Supplements, with more to come. The literature is covered through 1909 in the Main Series (33 parts), through 1919 in the First Supplement, through 1929 in the Second Supplement, and will be covered through 1949 in the Third Supplement.

Beilstein's *Handbuch* gives, so far as recorded in the literature, concise particulars on the following topics: *historical; occurrence; formation;* methods of *preparation;* physical properties (including generalities on salt formation); *chemical behavior* (under heat, electricity, oxidation, reduction, halogenation, etc., inorganic reagents, and reactions with other organic compounds); *biochemical behavior; uses; analytical* (identification, purity tests, quantitative determination, etc.); addition compounds and salts.

As guides to the consultation of this great work there are available:

1. Prager, B., *et al., System der organischen Verbindungen: ein Leitfaden für die Benützung von Beilsteins Handbuch der organischen Chemie* (Berlin: Springer, 1929), 246 pp. This book is not an index, but gives the system of classification and the system numbers of groups of compounds employed in *Beilsteins Handbuch.*

2. Richter, F., *Kurze Anleitung sur Orientierung in Beilsteins Handbuch der organischen Chemie* (Berlin: Springer, 1936), 23 pp. Affords a summary view of the division by functional groups, etc.

3. Huntress, E. H., *Brief Introduction to the Use of Beilstein* (N.Y.: Wiley, 1938). Provides a very helpful guide in English.

Adams, R. (ed.), *Organic Reactions* (N.Y.: Wiley, 1942——). Ten volumes.

Faraday, J. E., *Encyclopedia of Hydrocarbon Compounds* (N.Y.: Chemical Pub. Co., 1946——). Ten volumes published, others to come.

Heilbron, I., *Dictionary of Organic Compounds* (2nd ed., N.Y.: Oxford, 1953), 4 vols. This is the most extensive catalog of organic compounds in the English language that has been published in completed form. Contains the formula, physical data, and characteristic reactions, together with the melting point of identification derivatives and many references.

Müller, E. (ed.), *Methoden der organischen Chemie* (*Houben-Weyl*) (Stuttgart: Thieme, 1952——). To be completed in about twelve volumes. An important source for identification work.

Organic Syntheses (N.Y.: Wiley, 1921——). An annual compilation, cumulated every ten years.

Weissberger, A. (ed.), *The Chemistry of Heterocyclic Compounds* (N.Y.: Interscience, 1950——). To be completed in about thirty volumes.

Books of More Limited Scope

Cheronis, N. D., *Micro and Semimicro Methods,* Vol. VI of Weissberger (ed.), *Technique of Organic Chemistry* (2nd ed., N.Y.: Interscience, 1957).

Dreisbach, R. R., *Physical Properties of Organic Compounds:* Advances in Chemistry Series, Vols. 15 and 22 (Washington, D. C.: American Chem. Soc. 1955 and 1957).

Hodgman, C. D. (ed.), *Handbook of Chemistry and Physics* (43rd ed., Cleveland: Chemical Rubber Co., 1961).

Jordan, T., *Vapor Pressure of Organic Compounds* (N.Y.: Interscience, 1953).

Kempf, R., and F. Kutter, *Schmelzpunkt, Tabellen zur Organischen Molekular-Analyse* (Ann Arbor, Mich.: Edwards Bros., 1944).

Lange, N. A. (ed.), *Handbook of Chemistry* (10th ed., N.Y.: McGraw-Hill, 1961).

Ralston. A. W., *Fatty Acids and Their Derivatives* (N.Y.: Wiley, 1948).

Suter, C. M., *The Organic Chemistry of Sulfur* (N.Y.: Wiley, 1944).

Timmermans, J., *Physico-Chemical Constants of Pure Organic Compounds* (Princeton: Elsevier, 1950).

Wagner, R. B., and H. D. Zook, *Synthetic Organic Chemistry* (N.Y.: Wiley, 1953).

Selected Books on Identification of Organic Compounds

Cheronis, N. D., and J. B. Entrikin, *Semimicro Qualitative Organic Analysis* (2nd ed., N.Y.: Interscience, 1957).

Feigl, F., *Spot Tests in Organic Analysis* (6th ed., Amsterdam: Elsevier, 1960).

Hodgman, C. D. (ed.), *Tables for Identification of Organic Compounds* (Cleveland: Chemical Rubber Co., 1960).

Huntress, H., *The Preparation, Properties, Chemical Behavior, and Identification of Organic Chlorine Compounds* (N.Y.: Wiley, 1948).

McElvain, S. M., *Characterization of Organic Compounds* (rev. ed., N.Y.: Macmillan, 1953).

Mulliken, S., *The Identification of Pure Organic Compounds* (N.Y.: Wiley, 1904–1922), 4 vols. This classic work is still useful. Two of the original volumes have been revised by E. H. Huntress: Volume I, compounds containing only carbon, hydrogen, and oxygen (N.Y.: Wiley, 1941); and the volume on chlorine compounds (see Huntress above).

Shriner, R. L., R. C. Fuson, and D. Y. Curtin, *The Systematic Identification of Organic Compounds* (4th ed., N.Y.: Wiley, 1956).

Smith, F. J., and E. Jones, *A Scheme of Qualitative Organic Analysis* (London: Blackie and Son, 1948).

Staudinger, H., *Introduction à l'analyse organique qualitative* (Paris: Dunod, 1958).

Veibel, S., *The Identification of Organic Compounds* (1st English ed., 4th Danish ed., Copenhagen: G. E. C. Gad, 1954).

Vogel, A. I., *Elementary Practical Organic Chemistry*, Part II, Qualitative Organic Analysis (N.Y.: Longmans, Green, 1957).

10

Preparation of Derivatives: I

Introduction

The preparation of derivatives is the last step in the rigorous and systematic identification of an unknown organic substance. The identification is regarded as rigorous and conclusive if (*a*) the physical constants, classification, and functional group tests fit the properties of the compound selected from the literature search; (*b*) the derivatives prepared from the unknown melt within 1–2° of the melting points given in the literature for the same derivatives of the compound selected as identical with the unknown; and (*c*) a mixture of the derivative prepared from the known and unknown shows no variation in their melting points.

Selection of Derivatives

A suitable derivative must fulfill certain requirements. For example, benzaldehyde may be oxidized to benzoic acid and reduced to benzyl alcohol and converted into phenylhydrazone, semicarbazone, 2,4-dinitrophenylhydrazone, and methone derivatives. However, not all of these are suitable derivatives. For example, the oxidation of 100–200 mg of benzaldehyde to benzoic acid will not yield a sufficient amount of the acid for melting-point determination. Similarly, reduction to benzyl alcohol is not advisable since the derivative is a liquid. The melting points of the remaining four derivatives are: phenylhydrazone, 158°; semicarbazone, 222° (also 233–5° when heated rapidly); 2,4-dinitrophenylhydrazone, 237°; and methone derivative, 193°. Since the determination of melting temperatures above 200° is subject to errors, the most suitable derivatives for benzaldehyde are the phenylhydrazone and the product with methone.

The requirements which suitable derivatives should fulfill may be summarized as follows:

1. The derivative should be a solid melting, if possible, above 50 and below 250°. If the derivative is an oil, it cannot be purified in small quantities; as a rule organic crystalline compounds melting below 50° do not crystallize well and in such cases they have a tendency to separate as oils. A derivative melting between 100–200° is preferable (other factors being equal) to a derivative melting much above 200°; the determination of melting points much above 200° is more difficult and requires considerable care in ascertaining the thermometer correction to be applied.

2. The derivative should have a melting point that is quite different from that of the original compound from which it is prepared; further, the melting point of the derivative should differ by more than 5° from the derivatives of closely related compounds. For example, let us assume that a sample of an unknown liquid, on the basis of tests, is tentatively identified as propionic acid. In the selection of the derivatives to be prepared, reference is made to Table 5A (page 355), and it is found that the anilide of propionic acid melts at 106°, whereas the anilide of isobutyric acid, a closely related acid and one of the possibilities, melts at 105°. On the other hand, the p-toluidide of propionic acid melts at 126°, whereas that of isobutyric melts at 107°; therefore the p-toluidide in this particular case is a much more suitable derivative than the anilide.

3. The reaction by which the derivative is made should be complete within 30 minutes, should be subject to few, if any, side reactions, and should afford a good yield.

4. The reagents used in the preparation of a derivative should be readily available.

5. The derivative should be readily purified; it should be slightly soluble in some common solvent in the cold and somewhat soluble at the boiling point of the solvent. More specifically, the ratio of the solubility at room temperature to that at the boiling point of the solvent should be more than 1:5.

Recommended Derivatives

A summary table of *most suitable derivatives* that have appeared in the literature is given under each class of compounds discussed in this chapter. In many cases this number is very large. Hence, a few (usually 3 to 4) have been selected and *are designated by an asterisk*. For these recommended derivatives the general preparative procedures have been checked and are described for semimicro quantities (usually 100–200 mg of the compound). In addition to the general method, an example of a specific preparation of a derivative is given, in order to illustrate the application of the general method to a specific case.

The student *should be warned* not to follow blindly "cook-book procedures" in the preparation and purification of a derivative. For example, let us assume that the unknown is tentatively identified as benzyl alcohol and it is desired to prepare the benzyl 3,5-dinitrobenzoate. The general method for the preparation should first be consulted (page 250), then the preparation of ethyl 3,5-dinitrobenzoate (page 251), and finally an outline for the preparation of the benzyl derivative should be made. For example, the solubility of the benzyl derivative should be expected to be less than that of the ethyl analog, and for this reason the volume of methanol or ethanol used to effect solution in the purification by crystallization must be greater than indicated in the procedure described for ethyl 3,5-dinitrobenzoate. In general, for the recrystallization of any derivative, do not follow blindly the volumes of the solvent indicated in the examples, but use the minimum amount of hot solvent to dissolve the sample.

Other Derivatives

If none of the recommended derivatives is found suitable, the longer treatise of the authors as well as other literature should be consulted for other derivatives. Most methods described in original papers on the preparation of derivatives for identification work are based on macro quantities, usually 1–5 g of the compound. The amount of the derivative yielded by macro quantities is so great that even a careless worker can perform two or three crystallizations and still have 100 mg of the derivative left. In the transition from macro to semimicro quantities, the amounts of reagents are reduced tenfold or more; consequently conditions must be chosen that will ensure: (*a*) completion of the reaction; (*b*) a minimum number of crystallizations to obtain a pure product; (*c*) exact quantities of solvent and conditions of crystallization; and (*d*) a sufficient yield of pure product for several melting-point determinations.[1]

**Precautions for Beginners in the Preparation
of Small Quantities of Derivatives**

The following precautions are recommended for beginners when they undertake to prepare 50–200 mg of a derivative:

1. The procedure should involve as few transfer operations as possible, and yield a product that requires the minimum number of purification steps.

2. The reaction vessel should be as small as possible, since the larger the vessel the greater is the loss from adsorption on the walls of the vessel.

[1] Check with the instructor whether the derivatizing reagents are available before planning the preparation of other derivatives.

3. The method for the isolation of the crude product and its subsequent purification should be carefully selected.

4. The purity of the derivatizing agent should be considered.

5. When there is doubt about the procedure a "trial run" should be made, using a sample of *the pure compound tentatively identified as the unknown*. This will indicate changes that should be made in the procedure and will also give the analyst practice that will be helpful in dealing with the preparation of the same derivative from the unknown.

ACETALS

The preparation of derivatives of acetals is based on their hydrolysis to aldehydes and alcohols, according to Eq. (10.1). The alcohol is then identi-

$$\text{RCH(OR')}_2 + \text{H}_2\text{O} \rightarrow 2\text{R'OH} + \text{RCHO} \tag{10.1}$$

fied by the preparation of the 3,5-dinitrobenzoate or other suitable derivative. For the aldehyde, either the semicarbazone or the 2,4-dinitrophenylhydrazone is prepared. A more detailed discussion is given in the following sections.

Caution: Where appreciable amounts of acetals are involved, the student should bear in mind that these compounds form peroxides like the ethers and therefore the precautions outlined on page 294 should be followed.

Hydrolysis of Acetals

The hydrolysis of acetals is best accomplished by heating with dilute acids; the lower acetals are usually hydrolyzed by boiling with 1–2N hydrochloric or sulfuric acid; for the hydrolysis of the higher acetals an organic solvent miscible with water and heating for 20–30 minutes or longer is necessary. One method is to use a 50 per cent solution of dioxane and water and, after the hydrolysis is complete, to neutralize the hydrolyzate and divide in half; one portion is used for the characterization of the aldehyde and the second portion for the derivatization of the alcohol. For semimicro quantities it is often more convenient to hydrolyze 100–200 mg quantities of the acetal separately for each characterization. For example, for the characterization of benzaldehyde dimethyl acetal 200 mg of acetal is required for the carbonyl component and 250 mg for the alcoholic part. In general, however, the hydrolysis of the acetal and subsequent preparation of the derivatives for the alcohol and the aldehyde depend on the nature of the compound and the hydrolysis products. Therefore the discussions of preparation of derivatives of alcohols (pages 247–57) and aldehydes (pages 257–67) should be consulted.

References

Bibliographic references on the derivatization of alcohols appear on pages 255–57 and on the derivatization of aldehydes on pages 265–67. However, the general methods for the derivatization of alcohols and aldehydes should be reviewed before any literature search is made.

ACID ANHYDRIDES AND ACID HALIDES

If the unknown under investigation has been tentatively identified as an acid anhydride or an acid chloride, the general directions given for the preparation of solid substituted amides (anilides or p-toluidides) from carboxylic acids should be consulted. For example, the acetylation of amino acids by acetic anhydride (page 232) serves, with few modifications, as a general procedure for the preparation of substituted amides from anhydrides. Similarly, the procedure described on pages 238–42 for the preparation of anilides and p-toluidides from carboxylic acids by conversion first to the acid chlorides serves as a general method derivatization. About 100–200 mg or less of the acid halide is dissolved in dry benzene and 500 mg of aniline or p-toluidine is added; the mixture is refluxed for 10 minutes, then treated as described in Example 10.5.

If the hydrolysis of the acid anhydride or halide gives an acid that is solid and slightly soluble in water (as, for example, phthalic anhydride), it may be used for identification. In such cases 100–200 mg of acid chloride is boiled with 5 ml of 5 per cent sodium carbonate solution for 20 minutes. If the acyl halide is reactive, the refluxing may be shortened to a few minutes. The solution is cooled, extracted with 5 ml of ether, and the aqueous layer is separated and acidified with dilute sulfuric acid to liberate the carboxylic acid. Other derivatives that may be considered are solid esters, which are formed by reaction of the acid halide or anhydride with phenols (see page 328).

References

See under Carboxylic Acids.

ACIDS

Amino Acids

The α-aminocarboxylic acids are nonvolatile compounds and generally melt above 200° with extensive decomposition; this property is assumed to be due to the dipolar structure of the solid state. Since the melting of the crystalline solid is accompanied by decomposition, there is no fixed temperature at which the solid and liquid phase coexist, but there is *a temperature range at which decomposition takes place;* further, this decomposition range depends on the *rate of heating* (see pages 47–49). Thus exists a considerable confusion in the literature as to the so-called "melting points" of amino acids. For example, D-glutamic acid has been reported to melt with decomposition at various temperatures from 198° to 225°, and L-tyrosine decomposes according to Fischer with rapid heating at 314–8° (corr.) and with slow heating at 290–5°; the same compound has been reported to decompose at 344°. In most cases if the compound is quickly heated it decomposes at a higher temperature than when heated slowly; the difference may be as much as 40–50°. This brief discussion shows that since the decomposition points are very unreliable, they cannot be used except to indicate a range of possibilities. Therefore it is advisable, in the determination of the melting point of a substance suspected to be an α-aminocarboxylic acid, to make determinations both with rapid and slow heating of the bath.

The reactions used to prepare solid derivatives of the amino acids are those that characterize the amino group. The most important derivatives are listed in Table 10.1 and discussed below in three sections: (*a*) *N*-acyl

TABLE 10.1

Derivatives for the Identification of Amino Acids

*2,4-Dinitrophenyl derivatives	Phthaloyl derivatives
*3,5-Dinitrobenzamides	*p*-Nitrobenzyloxycarbonyl derivatives
p-Toluenesulfonamides	Picrates and flavianates
Phenylureas and phenylhydantoins	Naphthalene-β-sulfonates
Picronolates	Phosphotungstates
β-Naphthalenesulfonamides	Phosphomolybdates
Benzamides	Copper salts
Acetamides and formamides	Dibenzofuran-2-sulfonates
Carbobenzoxy derivatives	

* Recommended derivative.

and *N*-aroyl derivatives; (*b*) *N*-ureido derivatives; (*c*) salts of complex acids. The asterisked recommended derivatives are the 3,5-dinitrobenzoyl, *p*-toluenesulfonyl, and 2,4-dinitrophenyl derivatives.

The chromatographic identification as described on page 233 should be considered in place of derivatization, with permission from the instructor.

N-Acyl and N-Aroyl Derivatives of Amino Acids

Discussion

The reaction of the amino acid with acyl and aroyl chlorides yields the substituted amides, according to Eq. (10.2). The following chlorides have

$$CH_3 \cdot C_6H_4SO_2Cl + H_2NCH_2COOH \rightarrow CH_3C_6H_4SO_2NHCH_2COOH + HCl \quad (10.2)$$

p-Toluenesulfonyl   Glycine   N-p-Toluenesulfonyl
   chloride   glycine

been used for derivatization: acetyl, formyl, benzoyl, 3,5-dinitrobenzoyl, benzenesulfonyl, p-toluenesulfonyl, β-naphthalenesulfonyl, and 4-nitro-toluene-2-sulfonyl. The usual procedure is to shake an alkaline solution of the amino acid with an equivalent amount of the acid chloride. The reaction in most cases takes place very slowly, often requiring 3–4 hours of mechanical shaking. The slow rate is probably due to the inability of the amino group to react in the charged (dipolar) form; the addition of alkali increases the concentration of the form having a free amino group, as in Eqs. (10.3) and (10.4). As a first trial the use of 3,5-dinitrobenzoyl chloride is

$$\overset{+}{N}H_3CH_2CO\overset{-}{O} + NaOH \rightarrow NH_2CH_2CO\overset{-}{O} + \overset{+}{Na} + HOH \quad (10.3)$$

$$H_2NCH_2CO\overset{-}{O} + RCOCl \rightarrow RCONHCH_2CO\overset{-}{O} + HCl \quad (10.4)$$

recommended wherever possible, since the reaction requires only a few minutes of shaking.

The 2,4-dinitrophenyl derivatives are formed by the action of 2,4-dinitrochlorobenzene or 2,4-dinitrofluorobenzene and are preferred for chromatographic work. It should be noted that other groups (besides the amino), such as the hydroxyl, thiol, and imidazole, form similar derivatives. Thus tyrosine may form either the O- or the N- or the ON-di-2,4-dinitrophenyl derivatives.

General Method for the Preparation of Benzoyl and 3,5-Dinitrobenzoyl Derivatives

One millimole of the amino acid is dissolved in 2.5–3 ml of 1N sodium hydroxide solution in a 6-inch tube. One millimole of benzoyl or 3,5-dinitrobenzoyl chloride is added in the case of a monoamino acid, and 2 millimoles in the case of a diamino acid. The tube is stoppered and shaken vigorously for 2 minutes and then at intervals for 15–30 minutes. The reaction mixture is acidified with dilute hydrochloric acid to pH 4–5, using Congo red or Universal indicator. The crystals are filtered and washed with 25 per cent methanol. The derivative can be crystallized by being dissolved in hot alcohol, and adding water cautiously until permanent cloudiness results.

The diamino acids usually react with 2 moles of the chloride forming bis(benzoyl or 3,5-dinitrobenzoyl) derivatives. In general, the dicarboxylic acids react at a much slower rate than the monoamino monocarboxylic acids. A few amino acids, for example, tyrosine, form benzoyl derivatives, but not 3,5-dinitrobenzoates.

Note: If excess of aroyl halide is used or if the reaction is incomplete on acidification, free benzoic or 3,5-dinitrobenzoic acid will precipitate and contaminate the derivative. Therefore, if the melting point of the derivative after one crystallization is several degrees below the expected value, the derivative should be treated with sodium carbonate solution as described in Example 10.9. In some cases better results are obtained if 3,5-dinitrobenzoyl chloride is dissolved in benzene before addition to the amino acid solution.

General Method for the Preparation of p-Toluenesulfonyl Derivatives

One millimole of the amino acid, 3 ml of 1*N* sodium hydroxide solution, and 250 mg of *p*-toluenesulfonyl chloride dissolved in 2 ml of ether are placed in a small bottle with a well-ground glass stopper, the top of which is flat so that it may be fastened with wire. The stopper is slightly greased, fitted in the bottle, and wired across the top to make a perfectly tight joint. The bottle is shaken manually for a few seconds every 5–10 minutes over a period of 5 hours, or mechanically for 2–3 hours. The ethereal layer is separated, and the aqueous layer is acidified to pH 4–5, using Congo red or Universal indicator; the solution is cooled for 1 hour. The derivative is separated and crystallized from an alcohol-water mixture.

Note: The original article by McChesney and Swann (see References) should be consulted for the preparation of the derivatives of the dicarboxylic acids, tyrosine, and alanine. The following amino acids yield oils that do not crystallize: glutamic, aspartic, arginine, lysine, trytophane, and proline.

General Method for the Preparation of Acetyl Derivatives

The addition of the anhydride and alkali in two portions and control of the pH and temperature as described below are necessary to prevent racemization. Two millimoles of the amino acid (about 200 mg) is suspended in 1 ml of water in a 6-inch microtube provided with a solid rubber stopper. The tube is cooled to about 10° and 5.0 millimoles of acetic anhydride is added in two portions over a period of 5 minutes, together with 2.0 ml of 6*N* sodium hydroxide solution, also added in two portions immediately after the anhydride. The tube is shaken vigorously after each addition; the temperature is kept between 10° and 15°, and the pH between 8.0 and 10.0; a drop of Universal indicator is added to the solution before addition of the reagents. The total amount of added alkali should be 12 millimoles. The mixture is shaken at frequent intervals during the introduction of the reagents and for a few minutes longer. It is then acidified to pH 3–4 with 6*N* hydrochloric acid and placed in a cold bath for 4–5 hours or in a refrigerator overnight. The tube is centrifuged and the supernatant liquid is transferred to another tube; the crystals are washed twice with 0.4 ml of ice water and the washings are combined with the supernate. The crystals are dried at 100°, while the combined supernates are evaporated to dryness by passing warm air over the solution. This residue is extracted with 1–2 ml of hot water, and upon cooling an additional amount of the acetyl derivative is obtained. The combined yield is about 1.5–1.8 millimoles.

An alternative method employs the sodium salt of the amino acid and 5–6 millimoles of alkali.

General Method for the Preparation of 2,4-Dinitrophenyl Derivatives

The reagent used for the preparation of the 2,4-dinitrophenyl derivatives is 2,4-dinitrofluorobenzene, which reacts at room temperature. 2,4-Dinitrochlorobenzene may be used in some cases if the reaction mixture can be heated to boiling.

The use of 2,4-dinitrofluorobenzene is illustrated by the preparation of the derivative from phenylalanine. A solution of 200 mg of phenylalanine is prepared in 5 ml of water and 400 mg of sodium bicarbonate in a small flask. To this is added a solution of 400 mg (0.28 ml) of 2,4-dinitrofluorobenzene in 10 ml of ethanol. The flask is stoppered and shaken at room temperature for 2 hours. The ethanol is removed by concentration under reduced pressure and the aqueous solution is extracted with ether to remove the excess reagent. The mixture is then acidified, and the oil that separates out soon solidifies. The derivative is crystallized twice from a methanol–water mixture. The yield is 270 mg of crystals melting at 186°

Note: When 2,4-dinitrochlorobenzene is used, the mixture is refluxed for 4 hours, then treated as described above. When the oily derivatives do not crystallize on standing, they are extracted with CHCl₃, dried with anhydrous Na₂SO₄, and the solvent removed.

Chromatographic Detection of Amino Acids

For laboratory practice to acquire experience in the chromatographic techniques, the separation of two amino acids as described in this section is recommended. As experience is acquired, the chromatographic separation and detection of a more complex synthetic amino acid mixture (unknown) may be undertaken. Other experiments in the detection of small quantities of aldehydes and ketones are described on page 264.

Example 10.1: Separation of Microgram Quantities of Two Amino Acids

1. Place in a test tube 0.5 ml each of glycine and tyrosine solutions containing about 1–2 mg per milliliter.

2. Prepare several pipets with long capillary tipes of varying diameters

(Appendix, Fig. A.1). Practice the application of sample by applying droplets on dots made on filter paper until the spot produced by a single droplet is 1–2 mm in diameter. Select the pipet that gives the best results.

3. Prepare 4–5 strips of paper for 6-inch tubes as described on pages 165–66. Do not handle the strips with the fingers but hold them lightly by the edges. Improper handling causes spots to develop. Select three strips on which to apply droplets of the amino acid mixture.

4. Place 1 droplet on the first strip, 2 droplets on the second, and 3 droplets on the third, being very careful to dry each spot before superimposing the next droplet.

5. Charge three dry 6-inch Pyrex tubes with 0.5 ml of a mixture of *tert*-butanol (70 per cent) and water (30 per cent) and place the strips in the tubes as directed on page 166. Develop for an hour or until the solvent has ascended ¾ of the way up.

6. Take out the strips and mark the solvent front with a pencil and hang the strips on clips (Figure 7.4) to dry in air.

7. Spray or dip the strips with a solution of 0.2 per cent ninhydrin in ethanol, using an atomizer if spraying is used. Hang the strips to dry.

8. When the spots are clearly visible, remove the strips and examine them. The distance from the dot at the lower end of the strip (where the sample was placed) to the pencil line, indicating the solvent front, is measured. The distance from the original dot to the center of the spot is also measured in this manner, the R_f value is thereby calculated (see pages 158–59). The difference between the R_f values of the three strips should not be larger than 0.01. Generally the best strip is the one with the most well-defined spot. There should be a minimum of diffusion outward (absence of "tails" and "beards"). Examination of the three strips made with 1, 2, and 3 droplets will reveal the best method of application to be used in future trials.

9. Three other solvent systems useful for the separation of amino acids are: (a) 1-butanol, 6 ml, glacial acetic acid, 1.5 ml, and water, 2.5 ml; (b) phenol, 8 g, water, 2 ml; (c) 1-butanol, 6 ml, pyridine, 6 ml, and water, 6 ml. Solvent mixture (a) gives very compact spots, and for a large number of compounds is more effective than *tert*-butanol–water.

Example 10.2: Separation and Detection of Two Amino Acids Containing Sulfur

1. Place in a test tube 0.5 ml each of cystine hydrochloride and methionine hydrochloride solutions containing 1 mg per milliliter.

2. Prepare 4 strips of paper for 6-inch tubes. Pass each paper spot twice over the mouth of a bottle of ammonia to convert the hydrochloride to the free amino acid. Develop as described in the above example, using the

same type of solvent mixture (either *tert*-butanol–water or 1-butanol–acetic acid–water).

3. Spray two of the strips (after evaporation of the solvent) with ninhydrin solution and the other two with a solution of 0.1 per cent iodoplatinate reagent prepared by mixing equimolar quantities of chloroplatinic acid and potassium iodide.

References

N-Acyl and N-Aroyl Derivatives

Acetyl and formyl derivatives: Bergman and Zervas, *Biochem. Z.* **203**, 288 (1928); Chattaway, *J. Chem. Soc.*, 2405 (1931); Curtius, *Ber.* **17**, 1665 (1884); Fischer, *Ber.* **39**, 2330 (1906); Herbst and Shemin, *Organic Syntheses,* Vol. 19 (N.Y.: Wiley, 1939), p. 4; Kraut and Hartmann, *Ann.* **133**, 105 (1865); Reimschneider and Wiegand, *Monatsch* **86**, 201 (1955); Rochwen, *J. Org. Chem.* **18**, 53 (1953).

Benzoyl derivatives: Fischer, *Ber.* **32**, 2451 (1899); Fischer and Bergell, *Ber.* **35**, 3779, 3784 (1902); **39**, 597 (1906).

Carbobenzoxy derivatives: Bergman, Zervas, and Ross, *J. Biol. Chem.* **111**, 245 (1945); Newberger and Sanger, *Biochem. J.* **37**, 515 (1943).

3,5-Dinitrobenzoyl derivatives: Saunders, *Biochem. J.* **28**, 580 (1934); *J. Chem. Soc.*, 1397 (1938); Saunders *et al., Biochem. J.* **36**, 368 (1942); Town, *Biochem. J.* **35**, 578 (1941).

2,4-Dinitrophenyl derivatives: Sanger, *Biochem. J.* **39**, 507 (1945); Levy, *Nature* **174**, 126 (1954).

β-Naphthalenesulfonyl derivatives: Bergman and Stein, *J. Biol. Chem.* **129**, 609 (1939); Fischer and Bergell, *Ber.* **35**, 3779, 3784 (1902); **39**, 597 (1906).

p-Nitrobenzyloxycarbonyl derivatives: Gish and Carpenter, *J. Am. Chem. Soc.* **75**, 950 (1953).

Phthaloyl derivatives: Billman and Hartig, *J. Am. Chem. Soc.* **70**, 1473 (1948).

p-Toluenesulfonyl derivatives: Fischer and Bergell, *Ber.* **35**, 3779, 3784 (1902); **39**, 597 (1906); McChesney and Swann, *J. Am. Chem. Soc.* **59**, 1116 (1937).

Dibenzofuran-2-sulfonates: Dunbar and Ferrin, *Microchem. J.* **4**, 167 (1960).

Chromatographic Detection

Block, R. J., E. L. Durrum, and G. Zweig, *A Manual of Paper Chromatography and Paper Electrophoresis* (2nd ed., N.Y.: Academic Press, 1955).

Block, R. J., R. LeStrange, and G. Zweig, *Paper Chromatography* (N.Y.: Academic Press, 1952).

Boissonnas, *Helv. Chim. Acta* **33**, 1966 (1950).

Cassidy, H. G., *Adsorption and Chromatography,* Vol. V of Weissberger (ed.), *Technique of Organic Chemistry* (N.Y.: Interscience, 1951).

Consden *et al., Biochem. J.* **38**, 224 (1944); **40**, 590 (1947); **46**, 8 (1950).

Cifonelli and Smith, *Anal. Chem.* **27**, 1501 (1955). Detection of amino acids on paper chromatograms.

Decker and Riffart, *Chem-Ztg.* **74**, 261 (1950). R_f values.

Dent, *Biochem. J.* **43**, 169 (1948).

Lacourt *et al., Anal. Chim. Acta* **14**, 100 (1956).

Lederer, M., and E. Lederer, *Chromatographic Separation of Amino Acids* (Houston and N.Y.: Elsevier, 1953).

Levy, *Nature* **174**, 126 (1954).

Martin and Synge, *Biochem. J.* **35**, 1358 (1941).

Peck and Gale, *Anal. Chem.* **24**, 118 (1952).

Rockland *et al., Anal. Chem.* **23**, 1142 (1951); *Science* **109**, 539 (1949).

Subramanian *et al., J. Sci. Ind. Research (India)* **14C**, 56 (1955).

Underwood and Rockland, *Anal. Chem.* **26**, 1553 (1954).

Mills, L., in I. Smith (ed.), *Chromatographic and Electrophoretic Techniques* (N.Y., Interscience, 1960), pp. 143–65.

Carboxylic Acids

Table 10.2 lists a large number of derivatives that have been described in the literature for the derivatization of the carboxylic functions. The references at the end of this chapter further indicate the tremendous interest in the characterization of carboxylic acids.

Generally, the derivatives listed in Table 10.2 fall into four classes: (a) amides and substituted amides, such as anilides, *p*-bromoanilides, *p*-toluidides, and the like, which are formed by reaction of the carboxyl group with ammonia or amines; (b) crystalline esters, such as the *p*-nitrobenzyl, phenacyl, *p*-bromophenacyl, and the like, formed by the reaction of the arylalkyl halides and the sodium salts of the acids; (c) salts of the carboxylic acids with such bases as phenylhydrazine, 2,4-dinitrophenylhydrazine, benzylamine, piperazine, and the like; (d) a variety of derivatives formed by reaction of the carboxyl group with various functions.

The discussion that follows deals only with the asterisked derivatives listed in Table 10.2. In general, if the carboxylic compound is in the anhydrous condition or can be readily obtained in this form (for example, by evaporation of the neutralized aqueous solution of the acid), the preparation of the *p*-toluidide, or anilide is recommended. If the carboxylic compound is in aqueous solution or if the water cannot be readily removed, derivatization by means of one of the sparingly soluble esters (*p*-nitrobenzyl and the like) is recommended.

Anilides, *p*-Toluidides, and Other Substituted Amides of Carboxylic Acids

Discussion

The formation of amides, anilides, *p*-bromoanilides, *p*-toluidides, and naphthylamides is illustrated in Eqs. (10.5) and (10.6).

$$CH_3(CH_2)_2COOH \xrightarrow{SOCl_2} CH_3(CH_2)_2COCl + 2C_6H_5NH_2 \longrightarrow$$

$$\underset{\text{Aniline}}{CH_3(CH_2)_2CONHC_6H_5} + C_6H_5NH_3{}^+HCl^- \qquad (10.5)$$

$$\underset{\substack{\text{\textit{n}-Butyranilide}\\(\textit{N}\text{-Phenyl-\textit{n}-butyramide})}}{}$$

or

$$\underset{\text{\textit{p}-Toluidine}}{CH_3CH_2COOH + CH_3C_6H_4NH_2} \xrightarrow{180°} \underset{\substack{\text{Propiono-\textit{p}-toluidide}\\(\textit{N}\text{-\textit{p}-Tolylpropionamide})}}{CH_3CH_2CONHC_6H_4CH_3} + H_2O \qquad (10.6)$$

TABLE 10.2

Derivatives for the Identification of Carboxylic Acids

Amides	Phenacyl derivatives
*Anilides	p-Chlorophenacyl derivatives
*p-Bromoanilides	p-Bromophenacyl esters
*p-Toluidides	p-Phenylphenacyl esters
*α-Naphthylamides	*p-Nitrobenzyl esters
2,4-Dinitrophenylhydrazides	Diazomethane derivatives
S-Benzylisothiouronium salts	Menthyl esters
α-Bromo-β-naphthylamidides	4-Phenylazophenacyl esters
o-Bromo-p-toluidides	Acid esters of dibasic acids
N-Acyl-p-amidoazobenzols	Lactone derivatives
p-Acylaminobenzoic acids	Bis-(p-dimethylaminophenylureides)
3-Acylaminodibenzylfuranes	Monoureides
N-Acylanthranilic acids	Monothioureides
N-Acylcarbazoles	p-Chlorobenzylisothiouronium salts
N-Acyl-2-nitro-p-toluidides	p-Bromobenzylisothiouronium salts
N-Acylphenothiazines	Carbodiimide derivatives
N-Acylsaccharins	Benzylammonium salts
Diethanolamides	α-Phenylethylammonium salts
Dimethylamides	S-(p-Nitro)benzylthiouronium salts
Diphenylamides	S-(α-Naphthyl)methylthiouronium salts
Dodecylamides	Phenylhydrazonium salts
N-Ethanolamides	Piperazonium salts
N-Methylamides	Octadecyl ammonium salts
N-Isopropanolamides	Dodecyl ammonium salts
p-Hydroxyanilides	Phenylmercuric salts
2-Methyl-5-isopropyl anilides	p-Tolylmercuric salts
o-, m-, and p-Nitroanilides	Triphenyl lead salts
Octadecylamides	Hydrazides
o- and p-Phenetides	Phenyl hydrazides
Thiophenylamides	N-Acyl-2-acylcarbazoles
o-Toluidides	2,8-Diacylcarbazoles
2,4,6-Tribromoanilides	2-Alkylbenzimidazole picrates
p-Xenylamides	2-Alkylbenzimidazoles
m-Xylides	Hydroxamic acids
N-Phenylimides of dibasic acids	Tetraphenyl stibonium salts

* Recommended derivative.

As shown in Eq. (10.5), the acid is first converted to the acid chloride by treatment with thionyl chloride and then treated with ammonia or the amine. The direct reaction of the carboxylic acid with the amine is possible in some cases, as shown in Eq. (10.6), and for the toluidides of the lower fatty acids the method is described in Example 10.3.

General Method for the Preparation of p-*Toluidides and Anilides*

Of the substituted amides the p-toluidides are recommended. They are easily prepared and have desirable crystallizing properties. If the carboxylic acid contains less than 8 carbon atoms, the p-toluidide may be conveniently prepared by heating the acid with an excess of p-toluidine at 180–200°. The reaction mixture is extracted with dilute hydrochloric acid to remove the excess of the base, and with dilute sodium hydroxide to remove traces of unreacted acid, and is then crystallized from an alcohol–water mixture. The toluidides of the higher fatty acids and of the aromatic carboxylic acids are prepared by first converting the acid to the acid chloride and then adding an excess of p-toluidine. The same method is used for the preparation of anilides and p-bromoanilides.

The acid chlorides are conveniently prepared by heating under reflux 1 millimole of the acid with 1.1 millimoles of pure thionyl chloride at about 70–78° for about 30 minutes. If the acid contains water, it is neutralized with sodium carbonate and evaporated to a small volume and then placed in the reaction tube and evaporated to dryness. In such cases the amount of thionyl chloride is slightly increased (1.3 millimoles/millimole of acid). If the acid is unsaturated, there is danger of the reaction between the hydrogen chloride evolved from the thionyl chloride and the unsaturated linkages. This tendency can be kept at a minimum if the acid is first dissolved in dry benzene and then the thionyl chloride is added. Another limitation of the use of thionyl chloride is with dicarboxylic acids in which the carboxyl groups are separated by two or three carbon atoms when anhydrides are formed; however, these will react readily with the amines to form the desired derivative, but the yield will be lowered.

After the formation of the acid chloride is completed, dry benzene and an excess of the amine are added and the mixture is refluxed. The solution is then extracted with dilute hydrochloric acid to remove the excess of amine; the benzene solution is evaporated; and the crude substituted amide (anilide, p-toluidide, or p-bromoanilide) is purified by crystallization.

Example 10.3: Direct Preparation of the p-*Toluidide of a Monocarboxylic Acid: Propiono-p-toluidide*

Place in an 8-inch test tube 0.3 ml of propionic acid and 1.2 g of p-toluidine. Immerse the tube in an oil bath and raise the temperature slowly over a period of 10 minutes to 190°. Keep the temperature at 190–200° for 30 minutes; then remove the tube and allow to cool. Add 7 ml of 5 per cent hydrochloric acid and wash down the inner sides of the tube, by means of a pipet dropper, with 1 ml of methanol. Heat nearly to boiling and set in cold running water to cool for 10–15 minutes, shaking the tube from time to time. Filter; wash the solid with a mixture of 2 ml of 1–2N hydrochloric acid and 3 ml of water, then with 5 ml of water, followed by 5 ml of 2 per cent sodium hydroxide solution, and finally twice with 3 ml of water. By means of the spatula remove the solid, together with filter paper, and place in an 8-inch test tube. Add about 3 ml of methanol and heat nearly to boiling until the crystals are completely dissolved.[2] Add as much charcoal as can be placed on one half of the spatula blade and heat for a few seconds; then set aside. Place the perforated porcelain disc inside the funnel; set the filter paper on top of the porcelain disc and wet it with 3–4 drops of water. Fit the funnel into an 8-inch tube with side arm and start light suction. Inspect the filter paper and see that it fits tightly on the sides of the funnel. Add to the filter funnel 4–5 drops of methanol and then drain the receiving tube. Heat the solution of the derivative to boiling and pour slowly into the filter. The filtrate should be clear. If a minute amount of charcoal has passed through the filter paper, it may be overlooked, since the compound is to be recrystallized. About 1 ml of methanol is poured dropwise down the sides of the tube that was used to prepare the solution and, after rotating and heating for a few seconds, the washings are added into the filter. The suction is discontinued and 1.5–2 ml of water is added by means of the pipet dropper to the filtrate. The tube is heated until the cloudiness disappears and is then cooled for 15–20 minutes. The crystals are filtered and washed three times with 3–4 ml of water. The yield of dry crystals is 75–100 mg; the melting point is 122–123°. In practice, the crystals are recrystallized immediately after filtration from 3 ml of methanol and 1.5 ml of water, using the procedure described. The yield of pure propiono-p-toluidide is 50–70 mg, melting at 124–125°.

Note: The solubility of p-toluidides decreases with an increase of the molecular weight of the acid. Thus, by using 0.3 ml of n-caproic acid with the procedure described, and the same number of crystallizations, about 110–130 mg of n-capro-p-toluidide is obtained, melting at 74–75°. However, as the molecular weight increases, there is a tendency for the p-toluidides to separate as oils. The p-toluidide of n-caprylic acid illustrates this tendency. On heating with dilute hydrochloric acid the reaction mixture of 0.3 ml of acid and 1.2 g of p-toluidine, an oil separates that does not crystallize on cooling. If, during cooling, the tube is shaken frequently and the inner sides of the tube are scratched with a glass rod, the oil crystallizes into a dark mass. After washing as described, it is dissolved in 5 ml of hot ethanol, treated with charcoal, and filtered; the filtrate is then diluted with 2 ml of water. The crystals that separate on cooling are crystallized again from 5 ml of ethanol and 3 ml of water. The yield is about 200–220 mg of n-caprylo-p-toluidide, melting at 70°.

The preparation of the derivatives of fatty acids having 5–8 carbon atoms may be accomplished by using 0.1–0.2 ml of the sample. The dry sodium salt of the carboxylic

---

[2] A reminder to the beginner of the caution given on page 227. "Do not follow blindly the volumes of the solvent stated in the illustrative examples, but use the minimum amount of hot solvent to dissolve the sample."

acid may be used instead of the free acid. In such cases equal amounts of sodium salt and concentrated hydrochloric acid are placed in the tube. The toluidine is added and the mixture is heated gradually until a temperature of 140° is reached; it is kept at this temperature for 15 minutes. The temperature is then raised to 190–200° and maintained for 30 minutes. This method is effective when aqueous solutions of the lower carboxylic acids are involved. Since it is difficult to extract small amounts, it is more appropriate to neutralize the solution carefully with sodium carbonate and evaporate the salt solution to dryness, using this salt mixture for the preparation of the derivative.

Example 10.4: Direct Preparation of the Toluidide of a Dicarboxylic Acid: Adipobis-p-toluidide

The procedure described in Example 10.3 is followed, using 200 mg of adipic acid and 1 g of p-toluidine. Add to the cooled reaction mixture 7 ml of 5 per cent hydrochloric acid and heat nearly to boiling. Cool slightly and pour the contents of the tube into a mortar. Grind the mass of crystals until no lumps remain. Replace the mixture in the 8-inch tube and heat. Filter and wash as described in Example 10.3. Dissolve in sufficient amount of hot ethanol (10–15 ml), and, after treating with charcoal, filter. Add 1 ml of water and cool. The solid that separates is crystallized, after filtration, from 12–15 ml of ethanol. About 150 mg of the pure crystals, melting at 239°, is obtained.

Note: Dicarboxylic acids may react with only one mole of the base to give the mono-p-toluidide. This tendency is particularly manifested when the dicarboxylic acid forms an anhydride upon heating. For example, phthalic and maleic acids give the mono-p-toluidide when they are heated with p-toluidine. The literature value for the melting point of this bis-p-toluidide is 241°; however, the authors have not been able to obtain values above 239° (see discussion in Chapter 3 on the melting-point data).

Example 10.5: Preparation of a p-Toluidide through the Preparation of the Acid Chloride: Benzo-p-toluidide

Place in a 25-ml tube or flask, fitted with a microcondenser, 180 mg of benzoic acid. Lift the cork holding the microcondenser and add, by means of the pipet dropper, 1.2 ml of thionyl chloride. Heat in a water bath at 75–80° for 30 minutes. Lift the cork and add a solution of 500 mg of p-toluidine in 25 ml of dry benzene. (**Caution:** *Flames should not be in the vicinity.*)

Replace the microcondenser and reflux gently for 15 minutes. Cool, add 5 ml of water, and transfer the mixture into a small separatory funnel or into an 8-inch tube provided with separatory stopper. Wash the distilling tube with 1 ml of ethanol and add to the mixture in the separatory vessel. Shake the benzene–water mixture gently to ensure thorough mixing but to avoid formation of an emulsion. Remove the aqueous layer and wash the benzene solution successively with 5 ml of 1–2N hydrochloric acid, 5 ml of 5 per cent sodium hydroxide, and 5 ml of water. Allow sufficient time after shaking for complete separation of the two immiscible layers.

Pour the benzene solution into an evaporating dish and evaporate cautiously over a water bath. Wash the separatory vessel with 1–2 ml of ethanol and add the washings to the evaporating dish. Add to the residue 5 ml of ethanol and warm until it is completely dissolved. Add charcoal and filter by suction. Wash the dish, by means of the pipet dropper, with 1–2 ml of ethanol and pour the washings through the filter. Add 2–3 ml of water slowly to the filtrate and heat the tube until the cloudiness disappears. Cool and filter the crystals; wash the crystals twice with 3 ml of water, and dissolve directly in 5 ml of hot ethanol. Filter and add 2 ml of water. Cool and filter the toluide. About 180 mg of pure benzo-*p*-toluidide, melting at 157–158°, is obtained.

Note: This method illustrates the preparation of *p*-toluidides, anilides, *p*-bromo-anilides, and other substituted amides by reaction of the amine with the acid chloride. The amount of carboxylic acid can be reduced to 0.5 millimole or less (about 50–60 mg for most common acids) with proportionate amounts of other reagents with fairly good results.

Example 10.6: Preparation of an Anilide through the Preparation of the Acid Chloride: n-Butyranilide

Place in an 8-inch tube 150 mg of butyric acid and use the procedure as described in Example 10.5, except that a solution of 0.5 ml of aniline dissolved in 25 ml of dry benzene is substituted for *p*-toluidide. The residue, after evaporation of benzene, is dissolved in 4 ml of methanol and 2 ml of water. The crystals that separate after the first crystallization are dissolved in 5 ml of hot methanol and filtered; to the hot filtered solution 4 ml of water is added. The yield of pure *n*-butyranilide, melting at 94–95°, is about 70–75 mg.

Note: The solubility of the anilides of aryl carboxylic acids is much less than of those of the aliphatic; hence the same variation in the amount of alcohol and water for crystallization is observed as in the *p*-toluidides. For the preparation of *p*-bromoanilides, the same procedure is followed as for the preparation of benzo-*p*-toluidide (Example 10.5). The preparation of *p*-bromoanilides of the lower carboxylic acids is preferred over the anilides if the quantities of the carboxylic acid available are small.

As stated on page 238, the sodium salt of the carboxylic acid may be used instead of the free acid for the preparation of the substituted amides. In cases where it is required to prepare the acid chloride and then react it with the arylamine, the dry salt is treated with the thionyl chloride as described under the preparation of benzo-*p*-toluidide (Example 10.5), using the same quantities of salt instead of the acid. Dicarboxylic acids, such as succinic, glutaric, and maleic, which form anhydrides easily, are likely to give the anhydride rather than the acid chloride by reaction of the salt with thionyl chloride; in such cases the resulting anhydride may form the monosubstituted amide instead of the diamide. For example, maleic acid may yield under such conditions the mono-*p*-toluidide instead of the di-*p*-toluidide. On the other hand, aromatic acids that contain a negative substituent in the position para to the carboxyl group (*p*-chlorobenzoic, *p*-bromobenzoic, *p*-hydroxybenzoic, and the like), do not react easily with thionyl chloride. In such cases it is often possible to convert the carboxylic acid to the acid chloride by the use of phosphorus pentachloride.

General Methods for the Preparation of Other Substituted Amides

The preparation of other substituted amides, such as α-naphthylamides, β-naphthylamides, diphenylamides, o-bromo-p-toluidides, and the like, is similar to the procedures described for the p-toluidides and anilides. When the boiling point of the carboxylic acid and the amine is high, direct heating may be employed instead of converting the acid to the halide and then reacting it with the amine.

Example 10.7: Preparation of the β-Naphthylamide of Palmitic Acid: N-β-Napthylpalmitamide

Place in an 8-inch test tube, fitted with a microcondenser, 1 millimole (256 mg) of palmitic acid. Lift the cork holding the microcondenser and add, by means of a pipet dropper, 0.8 ml of thionyl chloride. Reflux in an oil or water bath at 85–90° for 30 minutes. While the test tube is still in the bath, the condenser is removed and the unreacted thionyl chloride is removed with the aid of a vacuum (or by increasing the temperature and driving off the excess thionyl chloride). Lift the cork and add a solution of 5 millimoles (572 mg) of β-naphthylamine dissolved in 25 ml dry benzene (**Caution**: *Benzene is flammable*). Replace the microcondenser and reflux gently for another 30 minutes. Cool, filter, and add to the solution 5 ml of water, and transfer the mixture to a small separatory funnel. Shake the benzene–water mixture gently to ensure thorough mixing but to avoid formation of an emulsion. Remove the aqueous layer (the separation may also be done with an ordinary capillary pipet from a test tube) and wash the benzene solution successively with 5 ml of 5 per cent hydrochloric acid, 5 ml of 5 per cent sodium hydroxide solution, and 5 ml of water. Allow sufficient time after shaking for the complete separation of the two immiscible layers. If crystallization does not occur, pour the benzene solution into an evaporating dish and cautiously evaporate over a water (steam) bath. Wash the separatory funnel with 1–2 ml of ethanol and add the washings to the evaporating dish. Add to the residue 5 ml of ethanol and warm until it is completely dissolved. Add charcoal and filter by suction. Wash the dish, by means of the pipet dropper, with 1–2 ml of ethanol and pour the washings through the filter. Add 2–3 ml of water slowly to the filtrate and heat the tube until cloudiness disappears. Cool and filter; wash the crystals twice with 3 ml of water, and dissolve directly in 5 ml of hot ethanol. Filter and add 2 ml of water. Cool and filter. The yield of crystals is about 80 per cent of the theoretical. The crystals melt at 109–110°.

Solid Esters of Carboxylic Acids

General Method

Eqs. (10.7) and (10.8) show the preparation of p-nitrobenzyl and phenacyl esters of carboxylic acids that may be used for characterization.

$$CH_3COONa + O_2N\langle\underline{\quad}\rangle CH_2Cl \rightarrow CH_3COOCH_2C_6H_4NO_2 + NaCl \qquad (10.7)$$

p-Nitrobenzyl chloride p-Nitrobenzyl acetate

$$CH_3COONa + \langle\underline{\quad}\rangle COCH_2Cl \rightarrow CH_3COOCH_2COC_6H_5 + NaCl \qquad (10.8)$$

Phenacyl chloride   Phenacyl acetate

A great amount of caution, however, should be exercised in the use of these reactions by the beginner. All of the phenacyl halides have lachrymatory properties and, further, cause blisters on contact with the skin. Handling of crystalline esters may cause an irritation between the fingers, probably due to the small amounts of halide adhering to the crystals of the esters. Nevertheless, the preparation of the p-nitrobenzyl and the phenacyl esters as derivatives is very valuable when the acids cannot be easily separated from aqueous solutions, as is often the case in the hydrolysis of the esters of lower carboxylic acids.

For the preparation of the p-nitrobenzyl and phenacyl esters, 1 millimole of the carboxylic acid is converted to the sodium salt by neutralization with dilute sodium hydroxide (5 per cent) or sodium carbonate (5 per cent) and is then heated for an hour or more with an alcoholic solution of 0.9 millimole of the halide in 5–8 ml of alcohol. It is essential to avoid the use of an excess of the halide, since it cannot be easily removed from the ester. The esters are recrystallized from alcohol; their separation from solution is rather slow and sufficient time should be allowed for the oils that often separate to change into crystalline state. The formation of p-nitrobenzyl and substituted phenacyl esters is slow and refluxing of semimicro quantities should be continued for 1.5–2 hours or more to ensure complete reaction. For example, by using p-nitrobenzyl chloride and quantities as outlined above and refluxing for one hour, the following results were obtained (in each case the temperature listed after the name of the acid is the melting point of the p-nitrobenzyl ester after one crystallization; the temperature in the parentheses is the melting point listed in the literature): acetic, 72° (78°); succinic, 71° (88°); citric, 74° (102°); benzoic, 65° (89°). Since pure p-nitrobenzyl chloride melts at 71°, it is evident that after one hour the reaction was only partially completed. For these reasons the preparation of p-nitrobenzyl and substituted phenacyl esters is to be undertaken only if other derivatives are not suitable.

Example 10.8: Preparation of p-*Nitrobenzyl Salicylate*

Place in an 8-inch test tube 200 mg of salicylic acid and then add a drop of phenolphthalein and 2–3 drops of 5 per cent sodium carbonate. Warm the tube over a free flame and continue the addition of carbonate dropwise until the acid has been neutralized and the color of the solution is just pink. Warm the solution in order to be certain that all the acid has reacted. Add 2 drops of 5 per cent hydrochloric acid so that the pink color of the solution is discharged. Add 250 mg of p-nitrobenzyl chloride or p-nitrobenzyl bromide, 8 ml of alcohol, and a small boiling stone. (**Caution:** *Handle* p-*nitrobenzyl halides with care. Replace the stopper of the bottle immediately and avoid contact with the skin.*) Arrange tube for reflux (page 17) and boil gently for 1.5 hours. Cool and add 1 ml of water and scratch the sides of the tube. After 20 minutes filter the ester and wash first with 4 ml of 5 per cent sodium carbonate and then twice with 4 ml of water. Dissolve crystals in 8–10 ml of hot alcohol, filter, and add water to filtrate dropwise until a cloudiness appears. Heat tube until cloudiness disappears and then cool, scratching the sides of the tube with a glass rod. Cool for 15 minutes; then filter crystals. Wash with 1–2 ml of 50 per cent methanol and dry on paper disc. Avoid handling or contact of the crystals with the skin. The yield is 110 mg; the melting point, 97–98°.

Note: p-Bromophenacyl esters are prepared in the same manner. From 180 mg of benzoic acid and 400 mg of p-bromophenacyl bromide (2.5 hours heating) about 150–180 mg of the derivative (m.p. 118–119°) was obtained.

References

Amides and Substituted Amides

N-Acyl-p-aminoazobenzols: Escher, *Helv. Chim. Acta* **12**, 27 (1929).

Amides: Assano, *J. Pharm. Soc. Japan* **480**, 97 (1929); DeConno, *Gazz. chim. ital.* **47**, I, 93 (1917); Mitchell and Reid, *J. Am. Chem. Soc.* **53**, 1879 (1931); Swera *et al., J. Am. Chem. Soc.* **71**, 2215, 3017 (1949).

Amides and p-toluidides (microscopic identification): Dunbar and Moore, *Microchem. J.* **3**, 479 (1959).

Anilides: Barnicoat, *J. Chem. Soc.,* 2927 (1927); Blodinger and Anderson, *J. Am. Chem. Soc.* **74**, 5514 (1952); Carre and Libermann, *Compt. rend.* **194**, 2218 (1932); *Bull. soc. chim.* **53**, 293 (1953); DeConno, *Gazz. chim. ital.* **47**, I, 93 (1917); Hann and Jamieson, *J. Am. Chem. Soc.* **50**, 1442 (1928); Hardy, *J. Chem. Soc.,* 398 (1936); Robertson, *J. Chem. Soc.* **93**, 1033 (1908); **115**, 1210 (1919); Shah and Deshpande, *J. Univ. Bombay* (*2 pt*) **2**, 125 (1933); *C.A.* **28**, 6127.

Benzidinediamides: Buu-Hoi, *Bull. soc. chim. France* **12**, 587 (1945).

Benzylamides: Dermer and King, *J. Org. Chem.* **8**, 168 (1943); Stafford *et al., Anal. Chem.* **21**, 1454 (1949).

p-Bromoanilides: Bryant, *J. Am. Chem. Soc.* **60**, 1394 (1938); Bryant and Mitchell, *J. Am. Chem. Soc.* **60**, 2748 (1938); Houston, *J. Am. Chem. Soc.* **62**, 1303 (1940); Kuehn and McElvain, *J. Am. Chem. Soc.* **53**, 1173 (1931); Robertson, *J. Chem. Soc.* **93**, 1033 (1908); **115**, 1210 (1919).

α-Bromo-β-naphthylamides: Robertson, *J. Chem. Soc.* **93**, 1033 (1908); **115**, 1210 (1919).

o-Bromo-p-toluidides: Robertson, *J. Chem. Soc.* **93**, 1033 (1908); **115**, 1210 (1919).

Diamides of p,p'-diaminodiphenylmethane: Ralston and McCorkle, *J. Am. Chem. Soc.* **61**, 1604 (1935).

Diethanolamides: D'Alelio and Reid, *J. Am. Chem. Soc.* **59**, 109 (1937).

Dimethylamides: Kirsavov and Zolotov, *Zhur. Obshchei Khim.* (*J. Gen. Chem.*) *U.S.S.R.* **21**, 1166 (1951); Mitchell and Reid, *J. Am. Chem. Soc.* **53**, 1879 (1931); Prelog, *Collection Czechoslov. Chem. Commun.* **2**, 712 (1930).

Dodecylamides, octadecylamides: Hunter, *Iowa State Coll. J. Sci.* **15**, 223 (1941).

p-Hydroxyanilides: DeConno, *Gazz. chim. ital.* **47**, I, 93 (1917).

2-Methyl-5-isopropylanilides: Hann and Jamieson, *J. Am. Chem. Soc.* **50**, 1442 (1928).

α- and β-Naphthylamides: De Conno, *Gazz. chim. ital.* **47**, I, 93 (1917); Robertson, *J. Chem. Soc.* **93**, 1033 (1908); **115**, 1210 (1919); Shah and Deshpande, *J. Univ. Bombay* (*2 pt*) **2**, 125 (1933); *C.A.* **28**, 6127.

o-, m-, and p-Nitroanilides: Shah and Deshpande, *J. Univ. Bombay* (*2 pt*) **2**, 125 (1933); *C.A.* **28**, 6127.

o- and p-Phenetides: DeConno, *Gazz. chim. ital.* **47**, I, 93 (1917).

o-Phenyldiamides: Shah and Deshpande, *J. Univ. Bombay* (*2 pt*) **2**, 125 (1933); *C.A.* **28**, 6127.

N-Phenylamides of dibasic acids: Laurent and Gerhardt, *Ann. chim. Phys.* (*3*) **24**, 188 (1848).

Thiophenylamides: Shah and Deshpande, *J. Univ. Bombay* (*2 pt*) **2**, 125 (1933); *C.A.* **28**, 6127.

o- and p-Toluidides: DeConno, *Gazz. chim. ital.* **47**, I, 93 (1917); Robertson, *J. Chem. Soc.* **93**, 1033 (1908); **115**, 1210 (1919).

2,4,6-Tribromoanilides: Robertson, *J. Chem. Soc.* **93**, 1033 (1908); **115**, 1210 (1919).

p-Xenylamides: Kimura and Nihayashi, *Ber.* **68B**, 2028 (1935).

m-Xylides: DeConno, *Gazz. chim. ital.* **47**, I, 93 (1917).

Esters

Acid esters of dibasic acids: Cazeneuve, *Bull. soc. chim. France* **9**, 90 (1893); Veibel and Lillelund, *Acta Chem. Scand.* **8**, 1954; Walker, *J. Chem. Soc.* **61**, 1088 (1892).

Lactone derivatives: Meyer, *Monatsh.* **20**, 717 (1899).

Methyl esters using diazomethane: Herzig and Wenzel, *Monatsh.* **229**, 22 (1901).

Menthyl esters: Brauns, *J. Am. Chem. Soc.* **42**, 1478 (1920); Tchugaeff, *Ber.* **31**, 360 (1898).

p-Nitrobenzyl esters: Blike and Smith, *J. Am. Chem. Soc.* **51**, 1947 (1929); Kelly and Segura, *J. Am. Chem. Soc.* **56**, 2497 (1934); Lyons and Reid, *J. Am. Chem. Soc.* **39**, 701, 1727 (1917); Rather and Reid, *J. Am. Chem. Soc.* **43**, 629 (1921); Reid, *J. Am. Chem. Soc.* **39**, 124, 701, 1727 (1917).

Phenacyl, p-chlorophenacyl, and p-bromophenacyl esters: Berger, *Acta Chem. Scand.* **10**, 638 (1956); Chen and Shih, *Trans. Sci. Soc. China* **7**, 81 (1931); Erickson et al., *J. Am. Chem. Soc.* **73**, 5301 (1951); Hann et al., *J. Am. Chem. Soc.* **52**, 818 (1930); Harmon and Marvel, *J. Am. Chem. Soc.* **54**, 2515 (1932); Kelly et al., *J. Am. Chem. Soc.* **54**, 4444 (1932); Kananiwa and Isono, *Ann. Rept. Fac. Pharm. Kanazawa Univ.* **2**, 30 (1952); Kimura, *J. Soc. Chem. Ind. Japan* **35**, Sb, 221 (1932); Lundquist, *J. Am.*

Chem. Soc. **60**, 2000 (1938); Price and Griffith, *J. Am. Chem. Soc.* **64**, 2884 (1942); Reid *et al., J. Am. Chem. Soc.* **41**, 4175 (1919); **42**, 1043 (1920); **52**, 818 (1930); **53**, 1172 (1931); **54**, 2101 (1932); Wilson, *J. Am. Chem. Soc.* **67**, 2161 (1945).

4-Phenylazophenacyl esters: Masuyama, *J. Chem. Soc. Japan, Pure Chem. Sect.* **71**, 402 (1950); Mowrey and Frode, *J. Am. Chem. Soc.* **63**, 2281 (1941); Viogue and LaMaza, *Grasas y aceites (Seville, Spain)* **8**, 19 (1957).

p-Phenylphenacyl esters: Drake *et al., J. Am. Chem. Soc.* **52**, 3717 (1930); **54**, 2059 (1932); **58**, 1502 (1936); Erickson *et al., J. Am. Chem. Soc.* **73**, 5301 (1951); Ford, *Iowa State Coll. J. Sci.* **52**, 818 (1930); Hawkins *et al., Anal. Chem.* **28**, 1975 (1956); Kass *et al., J. Am. Chem. Soc.* **64**, 1061 (1942); Kelly *et al., J. Am. Chem. Soc.* **58**, 1502 (1936).

Salts and Other Derivatives

N-Acyl-2-acylcarbazoles and 2,8-diacylcarbazoles: Ford, *Iowa State Coll. J. Sci.* **12**, 121 (1937); Gilman and Ford, *Iowa State Coll. J. Sci.* **13**, 135 (1939).

N-Acylsaccharins: Stephen and Stephen, *J. Chem. Soc.* **1957**, 492.

2-Alkylbenzimidazoles: Pool, Harwood, and Ralston, *J. Am. Chem. Soc.* **59**, 178 (1937); Seka and Mueller, *Monatsh.* **57**, 95 (1931).

2-Alkylbenzimidazole picrates: Brown and Campbell, *J. Chem. Soc.,* 1699 (1937).

Benzylammonium and α-phenylethylammonium salts: Boudet, *Bull. soc. chim. France* **1948**, 390; Buehler, Carson, and Edds, *J. Am. Chem. Soc.* **56**, 1759 (1934); Buehler *et al., J. Am. Chem. Soc.* **57**, 2181 (1935).

S-Benzylisothiouronium salts: Bolliger, *Helv. Chim. Acta* **34**, 916 (1951); Chambers and Sherer, *Ind. Eng. Chem.* **16**, 1272 (1924); Donleavy, *J. Am. Chem. Soc.* **58**, 1004 (1936); Friediger and Pederson, *Acta Chem. Scand.* **8** (1954); Kass *et al., J. Am. Chem. Soc.* **64**, 1061 (1942); *Am. J. Biol. Chem.* **192**, 301 (1951); Veibel and Lillelund, *Bull. soc. chim. (5)* **5**, 1153 (1938); Veibel and Ottung, *Bull. soc. chim. (5)* **6**, 1434 (1939).

S-Benzylisothiouronium salts (melting points): Berger, *Acta Chem. Scand.* **8**, 427 (1954); Walker, *J. Chem. Soc.,* 1999 (1949).

S-(*p*-Bromo-) and *S*-(*p*-chloro-)benzylisothiouronium salts: Dewey and Shasky, *J. Am. Chem. Soc.* **63**, 3526 (1941); Dewey and Sperry, *J. Am. Chem. Soc.* **61**, 3251 (1939).

Bis-(*p*-dimethylaminophenylureides): Breusch and Ulusoy, *Arch. Biochem.* **11**, 489 (1946).

2,4-Dinitrophenylhydrazides: Cherezo and Olay, *Anales soc. españ. fís. y quím.* **32**, 1090 (1934); Gilman and Ford, *Iowa State Coll. J. Sci.* **13**, 135 (1939).

2,4-Dinitrophenylhydrazone: Hawkins *et al., Anal. Chem.* **28**, 675 (1956).

Hydrazides: Hanus and Vorishek, *Collection Czechoslov. Chem. Commun* **1**, 223 (1929); Kyame *et al., J. Am. Oil Chemists' Soc.* **24**, 332 (1947); Pajari, *Fette u, Seifen* **51**, 347 (1944); Paschke and Wheeler, *J. Oil Chemists' Soc.* **26**, 637 (1949); Sah, *Rec. trav. chim.* **59**, 1046 (1940) (in English).

Hydrazides and phenylhydrazonium salts: Stempel and Schaffel, *J. Am. Chem. Soc.* **64**, 470 (1942).

Hydroxamic acids: Inoue, Noda, *et al., J. Agr. Chem. Soc. Japan* **23**, 294, 368 (1950); Inoue and Yukawa, *J. Agr. Chem. Soc. Japan* **16**, 504, 510 (1940); **17**, 411, 491, 771 (1941); **18**, 415, 875 (1942); *Bull. Agr. Chem. Soc. Japan* **16**, 100 (1940); **17**, 44, 89 (1941) (in English); **17**, 59 (1941); **18**, 33, 72 (1942) (in English).

Monoureides and monothioureides: Jacobson, *J. Am. Chem. Soc.* **58**, 1984 (1936); Stendahl, *Compt. rend.* **196**, 1810 (1933).

S-(α-naphthyl)methylisothiouronium salts: Bonner, *J. Am. Chem. Soc.* **70**, 3508 (1948).

S-(*p*-nitro)benzylisothiouronium salts: Rupe and Zweidler, *Helv. Chim. Acta* **23**, 1025 (1940).

Octadecyl and dodecyl ammonium salts: Hunter, *Iowa State Coll. J. Sci.* **15**, 223 (1941).

Phenylhydrazides: Brauns, *J. Am. Chem. Soc.* **42**, 1478 (1920); Cheronis and Cohn, *Mikrochim. Acta*, 925 (1956); Shah and Deshpande, *J. Univ. Bombay (2 pt)* **2**, 125 (1933); Van Alpen, *Rec. trav. chim.* **44**, 1064 (1925); Veseky and Haas, *Chem. Listy.* **21**, 351 (1927).

Phenylmercuric and *p*-tolylmercuric salts: Ford, *Iowa State Coll. J. Sci.* **12**, 121 (1937); Gilman and Ford, *Iowa State Coll. J. Sci.* **13**, 135 (1939).

Piperazonium salts: Pollard and Adelson, *J. Am. Chem. Soc.* **56**, 1759 (1934).

Tetraphenyl stibonium salts: Affsprung and May, *Anal. Chem.* **32**, 1164 (1960).

Triphenyl lead salts: Ford, *Iowa State Coll. J. Sci.* **12**, 121 (1937); Gilman and Ford, *Iowa State Coll. J. Sci.* **13**, 135 (1939).

Ureids: Schmidt *et al., Ber.* **71**, 1933 (1938); **73**, 286 (1940); Zetzsche *et al., Ber.* **71B**, 1088, 1516, 2095 (1938); **72**, 1599 (1939); **72B**, 1735, 2095 (1939); **73B**, 465, 1114 (1940); **74B**, 183 (1941); **75B**, 100 (1942).

Critical discussion of derivatives: Veibel, *Chim. Anal.* **41**, 12, 49 (1959).

Chromatographic Procedures

Airan *et al., Anal. Chem.* **25**, 659 (1953).

Asselineau, *Bull. soc. chim. France* **1952**, 884

Boldingh, *Rec. trav. chim.* **69**, 247 (1950).

Brown, *Biochem. J.* **47**, 598 (1950); *Nature* **166**, 66 (1950).

Buch *et al., Anal. Chem.* **24**, 489 (1952).

Cheronis and Cohn, *Mikrochim. Acta*, 925 (1956).

Denison and Phares, *Anal. Chem.* **24**, 1628 (1952).

Fink and Fink, *Proc. Soc. Exptl. Biol. Med.* **70**, 654 (1949).

Hiscox and Berridge, *Nature* **166**, 522 (1950).

Inoue *et al., Bull. Agr. Chem. Soc. Japan* **19**, 214 (1955); *J. Japan Oil Chemists' Soc.* **5**, 16 (1956).

Italman, *Progress in the Chemistry of Fats*, Vol. I (N.Y.: Academic Press, 1952), p. 104. Review.

Kennedy and Barker, *Anal. Chem.* **23**, 1033 (1951).

Lugg and Overell, *Australian J. Sci. Research* **1**, 98 (1948).

Reid and Lederer, *Biochem. J.* **50**, 60 (1951).

Sataki and Seki, *J. Jap. Chem.* **4**, 557 (1950).

Smith, I. (ed.), *Chromatographic and Electrophoretic Techniques* (N.Y.: Interscience, 1960), Chs. 14–16.

Thompson, *Australian J. Sci. Research* **B4**, 180 (1951).

ALCOHOLS

Table 10.3 summarizes the most important derivatives that have been described in the literature for the derivatization of compounds having the hydroxyl-group function. These compounds include alcohols, glycols, polyhydroxy compounds, and phenols. The last-named, however, although they are derivatized by many of the reagents described in this section, are discussed separately, since the presence of aromatic structures imparts a number of additional reaction properties by which they may be derivatized. The discussion of phenols appears on pages 328–32.

TABLE 10.3

Derivatives for the Identification of Alcohols,
Glycols, and Polyhydroxy Compounds

*α-Naphthylurethans	m-Nitrophenylurethans
*3,5-Dinitrobenzoates	o-Nitrophenylurethans
*p-Nitrobenzoates	p-Nitrophenylurethans
3-Nitrophthalates	Phenylurethans
Benzoates	p-Xenylurethans
p-Phenylazobenzoates	4-Phenylazophenylurethans
Trinitrobenzoates	Hydrogen phthalates
Aryloxyacetic acids	Tetrachlorohydrogenphthalates
Allophanates	Anthranilates
Acetates	Trityl ethers
p-Nitrophenyl acetates	2,4-Dinitrophenyl ethers
p-Anisylurethans	Pseudosaccharin ethers
p-Chlorophenylurethans	Xanthates
3,5-Dinitrophenylurethans	S-Benzylisothiouronium derivatives
3,5-Dinitro-4-methylphenylurethans	2,4-Dinitrophenylhydrazones by oxidation
p-Iodophenylurethans	to aldehydes or ketones
2,4,6-Trinitrobenzoyl esters	p-Fhenylazobenzoyl esters

* Recommended derivative. Literature references to all derivatives listed in this table are given on pages 255–57.

The asterisked derivatives in Table 10.3 are those that are recommended for a first trial. It will be noted that the preparation of suitable derivatives for alcohols is based to a large extent on the following two reactions: (a) formation of sparingly soluble esters with aromatic acids, preferably nitro aromatic acid chlorides, as shown by Eq. (10.9); and (b)

$$R \cdot OH + Ar(NO_2)_xCOCl \longrightarrow Ar(NO_2)_xCOOR + HCl \qquad (10.9)$$

Alcohol   Nitro aromatic acid chloride   Nitro aromatic ester

reaction of the hydroxyl compound with an isocyanate to form a urethan, which also can be called a carbamate or an ester of carbamic acid, NH_2COOR, as is represented in Eq. (10.10). Thus when ethanol reacts with

$$C_2H_5OH + C_{10}H_7NCO \longrightarrow C_{10}H_7NHCOOC_2H_5 \qquad (10.10)$$

Ethanol α-Naphthyl isocyanate α-Naphthylurethan
or
Ethyl-N-α-naphthylcarbamate
or
Ethyl-N-1-naphthalenecarbamate

α-naphthyl isocyanate, the product can be called either a urethan or a derivative of carbamic acid.

The discussion in the following section deals with 3,5-dinitrobenzoates and p-nitrobenzoates, α-naphthylurethans, and derivatives for glycols and polyhydric alcohols. The two derivatives recommended for small quantities are the *3,5-dinitrobenzoates* and *α-naphthylurethans*. The reagents for the preparation of these esters (3,5-dinitrobenzoyl chloride and α-naphthyl isocyanate) are commercially available. However, great care should be exercised in stoppering the bottles of these reagents, since exposure to air causes decomposition.

3,5-Dinitrobenzoates and *p*-Nitrobenzoates

Discussion

Eqs. (10.11) and (10.12) illustrate the formation of *p*-nitrobenzoates and 3,5-dinitrobenzoates, respectively. The alcohol is heated for several

$$NO_2 \cdot C_6H_4COCl + CH_3OH \rightarrow NO_2 \cdot C_6H_4COOCH_3 + HCl \qquad (10.11)$$

p-Nitrobenzoyl chloride   Methyl *p*-nitrobenzoate

$$(NO_2)_2C_6H_3COCl + C_2H_5OH \rightarrow (NO_2)_2C_6H_3COOC_2H_5 + HCl \qquad (10.12)$$

3,5-Dinitrobenzoyl chloride   Ethyl 3,5-dinitrobenzoate

minutes with the acid chloride in a dry vessel and then water is added to separate the solid ester, which is filtered and purified. In many macro procedures an excess of alcohol is employed in order to convert all of the acid chloride to the ester. Therefore, it is assumed that no chloride will be left to form nitrobenzoic acid when water is added after the reaction is complete. It has been shown[3] that a large excess of alcohol heated with 3,5-dinitrobenzoyl chloride does not prevent the formation of impurities, which are essentially 3,5-dinitrobenzoic acid and its anhydride. Therefore, only a small excess of alcohol is employed and the product is thoroughly pulverized and extracted with dilute sodium carbonate solution to remove the acid and anhydride formed and then recrystallized. The *p*-nitrobenzoates are formed and purified in the same manner. When the amount of hydroxy compound is less than 100 mg, it is advisable to heat it with the acid chloride in the presence of an inert solvent such as isopropyl ether, dry benzene, pyridine, or other.[4] In general, this method should be used whenever the formation of the ester is slow, as in the case of tertiary alcohols. For example, when 30 mg each of *n*-butyl, isobutyl, *sec*-butyl, and *tert*-butyl alcohols and 50 mg of 3,5-dinitrobenzoyl chloride were heated for 30 minutes, the yields of the dinitrobenzoates in milligrams were, respectively, 16, 20, 18, and 0.5. It is obvious that the yield of the ester from the tertiary alcohol is insufficient for the ordinary methods of purification and melting-point determination. In such cases it is advisable to boil the alcohol for an hour or two with the dinitrobenzoyl chloride and pyridine dissolved in isopropyl ether.[4] In the absence of a proton acceptor there is

[3] Cheronis and Vavoulis, *Mikrochemie* 38, 428 (1952); Cheronis, *Micro and Semimicro Methods*, Vol. VI of Weissberger (ed.), *Technique of Organic Chemistry* (N.Y.: Interscience, 1957), p. 483.

[4] For details see Cheronis and Entrikin, *op. cit.*, p. 373.

a tendency for tertiary alcohols to form the halide, RX, and an olefin.

The 3,5-dinitrobenzoates form crystalline complexes with aromatic amines. The complexes with α-naphthylamine are particularly useful because they have characteristic colors from orange to red according to the radical in the alcohol. Further, they form well-defined crystals, having definite melting points and may be used to purify those 3,5-dinitrobenzoates that are not easily crystallized.

Although the presence of water is a disadvantage in the preparation of p-nitrobenzoates, 3,5-dinitrobenzoates, and other arylnitro esters, it is possible to prepare these derivatives from dilute aqueous solutions of alcohols. The acid chloride is dissolved in purified ligroin and then is shaken with a 5 per cent aqueous solution of the alcohol in the presence of sodium acetate. It is possible by this method to prepare derivatives from 500 mg of alcohol when it is admixed with 10 ml of water.

General Method for Preparation

If the alcohol has less than 5 per cent water, 1.5–2 millimoles is heated in a small test tube with 1 millimole of the reagent. Both p-nitrobenzoyl chloride and 3,5-dinitrobenzoyl chloride are susceptible to hydrolysis on storage from atmospheric moisture and their melting points should be checked; if the melting point is more than 3° lower than that recorded in the literature, the acid chloride should be crystallized from carbon tetrachloride.

The mixture of alcohol and reagent is heated cautiously over a small flame at the lowest temperature at which the mixture remains liquid. The lower alcohols are heated for 3–5 minutes and the higher alcohols for 10–15 minutes. The melt is allowed to solidify, after which the crystalline mass is broken up thoroughly with a microspatula without transfer; it is then shaken with 2 per cent sodium carbonate at 50–60° for about 10–30 seconds and filtered. Prolonged treatment with sodium carbonate solution tends to hydrolyze the ester. The crystalline mass is then washed thoroughly with water and crystallized by dissolving in hot methanol or ethanol and adding water to cloudiness. Usually only one crystallization is required. However, where the extraction with sodium carbonate is faulty, two or three crystallizations may be necessary. From 1 millimole of the reagent the yield of pure derivative after one crystallization is 150–200 mg.

When the alcohol is not reactive or when a polyhydric alcohol is derivatized, the use of solvent is of advantage. Isopropyl ether or n-butyl ether may be used as the solvent. The mixture of reagent and alcohol is refluxed for 0.5–1 hour and then washed with sodium carbonate solution; the ethereal layer is separated and dried; then the solvent is evaporated to obtain the crude ester. This method, which entails several transfers, cannot normally be employed for quantities less than 50 mg. With tertiary alcohols the pyridine alone is employed as the solvent. One millimole of reagent is mixed with 1.5 millimoles of the alcohol in a tube containing 2 ml of pyridine. The mixture is refluxed 0.5–1 hour, cooled, and extracted with 5 ml of 1 per cent sulfuric acid, which removes the pyridine and separates a crystalline mass of the crude ester.

When the alcohol contains water, it is diluted so that 250–500 mg is contained in a volume of about 5–6 ml. The mixture is cooled to 0° and then shaken with a solution of 500 mg of the acid chloride in 2 ml of specially purified hexane (ligroin or petroleum ether) and 3 ml of benzene. The hexane is specially purified by washing, first, with concentrated H_2SO_4 and then with water, drying with anhydrous $CaCl_2$ or $CaSO_4$, and distilling. The mixture is well shaken, sodium acetate is added, and the temperature is kept below 5° with frequent shaking for 15–30 minutes. Alcohol-free ether is then added, the upper layer separated and washed, first with dilute sodium hydroxide, then with dilute hydrochloric acid, then with water. The solvent is then evaporated to obtain the solid ester.

Caution: For good results, the 3,5-dinitrobenzoyl chloride either must be prepared before use (see Cheronis and Entrikin[5]) or a sample from an unopened bottle of the commercial reagent should be employed. If the laboratory bottle containing the reagent has been left unsealed for some time, it is suggested to recrystallize the reagent from dry carbon tetrachloride and check its melting point. A product melting at 72–74° is satisfactory. If a reagent from an unsealed bottle is used without purification, it is suggested to use a slight excess and repeat the extraction twice with sodium carbonate.

Example 10.9: Derivatization of Ethanol: Ethyl 3,5-Dinitrobenzoate

Place in a 6-inch tube 400 mg of pure 3,5-dinitrobenzoyl chloride and 0.12 ml of 95 per cent ethanol. Heat for about 5 minutes by means of a microflame so that the melt at the bottom of the tube does not solidify. Avoid a hot flame. If there is much evidence of condensation on the sides of the tube 20–30 mm above the reaction mixture, the flame should be reduced or the tube raised.

Allow the melt to solidify. By means of a glass rod or a microspatula break and thoroughly pulverize the crystalline mass, so that *no lumps* remain. Add to the tube 5 ml of 2 per cent sodium carbonate and continue the grinding of crystals against the walls of the tube. Heat the mixture gradually to 50–60°, place a solid rubber stopper on the mouth of the tube,

[5] Cheronis and Entrikin, *Semimicro Qualitative Organic Analysis* (2nd ed., N.Y.: Interscience, 1957), pp. 374–75.

and shake for about 15 seconds. Filter the mixture and wash 3 times with 3–4 ml of water.

Place the crystals in the tube in which the derivative was prepared, add 15 ml of ethanol or methanol, and heat until solution is effected. Filter and add water to the filtrate until cloudiness appears; reheat until the cloudiness disappears. If the cloudiness persists near the boiling point of the solution, add alcohol dropwise. Cool for 10–15 minutes and filter. Wash twice with 3-ml portions of equal parts of alcohol and water, and dry on a paper drying disc. About 250–300 mg of pure ethyl 3,5-dinitrobenzoate, melting at 93°, is obtained.

Note: In the case of the alcohols having 6 or more carbon atoms, the heating of the reaction mixture should be prolonged to 10 minutes and, if the results are poor, the procedure described for the preparation of β-naphthyl 3,5-dinitrobenzoate (page 329) should be used. If the melting point of the dinitrobenzoate after crystallization is more than 2–4° below that recorded in the literature, process as follows: Dissolve the dinitro-benzoate (it need not be dry) in 5 ml of ethyl or isopropyl ether and wash the ethereal solution first with 3 ml of 2 per cent sodium hydroxide solution and then with 3 ml of water. Evaporate the ether and crystallize the residue once from an alcohol–water mixture. If this method is applied to a crude sample of ethyl 3,5-dinitrobenzoate melt-ing at 84–86°, a product is obtained melting at 92–93° without further crystallization. Commercially available chloride should be recrystallized from carbon tetrachloride unless the purity is specified. The stopper of the bottle in which the chloride is kept should be sealed with paraffin wax and exposure to air should be kept at a minimum.

α-Naphthylurethans

General Method for Preparation

As was pointed out in the introduction to the derivatization of alcohols, a urethan or carbamate is formed by the reaction of an isocyanate and a hydroxy compound, as shown by Eqs. (10.13) and (10.14). A ratio of

$$RNCO + R'C\!-\!OH \longrightarrow RNH\overset{\overset{\displaystyle O}{\|}}{C}\!-\!OR' \qquad (10.13)$$

Isocyanate   Hydroxy   Urethan or ester of
compound   substituted carbamic acid

$$C_{10}H_7NCO + C_3H_7OH \longrightarrow C_{10}H_7NHCOOC_3H_7 \qquad (10.14)$$

α-Naphthyl   1-Propanol   1-Propyl-α-naphthylurethan
isocyanate

1.25 millimoles of α-naphthyl isocyanate to 1 millimole of alcohol is used. The anhydrous alcohol and isocyanate are mixed in an appropriate 3-inch or 6-inch test tube and heated in a water bath at 60–70° for 10–15 minutes. The crude urethan solidifies on cooling and is pulverized in the reaction vessel by means of the microspatula. It is then extracted with a minimum amount of petroleum ether to remove the soluble impurities.[6] The first extract, which contains a considerable amount of the derivative, is set aside

[6] Cheronis, *op. cit.,* pp. 485–89, 497–99.

and the residue is extracted with a fresh amount of the solvent. The second extraction yields the pure derivative. The residue contains di-α-naphthylurea formed by the reaction of moisture present in the alcohol and the reaction vessel.

Attention should be paid to the filtration of the petroleum ether extract of the reaction mixture. These and other details are described in the following examples.

Example 10.10: n-Propyl-α-naphthylurethan

Heat a 6-inch test tube over a flame until all moisture has been driven off; then cork and allow to cool. By means of a pipet dropper, rapidly put 0.2 ml of 1-propanol and 0.25 ml of α-naphthyl isocyanate into the tube; cork the tube immediately. Place the tube in a water bath at 60–70° for 5 minutes. Remove the tube and add 4 ml of petroleum ether (b.p. 90–110°) or 4 ml of commercial heptane. Heat to boiling and filter through a funnel prepared as follows: Insert the perforated disc inside of the funnel and place upon it a disc of filter paper; add 2 drops of water and apply light suction; then adjust filter paper. Add 3 drops of methanol and apply light suction again. Finally, add 4 drops of petroleum ether and repeat the application of suction. Fit the filter into the mouth of a clean and dry 8-inch test tube with a side arm, apply suction, and add the hot solution. Set the filtrate aside and transfer the residue with the filter paper back into the reaction tube and add 8 ml of the solvent. Heat the tube cautiously and as the solvent boils lift the filter paper from the bottom of the tube by means of the microspatula so that it adheres to the sides of the tube about 5 mm above the liquid. In this manner as the vapors condense the material adhering to the paper is washed down. Prepare the funnel for a second filtration as described. Chill the second filtrate for a few minutes and scratch the sides of the tube by means of a glass rod. After 15 minutes, filter the crystals and crystallize from 5–6 ml of hydrocarbon. The yield is 80–90 mg of crystals, melting at 79–80°.

Note: It is necessary that the alcohol be free from water; otherwise a considerable amount of dinaphthylurea forms, with a corresponding decrease in the yield of the desired urethan. It is clear, therefore, that the preparation of the urethan should not be attempted unless it is known that the hydroxy compound is anhydrous. For secondary alcohols the duration of heating should be increased to 10 minutes. For example, from 0.2 ml of 2-butanol or 2-pentanol, 80–90 mg of the pure urethan is obtained. For smaller quantities of the hydroxy compound and for higher alcohols, the procedure described below is recommended.

Example 10.11: Cyclohexyl-α-naphthylurethan

Put 100 mg of cyclohexanol and 125 mg of α-naphthyl isocyanate into an 8-inch test tube dried as described in the preparation of n-propylurethan.

Fit a two-hole stopper into the test tube holding a microcondenser and a calcium chloride tube. Add 6 ml of petroleum ether and heat in a water bath at about 90° for 30 minutes. Prepare a funnel as described in Example 10.10 above, and filter the hot solution from the dinaphthylurea. Cool the solution for 10–15 minutes and then filter the crystals. Dissolve these in the minimum amount of hot petroleum ether and crystallize. The yield is 40–50 mg, melting at 128–129°.

Note: Tertiary alcohols react very slowly and yield but small amounts of urethans; dehydration of the alcohol occurs and results in increased formation of the dinaphthylurea. For example, 200 mg of *tert*-amyl alcohol with 0.25 ml of the isocyanate, treated as described in Example 10.11, gave 1.5 mg of a gummy substance that could not be purified for melting-point determination.

Benzoates of Polyhydroxy Compounds

The benzoates are useful for the derivatization of glycols, glycerol, and other polyhydroxy compounds. The same method may be used for the preparation of benzoates of alcohols. However, as a rule, these have too low melting points to be useful derivatives.

Example 10.12: Preparation of Glycerol Tribenzoate

Place in an 8-inch test tube 100 mg (3 drops) of glycerol and add 0.5 ml of benzoyl chloride. Select a solid rubber stopper that fits securely in the mouth of the tube. Add 5 ml of 10 per cent sodium hydroxide solution and shake vigorously for 1 minute and then intermittently for 5 minutes or until the solid derivative separates out. Allow to stand in a cold bath for 30 minutes, shaking the tube at intervals so that the lumps forming at the beginning break up into small granules; use a rod for this purpose if necessary. Add 5 ml of water, shake vigorously for a minute, and then filter. Wash twice with 5 ml of water and place on a drying disc.

The recrystallization of the tribenzoate of glycerol and of the dibenzoate of the glycols entails considerable difficulty, owing to the tendency of these esters (glycerides and glycol esters) to separate as oils. The following procedure gives fairly good results: Retain about 5 mg of the crystals and dissolve the rest in 10 ml of hot ethanol; then filter. Add 5 ml of water and reheat until the cloudiness disappears. Cool and add a few crystals from the lot set aside for seeding. Allow the tube to stand in the cold bath for 1–2 hours and then filter; wash twice with 2–3 ml of water and set aside to dry. The yield is 200–225 mg of crystals, melting at 75–76°. This derivative has also been reported to melt at 71–72°.

References

3,5-Dinitrobenzoates and Other Nitrobenzoates

3,5-Dinitrobenzoates: Adamson and Kenner, *J. Chem. Soc.* **1935**, 287; Bryant, *J. Am. Chem. Soc.* **54**, 3758 (1932); King, *J. Am. Chem. Soc.* **61**, 2383 (1939); Lipscomb, Malone, and Reid, *J. Am. Chem. Soc.* **51**, 3424 (1929); Reichstein, *Helv. Chim. Acta* **9**, 799 (1926).

3,5-Dinitrobenzoates (microscopic identification): Dunbar and Ferrin, *Microchem. J.* **3**, 65 (1959).

3,5-Dinitrobenzoates and other derivatives of glycol and glycerol: Boehm and Threme, *Pharmazie* **11**, 175 (1956); *C.A.* **51**, 6088d.

3,5-Dinitrobenzoates of hexanols: Sutter, *Helv. Chim. Acta* **21**, 1266 (1938).

o-Nitrobenzoates: Lowe, *J. Am. Chem. Soc.* **74**, 841 (1952).

p-Nitrobenzoates: Adamson and Kenner, *J. Chem. Soc.* **1935**, 287; Armstrong and Copenhaver, *J. Am. Chem. Soc.* **65**, 2252 (1943); Henstock, *J. Chem. Soc.* **1933**, 216; King, *J. Am. Chem. Soc.* **61**, 2383 (1939); Meisenheimer and Schmidt, *Ann.* **475**, 157 (1929).

2,4,6-Trinitrobenzoates: Laskowski and Adams, *Anal. Chem.* **31**, 148 (1959).

Urethan Derivatives

p-Anisylurethans: Brunner and Wohrl, *Monatsh.* **63**, 374 (1933).

3,5-Dinitrophenylurethans: Hoeke, *Rec. trav. chim.* **54**, 505 (1935).

α-Naphthylurethans: Bickel and French, *J. Am. Chem. Soc.* **48**, 747 (1926); French and Wirtel, *J. Am. Chem. Soc.* **48**, 1736 (1926); Neuberg and Kansky, *Biochem. Z.* **20**, 445 (1909).

m-Nitrophenylurethans: Hoeke, *Rec. trav. chim.* **54**, 505 (1935); Veibel *et al.,* *Dansk Tidsskr. Farm.* **14**, 241 (1940); **17**, 187 (1943).

p-Nitrophenylurethans: Shriner and Cox, *J. Am. Chem. Soc.* **53**, 1601, 3186 (1931); Van Hoogstraten, *Rec. trav. chim.* **51**, 414 (1932).

4-Phenylazophenylurethans: Masuyama and Hamada, *J. Chem. Soc. Japan, Pure Chem. Sect.* **70**, 198 (1949).

Phenylurethans: Dewey and Witt, *Ind. Eng. Chem., Anal. Ed.* **12**, 459 (1940); **14**, 648 (1942); Lanbling, *Bull. soc. chim.* (3) **19**, 771 (1898); McKinley *et al., Ind. Eng. Chem., Anal. Ed.* **16**, 304 (1944); Witten and Reed, *J. Am. Chem. Soc.* **69**, 2470 (1947).

Substituted azides and urethans: Sah *et al., Science Repts., Natl. Tsing Hua Univ.* (A) **3**, 109 (1935); *J. Chinese Chem. Soc.* **2**, 229 (1934); *Rec. trav. chim.* **58**, 453, 582, 591, 595, 1013 (1939); **59**, 238, 357 (1940).

p-Xenylurethans: Morgan and Hardy, *J. Soc. Chem. Ind.* **52**; *Chemistry and Industry II,* 519 (1933); Morgan and Pettet, *J. Chem. Soc.* **1931**, 1124.

Other Derivatives

Identification and detection of volatile alcohols and acids in biological materials: Friedemann and Brook, *J. Biol. Chem.* **123**, 161 (1938).

Oxidation of alcohols to aldehydes and preparation of the 2,4-dinitrophenylhydrazones: Duke and Wiman, *Anal. Chem.* **20**, 490 (1948).

Acetates of tertiary alcohols: Spassow, *Ber.* **70**, 1926 (1937); **75**, 779 (1942).

Allophanates: Lane, *J. Chem. Soc.* **1951**, 2764.

Aryloxyacetic acids: Hayes and Branch, *J. Am. Chem. Soc.* **65**, 1555 (1943); Koelsch, *J. Am. Chem. Soc.* **53**, 305 (1931).

256   IDENTIFICATION OF ORGANIC COMPOUNDS

S-Benzylisothiouronium derivatives: Bair and Suter, *J. Am. Chem. Soc.* **64**, 1978 (1942); Berge, *Acta Chim. Scand.* **8**, 427 (1954).

p-Bromomethylphthalimides: Maxera and Lemberger, *J. Org. Chem.* **15**, 1253 (1953).

2,4-Dinitrophenyl ethers: Bost and Nicholson, *J. Am. Chem. Soc.* **57**, 2368 (1935).

1-(2,4-Dinitrophenyl)-3-methyl-3-nitrourea: Van Ginkel, *Rec. trav. chim.* **61**, 149 (1942).

Hydrogen phthalates: Fessler and Shriner, *J. Am. Chem. Soc.* **58**, 1384 (1936); Goggans and Copenhaver, *J. Am. Chem. Soc.* **61**, 2909 (1939); Reid, *J. Am. Chem. Soc.* **39**, 1250 (1917).

Substituted hydrogen phthalates: Lawlor, *J. Ind. Eng. Chem.* **39**, 1419, 1424 (1947).

p-Nitrophenylacetyl derivatives: Ward and Jenkins, *J. Org. Chem.* **10**, 371 (1945).

3-Nitrophthalates: De Graef and Pierret, *Bull. soc. chim. Belges* **57**, 307 (1948); Dickinson, *J. Am. Chem. Soc.* **59**, 1094 (1937); Dickinson, Crosson, and Copenhaver, *J. Am. Chem. Soc.* **59**, 1094 (1937); Nicolet and Sacks, *J. Am. Chem. Soc.* **47**, 2348 (1925); Veraguth and Diehl, *J. Am. Chem. Soc.* **62**, 233 (1940).

p-Phenylazobenzoyl esters: Woolfolk and Taylor, *J. Org. Chem.* **22**, 827 (1957).

Pseudosaccharin ethers: Böhme and Opper, *Z. anal. Chem.* **139**, 255 (1953); Meadow and Reid, *J. Am. Chem. Soc.* **65**, 457 (1943).

Trityl ethers: Seikel and Huntress, *J. Am. Chem. Soc.* **25**, 495 (1942); Sobetay, *Compt. rend.* **203**, 1164 (1936).

Xanthates: Shupe, *J. Assoc. Official Agr. Chemists* **25**, 495 (1942); Whitmore and Lieber, *Ind. Eng. Chem., Anal. Ed.* **7**, 127 (1935).

Miscellaneous: Naphthylamine addition products of 3,5-dinitrobenzoates, Benfey *et al., J. Org. Chem.* **20**, 1777 (1955); Sutter, *Helv. Chim. Acta.* **21**, 1266 (1938). Alkyl 3,5-dihydroxybenzoates, Suter and Weston, *J. Am. Chem. Soc.* **61**, 531 (1939). Fluorenyl, triphenylmethyl, and iodobiphenyl substituted urethans, Witten and Reid, *J. Am. Chem. Soc.* **69**, 2470 (1947); Kawai and Tamura, *J. Chem. Soc. Japan* **52**, 77 (1931). S-Benzylisothiouronium deriv., Schotte and Veibel, *Acta Scand.* **7**, 1357 (1953). Derivs. of glycols, Nason and Manning, *J. Am. Chem. Soc.* **62**, 1635, 3136 (1940). Triiodobenzoates, O'Donnell *et al., J. Am. Chem. Soc.* **68**, 1865 (1946); **70**, 1657 (1948). 2,4-Dinitrophenyl derivs., Vozozhstov and Yacobson, *C.A.* **52**, 12784h. 4,6-Dinitroresorcinol esters, Vozozhstov and Yacobson, *C.A.* **52**, 1998i. 2,4,6-Trinitrobenzoates, Laskowski and Adams, *Anal. Chem.* **31**, 148 (1959). Anthranilates, Staiger, *J. Org. Chem.* **18**, 1427 (1953).

Chromatographic Procedures

Cerbulis, *Anal. Chem.* **27**, 1400 (1955). Chromatography of sugar alcohols and glycosides.

Hough, *Nature* **165**, 400 (1950). The separations of polyhydroxy compounds by paper chromatography and column chromatography.

Kariyone and Hashimoto, *Nature* **168**, 511 (1951). Potassium xanthates.

Lederer, M., and E. Lederer, *Chromatograpry* (Houston-N.Y.: Elsevier, 1953), p. 106. The separation of polyhydroxy compounds by paper chromatography and column chromatography.

Masuyama and Hamada, *J. Chem. Soc. Japan, Pure Chem. Sect.* **70**, 198 (1949). Chromatographic separation of aliphatic 4-phenylazophenylurethans.

Meigh, *Nature* **169**, 706 (1952). Microdetection of 3,5-dinitrobenzoates of alcohols by paper chromatography.

Momosa and Yamada, *J. Pharm. Soc. Japan* **71**, 980 (1951). 3,6-dinitrophthalates.

Rice *et al., Anal. Chem.* **23,** 195 (1951). Microdetection of 3,5-dinitrobenzoates of alcohols by paper chromatography.

Spayner and Phillips, *Anal. Chem.* **28,** 253 (1956). Separation of xanthates.

Sundt and Winter, *Anal. Chem.* **29,** 851 (1957). Separation of 3,5-dinitrobenzoates.

Woolfolk *et al., J. Org. Chem.* **20,** 391 (1956); **22,** 827 (1957). Separation and identification by means of *p*-phenylazobenzoyl chloride.

ALDEHYDES

The discussion in this chapter pertains to the preparation of derivatives suitable for the characterization of aldehydes; however, a large number of the derivatives and the procedures for their preparation apply equally to the characterization of ketones, ketoesters, ketoacids, and many of the polyfunctional compounds that have a carbonyl group. For this reason the discussion on the preparation of derivatives of ketones given on pages 318–21 is very brief.

TABLE 10.4

Derivatives for the Identification of Aldehydes

*2,4-Dinitrophenylhydrazones	Tolylsemicarbazones (*o, m, p*)
*Semicarbazones	Phenylsemicarbazones
*Dimethone or dimedone derivatives	1-Naphthylsemicarbazones
*Phenylhydrazones	2-Naphthylsemicarbazones
p-Nitrophenylhydrazones	3,5-Dinitrophenylsemicarbazones
Thiosemicarbazones	Dibromomethone derivatives
Oximes	*p*-Phenyl- and *p*-tolylthiosemicarbazones
Azines	4-*p*-Bromophenylthiosemicarbazones
Bromobenzohydrazones (*o, m, p*)	4-*p*-Chlorophenylthiosemicarbazones
2,4-Dinitromethylphenylhydrazones	Benzothiazoles
p-Carboxyphenylhydrazones	Benzothiazolines
Nitrobenzenesulfonhydrazones	Hydantoins
β-Nitrobenzohydrazones	Aminomorpholines
Diphenylhydrazones	Hydrazinobenzoic acid derivatives
2-Naphthylhydrazones	5-(1-Phenylmethyl)semioxazide derivatives
p-Chlorobenzohydrazones	1,3-Cyclohexadione derivatives
m-Chlorobenzohydrazones	1,2-Bis(*p*-methoxybenzylamino)-ethane
Nitroguanylhydrazones	derivatives
α-(2,4-Dinitrophenyl)-α-methylhydrazones	Solid alcohols obtained on reduction
Xenylsemicarbazones	Solid carboxylic acids obtained on oxidation

* Recommended derivative.

Table 10.4 lists the majority of compounds proposed in the literature for the derivatization of carbonyl compounds. Those asterisked are the ones that should be first considered and a detailed discussion of their preparation and purification is given in the following sections. All factors should be considered in selecting the particular derivative to be prepared. For example, the recommended (asterisked) derivatives for glyoxal are semi-carbazone (m.p., 279°), 2,4-dinitrophenylhydrazone (m.p., 328°), phenyl-hydrazone (m.p., 180°), and methone (m.p., 228°). As a first trial the phenylhydrazone should be selected since it usually forms readily and its melting point is below 200°, but if it is found unsuitable owing to decomposition on purification (that is, if the melting point consistently differs by 5–10° from the value listed in the literature), the methone derivative should be selected.

Phenylhydrazones

Discussion

The reaction of phenylhydrazine with an aldehyde yields a phenylhy-drazone, according to Eq. (10.15). Since phenylhydrazine is easily available

$$\underset{\text{Phenylhydrazine}}{RC{=}O + H_2NNHC_6H_5} \longrightarrow \underset{\text{Phenylhydrazone}}{RC{=}NNHC_6H_5 + H_2O} \tag{10.15}$$

and the derivatives form readily, the use of phenylhydrazones, particularly for the aryl carbonyl compounds, is advisable. For the preparation of the derivatives, the carbonyl compound is dissolved in methanol or ethanol and heated with phenylhydrazine base and a small amount of acetic acid; the phenylhydrazone separates even while the solution is hot. After filtration, the derivative should be dried rapidly and the melting point determined at once, since phenylhydrazones as a rule undergo slow decomposition when dried in air. In general, when derivatives of phenylhydrazine are involved, it is recommended that the product be crystallized immediately and dried as rapidly as possible for melting-point determination.

A number of carbonyl compounds fail to give stable phenylhydrazones even when the derivatives are prepared from pure reagents and with the utmost care. For example, cyclohexanone and acetophenone yield deriva-tives that melt 5–10° below the melting point of the pure compound; even after crystallization the derivatives undergo change in the melting point when dried in a desiccator.

General Method for Preparation

About 1 millimole of the carbonyl compound is dissolved in 4–5 ml of methanol or ethanol, and 1.05 millimoles of phenylhydrazine base is added, followed by an equivalent amount of glacial acetic acid. The mixture is refluxed for 3–15 minutes, depending on the reactivity of the carbonyl group. The mixture is cooled and water is added dropwise until a cloudi-ness indicates the separation of the derivative. The mixture is then chilled until precipitation is complete and the phenylhydrazone is filtered, washed, and purified as usual. The actual yield of the pure derivative varies from

40–50 per cent of the theoretical. The following example illustrates the preparation of the phenylhydrazone from an aromatic aldehyde.

Example 10.13: Phenylhydrazone of an Aryl Aldehyde: Piperonal Phenylhydrazone

Place in an 8-inch tube 150 mg of piperonal and 5 ml of methanol. Heat for a few seconds to effect solution; then add 0.1 ml of phenylhydrazine. (**Caution:** *Use care in handling the reagent.*) Boil the mixture for 1 minute and add 1 drop of glacial acetic acid and boil gently for 3 minutes. Add dropwise 1.5 ml of water until a permanent cloudiness results. Cool, filter the crystals, and wash with 1 ml of water containing 1 drop of acetic acid. Recrystallize the product immediately by dissolving it in 3 ml of hot methanol. Add 0.5 ml of water to the hot solution, cool, and scratch the sides of the tube, if crystals do not separate readily. Filter, wash with a few drops of 50 per cent methanol, and dry rapidly by pressing the crystals first between filter paper before spreading on the paper filter disc. Determine the melting point as soon as the crystals are dry. The yield is about 80–100 mg of the product, melting at 99–100°.

Note: Other aromatic aldehydes, such as vanillin and *m*-nitrobenzaldehyde, yield phenylhydrazones readily by the same method as described for piperonal.

2,4-Dinitrophenylhydrazones

Discussion

The formation of 2,4-dinitrophenylhydrazones is shown by Eq. (10.16). The preparation of 2,4-dinitrophenylhydrazones is advisable for semi-

$$CH_3CHO + H_2NNHC_6H_3(NO_2)_2 \longrightarrow CH_3C{=}NNHC_6H_3(NO_2)_2 + H_2O \quad (10.16)$$
2,4-Dinitrophenylhydrazine   Ethanal 2,4-dinitrophenylhydrazone

micro quantities. In most cases it is possible to start with as little as 20–30 mg of the carbonyl compound and obtain a sufficient quantity of the pure derivative for several determinations of the melting point. In some cases the use of nitro-substituted phenylhydrazones may be limited by the high melting points of the derivatives; in such instances the use of phenlyhydrazones may be found suitable. Another limitation of the 2,4-dinitrophenylhydrazones is that they are not very satisfactory for α-hydroxy aldehydes ketones, and sugars, and a number of substituted aldehydes and ketones. In such cases the initial products often undergo secondary reactions; for example, the derivative of an α-hydroxy aldehyde may split off a molecule of water. The evaluation of the melting points of the 2,4-dinitrophenylhydrazones often leads to difficulties on account of the tendency of this group to polymorphism, and also because of the possibility of geometrical isomers from the C=N linkage. The differences in the melting points of the modifications may be 1–2°, as for the polymorphs of propanal and butanal, or as much as 24°, as in the case of furfural. In general, it is advised to determine the melting point of the 2,4-dinitro derivative, allow the melt to solidify, and then redetermine the melting point. As shown in Example 10.17, the 2,4-dinitro derivatives may be used for the chromatographic identification of microgram quantities of aldehydes and ketones.

General Method for Preparation

The ratio of the reactants is 0.5–0.6 millimole of the carbonyl compound to 0.4 millimole of the nitrophenylhydrazine, which is about 40 mg. The nitrophenylhydrazine is placed in a tube containing 4 ml of methanol and 0.1 ml of 6N hydrochloric acid, and dissolved by heating the tube in a water bath. Another method of preparing a stable solution is to pass hydrogen chloride gas through 50 ml of methanol until 3.5 g has been absorbed; about 2 g of 2,4-dinitrophenylhydrazine is added and the mixture is shaken until the solid dissolves. The solution contains 40 mg/ml.

The carbonyl compound (0.5–0.6 millimole) dissolved in 1 ml of methanol is added to an amount of the reagent containing 40 mg and the mixture is heated in the water bath (50–60°) for 1–2 minutes and then allowed to stand for 15–30 minutes, depending on the reactivity of the carbonyl compound. The derivatives often separate out on cooling, but it is advisable to add water dropwise to cloudiness. The derivatives are purified by crystallization from a solvent pair, usually alcohol and water. When the solubility in boiling methanol or ethanol is low, dioxane, ethyl acetate, toluene, or xylene may be used.

For carbonyl compounds that are sensitive the use of dimethyl ether of diethylene glycol (diglyme) as a solvent has been recommended.[7]

Example 10.14: Preparation of Formaldehyde 2,4-Dinitrophenylhydrazone

Place in an 8-inch tube 4 ml of methanol, 40 mg of 2,4-dinitrophenylhydrazine, and 0.1 ml of 6N HCl. Boil for a minute to effect solution. If all the solid does not dissolve, remove the flame and, after 1 minute, pour the clear solution into another test tube. Add to the clear solution 0.4 ml of 40 per cent aqueous formaldehyde solution and heat to boiling, then allow to stand for 5 minutes. If no precipitate appears, add water drop by drop to cloudiness. Filter the crystals and recrystallize from a mixture of 3 ml of methanol and 1 ml of water. The yield is about 20 mg of crystals, melting at 166–167°.

[7] Shine, *J. Org. Chem.* **24**, 252 (1959); *J. Chem. Educ.* **36**, 575 (1959); *J. Am. Chem. Soc.* **24**, 1790 (1959).

Note: For aromatic carbonyl compounds, the amount of the materials (carbonyl compound, solvent, and reagent) may be reduced to one-half. When 50 mg of benzaldehyde is used, 20 mg of the pure derivative is obtained after crystallization from 4 ml of methanol; *o*-chlorobenzaldehyde yields 25 mg of the pure derivative. With piperonal it is possible to obtain a pure derivative with 25 micrograms.

Semicarbazones

Discussion

The formation of semicarbazones from carbonyl compounds is represented by Eq. (10.17). Generally, carbonyl compounds react rapidly

$$C_6H_5CHO + H_2NNHCONH_2 \longrightarrow C_6H_5CH{=}NNHCONH_2 + H_2O \qquad (10.17)$$

Benzaldehyde   Semicarbazide   Benzaldehyde
                                      semicarbazone

with semicarbazide to yield crystalline derivatives, so that in many cases the separation of crystals begins upon warming of the reaction mixture. However, this fact should not be taken as evidence that all carbonyl compounds react readily with semicarbazide for, with the lower aldehydes, the time required for complete reaction is several days. For example, a mixture of formaldehyde and semicarbazide does not yield a crystalline derivative even after 10 days; if sodium acetate is omitted from the reaction mixture, an amorphous polycondensation product of formaldehyde and semicarbazide is formed. Acetaldehyde forms a soluble semicarbazone that is not easily isolated. Complications may also arise on prolonged standing or heating in order to complete the reaction, such as the formation of acetylsemicarbazones and hydrazodicarbonamides. Therefore, it is advisable not to attempt the preparation of semicarbazones of aldehydes with less than five carbon atoms.

Semicarbazones from unsymmetrical carbonyl compounds are capable of existing in two stereoisomeric forms. Thus a pure isomer may undergo a rearrangement under the effect of heating and be converted to a mixture that melts at a lower temperature than either of the pure isomers. Hence it is advisable, when the melting point of the derivative is lower or higher than that listed in the literature, that the same derivative be prepared from a pure sample of the compound tentatively identified as identical with the unknown and its melting point determined in the same apparatus using the same rate of heating.

General Method for Preparation

The semicarbazones are formed when the carbonyl compounds are heated with aqueous solutions of the semicarbazide hydrochloride buffered with sodium acetate. The ratio employed is 3 millimoles of semicarbazide hydrochloride in 2 ml of water buffered with about 4 millimoles of sodium acetate to 2.75 millimoles of the carbonyl compound. The mixture is heated for 10–15 minutes and then allowed to stand. In the cases of the higher carbonyl compounds the separation of the derivative begins immediately, but in most cases the reaction is complete after 1 hour. The crude derivative is filtered and crystallized from the minimum amount of water. When the solubility of the semicarbazone in water is low (as in the case of the semi-carbazones of the aromatic carbonyl compounds), the derivative is crys-tallized from methanol or ethanol and water. When the semicarbazone is sparingly soluble in hot alcohol, purification may be effected by extracting the derivative with a small amount of the hot solvent.

Example 10.15: Preparation of the Semicarbazone of an Aldehyde: Benzaldehyde Semicarbazone

Place in an 8-inch tube 100 mg of semicarbazide hydrochloride, 150 mg of sodium acetate, 1 ml of water, and 1 ml of alcohol. Add 0.1 ml of benz-aldehyde and heat tube in a water bath at 70° for 10 minutes. Add 2 ml of water and cool. Filter the crystals and wash with two 1-ml portions of water. Recrystallize from a mixture of 6 ml of methanol and 2 ml of water. The yield is 80 mg, melting at 221–222°.

Note: The semicarbazones of aromatic aldehydes have melting points usually above 200° and care must be taken to apply proper thermometer corrections.

Dimethone Derivatives

Discussion

The reaction of an aldehyde with dimethylcyclohexanedione, commonly known as "methone" reagent (also called in the literature dimethyldihy-droresorcinol), is represented by Eq. (10.18). One mole of aldehyde con-

$$ (10.18) $$

Octahydroxanthene derivative

denses with 2 moles of the reagent, and therefore derivatives are often named with the prefix of the aldehyde and the ending dimethone or "dimedone," for example, formaldimethone and acetaldimethone. The reaction is not given by ketones and is helpful for detecting traces of the aldehydes. The dimethones are more suitable for the lower aldehydes than are the 2,4-dinitrophenylhydrazones.

The methone derivatives of most aldehydes can be made to undergo cyclization to give octahydroxanthenes. The usual method is to heat the methone derivative with acetic anhydride or with alcohol containing a small amount of hydrochloric acid. The cyclization usually requires 5 minutes, and the yield is nearly quantitative. In most cases the melting points of the new derivatives (octahydroxanthenes) differ by more than 15° from those of the methones; therefore it is possible to prepare two different derivatives through the reaction. The methone is first prepared, and after the determination of the melting point it is then converted by cyclization to the xanthene derivative.

General Method for Preparation

A ratio of 2.1 millimoles of the reagent to 1 millimole of the aldehyde is employed. The reagent is dissolved in 3–4 ml of 50 per cent aqueous methanol or ethanol; the carbonyl compound is added and the mixture is heated to boiling for about 30 seconds and then allowed to crystallize for 3–4 hours. Prolonged heating is avoided in order to prevent the cyclization to the xanthene derivative. The crystals are filtered and crystallized by dissolving in the minimum amount of alcohol and then adding water until cloudiness appears.

Example 10.16: Preparation of Dimethone of Butanal

Place in a test tube 300 mg of methone, 3 ml of a 50 per cent methanol–water mixture, 50 mg of butanal, and 1 drop of piperidine. Heat in a water bath under reflux for 10 minutes. If the solution is clear at this point, add water dropwise until a cloudiness appears, then cool the mixture. Filter and wash the crystals with two 1-ml portions of 30 per cent methanol. The crystals melt at 134–135°; the yield is 120–130 mg. If crystallization is necessary, the solid is dissolved in 1 ml of methanol and water is added dropwise until a permanent cloudiness results; the solution is warmed until clear and then cooled. About 60 mg of the methone derivative are obtained and additional amounts of crystals separate out from the filtrate on standing.

Note: The preparation of methone derivatives is recommended when only a small amount of the aldehyde is available. For example, 1 drop of butanal is added to a solution of 50 mg of methone, dissolved in 0.5 ml of methanol in a small test tube, and allowed to stand for 4 hours. A sufficient amount of the derivative is obtained for several determinations of the melting point. The methone derivatives separate out slowly from solutions; a cloudy filtrate is an indication that crystallization is incomplete; in such cases the mixture is corked and allowed to stand in the cold overnight. For the cyclization of the methone derivative, 50–100 mg of the crystals is dissolved in 2–4 ml of a hot 80 per cent methanol–water mixture; 1 drop of concentrated hydrochloric acid solution is added and the solution is heated under reflux for 5 minutes. Water is added dropwise until a cloudiness appears. The xanthene derivative separates out on cooling. The values of the melting points of the xanthene derivatives do not appear in the tables of this text, but can be found by consulting the original article cited in the bibliography section.

Example 10.17: Chromatographic Separation and Detection of Small Quantities of Aldehydes

Place in a 6-inch tube 0.1 ml each of an alcoholic solution of an aliphatic aldehyde (propanal or butanal) and an aromatic aldehyde (benzaldehyde or piperonal) containing 0.1 mg per milliliter. Add 1 ml of a solution (alcoholic) of 2,4-dinitrophenylhydrazine hydrochloride containing 0.1 mg per milliliter. The tube is securely closed with a tinfoil-lined stopper and placed in an oven at 55° for 24 hours. Alternately, the tube can be heated at 50–55° in a water bath for 2–3 hours and then allowed to stand for a week. The tube is opened and diluted with 9 ml of methanol. By means of pipets, 10 μl and 20 μl are placed on separate strips which fit 6- or 8-inch tubes and then developed by the ascending technique (pages 165–66), using a 95:5 hexane–ether mixture saturated with water. The time required for the solvent front to reach near the top is about 30–45 minutes. The strips are removed and the solvent front is marked lightly by a pencil. After evaporation of the solvent, the strips are sprayed with a 5 per cent solution of potassium hydroxide in methanol. The spot caused by the excess reagent will be found at the point of origin ($R_f = 0$). The spot due to the aliphatic aldehyde will be found near the top, and the aromatic aldehyde near the middle. Determine accurately the R_f value of each.

References

2,4-Dinitrophenylhydrazones

Derivatives of aminocarbonyl compounds: Johnson, *J. Am. Chem. Soc.* **73**, 5888 (1951).

X-ray identification and crystallography of aldehydes and ketones as 2,4-dinitrophenylhydrazones: Clark, Kaye, and Parks, *Ind. Eng. Chem., Anal. Ed.* **18**, 310 (1946).

2,4-Dinitrophenylhydrazones: Allen, *J. Am. Chem. Soc.* **52**, 2955 (1930); Allen and Richmond, *J. Org. Chem.* **2**, 222 (1937); Brady, *Analyst* **51**, 77 (1926); *J. Chem. Soc.*, 756 (1931); Campbell, *Analyst* **61**, 391 (1936); Castillo, *Farm. mod.* **47**, 640 (1936); Iddles, Low, Rosen, and Hart, *Ind. Eng. Chem., Anal. Ed.* **11**, 102 (1939); Johnson, *J. Am. Chem. Soc.* **75**, 2720 (1953); Neuberg and Grauer, *Anal. Chim. Acta* **7**, 238 (1952); Perkins and Edwards, *Am. J. Pharm.* **107**, 208 (1935); Purgotti, *Gazz. chem. ital.* **24**, I, 555 (1894); Strain, *J. Am. Chem. Soc.* **57**, 758 (1935).

Microscopical identification and polymorphism: Allen and Richmond, *J. Org. Chem.* **2**, 222 (1937); Braddock *et al., Anal. Chem.* **25**, 301 (1953); Brandstatter, *Mikrochim. Acta* **32**, 33 (1944); Marovic, *Mikrochim. Acta* **32**, 6 (1944); Sandulesco, *Helv. Chim. Acta* **19**, 1095 (1936).

Other Substituted Hydrazones

o, m, p-Bromobenzohydrazones: Kao, *J. Chinese Chem. Soc.* **4**, 69 (1936); *Science Repts., Natl. Tsing Hua Univ.* **4**, 62 (1936); Wang, *Science Repts., Natl. Tsing Hua Univ.* (A) **3**, 279 (1935).

p-Carboxyphenylhydrazones: Viebel *et al., Dansk Tidsskr. Farm.* **14**, 184 (1940); *Acta Chem. Scand.* **1**, 54 (1947); **2**, 545 (1948); *C.A.* **42**, 434 (1948); **43**, 5764 (1949).

o and *m*-Chlorobenzohydrazones: Sah and Wu, *Science Repts., Natl. Tsing Hua Univ.* **3**, 443 (1936); Sun and Sah, *Science Repts., Natl. Tsing Hua Univ.* (A) **2**, 359 (1934).

p-Chlorobenzohydrazones: Sah and Wang, *J. Chinese Chem. Soc.* **14**, 39 (1946); Shih and Sah, *Science Repts., Natl. Tsing Hua Univ.* (A) **2**, 353 (1934).

Dimethylhydrazones: Wiley *et al., J. Org. Chem.* **22**, 204 (1957).

2,4-Dinitromethylphenylhydrazones: Vis, *Rec. trav. chim.* **58**, 387 (1939).

α-(2,4-Dinitrophenyl)-α-methylhydrazones: Blanksma and Wackers, *Rec. trav. chim.* **55**, 655 (1936).

Diphenylhydrazones: Maurenbrecher, *Ber.* **39**, 3583 (1906).

β-Naphthylhydrazones: Chen and Sah, *Science Repts., Natl. Tsing Hua Univ.* **4**, 62 (1936); Lei, Sah, and Kao, *Science Repts., Natl. Tsing Hua Univ.* (A) **2**, 335 (1934).

o, m, and *p*-Nitrobenzenesulfonhydrazones: Cameron and Storrie, *J. Chem. Soc.,* 1330 (1934).

3-Nitrobenzohydrazones: Chen, *J. Chinese Chem. Soc.* **3**, 251 (1935); Meng and Sah, *Science Repts., Natl. Tsing Hua Univ.* (A) **2**, 347 (1934); Strain, *J. Am. Chem. Soc.* **57**, 758 (1935).

Nitroguanylhydrazones: Whitmore, Revukas, and Smith: *J. Am. Chem. Soc.* **57**, 706 (1935).

p-Nitrophenylhydrazones: Bamberger, *Ber.* **32**, 1806 (1899); **34**, 546 (1901); Petit, *Bull. soc. chim. France* **1948**, 141.

Semicarbazones

3,5-Dinitrophenylsemicarbazones: Sah and Tao, *J. Chinese Chem. Soc.* **4**, 506 (1936).

α-Naphthylsemicarbazones: Sah and Chiang, *J. Chinese Chem. Soc.* **4**, 496 (1936).

Phenylsemicarbazones: Sah and Ma, *J. Chinese Chem. Soc.* **2**, 32 (1934).

p-Phenylthiosemicarbazones, *p*-tolylthiosemicarbazones, 4-*p*-bromophenylthiosemicarbazones, and 4-*p*-chlorophenylthiosemicarbazones: Tisler, *Z. Anal. Chem.* **149**, 164 (1956); **150**, 345 (1956); *Proc. XV Intern. Congr. Pure Appl. Chem.* 1–29; *Z. Anal. Chem.* **151**, 187 (1956); **155**, 186 (1957).

Semicarbazones: Angla, *Ann. chim. anal. chim. appl.* **22**, 10 (1940); Michael, *J. Am. Chem. Soc.* **41**, 417 (1919); Shriner and Turner, *J. Am. Chem. Soc.* **52**, 1267 (1930); Thiele and Stange, *Ber.* **27**, 31 (1894); Zelinsky, *Ber.* **30**, 1541 (1897).

Semicarbazones (microscopic identification): Dunbar and Aaland, *Microchem. J.* **2**, 113 (1958).

Semicarbazones (reliability of melting points): Veibel, *Bull. soc. chim.* (4) **41**, 1410 (1947).

Thiosemicarbazones: Busch, *J. pract. Chem.* **124**, 301 (1930); Duval and Xuong, *Mikrochim. Acta* **1956**, 747; Freund and Schander, *Ber.* **29**, 2501 (1896); **25**, 2602 (1902); Kitamura, *J. Pharm. Soc. Japan* **57**, 51 (1937).

m-Tolylsemicarbazones: Sah, Wang, and Kao, *J. Chinese Chem. Soc.* **4**, 187 (1936).

o-Tolylsemicarbazones: Lei, Sah, and Shih, *J. Chinese Chem. Soc.* **3**, 246 (1935).

p-Tolylsemicarbazones: Sah and Lei, *J. Chinese Chem. Soc.* **2**, 167 (1934).

p-Xenylsemicarbazones: Sah and Kao, *Rec. trav. chim.* **58**, 459 (1939).

Methones

Dibromomethones: Voitila, *Suomen Kemistilehti*. **10B**, 14 (1937).

Methone derivatives: Klein and Linser, *Mikrochemie, Pregl Festscht.*, 204 (1929); Volander, *Z. anal. Chem.* **77**, 245 (1929); *Z. angew Chem.* **42**, 46 (1929); Weinberger, *Ind. Eng. Chem., Anal. Ed.* **3**, 365 (1931).

Methone derivatives (and their cyclization to xanthenes): Horning and Horning, *J. Org. Chem.* **11**, 955 (1946).

Other Derivatives

Aldehydes in cigarette smoke: Touey, *Anal. Chem.* **27**, 1788 (1955).

Azines: Braun and Mosher, *J. Am. Chem. Soc.* **80**, 2749, 3048 (1958).

Benzothiazoles and benzothiazolines: Lankelma and Sharnoff, *J. Am. Chem. Soc.* **53**, 2654 (1931).

Benzylidine aminomorpholines: Dugan and Handler, *J. Am. Chem. Soc.* **64**, 522 (1942).

1,3-Cyclohexadione derivatives: King and Felton, *J. Chem. Soc.*, 1371 (1948).

1,2-Bis(*p*-methoxybenzylamino)ethane derivatives: Billman *et al., J. Org. Chem.* **17**, 1375 (1952).

Mercaptals: Ritter and Lover, *J. Am. Chem. Soc.* **74**, 5576 (1952).

5-(1-Phenylethyl)semioxamazide derivatives: Leonard and Boyer, *J. Org. Chem.* **15**, 42 (1950).

Miscellaneous: Deriv. of bisulfite addition products, Von Wacek and Kratzl, *Ber.* **76**, 1209 (1943); Adams and Gerber, *J. Am. Chem. Soc.* **71**, 522 (1949); *p*-Dimethyl-aminoanils, Utzinger and Regenass, *Helv. Chim. Acta* **37**, 1901 (1954). Carboethoxy-, carbomethoxy-, and carboxyphenylhydrazones, Rabjohn and Barnstorff, *J. Am. Chem. Soc.* **75**, 2259 (1953); Zellner, *Monatsh.* **80**, 330 (1951). Iodo-, methyl-, and nitro-substituted benzohydrazones, Sah *et al., J. Chinese Chem. Soc.* **14**, 24, 31, 45 (1946); *Rec. trav. chim.* **59**, 349 (1949); Gaudemaris and Dubois, *Bull. soc. chim. France*, **1950**, 63. 2,4-Dinitrophenyl-, *α*-, and *β*-naphthylsemicarbazones, McVeigh and Rose, *J Chem. Soc.*, 713 (1945); Sah and Tao, *J. Chinese Chem. Soc.* **4**, 501 (1936). Thiosemi-carbazones and L-menthylsemicarbazones, Sah and Daniels, *Rec. trav. chim.* **69**, 1545 (1950); Woodward *et al., J. Am. Chem. Soc.* **63**, 120 (1941). *α*- and *β*-Naphthyl- and 5-phenylaminosemioxamazides, Sah *et al., J. Chinese Chem. Soc.* **14**, 39, 101 (1946). Substituted hydantoins, Henze and Speer, *J. Am. Chem. Soc.* **64**, 522 (1942). *N*-Methyl-*β*-carbohydrazidopyridinium *p*-toluenesulfonate deriv., Allen and Gates, *J. Org. Chem.* **6**, 596 (1941); Derivatives of cyclic aldehydes with 4-aminoantipyrine, Manns and Pfeifer, *Mikrochim. Acta* **1958**, 630; 1,3-Bis(*p*-chlorobenzyl) tetrahydroimidazoles, Chen and Billman, *C.A.* **52**, 16340.

Chromatographic Procedures

Cheronis and Levey, *Microchem. J.* **228** (1957).

Kirchner and Keller, *J. Am. Chem. Soc.* **72**, 1867 (1950).

Meigh, *Nature* **170**, 579 (1952).

Rice *et al., Anal. Chem.* **23**, 195 (1951).

Optical and Infrared Methods

Davidson and Christie, *J. Chem. Soc.* **1955**, 3389.
Grammaticakis, *Bull. soc. chim. France* **7**, 527 (1940); **8**, 38 J27 (1941).
Jones *et al., Anal. Chem.* **28**, 191 (1956).
Struck, *Mikrochim. Acta* **1956**, 1277.
Wiley *et al., J. Org. Chem.* **22**, 204 (1957).

AMIDES, IMIDES, AND UREAS

This group (amides, imides, and ureas) includes unsubstituted amides, such as acetamides and benzamides; substituted amides, such as anilides, *o*- and *p*-toluidides; imides, such as succinimides, phthalimides, and substituted imides; ureas and their many derivatives, such as the mono- and di-substituted ureas and ureids. Table 10.5 summarizes the few types of derivatives that may be employed for their characterization.

TABLE 10.5

Derivatives for the Identification of Amides, Imides, and Ureas

*Hydrolysis and derivatization of the hydrolytic products	Oxalates
*N-Xanthylamides	N-Acylphthalimides
*Mercury derivatives	p-Nitrobenzyl derivatives

* Recommended derivative.

By far the best method of preparing derivatives for this group of compounds is hydrolysis and identification of hydrolytic products. For example, the identification of the amides, $RCONH_2$, and substituted amides, $RCONHR'$, such as anilides, toluidides, and the like, is best accomplished by hydrolysis, as represented by Eqs. (10.19) and (10.20). The hydrolytic

$$RCONH_2 + H_2O + HCl \rightarrow RCOOH + NH_4Cl \qquad (10.19)$$

$$RCONHR' + NaOH \rightarrow RCOONa + R'NH_2 \qquad (10.20)$$

products are then characterized by methods suitable for identification of carboxylic acids and amines.

The formation of xanthylamides by reaction with xanthydrol can be applied to the direct derivatization of about twenty-five to thirty amides including urea and several ureids and substituted ureas. The formation of mercury salts has been applied to the derivatization of fifteen amides.

Characterization by Hydrolysis

Discussion

The hydrolysis of the simpler amides and substituted amides is effected by boiling the compound with 6N hydrochloric acid or with a 10–20 per cent solution of sodium hydroxide. Alkaline hydrolysis is usually faster than acid hydrolysis. Amides and substituted amides that are resistant to hydrolysis when boiled with aqueous solutions of acids or bases may be hydrolyzed by heating with 100 per cent phosphoric acid. Another method for the hydrolysis of resistant amides is to heat the compound at 200° in a 20 per cent solution of potassium hydroxide in glycerol.

If acid hydrolysis is used, the reaction mixture is made alkaline in order to separate the amine. If the original compound is an amide, the ammonia is identified by its odor or it may be distilled into a receiver containing dilute hydrochloric acid and then tested with Nessler's solution or as in test 6.5. In the case of substituted amides, the amine that separates out may be isolated either by filtration, if it is a solid, or by extraction with ether or by distillation. Two or three extractions with ether will remove the amine from the reaction mixture. The ether extract is shaken with a small amount of hydrochloric acid. The amine, through salt formation, passes into the aqueous layer, which is then separated, carefully neutralized, and treated with an acid chloride as described in Examples 10.21 and 10.22.

After removal of the ammonia or amine the remaining alkaline solution is evaporated to a small volume and carefully neutralized. This solution is used for the preparation of the p-nitrobenzyl ester, as described in Example 10.8. If the acid is a solid, the concentrated alkaline solution is acidified and cooled, and the carboxylic acid that separates out is removed by filtration.

The general method for the hydrolysis of amides and derivatization of the hydrolytic products is illustrated in the following example describing the characterization of accetanilide.

Example 10.18: Hydrolysis and Characterization of Acetanilide

Place in a small distilling tube, fitted with a reflux condenser and with a side arm leading into a receiving tube, 5 ml of 10 per cent sodium hydroxide solution and 300 mg of acetanilide. The receiving tube contains 2 ml of 10 per cent hydrochloric acid, and the delivery tube reaches to about 20 mm above the surface of the acid. Add 2 boiling stones to the distilling tube and boil gently for 10–15 minutes. Remove the reflux condenser from the distilling tube and place it into the receiving tube (Fig. 2.13). Distill the reaction mixture until about 3 ml of distillate has been collected in the receiving tube containing the acid. Save the residue in the distilling tube for identification of the acid.

Add to the tube containing the distillate 5 ml of 10 per cent sodium hydroxide and 0.2 ml of benzenesulfonyl chloride and proceed according to the directions given in Example 10.22.

Pour the alkaline residue remaining in the distilling tube into an 8-inch tube. Wash the distilling vessel with 1–2 ml of water and add the washings to the mixture in the 8-inch tube. Add a drop of phenolphthalein and then cautiously add dilute hydrochloric acid until the color is just discharged and prepare the p-nitrobenzyl acetate as described in Example 10.8.

Note: Certain acyl-substituted amides, particularly those containing nitro and halogen groups, are resistant to hydrolysis. In such cases use is made of 100 per cent phosphoric acid About 300–400 mg of the amide is placed in a tube containing a mixture of 1.0 g of 85 per cent phosphoric acid and 400 mg of phosphorus pentoxide. The mixture is boiled gently with a very small flame for 1 hour under reflux to effect hydrolysis. The hydrolytic products are treated in the manner outlined.

Xanthyl Derivatives of Amides

Discussion

The reaction of amides with xanthydrol takes place according to Eq. (10.21). The N-xanthylamides (or 9-acylamidoxanthenes) crystallize

$$\text{Xanthydrol} + RCONH_2 \longrightarrow N\text{-Xanthylamide} + H_2O \qquad (10.21)$$

readily when xanthydrol is heated with the amide in the presence of acetic acid. The derivatives are purified by crystallization from aqueous dioxane or alcohol mixtures. The N-substituted amides and a number of other amides (oxamides, trichloroacetamides, salicylamides) do not form derivatives. Urea forms a derivative readily, whereas the substituted ureas and ureids (barbiturates) react more slowly so that heating for 30–45 minutes may be necessary.

The general method for the formation of xanthyl derivatives is to dissolve 2 millimoles (400 mg) of xanthydrol in 5 ml of glacial acetic acid and, after adding 1–1.5 millimoles of the amide, warm the mixture for 10–30 minutes; upon cooling the xanthylamide separates out. The crystals are filtered and crystallized from a mixture of 70 per cent dioxane or alcohol and 30 per cent water.

When the amide is not soluble in acetic acid it is first dissolved in 2 ml of ethanol and this is added to the acetic acid. After heating, 1 ml of water is added and the mixture is cooled.

Example 10.19: Xanthyl Derivative of Acetamide

Place 400 mg of xanthydrol and 4 ml of glacial acetic acid in a 6-inch tube. Shake at intervals for 5 minutes until the xanthydrol dissolves. If an oil separates, decant the clear supernatant solution into another tube. Add to the xanthydrol solution 100 mg of acetamide and place in a beaker of water heated to 80–85° for 15–20 minutes, then cool for 30 minutes. Filter off the solid and recrystallize it from a mixture of 70 per cent dioxane and 30 per cent water. The yield is 100–120 mg of crystals, melting at 238–240° uncorrected.

Mercuric Salts and Other Derivatives of Amides

Mercuric Salts

The formation of a salt through the weak acidic properties of the amides is represented by Eq. (10.22). The mercury salts are formed by

$$2RCONH_2 + HgO \longrightarrow (RCONH)_2Hg + H_2O \qquad (10.22)$$
$$\text{Mercuric salt}$$

heating a mixture of mercuric oxide and the amide in alcohol; an alternate method consists in heating a mixture of the oxide and the amide to the melting point of the amide. The melting points of fifteen mercuric derivatives have been described; most of these are above 200°.

The general method is illustrated by the derivatization of benzamide described in Example 10.20.

Example 10.20: Mercuric Derivative of Benzamide

Place in an 8-inch tube arranged for reflux 300 mg of benzamide, 400 mg of finely powdered mercuric oxide, and 4 ml of methanol. Add a boiling stone and boil gently for 30 minutes. Filter the hot mixture with suction. Cool the filtrate in an ice-water mixture for 10–15 minutes. Filter off the solid and wash with 1–2 ml of cold methanol. The yield is 90–100 mg of crystals, melting at 222° uncorrected.

Note: An alternate method for the preparation of the mercury derivatives is to place the mixture of amide and mercuric oxide in an 8-inch tube and then heat by means of a small flame until the amide melts and the reaction begins. If the yellow color is not discharged, more amide is added in small portions until the color is removed completely. The mixture is heated for a few minutes and then allowed to cool. About 3–4 ml of alcohol is added and the mixture is heated until the solid dissolves; the solution is allowed to crystallize.

Since the derivatives of aliphatic amides are soluble in cold ethanol and methanol, the amount of solvent should be regulated. On the other hand, the derivatives of aromatic amides have low solubilities and in certain cases purification is effected by leaching out the unreacted amide with boiling alcohol.

References

Derivatives of Amides, Imides, and Ureas

Barbiturates derivatized directly: Castle and Poe, *J. Am. Chem. Soc.* **66**, 1440 (1944). Identification of barbiturates: Bush, *Microchem. J.* **5**, 73 (1961).

Hydrolysis of amides with phosphoric acid: Dehn and Jackson, *J. Am. Chem. Soc.* **55**, 4285 (1933).

Hydrolysis of arylamides without affecting alkoxyl groups: MacGregor and Wilson, *J. Soc. Dyers Colourists* **55**, 449 (1939).

Potassium salts of imides reaction with alkyl halides to form *N*-alkylimides: Gabriel and co-workers, *Ber.* **20**, 2224 (1887); **21**, 566 (1888); **22**, 2220 (1889); **25**, 3056 (1892); **26**, 2197 (1893); **35**, 3805 (1902).

Saponification of amides and nitriles: Olivier, *Rec. trav. chim.* **46**, 600 (1927).

Urethanes and isocyanates converted to sulfamic acids: Bieber, *J. Am. Chem. Soc.* **75**, 1405 (1953).

N-Acylphthalimides: Evans and Dehn, *J. Am. Chem. Soc.* **51**, 3651 (1929).

m-Bromobenzazides for identification of amides: Sah and Chang, *Rec. trav. chim.* **58**, 8 (1939).

Hydrazides for amides and ureas: Sah, *Rec. trav. chim.* **59**, 1036 (1940).

Mercury derivatives of amides: Williams *et al., J. Am. Chem. Soc.* **64**, 1738 (1942).

Oxalates of amides: McKenzie and Rawles, *Ind. Eng. Chem., Anal. Ed.* **12**, 737 (1940).

N-Xanthylamides: Adriani, *Rec. trav. chim.* **35**, 180 (1916); Kny-Jones and Ward, *Analyst* **54**, 574 (1929); Phillips and Frank, *J. Org. Chem.* **9**, 9 (1944); Phillips and Pitt, *J. Am. Chem. Soc.* **65**, 1355 (1943).

Xanthyl derivatives of ureas, monosubstituted ureas, and ureides; *p*-nitrobenzyl chloride or bromide for derivatives of barbituric acids: Jespersen *et al., Dansk Tidsskr. Farm.* **8**, 212 (1935); *C.A.* **29**, 3459 (1935).

AMINES

Table 10.6 summarizes the derivatives proposed in the literature for the characterization of primary, secondary, and tertiary amines. The majority of these are: (*a*) substituted amides formed by the reaction of amines with acyl or aroyl halides; (*b*) substituted thioureas formed with isothiocyanates; (*c*) salts with acids.

Benzamides of Amines

The acetyl derivatives of alkylamines, although readily formed, have low melting points and are not useful as derivatives. Some arylamines form suitable derivatives, but in most cases the benzoyl derivatives are preferable.

A substituted benzamide is formed by reaction of the amine with benzoyl chloride, as shown by Eq. (10.23). The best procedure for benzoyla-

$$C_2H_5NH_2 + C_6H_5COCl \longrightarrow C_6H_5CONHC_2H_5 + HCl \qquad (10.23)$$

Ethylamine   Benzoyl chloride   *N*-Ethylbenzamide

tion is to suspend the amine in an aqueous alkaline solution and add the aroyl chloride in small amounts with vigorous shaking, keeping the mixture cold; this method is often called the *Schotten-Baumann method* or *reaction.*

TABLE 10.6

Derivatives for the Identification of Amines

*Benzamides	*Salts*
*Benzenesulfonamides	
*Phenylthioureas	3,5-Dinitrobenzoates
*Picrates	Arylsulfonates
*Quaternary ammonium salts	2,4-Dinitrobenzoates
	β-Resorcylates
Substituted ureas and thioureas	Acetates and benzoates
	p-Nitrophenyl acetates
α-Naphthylthioureas	Ethyl sulfone-bis acetates
α-Naphthylureas	Picramides
p-Iodo-, p-bromo-, m-bromophenylureas	2-Nitro-1,3-indandiones
β-Naphthylureas	Dibenzofuran-2-sulfonates
p-Chlorophenylureas	Tetraphenyl borates
3,5-Dinitrophenylureas	
p-Nitrophenylureas	*Miscellaneous derivatives*
3,5-Dinitro-4-methylphenylureas	
N-Nitro-N'-2,4-dinitrophenylureas	Alkanolamine hydrochlorides
P-Chlorobenze thioureas	2-Isonitrosocyclohexane derivatives
Phenyl and o-tolyl thioureas	2,4,5-Trinitrotoluene derivatives
Xenylthioureas	N-(2-Naphthyl)nitroamines
	Nitrophenacyl derivatives
Substituted amides	N-Substituted dinitroanilines
	N-Substituted trinitroanilines
Acetamides	
p-Bromo- and m-nitrobenzenesulfonamides	
Benzylsulfonamides	
Methanesulfonamides	
2,4-Dinitro- and o-nitrobenzenesulfenylamides	
3-Nitrophthalimides	
p-Phenylazobenzamides	

* Recommended derivative.

General Method for Preparation

About 1 millimole of the amine is suspended in an aqueous solution of sodium hydroxide (0.6–1.0 ml of 10 per cent solution), and 2–3 millimoles of benzoyl chloride is added in small amounts, with vigorous shaking, while the mixture is kept cold. After about 5–10 minutes of shaking, the reaction mixture is carefully neutralized to about pH 8.0; this ensures the separation of the derivatives of the primary amines, $RNHCOC_6H_5$, which are somewhat soluble in strong alkaline media, owing to the presence of an aminohydrogen atom. The N-substituted amides separate out usually in lumpy or granular masses and are filtered, washed thoroughly with water, and crystallized from an alcohol–water mixture.

For the preparation of *p*-nitrobenzoyl or 3,5-dinitrobenzoyl derivatives the aroyl chloride is dissolved in 2–3 ml of dry benzene, and after addition of the amine it is shaken with 1 ml of 10 per cent sodium hydroxide for 10–15 minutes. The crude derivative is separated by evaporating the benzene solution. If this procedure is not satisfactory, the amine and aroyl chloride are mixed with 2 ml of pyridine and then boiled for 30 minutes. The *N*-substituted *p*-nitrobenzamides and 3,5-dinitrobenzamides have melting points above 200° and, therefore, are not recommended as derivatives unless other more suitable ones cannot be prepared.

Example 10.21: Benzoylation of Ethylamine

In an 8-inch tube, provided with a solid rubber stopper, place 0.4 ml of an aqueous (33 per cent) solution of ethylamine; add, by means of a pipet dropper, 0.6 ml of benzoyl chloride and then 6 ml of 10 per cent solution of sodium hydroxide. Stopper the tube and shake for 1 minute, then at intervals over a period of about 5 minutes. After each shaking, carefully release the rubber stopper (*preferably in a hood, since the vapors of benzoyl chloride have lacrymatory properties*). The oil that separates at first soon crystallizes in shiny plates. Cool and filter the crystals; wash twice with water. Neutralize the filtrate cautiously to about pH 7–8 in order to precipitate an additional amount of the derivative dissolved by the excess of alkali. Dissolve the combined solid in 1–2 ml of boiling methanol and filter; wash the tube with 0.5 ml methanol and add the washings to the filtrate. To the filtrate add 3 ml of water and cool for 30 minutes. If crystals do not separate within 5 minutes, scratch the inner side of the tube by means of a glass rod. Filter and wash twice with 1 ml of water. The yield is 40–50 mg, melting at 70–71°.

Note: The benzoyl derivatives of the primary amines have a tendency to dissolve in excess of alkali; it is advisable, therefore, to neutralize the filtrate with dilute hydrochloric acid. It is evident, however, that the precipitated derivative from the filtrate will contain a small amount of benzoic acid resulting from the hydrolysis of the reagent; the amount of benzoic acid that is coprecipitated can be kept to a minimum if on neutralizing with dilute acid the pH is adjusted (with the aid of Universal indicator or Hydrion paper) to about 8.

Ethyl-*p*-aminobenzoate and anthranilic acid are benzoylated easily by the same method. The reaction mixture, however, should be cautiously neutralized with dilute hydrochloric acid in order to precipitate completely the derivative. From 100 mg of ethyl-*p*-aminobenzoate, 120–130 mg of pure benzoyl derivative is obtained, having a melting point of 147–148°; similarly, 100 mg of anthranilic acid yields 110–120 mg of the benzoyl derivative, melting at 180–181°.

Sulfonamides of Amines

Discussion

The formation of various substituted sulfonamides is represented in Eqs. (10.24) and (10.25). Many substituted benzenesulfonamides have been

$$
\left.\begin{array}{l} \underset{\text{Aniline}}{C_6H_5NH_2} \\[2mm] \underset{\text{Methylaniline}}{C_6H_5NHCH_3} \end{array}\right\} + 2C_6H_5SO_2Cl \longrightarrow \left[\begin{array}{l} \longrightarrow \underset{\text{N-Phenylbenzenesulfonamide}}{C_6H_5SO_2NHC_6H_5} + HCl \\[2mm] \longrightarrow \underset{\text{N-Methyl-N-phenylbenzenesulfonamide}}{C_6H_5SO_2N(CH_3)C_6H_5} + HCl \end{array}\right. \quad (10.24)
$$

$$
\underset{\text{Dimethylamine}}{(CH_3)_2NH} + \underset{\text{p-Toluenesulfonchloride}}{CH_3C_6H_4SO_2Cl} \longrightarrow \underset{\text{N,N-Dimethyl-p-toluenesulfonamide}}{CH_3C_6H_4SO_2N(CH_3)_2} + HCl \quad (10.25)
$$

described in the literature as suitable derivatives for amines; the following partial list is arranged in decreasing order of utility for semimicro work: *p*-toluenesulfonamides, *p*-bromobenzylsulfonamides, *m*-nitrobenzenesulfonamides, and methanesulfonamides.

A reagent related to the aryl sulfonyl chlorides for the derivatization of amides is the aryl sulfenyl chloride, Ar—SCl, which on reaction with amines forms amides of the sulfenic acid, according to Eq. (10.26). The advantage

$$
ArSCl + 2R_2NH \longrightarrow \underset{\text{Sulfenamide}}{ArSNR_2} + R_2NH \cdot HCl \quad (10.26)
$$

of these reagents is that the reaction takes place rapidly at room temperature. The *o*-nitrobenzenesulfenyl chloride is prepared by the action of chlorine on *o,o'*-dinitrodiphenyl disulfide, whereas the 2,4-dinitrobenzenesulfenyl chloride is prepared by a similar method from 2,2',4,4'-tetranitrodiphenyldisulfide.[8]

General Method for Preparation

The benzenesulfonamides and *p*-toluenesulfonamides are prepared by mixing 1 millimole of the amine, 1.2–1.5 millimoles of the sulfonyl chloride, and 4–5 ml of 10 per cent sodium hydroxide solutions and shaking the mixture intermittently for 3–5 minutes. If the crystals do not separate out at once, the reaction mixture is carefully acidified with 6N hydrochloric acid to pH 6.0. The granular mass of crystals is filtered, washed thoroughly with water, and crystallized from an alcohol–water mixture.

Most of the other substituted sulfonamides (*p*-bromobenzene-, *p*-nitrobenzene-, *m*-nitrobenzene-, *α*-naphthyl-, benzyl- and methanesulfonamides), are prepared by heating for 5–10 minutes a mixture of 1 millimole of the chloride, 2.1 millimoles of the amine, and 4 ml dry benzene and then allowing the mixture to cool. The amine hydrochloride that separates out is removed by filtration or centrifugation and the clear solution is evaporated

---

[8] For methods see citations of original papers in the reference section.

to obtain the crude sulfonamide, which is purified by crystallization from alcohol. For a discussion of the use of benzenesulfonchloride for the separation of primary, secondary, and tertiary amines (often called Hinsberg's method), the reader is referred to the larger treatise of the authors.[9]

Example 10.22: Preparation of the Benzenesulfonyl Derivative of Aniline

Place in an 8-inch test tube 0.1 ml of aniline, 0.2 ml of benzenesulfonyl chloride, and 5 ml of 10 per cent sodium hydroxide solution. Stopper the tube by means of a solid rubber stopper and shake at frequent intervals for 3 minutes. Remove the stopper and warm the tube; then shake again for 1 minute. Cool in running water and then add carefully dilute hydrochloric acid solution to neutralize the excess of sodium hydroxide. Filter the crystals and wash twice with 3 ml of water. Dissolve the derivative in 8 ml of hot methanol, filter, and add 3 ml of water to the filtrate. The product melts at 109–110° and requires an additional recrystallization to give a melting point of 111–112°. The yield of the pure sulfonamide is about 140–150 mg.

Substituted Thioureas of Amines

Discussion

Substituted thioureas are formed by reaction of amines with isothiocyanates, as shown by Eqs. (10.27) and (10.28).

$$CH_3(CH_2)_4NH_2 \ + \ C_6H_5NCS \ \longrightarrow \ C_6H_5NHCSNH(CH_2)_4CH_3 \quad (10.27)$$

n-Amylamine   Phenylisothiocyanate   *sym-n*-Amylphenylthiourea

$$(C_4H_9)_2NH \ + \ C_6H_5NCS \longrightarrow C_6H_5NHCSN(C_4H_9)_2 \quad (10.28)$$

Di-*n*-butylamine   *sym*-Phenyl-di-*n*-butylthiourea

General Method for Preparation

For the preparation of thioureas a solution of 1 millimole of the isothiocyanate in 1–2 ml of alcohol is mixed with 1.1–1.3 millimoles of the amine and the mixture is gently refluxed for 5–10 minutes. To the warm mixture water is added dropwise until a permanent cloudiness occurs, and the mixture is chilled. The derivative usually separates out as an oil, which solidifies when it is carefully rubbed by means of a glass rod against the walls of the vessel. The solid is dried rapidly by pressing between filter paper and extracted with 1–2 ml of petroleum ether (commercial hexane or heptane). A minute amount of the crude is saved for "seeding" and the rest is crystallized by dissolving in hot alcohol, adding water to cloudiness, seeding with a crystalline aggregate of the crude, and scratching the inner walls of the vessel.

[9] Cheronis and Entrikin, *op. cit.*, p. 412.

Example 10.23: Preparation of a Substituted Phenylthiourea from Methylamine: N-Methyl-N'-phenylthiourea

Place 0.15 ml (7 drops) of phenylisothiocyanate in an 8-inch test tube; add 2 drops of methylamine solution (33 per cent) and 1 ml of methanol so that a clear solution results. Heat for 10 minutes at 60–70°; if the tube is partially immersed in the water bath, the alcohol that boils off from the mixture condenses on the sides of the tube. If the boiling is brisk, either a reflux condenser is used or an additional 1 ml of methanol is added after about 5 minutes of heating.

Add to the reaction mixture, while it is still hot, 1.5 ml of water and cool the tube in tap water; by means of a glass rod scratch the inner sides of the tube until the oil that separates at first begins to crystallize. Allow the tube to stand in the cold bath for 10 minutes; then filter. Wash first with 2 ml of water to which 1 drop of 6N hydrochloric acid has been added, then with 1 ml of plain water. Transfer the solid into the 8-inch reaction tube and dissolve in about 1.5 ml of hot methanol. Filter and add to the hot solution 0.7 ml of water. Stir and scratch the inner sides of the tube until crystallization begins. Filter and wash the crystals with 1 ml of 25 per cent methanol. The yield is 110 mg of crystals, melting at 112–113°.

Note: As stated in the discussion of the general method, the main disadvantage of the thioureas is that these derivatives often separate as oils from the reaction mixture and are, at times, slow in crystallizing out. The derivatives of the lower alkylamines are quite soluble in alcohol and hence water must be added for separation. The addition of water invariably causes the separation of the thiourea as an oil. Therefore, in the purification of the crude thiourea, it is advisable to save a small amount of crystals so that a minute amount may be used to seed the oily mixture that separates from the filtered solution of the derivative upon cooling.

Quaternary Ammonium Salts of Amines

Discussion

A number of amine salts from primary, secondary, and tertiary amines are readily formed even if the amines have even small K_b's. For example, the hydrochlorides can be prepared readily by passing dry hydrogen chloride gas into a solution of the amine in ether, as shown by Eq. (10.29). The

$$\left.\begin{array}{l} RNH_2 \\ R_2NH \\ R_3N \end{array}\right\} + HCl \longrightarrow \begin{array}{l} [RNH_3]^+Cl^- \\ [RNH_2]^+Cl^- \\ [RNH]^+Cl^- \end{array} \qquad (10.29)$$

salts are filtered, washed with ether, and dried in a desiccator. However, the melting points of hydrobromides and hydrochlorides are not particularly suitable since these salts melt with decomposition at temperatures that are often dependent on the rate of heating The salts are important mainly as derivatives of tertiary amines. These compounds do not have amino hydrogen atoms; hence the reactions by which derivatives may be made are restricted to the formation of salts. Among the most useful of the salts are the picrates. These are easily formed by boiling the amine with a saturated solution of picric acid in methanol. The picrates may be purified by crystallization without appreciable decomposition.

Another type of derivative is formed by tertiary amines by the acceptance of an organic radical to form quaternary ammonium salts, as represented by Eqs. (10.30) and (10.31). The most useful quaternary salt types

$$R_3N : + R'^+X^- \longrightarrow [R_3N : R']^+ + X^- \tag{10.30}$$

$$\underset{\text{Dimethylaniline}}{C_6H_5N(CH_3)_2} + CH_3I \longrightarrow \underset{\text{Phenyltrimethylammonium iodide}}{[C_6H_5N(CH_3)_3]^+I^-} \tag{10.31}$$

of derivatives are those formed with methyl iodide, methyl-p-toluenesulfonate, and benzyl chloride. The quaternary methyl-p-toluenesulfonates are useful in the case of nitrogen cyclic compounds and less useful with other types of amines.

General Method for Preparation of Picrates of Tertiary Amines

One millimole of the tertiary amine dissolved in 5 ml of alcohol is mixed with 2–3 ml of a hot saturated solution of picric acid in methanol, and is then allowed to cool. The crystals that separate are filtered, washed with a small amount of methanol, and dried. If crystals do not separate on cooling the solution is concentrated to about 4 ml (see page 16). The picrates may be recrystallized from methanol or ethanol; however, in many cases no purification is necessary.

Example 10.24: Preparation of Dimethylaniline Picrate

Place into an 8-inch test tube 3 ml of a saturated solution of picric acid in methanol and 0.1 ml of N,N-dimethylaniline. Add 5 ml of methanol, boil for a few minutes using a reflux condenser, and allow to cool. Filter the crystals and wash twice with 1 ml of methanol. The yield is 180–200 mg of crystals, melting at 162–163°.

Preparation of Quaternary Salts of Tertiary Amines

The most suitable quaternary salts are the methiodides and methyl-p-toluenesulfonates. The general method is to mix equimolar quantities of the tertiary amine and the reagent and heat for a few minutes; the use of isopropyl ether or benzene facilitates the handling of small quantities of material. The following examples illustrate details of the general procedure.

Example 10.25: Preparation of the Methiodide of Tri-n-butylamine: Methyltri-n-butylammonium Iodide

Place 0.1 ml of tri-*n*-butylamine and 0.1 ml of methyl iodide in a test tube and add 1 ml of isopropyl ether. Heat under reflux for 5 minutes; then cool. Filter and wash the crystals with 1 ml of isopropyl ether. The yield is 180 mg of crystals, melting at 179–180°.

Example 10.26: Preparation of Pyridine Methyl-p-toluenesulfonate

Pour into a 6-inch test tube 0.2 ml of pyridine, 300 mg of methyl-*p*-toluenesulfonate, and 1 ml of isopropyl ether. Boil (under reflux) in a water bath for 20 minutes; then cool. Pour off the ether from the crystals and add 1 ml of methanol. Heat until a solution results and then add 5 ml of ethyl acetate. Cool and filter the product; wash with 1 ml of ethyl acetate and dry a small amount of crystals by pressing between filter paper; dry in the air for 2–3 minutes and then determine the melting point. If the melting point is below 139°, the product is recrystallized. The yield is 175–200 mg of crystals.

Note: The preparation of the methyl-*p*-toluene sulfonate is undertaken if the picrate is found unsuitable. In the case of pyridine, the picrate is easily prepared and purified.

References

Substituted Thioureas

4-Biphenylthioureas, β-naphthylthioureas, *p*-chlorophenylthioureas, *p*-xenylthioureas: Brown and Campbell, *J. Chem. Soc.*, 1699 (1937).

α-Naphthylthioureas: Suter and Moffett, *J. Am. Chem. Soc.* **55**, 2496 (1933).

Phenylthioureas and *o*-tolylthioureas: Brown and Campbell, *J. Chem. Soc.*, 1699 (1937); Fry, *J. Am. Chem. Soc.* **35**, 1544 (1913); Whitmore and Otterbacher, *J. Am. Chem. Soc.* **51**, 1909 (1929).

p-Xenylthioureas: Brewster and Honer, *Trans. Kansas Acad. Sci.* **40**, 101 (1937).

p-Chlorobenzyl thioureas: Tišler, *Z. Anal. Chem.* **165**, 272 (1959).

Substituted Amides

Benzamides, *p*-nitrobenzamides, benzenesulfonamides, *p*-toluenesulfonamides, and acetamides of long-chain amides: Sasin *et al.*, *J. Am. Oil Chemists' Soc.* **34**, 358 (1957).

Benzylsulfonamides: Marvel and Gillespie, *J. Am. Chem. Soc.* **48**, 2943 (1926).

p-Bromobenzenesulfonamides and *m*-nitrobenzenesulfonamides: Marvel *et al.*, *J. Am. Chem. Soc.* **45**, 2696 (1923); **47**, 166 (1925).

2,4-Dinitrobenzenesulfenamides: Billman *et al.*, *J. Am. Chem. Soc.* **63**, 1920 (1941).

Methanesulfonamides: Marvel *et al.*, *J. Am. Chem. Soc.* **51**, 1272 (1929).

N-Substituted 2,4-dinitroanilines using 2,4-dinitrochlorobenzene: Van Der Kam, *Rec. trav. chim.* **45**, 722 (1926).

N-Substituted 2,4,6-trinitroanilines using picryl chloride: Mulder, *Rec. trav. chim.* **25**, 108 (1906); Van Romburgh, *Rec. trav. chim.* **2**, 103 (1883); **4**, 189 (1885).

o-Nitrobenzenesulfenamides: Billman and O'Mahoney, *J. Am. Chem. Soc.* **61**, 2340 (1939).

p-Phenylazobenzamides: Woolfolk and Roberts, *J. Org. Chem.* **21**, 436 (1956); **22**, 827 (1957).

p-Phenylazobenzenesulfonamides: Woolfolk, Reynolds, and Mason, *J. Org. Chem.* **24**, 1445 (1959).

Sulfenamides: Billman *et al., J. Am. Chem. Soc.* **61**, 2340 (1939); **63**, 1920 (1941).

Salts

Amine salts of anthraquinone sulfonic acids: Perkin and Sewell, *J. Soc. Chem. Ind. (London)* **42**, 27T (1923).

Amine salts of arylsulfonic acids: Foster and Keyworth, *J. Soc. Chem. Ind.* **43**, 165T (1924); **43**, 299T (1924); Keyworth, *J. Soc. Chem. Ind.* **43**, 341T (1924); **46**, 20T (1927); **46**, 397T (1927); Noller and Liang, *J. Am. Chem. Soc.* **54**, 670 (1932).

Amine salts of 3,5-dinitrobenzoic and 2,4-dinitrobenzoic acids: Buelhler *et al., Ind. Eng. Chem., Anal. Ed.* **5**, 277 (1933); **6**, 351 (1934).

β-Resorcylates: Wilson *et al., Anal. Chem.* **23**, 1032 (1951).

Dibenzofuran-2-sulfonates (microscopic identification): Dunbar and Ferrin, *Microchem. J.* **4**, 163 (1960).

Other Derivatives

Acetylation of amines: Chattaway, *J. Chem. Soc.*, 2495 (1931); Kaufmann, *Ber.* **42**, 3480 (1909); Raiford *et al., J. Am. Chem. Soc.* **46**, 205 (1924).

Benzoylation of amines: Henstock, *J. Chem. Soc.* 216 (1933); Menalda, *Rec. trav. chim.* **49**, 967 (1930).

Detection of aniline vapors in air: Riehl and Heger, *Anal. Chem.* **27**, 1768 (1955).

Separation of amines by the Hinsberg reaction: Bell, *J. Chem. Soc.*, 2787 (1929); 1072 (1930); Herzog and Hancu, *Ber.* **41**, 636 (1908); Hinsberg, *Ber.* **23**, 2962 (1890); Ssolonina, *Centralblatt II*, 848 (1897); *II*, 867 (1898).

Alkanolamine hydrochlorides: Jones, *J. Assoc. Official Agr. Chem.* **27**, 467 (1944).

N-(Arylaminomethyl)phthalimides: Winstead and Heine, *J. Am. Chem. Soc.* **77**, 1913 (1955).

Diliturates (nitrobarbiturates). Optical properties of: Plein and Dewey, *Ind. Eng. Chem., Anal. Ed.* **15**, 534 (1943); **18**, 575 (1946).

N-(2-Naphthyl) nitroamines using 2-naphthol. Borodkin and Burmistrov: *J. Gen. Chem. (U.S.S.R.)* **17**, 63 (1947).

2-Nitro-1,3-indandione: Christensen *et al., Anal. Chem.* **21**, 1573 (1949); Wanag and Dombrowski, *Ber.* **75B**, 82 (1942); Wanag and Lode, *Ber.* **70**, 547 (1937); **69B**, 1066 (1936).

p-Nitrophenylacetamides of aromatic amines and color phenomena: Smirnov, *J. Gen. Chem. (U.S.S.R.)* **20**, 733 (1950).

p-Nitrophenylacetyl chloride: Ward and Jenkins, *J. Org. Chem.* **10**, 371 (1945).

3-Nitrophthalimides: Alexander and McElvain, *J. Amer. Chem. Soc.* **60**, 2285 (1938).

Picramides: Linke *et al., Ber.* **65**, 1282 (1932).

Substituted 2-isonitrosocyclohexane derivatives: Birch, *J. Chem. Soc.*, 314 (1944).

Sulfon-bis-acetamides: Alden and Houston, *J. Am. Chem. Soc.* **56**, 413 (1934).

2,4,5-Trinitrotoluene derivative: Barger, *Biochem. J.* **12**, 402 (1918).

Miscellaneous derivatives: Reineckates for tertiary amines, Aycock *et al., J. Am. Chem. Soc.* **73**, 1351 (1951). Quaternary salts of tertiary amines, Marvel *et al., J. Chem. Soc.* **51**, 3638 (1929). Compounds of amines with phenols and *p*-nitrobenzyl halides, Buehler *et al., J. Am. Chem. Soc.* **54**, 2398 (1932); Lyons, *J. Am. Pharm. Assoc.* **21**, 224 (1932). Ethylalkylaminomethylenemalonates, Lappen, *J. Chem. Educ.* **28**, 126

(1951). Halogenated quinone deriv., Buu-Hoi *et al., Rec. trav. chim.* **71,** 1059 (1952). Tetraphenyl borates of amines, Crane, *Anal. Chem.* **28,** 1794 (1956); Wendlandt, *Chemist Analyst* **47,** 6 (1958). Derivatives with isatoic anhydride, Staiger, *J. Org. Chem.* **18,** 1427 (1953).

Chromatographic Procedures

Baker *et al., J. Chem. Soc.* **1952,** 3215.
Bremmer and Kenten, *Biochem. J.* **49,** 651 (1951).
Burmistrov, *Zhur. Anal. Khim.* **5,** 39 (1950).
Cheronis, N. D., and J. B. Entrikin, *Semimicro Qualitative Organic Analysis* (2nd ed., N.Y.: Interscience, 1957), pp. 422–23.
Cuthbertson and Ireland, *Biochem. J.* **34,** 52 (1952).
Ekman, *Acta Chem. Scand.* **2,** 383 (1948).
James *et al., Biochem. J.* **52,** 238, 242 (1952).
Kariyoni and Hashiomot, *Nature* **168,** 739 (1951).
Roche and Lafon, *Bull. soc. chim. biol.* **33,** 1437 (1951).
Vitte and Boussemart, *C.A.* **45,** 7299f (1951).
Walker *et al., Australian J. Sci.,* **3,** 84 (1950).
Wickström and Salvesen, *J. Pharm. Pharmacol.* **4,** 631 (1952).
Woolfolk and Robert, *J. Org. Chem.* **21,** 436 (1956).

CARBOHYDRATES

Table 10.7 summarizes the various types of derivatives that are most commonly employed for the derivatization of carbohydrates. It should be

TABLE 10.7

Derivatives for the Identification of Carbohydrates

*Substituted phenylhydrazones	Benzimidazole derivatives
*Osazones	Tosyl esters
*Osotriazoles	Trityl ethers
Acetates and Benzoates	Thiobenzhydrazones
Azoates	

  * Recommended derivative.

pointed out that for the most part these derivatives are applicable to relatively simple sugars—that is, monosaccharides—and to a lesser degree to disaccharides.

In general, although the sugars undergo a large variety of reactions, their derivatization is a rather difficult matter, particularly when the sample contains traces of related compounds; the present discussion, therefore, is limited to the derivatization of pure sugars. The reader is referred to standard works on sugars for their separation and identification in mixtures. The specific rotation of sugars and their derivatives is discussed on pages 61–62.

Substituted Phenylhydrazones, Osazones, and Osotriazoles

Discussion

The formation of phenylhydrazones and substituted phenylhydrazones is represented in Eq. (10.32). The formation of hydrazones is accomplished

$$
\begin{array}{c}
\text{H} \\
\text{C}{=}\text{O} \\
| \\
\text{HC—OH} + \text{H}_2\text{NNHC}_6\text{H}_5 \\
| \\
\text{R}
\end{array}
\longrightarrow
\begin{array}{c}
\text{H} \\
\text{C}{=}\text{NNHC}_6\text{H}_5 \\
| \\
\text{HC—OH} \qquad + \text{H}_2\text{O} \\
| \\
\text{R}
\end{array}
\qquad (10.32)
$$

Hexose   Phenylhydrazine   Hexose phenylhydrazone

by treatment of the sugar with a little more than an equimolecular quantity of phenylhydrazine. The sugar is dissolved in a small quantity of water (100 mg/ml), and the required amount of phenylhydrazine in an equal volume of 50 per cent acetic acid is added; the mixture is allowed to stand for 24 hours in the cold. In the case of substituted phenylhydrazines a small amount of alcohol is used in the reaction mixture; for example, 100 mg of the sugar dissolved in 1 ml of water is mixed with 100 mg of *p*-nitrophenylhydrazine hydrochloride suspended in 1 ml of methanol. After standing, the hydrazones are filtered, washed with water, and crystallized by first dissolving in the minimum amount of hot methanol or ethanol and then precipitating by the cautious addition of water.

The hydrazones may be reconverted to the original sugar by reacting them with benzaldehyde or formaldehyde; the hydrazone of the aldehyde separates, leaving the sugar in solution. This property and the fact that the hydrazones of the various sugars separate with varying speeds make them useful in the separation of mixtures; the hydrazones are filtered off as they are formed and then are converted to the original sugar by treatment of the hydrazone with an aldehyde.

Among the most important substituted phenylhydrazines used for the preparation of derivatives are *p*-nitrophenylhydrazine (also the *m*- and *o*-isomers), p-bromophenylhydrazine, methylphenylhydrazine, diphenylhydrazine, and *β*-naphthylhydrazine.

The osazones are formed by reacting sugars with excess of phenylhydrazine, as represented by Eq. (10.33). The reaction takes place rapidly,

$$
\begin{array}{c}
\text{H} \\
\text{C}{=}\text{O} \\
| \\
\text{HC—OH} + 3\text{C}_6\text{H}_5\text{NHNH}_2 \\
| \\
\text{R}
\end{array}
\longrightarrow
\begin{array}{c}
\text{H} \qquad\qquad + 2\text{H}_2\text{O} \\
\text{C}{=}\text{NNHC}_6\text{H}_5 + \text{NH}_3 \\
| \\
\text{C}{=}\text{NNHC}_6\text{H}_5 + \text{C}_6\text{H}_5\text{NH}_2 \\
| \\
\text{R}
\end{array}
\qquad (10.33)
$$

Hexose   Phenylhydrazine   Hexose osazone

as compared with the formation of hydrazones, when a solution of the sugar is warmed with excess of the hydrazine. The mechanism of the reaction is assumed to be, first, the formation of the hydrazone, followed by oxidation of an adjacent carbon atom to the carbonyl stage and subsequent reaction with the hydrazine to produce the osazone. In the preparation of osazones the sugar solution is mixed with a solution of phenylhydrazine acetate, or phenylhydrazine hydrochloride and sodium acetate, in a tube and then heated in boiling water for 30 minutes. The time required for the formation of osazone may be used as additional evidence in the characterization of the unknown sugar, *provided the sample is a pure substance* and not contaminated with small amounts of other sugars. The exact time required for the appearance of the osazone, after the tube is immersed in boiling water, depends on several factors, such as the amount of sugar, reagent, pH of the solution, and the amount of solvent. Generally, however, the following descending ease of formation of phenylosazones is observed: fructose, sorbose, glucose, xylose, rhamnose, arabinose, galactose; sucrose undergoes hydrolysis and slowly forms (after about 20 minutes) a small amount of the glucosazone. The osazones of maltose and lactose are soluble in the hot solution and separate out only on cooling. It should be again noted that the time of formation of osazone is of value only in the case of pure sugars. The presence of impurities of other osazones greatly influences the rate of crystallization.

The purification of osazones must be undertaken immediately after they have been filtered and washed with cold water. A small amount is kept for the determination of the melting point, and the balance is dissolved in the minimum amount of hot methanol or ethanol and then precipitated cautiously by addition of water. It is not recommended to dry osazones in air. The phenyl-D-glucosazone does not show appreciable change when dried in air, but a number of other phenylosazones show extensive reduction of the melting point.

The melting points of the osazones are not to be considered with the same regard as the melting points of other derivatives because the melting points are really decomposition points that vary greatly, depending on the rate of heating. For example, the melting point of phenyl-D-glucosazone listed in the literature is 210°. For a given sample of the pure derivative the melting point of 210° will be observed if the rate of heating is 40–60° per minute. If the temperature is raised 8–10° per minute (which is regarded as very rapid in the usual practice near the melting point of a substance), the observed melting point will be below 200° and usually between 194–198°. It is obvious that under these conditions reproducibility of observations requires great care. The osazone of the sugar suspected to be the unknown should always be prepared for comparison.

Another limiting factor in the use of osazones for the characterization of sugars is the fact that a number of isomeric sugars give the same osazone.

The following serve as examples: D-glucose, D-mannose, and D-fructose; D-arabinose and D-ribose; D-xylose and D-lyxose. In addition, the corresponding sugars of the L-series, which have the same configuration beyond the second carbon atom, yield the same osazone, differing from the D-osazone only in the direction of the rotation.

Methylphenylhydrazine is useful for differentiating between aldoses and ketoses, which yield the same osazone. For example, D-fructose reacts readily with methylphenylhydrazine to form a characteristic osazone, whereas D-glucose and D-mannose do not.

The osazone may be converted to the osotriazole, as represented by Eq. (10.34).[10] In this reaction one of the phenylhydrazine groups is converted to

$$
\begin{array}{ccc}
& HC{=\!=}N & \\
& | \quad {\Large>}NC_6H_5 & \\
H & & \\
C{=\!=}NNHC_6H_5 & C{=\!=}N & \\
| & \xrightarrow{Cu^{++}} \quad | & \\
C{=\!=}NNHC_6H_5 & (HCOH)_3 & + C_6H_5NH_2 \quad\quad (10.34) \\
| & | & \\
(HCOH)_3 & CH_2OH & \\
| & & \\
CH_2OH & &
\end{array}
$$

Phenyl-D-glucosazone   Phenyl-D-glucosotriazole

aniline with formation of a ring containing three nitrogen (triazo) atoms. The phenylosotriazoles differ from the osazones by their sharp melting points, and hence they are recommended for confirmation of the identity of phenylosazones. The preparation of the phenylosotriazoles from 100 mg of the osazone is feasible.

General Method for Preparation of Substituted Hydrazones of Sugars

The ease of preparation of the hydrazone varies; for example, at room temperature fructose yields an impure phenylhydrazone but a relatively pure methylphenylhydrazone, whereas the action of glucose is the reverse. Similarly, the phenylhydrazone of mannose and the o-tolylhydrazone of galactose form readily and may be used for the characterization of these sugars. About 1 millimole of sugar and 2 millimoles of solid sodium acetate, dissolved in 2 ml of water, are mixed with 0.95 millimoles of the substituted hydrazine hydrochloride, dissolved in 2 ml of methanol. The mixture is allowed to stand 24–48 hours, then cooled; the crystals of the hydrazone are filtered and crystallized from 95 per cent ethanol.

Note: In preparing phenylhydrazones the amount of the base should be reduced to 0.9 millimoles per millimole of sugar. For pentoses benzylphenylhydrazine hydrochlo-

[10] Hudson *et al., J. Am. Chem. Soc.* **66,** 735 (1944); **67,** 939 (1945); **68,** 1769 (1946); **69,** 1050, 1461 (1947).

ride is useful. The benzylphenylhydrazone of arabinose (m.p., 174°) and xylose (m.p., 99°) form readily.

If crystals of the hydrazone do not separate out readily, water is added dropwise to turbidity and the mixture is cooled.

Example 10.27: Glucose p-Nitrophenylhydrazone

Place in a 6-inch tube 100 mg of p-nitrophenylhydrazine hydrochloride and 1 ml of methanol; shake for a few seconds and then add 100 mg of glucose, 150 mg of powdered sodium acetate, and 1 ml of water. Cork tube and shake gently so as to mix its contents; allow to stand overnight. Add 2 ml of water and filter the crystals; wash with water and crystallize from 5 ml of 95 per cent ethanol. The yield is 125 mg of crystals, which melt at 189–190°.

General Method for Preparation of Osazones and Conversion to Osotriazoles

Osazones are formed with excess of phenylhydrazine. The general methods for the preparation and purification of osazones and osotriazoles are illustrated in the examples given below.

Example 10.28: Rate of Osazone Formation

Place 100 mg of the sugar, 100 mg of sodium acetate, and 2 ml of water in a 6-inch tube. Add 5–6 drops (0.2 g) of phenylhydrazine and 6–7 drops (0.12 g) of glacial acetic acid. Close the tube loosely with a cork and set it in a 600-ml beaker half-filled with water that is already boiling. Note the time required for the appearance of the osazone. When the sample of sugar is pure, the following intervals of time (in minutes) are observed as measured from the moment the tube is immersed in the boiling water to the appearance of the osazone: mannose, 0.5–1; fructose, 1–2; glucose, 4–5; xylose, 6–8; arabinose, 9–10; galactose, 14–16; sucrose, 20–30 (by hydrolysis); and lactose and maltose, on cooling.

Note: Phenylhydrazine hydrochloride may be used in place of the base; in such a case, for 100 mg of sugar use 200 mg of phenylhydrazine hydrochloride, 300 mg of sodium acetate, and 2 ml of water. On standing the salt undergoes decomposition and darkens. It may be purified by crystallization from hot water; the salt is dissolved in the minimum amount of boiling water; the tarry impurities remain undissolved. A small amount of charcoal is added, and the hot solution is filtered rapidly. The solution is cooled, and concentrated hydrochloric acid is added so that the volume of the solution increases by one-third. After an hour the cold mixture is filtered, and the crystals are washed with ice water and dried.

Example 10.29: Preparation of Phenyl-D-glucosazone

Place in an 8-inch tube 100 mg of glucose, 100 mg of sodium acetate, 6 drops (0.21 g) of phenylhydrazine, 2 ml of water, and 7 drops (0.13 g) of glacial acetic acid. Immerse the tube in boiling water for 30 minutes. Add

5 ml of water and cool. Filter with suction and wash the tube and crystals—first with 2 ml of water to which 2 drops of acetic acid have been added and then twice with 3 ml of water. Remove a small amount of the osazone and dry in a vacuum desiccator. Transfer the rest of the crystals to the reaction tube and add 20 ml of methanol. Heat to boiling, adding more alcohol until practically all the osazone has dissolved. Filter and add 2–3 ml of water to the filtrate and cool in an ice-water mixture. Filter the crystals and wash twice with 2 ml of 25 per cent methanol. Dry in a vacuum desiccator. The yield is 90–100 mg. The melting point of the crystals, when determined by the capillary method in an oil bath with a temperature rise of 40–60° per minute, is 209–210° (corrected). The unrecrystallized material usually melts at 207–208°. The identity of phenyl-D-glucosazone is confirmed by conversion to the glucosotriazole as directed in Example 10.30.

Note: The formation of an osazone from an aldose (such as glucose) leads to the elimination of asymmetry in the α-carbon atom (to the carbonyl), and hence it yields the same osazone as the related ketose, D-fructose. The formation of osazone from fructose takes place with greater ease and a slightly better yield. Some authors recommend the use of pyridine in conjunction with alcohol for crystallization of the osazone. No advantage has been found for this solvent pair. When the osazone is required for determination of the optical rotation, it should be washed with acetone before crystallization. This operation is best accomplished just before the osazone is removed from the filter funnel for crystallization. About 2–3 ml of acetone is added to the crystals, and after a minute suction is applied and the washing repeated.

Many of the osazones can be identified by microscopical examination of the crystals. The original article cited in the bibliography section should be consulted.

Example 10.30: Preparation of Phenyl-D-glucosotriazole

Place 100–110 mg of glucosazone in an 8-inch tube; add 9 ml of water, 2 drops of 6N sulfuric acid, 300 mg of copper sulfate (pentahydrate), 6 ml of isopropyl alcohol, and 2 boiling stones. The mixture is boiled for 1 hour under reflux. The yellowish green solution is poured into an evaporating dish and concentrated over a water bath to a volume of 3–4 ml. Cool the dish in ice water and filter the granular crystals; dissolve the crude material in 12–14 ml of boiling water; add charcoal and filter. Cool the solution overnight in an icebox. Filter the derivative and wash twice with 1 ml of water. The yield is 16–18 mg of crystals, melting at 193–194°. To recrystallize the glucosotriazole, place the crystals in a 6-inch tube; add 1 ml of 95 per cent ethanol and heat to boiling. Add 1 ml of water to the clear solution and cool for 2 hours in an ice–salt mixture. Filter the crystals and wash with 0.5 ml of water. The yield is 13–14 mg of crystals, melting at 195–196°.

Note: In the above example 0.5 millimole of the osazone was used. It is recommended that a beginner start with 1 millimole of the osazone and proportionate amounts of the other reagents.

Specific Rotation of Carbohydrates and Their Derivatives

The specific rotation of sugars and their osazones, hydrazones, and azoates is often used as a means of identification, since it consists simply in the determination of the rotation of a solution of known concentration. However, the sample must be pure. The rotation must be measured under specified conditions as to quality of material and nature of solvent. For osazones a mixture of 40 per cent pyridine and 60 per cent alcohol is commonly used as a solvent; for azoates, alcohol-free chloroform is employed. The derivative must be of high purity. Special methods of purification are necessary when the original sample is an impure sugar. For example, glutose, for a long time, was considered a ketohexose sugar, that yielded a phenylosazone melting at 163–165°. However, it has been shown[11] to be a fructose anhydride mixture. The phenylosazone, which melts at 163–165°, is a mixture of glucosazone and of the osazone of methylglyoxal and may be recrystallized from alcohol and other osazone solvents without change in its melting point. When it is treated with dry acetone, the osazone of methylglyoxal dissolves, leaving the pure glucosazone. Therefore, it is recommended that osazones prepared from impure sugars be washed with acetone before final purification. The procedure for determination of optical rotation is described on page 61.

References

Armstrong and Armstrong, *The Carbohydrates* (N.Y.: Longmans, Green, 1934).

Bates *et al., Polarimetry and Saccharimetry and the Sugars* (Bur. Standards Circ. C440, 1942).

Bomer, Zuckenack, and Tillmans, *Untersuchungsmethoden,* 2nd part (Berlin: Springer, 1935).

Browne and Zerban, *Physical and Chemical Methods of Sugar Analysis* (3rd ed., N.Y.: Wiley, 1941). The most serviceable book in English on carbohydrates; contains a chapter on derivatives of sugars.

Jackson and Dehn, *Ind. Eng. Chem., Anal. Ed.* **6,** 382 (1934). Glycosides.

Micheel, *Chemie der Zucker und Polysaccharide* (Leipzig: Akademische Verlagsgesellschaft, 1939).

Tollens and Elsner, *Kurzes Handbuch der Kohlenhydrate* (4th ed., Leipzig: Barth, 1935).

Van der Haar, *Anleitung zum Nachweis zur Trennung und Bestimmung der Monosaccharide und Aldehydsäuren* (Berlin: Gebr. Borntraeger, 1920).

Vogel and Georg, *Tabellen der Zucker und ihrer Derivate* (Berlin: Springer, 1931).

Carbohydrate Tests

Devor, *J. Am. Chem. Soc.* **72,** 2008 (1950); *Anal. Chem.* **24,** 1626 (1951).

Feigl, *Spot Tests,* Vol. II (Houston and N.Y.: Elsevier, 1953).

Klein and Weissman, *Anal. Chem.* **25,** 771 (1953).

Steinmann, *Mikrochim. Acta* 1953/5, 537. Bibliography on micro tests.

[11] Sattler and Zerban, *Sugar* **39,** 12 (1944).

Hydrazones

Thiobenzhydrazones: Holmberg, *Arkiv Kemi* **4**, 33 (1952).
o-Tolylhydrazones of galactose, mannose: Fowweather, *Biochem. J.* **55**, 718 (1953).

Other Derivatives

Acetates: Hudson, *Ind. Eng. Chem.* **8**, 380 (1916).
Aldonic acids: Link *et al., J. Biol. Chem.* **150**, 345 (1943); **133**, 293 (1940); *J. Org. Chem.* **5**, 639 (1940).
Azoates: Coleman and McClosky, *J. Am. Chem. Soc.* **64**, 1401 (1942); **65**, 1588 (1943); Reich, *Biochem. J.* **33**, 1000 (1939).
Benzoates: Fischer and Noth, *Ber.* **51**, 321 (1918); Levene and Meyer, *J. Biol. Chem.* **76**, 513 (1928).
Osotriazoles: Hann and Hudson, *J. Am. Chem. Soc.* **66**, 735 (1944); **67**, 939 (1945); **68**, 1766 (1946); **69**, 1050, 1461 (1947).
Tosyl esters: Compton, *J. Am. Chem. Soc.* **60**, 395 (1938); Freudenberg *et al., Ber.* **55**, 929, 3233 (1922); **58**, 294 (1925); **59**, 714 (1926).
Trityl ethers: Helferich *et al., Ber.* **56**, 766 (1923); **58**, 872 (1925); Reynolds and Ewans, *J. Am. Chem. Soc.* **60**, 2559 (1938).
Derivatives of sugars: Pittenger, *Sugar Research Foundation Sci. Rept. Ser.* **5**, 51 (1947).

*Microdetection of Sugars by Paper Chromatography
and Infrared Spectra*

Bersin and Müller, *Helv. Chim. Acta* **35**, 475 (1952).
Boggs *et al., Nature* **166**, 520 (1950).
Cheronis, N. D., and J. B. Entrikin, *Semimicro Qualitative Organic Analysis* (2nd ed., N.Y.: Interscience, 1957), pp. 436–37.
Clark and Chianta, *Ann. N.Y. Acad. Sci.* **69**, 205 (1957). Bibliography of infrared spectra of carbohydrates.
Gordon *et al., Anal. Chem.* **28**, 849 (1956).
Hirst *et al., Nature* **163**, 177 (1949).
Hough *et al., Nature* **161**, 720 (1948); **164**, 1107 (1949); *J. Chem. Soc.*, 251 (1949).
Jeanes *et al., Anal. Chem.* **23**, 415 (1951).
Jermyn and Isherwood, *Biochem. J.* **44**, 402 (1949); **48**, 515 (1951).
Johanson, *Nature* **172**, 956 (1953).
Lemieux and Bauer, *Anal. Chem.* **26**, 920 (1954).
Partridge *et al., Nature* **158**, 270 (1946); **164**, 443, 479 (1949); *Biochem. J.* **42**, 238, 251 (1948).
Rafique and Smith, *J. Am. Chem. Soc.* **72**, 4634 (1950).
Sattler and Zerban, *Anal. Chem.* **24**, 826 (1952).
Schneider and Erlenmann, *C.A.* **46**, 381 (1952).
Smith, I. (ed.), *Chromatographic and Electrophoretic Techniques* (N.Y.: Interscience, 1960), p. 246.
Wiggins and Williams, *Nature* **170**, 279 (1952).
Wolfrom and Miller, *Anal. Chem.* **28**, 1037 (1956).

ESTERS

Table 10.8 lists the derivatives that have been proposed for the characterization of esters. There are two general approaches to the preparation

TABLE 10.8

Derivatives for the Identification of Esters

---

*Hydrolysis followed by derivatization of acidic and alcoholic components
*N-Benzylamides
p-Toluidides
Anilides
N-(β-Aminoethylmorpholides)
*3,5-Dinitrobenzoates
Hydrazides

---

* Recommended derivative.

of derivatives for the characterization of esters. One is to derivatize the ester directly, as, for example, react the ester directly with 3,5-dinitrobenzoyl chloride so as to obtain a derivative of the alcohol function and prepare the anilide or p-toluidide by reacting the ester with a Grignard reagent so as to obtain a derivative of the acid function. The other is to hydrolyze the esters and derivatize the resulting acids and alcohols. However, as will be discussed in the following sections, a judicious selection of the proper type of derivative and detailed procedure must be made with each case depending on the nature of the esters. Only the latter method will be discussed in this text; for the former the student is referred to the more extensive work of the authors.

In order to prepare derivatives for the final step of the identification of esters, it is necessary to hydrolyze the esters to their acidic and hydroxy components. The hydrolysis of semimicro quantities and isolation of 50–100 mg of an alcohol or an acid for derivatization require the utmost care. Even when the amount of ester used for hydrolysis is 2–3 g, the isolation of the hydrolytic products involves difficulties. The information obtained from the preliminary tests, boiling- or melting-point and refractive-index determinations, is of value in determining the probable nature of the ester; the type of hydroxy and acidic components present in the ester determines to a large extent the best method of procedure for the preparation of derivatives.

The most common esters are those of the lower alcohols having one to four carbon atoms. In the usual hydrolytic methods by aqueous alkali, the alcohol that is distilled after completion of hydrolysis contains a considerable amount of water and cannot be used for the preparation of nitrobenzoates or urethans. The preparation of benzoates through the Schotten-Baumann reaction is of little value for the lower alcohols, since these derivatives are chiefly liquids. On the other hand, the separation of anhydrous, or nearly so, methanol or ethanol from 5–10 ml of aqueous distillate is an extremely difficult operation requiring apparatus not available to the beginner. The general procedure for the separation of an alcohol is to saturate the distillate with potassium carbonate and then extract with ether (free from alcohol). This procedure gives fair results when the amount of ester is 5–10 g but is not satisfactory for the esters of lower alcohols, when the amount of ester hydrolyzed is less than 2 g.

Difficulties are also encountered in the identification of the acidic group of the esters. The common procedure is to evaporate the alkaline residue after distillation or extraction of the hydroxy compound and to use the sodium salt for the preparation of the p-toluidide, anilide, or other suitable derivative of the acid derived from the ester. The presence of excess alkali, however, complicates the preparation of such derivatives; as a consequence, even when one starts with 1 g of ester, little or no derivative of the acid is obtained with the use of such methods. This is particularly true in the case of the esters of the lower aliphatic carboxylic acids.

Consideration must be given to the effect of the excess alkali used in the hydrolysis upon labile functional groups such as are encountered in the β-keto esters, and in the esters of halogen acids; for example, alkali hydrolysis of ethyl acetoacetate will produce cleavage on the acetoacetic acid with the formation of acetone; similarly, alkaline hydrolysis of either α-chloro- or α-bromobutyrates will give rise to both crotonic and α-hydroxybutyric acids.

Finally, esters hydrolyze at vastly different rates. Most esters of alcohols with less than four carbon atoms hydrolyze when they are treated for 30 minutes or less with hot $6N$ sodium or potassium hydroxide solution. Esters boiling above 200° require from 1–2 hours for complete hydrolysis. The disappearance of the ester layer cannot be used as a criterion for completion of hydrolysis of compounds that are slightly miscible with water; as the hydrolysis proceeds, the hydroxy compound formed rises to the top and therefore, even at the completion of hydrolysis, there remains an immiscible layer.

From this brief discussion it is evident that the exact procedure to be followed in the hydrolysis of esters and identification of the hydroxy and acyl radicals depends on the nature of the ester. The saponification equivalent of the ester given in the Appendix (pages 433–34) is often an aid in the identification.

Derivatization of the Acidic Part of an Ester Without Hydrolysis

The acidic component of the ester may be identified by reaction with aqueous ammonia, benzylamine, or hydrazine hydrate, as represented by Eqs. (10.35), (10.36), and (10.37). The solubility of the amides renders the

$$RCOOR' + NH_3 \longrightarrow RCONH_2 + R'OH \qquad\qquad (10.35)$$
Amide

$$RCOOR' + C_6H_5CH_2NH_2 \longrightarrow RCONHCH_2C_6H_5 + R'OH \qquad (10.36)$$
N-Benzylamide

$$RCOOR' + HNHNH_2 \longrightarrow RCONHNH_2 + R'OH \qquad\qquad (10.37)$$
Hydrazide

reaction with ammonia unsuitable in most cases. The preparation of benzyl-amides is effected by boiling benzylamine with the ester as shown in Eq. (10.38). The reaction with benzylamine has been applied to the identification of about sixty-five acyl groups in esters using about 1 ml of the

$$RCOOR' + C_6H_5CH_2NH_2 \longrightarrow RCONHCH_2C_6H_5 + R'OH \qquad (10.38)$$

liquid ester. The esters that do not give good results with benzylamine owing to the nature of the hydroxy group can be converted by alcoholysis to the methyl esters. In general the N-benzylamides should be tried first, then the hydrazides and morpholides; the amides are suitable when the acidic group is complex.

General Method for Preparation of N-Benzylamides from Esters[12]

Place in a test tube arranged with a micro condenser for reflux 100 mg of powdered ammonium chloride, 1 ml of the ester (or 1 g if the ester is solid), 3 ml of benzylamine, and a small boiling stone. Boil gently for 1 hour and after cooling wash the reaction mixture with water to remove the excess of amine, and then adjust the pH to 5–6 with dilute hydrochloric acid. In most cases this produces crystallization of the benzylamide. If there is a considerable amount of unreacted ester the benzylamide does not crystallize. In such cases the contents of the tube are added into an evaporating dish and the tube washed with 1 ml of water; the washings are added to the dish. The mixture is boiled for a few minutes in order to volatilize the unreacted ester and then chilled. The crude benzylamide is extracted with 1 ml of ligroin and filtered, and then recrystallized from aqueous alcohol or acetone.

For the alcoholysis of esters higher than C_2 or C_3 in the alcohol radical of the ester, 1 ml of the ester is refluxed for 30 minutes with 5 ml of absolute methanol containing a trace of sodium methoxide (obtained by adding 100 mg of sodium metal to the 5 ml of absolute methanol). After refluxing, the excess of methanol is distilled off or evaporated, and the residue is reacted with benzylamine.

---

[12] Data on the melting points of N-benzylamides will be found in the original papers cited in the references at the end of this section.

General Method for Preparation of Hydrazides from Esters

Higher esters are converted to methyl esters by alcoholysis. About 1 ml of the ester and 1 ml of 85–90 per cent hydrazine hydrate are refluxed in a tube as described in the preceding paragraph for 10–15 minutes, the heat is removed, and absolute ethanol or methanol is added dropwise until a clear solution results. The mixture is refluxed for 1–2 hours. The mixture is poured into a small evaporating dish, and the alcohol is evaporated off. On cooling the residue, the crude hydrazide is obtained, which is recrystallized from aqueous alcohol.

Derivatization of the Acidic and Alcoholic Components After Hydrolysis

Discussion

If the acid is a solid it can be separated after hydrolysis. The ester is hydrolyzed by saponification with potassium or sodium hydroxide and if the alcohol is volatile the reaction is distilled and the distillate is used for the derivatization of the alcoholic component. If the acid is a solid the residue is acidified, the solid acid is isolated and purified, and its melting point is determined. This procedure is illustrated in Example 10.31, which deals with derivatization of the acidic and alcoholic components of butyl phthalate. If the acidic component is a liquid the residue can be derivatized directly by reaction with *p*-nitrobenzyl chloride to obtain the *p*-nitrobenzyl ester, or it may be evaporated and the sodium salt converted first to the acid chloride by means of thionyl chloride and then to the anilide or *p*-toluidide by reaction of the acid chloride with an arylamine. This is illustrated in the hydrolysis of isopropyl acetate (Example 10.32).

Example 10.31: Identification of Butyl Phthalate

Place in an 8-inch distilling tube 1.5 ml of diethylene glycol, 0.3 g (1 pellet) of potassium hydroxide, and 5 drops of water. Heat by means of a small flame until the pellet dissolves. Cool by means of tap water to room temperature and add 0.5 ml of the ester. Arrange tube with a reflux condenser and connect the outlet with a condensing setup (Figure 2.13). Heat to gentle boiling, shaking the tube from time to time; when the ester layer disappears (3–5 minutes), the reflux condenser is removed and 2 ml of dry pyridine are added. The tube is carefully heated until 2.2 ml of distillate have been collected. The distillate is used to prepare the 3,5-dinitrobenzoate, as directed in the general method of preparation when the sample to be derivatized contains water. The residue is diluted with 5 ml of water and then acidified with 6N sulfuric acid. The crystals that separate out are filtered and the melting point determined.

If a derivative of the acid is desired, the residue left in the distilling flask is diluted with 5 ml of water and 5 ml of ethanol and then neutralized to phenolphthalein with $6N$ sulfuric acid; it is then set aside to permit separation of potassium sulfate. The mixture is filtered and the clear filtrate is used for the preparation of p-nitrobenzyl ester as described in Example 10.8 (see also Example 10.32, Note).

Derivatization of the Alcoholic Component

There are two general methods for the derivatization of the alcoholic component. One is to hydrolyze the ester, separate the alcohol (or phenol), and derivatize it by the methods discussed on pages 248 and 328. The other method is to react the ester directly with 3,5-dinitrobenzoyl chloride to obtain the 3,5-dinitrobenzoate of the alcohol, as described in Example 10.32.

Examples of Complete Characterization of Esters

Example 10.32: Identification of Isopropyl Acetate

Place 300 mg of isopropyl acetate, 500 mg of 3,5-dinitrobenzoyl chloride, and 3 ml of pyridine in an 8-inch tube, provided with a condenser arranged for reflux. Add 2 small boiling stones and heat for 1.5–2 hours over a small flame so that the mixture boils gently. Cool and add a mixture of 1 ml of $6N$ sulfuric acid and 9 ml of water (10 ml of 3 per cent acid). Cool and shake the mixture vigorously; extract the mixture with 5 ml of isopropyl ether or 5 ml of ethyl ether (which has been washed with water, placed over calcium chloride and sodium, and then filtered to remove all traces of ethanol). Separate the ether layer and wash it, first with 5 ml of 2–3 per cent sulfuric acid, then with 4 ml of 2 per cent sodium hydroxide solution, and finally with 3 ml of water. Evaporate the ether layer from a small dish. Dissolve the residue in 5 ml of methanol, add a minute amount of charcoal, filter, and add to the filtrate 2 ml of water. About 60–70 mg of crystals separate out, melting at 116–118°. Dissolve the crystals in 4 ml of methanol and precipitate with 1.5 ml of water. Filter and dry the 3,5-dinitrobenzoyl ester of isopropyl alcohol. The yield is 30–45 mg of crystals, melting at 121–122°.

For the identification of the acidic part of the ester, a new portion of the ester is hydrolyzed. Place 300 mg of the ester, 0.3 g (1 pellet) of potassium hydroxide, 1 ml of water, and 1 boiling stone. Boil for 15 minutes and then distill off most of the water and finally dry the residue within the tube either in an oven or in the water bath. By means of a spatula or a glass rod pulverize the residue and convert the salt to the halide and then to the p-toluidide according to Example 10.3. The yield is 50–60 mg melting at 146–147°.

Note: The acidic component of the ester can be converted to the *p*-nitrobenzyl ester instead of the *p*-toluidide as follows. After boiling to hydrolyze the ester dilute to 2 ml with water; add 1 ml of 6*N* hydrochloric acid and 1 drop of phenolphthalein; if the solution is acid, add sodium hydroxide solution (5–10 per cent, until the color of the solution is just pink). Add 2 drops of 5 per cent hydrochloric acid so that the pink color is discharged. If the original solution is alkaline, add dilute hydrochloric acid (5–10 per cent) until the color of phenolphthalein just fades. Add 200 mg of *p*-nitrobenzyl bromide or chloride and 8 ml of methanol and a small boiling stone. (**Caution:** *Be careful in handling* p-*nitrobenzyl halides.*) Arrange a reflux condenser for the tube and boil gently for 1.5 hours. Cool and add 2–3 ml of water; by means of a glass rod scratch the inner sides of the tube. After 20 minutes filter the ester and wash first with 4 ml of 5 per cent sodium carbonate and then twice with 2 ml of water. Crystallize according to the method described in Example 10.8. The yield is 40 mg, melting at 77–78°.

References

Rapid saponification of esters by potassium hydroxide in diethylene glycol: Redemann and Lucas, *Ind. Eng. Chem., Anal. Ed.* **9**, 521 (1937).

Removal of acyl groups: Baltzly and Buck, *J. Am. Chem. Soc.* **63**, 2022 (1941).

Use of *N*-(β-aminoethyl) morpholine: Bost and Mullen, *J. Am. Chem. Soc.* **73**, 1967 (1951).

Anilides from esters: Hardy, *J. Chem. Soc.,* 398 (1936).

N-Benzylamides: Buehler and Mackenzie, *J. Am. Chem. Soc.* **59**, 421 (1937); Dermer and King, *J. Org. Chem.* **8**, 168 (1943).

3,5-Dinitrobenzoates: Renfrow and Chaney, *J. Am. Chem. Soc.* **68**, 150 (1946).

Hydrazides: Sah, *Rec. trav. chim.* **59**, 1036 (1940).

p-Toluidides from esters: Hardy, *J. Chem. Soc.* 398 (1936); Koelsch and Tannenbaum: *J. Am. Chem. Soc.* **55**, 3049 (1933).

ETHERS

Table 10.9 gives a summary of the few derivatives that may be prepared for characterization of ethers. Generally ethers are relatively inert

TABLE 10.9

Derivatives for the Identification of Ethers

*3,5-Dinitrobenzoates	Picrates
*Sulfonamides	Carbonyl compounds
*Bromo derivatives	Alkyl halides

* Recommended derivative.

compounds; the carbon–oxygen bond in ethers cannot be easily split and hence the preparation of derivatives from semimicro quantities involves some difficulties. In the case of aromatic ethers, it is possible by means of bromination, chlorosulfonylation, or other substitution reactions to prepare suitable solid derivatives that may be used for characterization. For the derivatization of aliphatic ethers, however, it is necessary to cleave the ether linkage in order to prepare a derivative of the resulting hydroxy compound. Aliphatic and aromatic ethers are, therefore, discussed separately.

Caution: In working with ethers bear in mind that they form *peroxides* easily, particularly when exposed to light and air. The peroxides detonate when heated, and hence the distillation of an appreciable amount of ether, which contains peroxide, involves the danger of an explosion if the distillation is allowed to proceed to dryness. The presence of peroxides in ethers is detected by means of starch iodide paper that has been moistened with dilute hydrochloric acid. The peroxides are removed by washing the ether with water containing a small amount of ferrous sulfate and dilute sulfuric acid. For small quantities of ethers, as, for example, 10 ml, washing with 2–3 ml of water containing 5 drops of 10 per cent solution of ferrous sulfate and 1 drop of sulfuric acid is sufficient.

Derivatives from Aliphatic Ethers

Discussion

A general method for obtaining derivatives from ethers is heating the sample in the presence of anhydrous zinc chloride and 3,5-dinitrobenzoyl chloride. The ether becomes cleaved to the hydroxy compound, which reacts with the acid chloride to give the 3,5-dinitrobenzoate. The first step in the cleavage (Eq. (10.39)) is assumed to be the formation of an alcohol and an olefin; the alcohol thus formed reacts with 3,5-dinitrobenzoyl chloride to give the ester and hydrogen chloride, as shown by Eq. (10.40); the latter converts some of the alcohol into an alkyl chloride, according to Eq. (10.41).

$$CH_3CH_2OCH_2CH_3 \xrightarrow{ZnCl_2} CH_3CH_2OH + CH_2{=}CH_2 \qquad (10.39)$$

$$CH_3CH_2OH + C_6H_3(NO_2)_2COCl \longrightarrow C_6H_3(NO_2)_2COOC_2H_5 + HCl \qquad (10.40)$$

$$CH_3CH_2OH + HCl \longrightarrow CH_3CH_2Cl + H_2O \qquad (10.41)$$

This method is not well suited for small quantities. Aside from the reactions that diminish the amount of 3,5-dinitrobenzoate, the chief difficulty of the method as applied to small quantities is that a number of aliphatic ethers boil at low temperatures and prolonged heating results in considerable losses of the compounds before cleavage is effected. The amount of 3,5-dinitrobenzoate obtained from a series of experiments with 0.5 ml of ethyl ether and isopropyl ether was often less than 10 mg, while in many runs no derivative at all was obtained. It has been observed that, if the zinc chloride is freshly fused and precautions are taken to dry thoroughly the tube and microcondenser, better yields are obtained. By the use of a sealed tube and heating under pressure, it is possible to prevent the loss of the ether and obtain a sufficient amount of the derivative.

General Method for Preparation of Alkyl 3,5-Dinitrobenzoates

The zinc chloride employed for cleavage is freshly fused and finely powdered. About 2 g of powdered commercial zinc chloride is fused in an iron crucible and then stirred with an iron rod while the melt cools. When the consistency becomes doughlike, the mass is pulverized with a small pestle to give a freely running fine powder that is still hot. The powder is transferred immediately to a dry bottle that has been previously warmed and tightly closed with a Bakelite screw cap. The small amount of iron oxide with which the zinc chloride is contaminated does not interfere with its activity.

A 6-inch tube is dried by being heated over a free flame and stoppered while hot. About 400–500 mg of zinc chloride is introduced, followed by 0.5 ml of ether and 250 mg of 3,5-dinitrobenzoyl chloride; the stopper is raised momentarily for the addition. A microcondenser is inserted after it has been thoroughly wiped off to remove any adhering moisture. The tube is immersed in a bath and heated at such a temperature that the ether does not condense much above the end of the microcondenser, which is about 10–15 mm from the bottom of the tube. The mixture is heated for 2 hours. The microcondenser is removed; the residue is heated in the bath until it is dry, and is then carefully pulverized with a rod. About 5 ml of 10 per cent sodium carbonate is heated separately at 60–70° and then added to the powdered mixture. A solid rubber stopper is inserted in the mouth of the tube and the contents are shaken for about 1 minute; then the remaining solid is stirred and crushed on the sides of the tube until no small lumps can be detected. The mixture is again heated to 60–70° and filtered by suction; the residue in the funnel is washed twice with 2 ml of sodium carbonate solution and twice with 2–3 ml of water. At this point the residue remaining in the filter should be a fine powder. If any lumps are present, the treatment with warm sodium carbonate solution is repeated. The residue is transferred along with the filter paper to a dry tube and extracted with 1.5–2 ml of boiling ethanol. The hot solution is filtered, the tube and residue are washed with an additional 1 ml of hot ethanol, and the filtrate is combined with the first extract. If no solid separates out, the filtrate is evaporated until the volume is about 2 ml, and water is added dropwise until the solution becomes cloudy. The tube is reheated until the solution is clear and is allowed to cool slowly. The crystals are filtered and dried and the melting point is determined. It is not uncommon for the melting point to be 5–10° below that of the pure derivative if the extraction is faulty. The yield varies from 5 to 50 mg. With low-boiling ethers, a yield of only 1 mg or less of the ester is not unusual. In such cases a closed tube (ignition tube) is used and the mixture is heated under pressure at 100° for 1 hour. It should be noted that this method is applicable only if both radicals in R_2O are identical.

Derivatives from Aromatic Ethers

Discussion

Several reactions can be employed to form suitable derivatives of aryl ethers. Eq. (10.42) shows the bromination of the aromatic ether; the com-

$$(CH_3)C_6H_4OCH_3 + Br_2 \longrightarrow (CH_3)C_6H_3BrOCH_3 + HBr \qquad (10.42)$$

o-Cresyl methyl ether   Monobromo-o-cresyl methyl ether

pound is dissolved in glacial acetic acid, alcohol, or chloroform and the required amount of bromine is added dropwise. The bromo-substituted ether is obtained either by addition of water or evaporation of the solvent. The extent of bromination depends on the groups already present: o-cresyl methyl ether forms a monobromo, whereas guiacol (1-methoxy-2-hydroxybenzene) forms a tribromo derivative. Ethers that have an unsaturated linkage undergo addition and substitution; thus anethole (p-propenylphenyl methyl ether) forms a monobromo dibromide.

Eq. (10.43) represents the formation of a molecular compound with

$$C_6H_5OCH_3 + C_6H_2(NO_2)_3OH \longrightarrow C_6H_5OCH_3 \cdot C_6H_2(NO_2)_3OH \qquad (10.43)$$

Anisole   Picric acid   Anisole picrate

picric acid. The picrates form readily by mixing equimolecular amounts of the ether and picric acid dissolved in the minimum amount of warm chloroform, and then allowing the mixture to stand for a short time. Although a number of the picrates are unstable on exposure to air, their preparation offers a relatively convenient method for identification of a number of aromatic ethers. Since halides, when treated with β-naphthol, form naphthyl ethers readily, which may be identified by the picrates, the method may be employed for the derivatization of a number of alkyl halides.

An aromatic ether is converted to a substituted benzenesulfonchloride through reaction with chlorosulfonic acid, according to Eq. (10.44). As

$$C_6H_5OCH_3 \xrightarrow{ClSO_2OH} C_6H_4(OCH_3)(SO_2Cl) \qquad (10.44)$$

Anisole   p-Methoxybenzenesulfonyl chloride

shown by Eq. (10.45), the sulfonyl chloride is converted by ammonolysis to

$$C_6H_4(OCH_3)SO_2Cl \xrightarrow{NH_3} C_6H_4(OCH_3)SO_2NH_2 + NH_4Cl \qquad (10.45)$$

p-Methoxybenzenesulfonamide

the substituted benzenesulfonamide, which may be identified through its melting point. The method is feasible when the preparation of a bromo derivative or a picrate does not afford satisfactory results.

Other reactions of aromatic ethers that may be used for the preparation of derivatives are nitration and oxidation. Oxidation is used in the case of ethers having side chains—as, for example, the cresyl ethers; these, by oxidation, yield alkoxybenzoic acids. With few exceptions both nitration and oxidation give poor yields of derivatives.

General Method for Preparation of Bromo Derivatives

Quantities as low as 1–5 mg can be brominated. The ether is dissolved in glacial acetic acid or chloroform, and a slight excess of bromine is added slowly while the reaction mixture is cooled. A solution of bromine in glacial acetic acid is convenient to handle. If glacial acetic acid is used as the solvent, the bromo compound is separated by the addition of water; if the bromo compound tends to separate as an oil, the solvent is evaporated off and the crude solid is crystallized from alcohol, petroleum ether, or isopropyl ether. The bromination of some ethers (anethole) is best accomplished in isopropyl ether. About 50 mg of the compound are added to 1 ml of isopropyl ether (or absolute ethyl ether) and cooled in an ice bath or tap water. To this is added over a period of 5 minutes a solution of 120 mg of bromine in 1 ml of ether. The mixture is allowed to stand in the cold for 5 minutes; then the crystals are filtered and crystallized from 4–5 ml of petroleum ether. The yield is about 75–90 mg of crystals melting at 107°.

For bromination in glacial acetic acid the ether is added to a cold 1 per cent solution of bromine in glacial acetic acid. The solution should contain slightly more bromine than the theoretical amount. For example, in the bromination of 100 mg of 2-naphthyl methyl ether (0.7 millimole) 6 ml of 1 per cent bromine solution is used since the monobromo derivative is formed. The bromine solution is first cooled in an ice bath and the ether is added. After a few minutes the tube containing the mixture is removed from the ice bath and allowed to stand at room temperature for 10 minutes. The derivative is precipitated by the addition of 10 ml of water and the crude crystals are separated and crystallized from an alcohol–water mixture. The yield from 100 mg of ether varies from 100–150 mg of the pure derivative.

General Method for Chlorosulfonylation of Ethers and Preparation of Sulfonamides

About 0.5 ml of the aromatic ether is dissolved in 2 ml of dry chloroform (in a clean tube) and cooled in an ice bath. About 1 g of chlorosulfonic acid is added dropwise over a period of 3–5 minutes. The tube is removed from the ice bath and allowed to stand at room temperature for 20–30 minutes. The contents of the tube are poured slowly into a small separatory funnel containing 5 ml of ice water. The tube is washed with 1 ml of chloroform and the washings are added to the funnel. The mixture is shaken gently and the chloroform layer is run into an 8-inch distilling tube. The chloroform is distilled off and, while the tube is still immersed in the water bath, 1 g of powdered ammonium carbonate and 5 ml of a concentrated solution of aqueous ammonia are added. The stopper of the distilling tube is replaced and the mixture is heated at 60° for 15 minutes and then at 80–90° for 10 minutes. The receiving tube is charged with water to absorb the ammonia given off during heating. The hot solution is filtered to remove any solid that has separated out; this may be crude sulfonamide and is crystallized separately. The clear filtrate is evaporated to dryness on the water bath; the residue is dissolved in 5–10 ml of boiling water; the solution is treated with charcoal, filtered, and cooled. Another method of purification consists in dissolving the crude sulfonamide in an alcohol–water mixture. If the melting point of the derivative is 5–10° below the value recorded in the literature, the sulfonamide may be contaminated with products of side reactions (sulfones and chlorinated compounds). The sulfonamide is dissolved by being heated gently in 5 ml of 5 per cent sodium hydroxide and any undissolved material is filtered off. The filtrate is acidified with dilute hydrochloric acid, and the sulfonamide is filtered and purified. The yield varies from 10 to 20 mg.

General Method for Preparation of Picrates of Phenolic Ethers

A number of phenolic ethers form picrates, the molecular compounds consisting of 1 molecule of ether to 1 to 2 molecules of picric acid. About 5 millimoles of the ether are dissolved in 5–6 ml of boiling chloroform and 5 millimoles of the picric acid are dissolved in 3 ml of boiling chloroform. The solutions are mixed, shaken, and then allowed to cool. The crystals are filtered and dried rapidly between filter papers; the melting point is determined immediately. The picrates should not be recrystallized or allowed to dry in air as a number of these decompose.

References

Chlorosulfonylation of ethers: Huntress and Carten, *J. Am. Chem. Soc.* **62,** 511 (1940).

Semimicro identification of ethers by catalytic conversion to carbonyl compounds: Sah, *Rec. trav. chim.* **58,** 758 (1939) (in English).

3,5-Dinitrobenzoates: Underwood, Baril, and Toone, *J. Am. Chem. Soc.* **52,** 4087 (1930).

Picrates: Andersen, *Acta Chem. Scand.* **8,** 157 (1954); Baril and Megrdichian, *J. Am. Chem. Soc.* **58,** 1415 (1936).

Picrates of naphthyl alkyl ethers: Dermer and Dermer, *J. Org. Chem.* **3,** 289 (1938).

Sulfonamides from aryl ethers: Huntress and Carten, *J. Am. Chem. Soc.* **62,** 603 (1940).

Miscellaneous derivatives: Naphthylamine adducts of 3,5-dinitrobenzoates from esters, Benfey *et al., J. Org. Chem.* **20**, 1777 (1955). Use of ethanolamine for esters, Rauscher *et al., Ind. Eng. Chem., Anal. Ed.* **22**, 923 (1940); *J. Am. Chem. Soc.* **70**, 438 (1948). *p*-Nitrobenzoates from ethers, Ward and Jenkins, *J. Org. Chem.* **10**, 371 (1945). *S*-Alkylisothiouronium picrates from alkoxyl compounds, Kratzle and Osterberger, *Monatsh.* **81**, 998 (1950). Hydrazides and phenylhydrazides from lactones, Darapsky *et al., J. pract. Chem.* (*2*) **147**, 150 (1936); Pummerer *et al., Ber.* **68**, 371 (1945); Seib, *Ber.* **60**, 1399 (1935).

11

Preparation of Derivatives: II

HALOGEN COMPOUNDS

Table 11.1 summarizes the derivatives that have been proposed for the characterization of alkyl, cycloalkyl, and aryl halides.

TABLE 11.1
Derivatives for the Identification of Halogen Compounds

*S-Alkyl isothiouronium picrates	N-Alkyl-3-nitrophthalimides
*Picrates of β-naphthyl ethers	N-Alkyl phthalimides
3,5-Dinitrobenzoates	N-Alkyl-p-bromobenzenesulfon-p-anisides
*Sulfonamides	N-Alkyl-p-toluenesulfontoluidides
*Nitro derivatives	Iodoso derivatives
Anilides, p-toluidides	6-Nitro-2-mercaptobenzothiazole derivatives
Alkyl mercuric halides	Piperidyl derivatives
Triiodophenyl ethers	Ethers of p-hydroxydiphenylamine
2,4-Dinitrothiophenol ethers	9-Alkyl fluorene carboxylic acids
Ethers of p-hydroxybenzoic acid	S-Alkyl-2-mercapto-4,5-dihydroglyoxali-
N-Alkyl tetrachlorophthalimides	nium salts

* Recommended derivative.

The discussion will be divided into derivatives suitable for: (a) alkyl and cycloalkyl halides, and (b) aryl halides. Of the asterisked derivatives the first two are suitable for alkyl and cycloalkyl halides; the sulfonamides and nitro compounds are suitable for aryl halides. For the preparation of anilides, p-toluidides, and alkyl mercuric salts by means of Grignard reagents obtained from the halides, the student is referred to the more extensive work of the authors. The majority of the non-asterisked derivatives are for alkyl and cycloalkyl halogen compounds.

The polyhalides are not treated separately. Except in the few instances in which derivatization is possible, the alkyl and cycloalkyl polyhalides are characterized by means of physical constants and functional group reactions. In general, aryl polyhalides are derivatized by the same procedures as the monohalides.

Alkyl and Cycloalkyl Halides

S-Alkyl Isothiouronium Picrates

Discussion

Eqs. (11.1) and (11.2) represent the formation of S-alkyl isothiouronium picrates. These molecular compounds, whenever they can be prepared,

$$C_2H_5Br + 2NH_2CSNH_2 \longrightarrow NH_2CS(NH)C_2H_5 + NH_2CSNH_3{}^+Br^- \qquad (11.1)$$

Thiourea S-Ethylisothiourea

$$NH_2CS(NH)C_2H_5 + C_6H_2(OH)(NO_2)_3 \longrightarrow NH_2CS(NH)C_2H_5 \cdot C_6H_2(OH)(NO_2)_3 \qquad (11.2)$$

Picric acid S-Ethylisothiouronium picrate

should be among the first derivatives to be tried. They are easily obtained in good yields by heating a mixture of the halide and thiourea. The mixture, dissolved in ethanol, is refluxed for 15–30 minutes; then a saturated alcoholic solution of picric acid is added and the vessel cooled. The picrate of the S-alkyl isothiouronium compound separates out. The adduct is recrystallized after filtration. If the organic halide is a chloride, a small amount of potassium iodide is added to the reaction mixture; the addition of potassium iodide increases the reactivity of the chloride, but it is not very effective with a number of chlorides. Therefore, the mixture of thiourea and halide should be heated for 2 or 3 hours. For example, n-butyl chloride and thiourea must be refluxed for 2 hours, and n-propyl chloride for 5 hours, in order to obtain satisfactory results. The number of halides from which S-isothiouronium picrates have been obtained is about 50; however, among these are the most common alkyl halides—that is, up to 10 carbon atoms.

General Method for Preparation

For the preparation of S-alkyl isothiouronium picrates 1 millimole of the alkyl bromide or iodide and 2 millimoles of thiourea are dissolved in 3 ml of alcohol and refluxed. The time required for heating varies from 20–30 minutes for the primary halides and 2–3 hours for the secondary. If the reaction is not complete, the yield is poor and the derivative melts 6–10° lower than the value in the literature. If the halide is a chloride, 100 mg of potassium iodide are first dissolved in 2–3 drops of water (heated to effect solution); then the alcohol, thiourea, and chloride are added in the order named and the mixture is heated. Benzyl chloride requires 30 minutes, while *n*-butyl and *n*-amyl chlorides require 1–2 hours. After heating the halide and thiourea, 0.5 ml of a saturated alcoholic solution of picric acid is added and the mixture is boiled for 5 minutes. On cooling, the *S*-alkyl isothiourea picrate crystallizes out and is purified by recrystallization from an alcohol–water mixture.

While good results are obtained from 1–2 millimoles of primary and secondary halides, the method must be modified for tertiary halides and the quantity of the halide to be derivatized should be increased to 5–10 millimoles. About 5 millimoles of the tertiary halide is mixed with a solution of 500 mg of thiourea in 10 ml of water and 1 ml of ethanol and the mixture is refluxed for 2–5 hours or until the alkyl halide layer has disappeared. The mixture is then poured into 100 ml of a 1 per cent aqueous solution of picric acid. The picrate separates out rapidly and after 15–20 minutes it is filtered and crystallized from an alcohol–water mixture.

Example 11.1: Preparation of S-*Ethylisothiouronium Picrate*

Place 300 mg of thiourea in a test tube and add 5 ml of ethanol or methanol and 0.2–0.25 ml of ethyl bromide. Attach to the tube a microcondenser and reflux the mixture for 15–20 minutes. Remove the flame and add 1 ml of a saturated alcoholic solution of picric acid. Boil the mixture for a short time (3–5 minutes), or until a clear solution results, and then cool in an ice-water mixture or in running cold water for 10 minutes. Filter the crystals and wash twice with 1 ml of 50 per cent alcohol. Save a very small amount (5 mg) of the derivative and crystallize the remainder. Use 2 ml of hot alcohol to effect solution of the picrate and add 1 ml of water to the filtered solution. Cool for 10 minutes and filter the crystals, washing twice with 1 ml of 50 per cent methanol. Dry the crystals in air or in a desiccator. The yield is 40–50 mg of crystals, melting at 187–188°.

Note: The same general method is used for the preparation of the alkyl isothiourea picrates from alkyl bromides and iodides. For chlorides the method is the same, except that potassium iodide is added to the reaction mixture to increase the reactivity of the halogen.

The crystallization of the alkyl isothiouronium picrates may be accomplished from hot solutions. When the conversion of the halide to the alkyl isothiouronium is incomplete, the melting point of the picrate after one crystallization is 6–10° lower than the value of the pure compound. In such cases it is often more time-saving to repeat the preparation of the derivative and increase the time of refluxing the halide–thiourea mixture to 30–40 minutes; in some cases (isopropyl bromide) 2–3 hours heating is necessary.

The yield of the picrate from 0.2–0.25 ml of alkyl bromides or iodides varies from 40 to 200 mg. Generally the higher primary alkyl halides give better yields than the lower halides or the secondary halides. For example, from 200 mg of the halide the following yields of the pure picrate were obtained: methyl iodide, 50 mg; *n*-propyl iodide, 60 mg; *n*-amyl bromide, 150 mg; and *sec*-amyl bromide, 50 mg.

Picrates of β-Naphthyl Ethers

An alkyl halide is converted to alkyl β-naphthyl ether by reaction with β-naphthol, according to Eq. (11.3). The alkyl β-naphthyl ether is con-

$$CH_3CH_2Br + \underset{\text{β-Naphthol}}{\text{[OH]}} \xrightarrow{\text{NaOH}} \underset{\text{Ethyl β-naphthyl ether}}{\text{[OC}_2\text{H}_5\text{]}} + \underset{\text{H}_2\text{O}}{\text{NaBr}} \qquad (11.3)$$

verted to the picrate as described on page 298 for the derivatization of aryl alkyl ethers.

For the preparation of the aryl alkyl ether 5 millimoles of the alkyl halide and slightly less than 5 millimoles (700 mg) of β-naphthol and 2 ml of 10 per cent sodium hydroxide are warmed for 30 minutes in a steam bath; about 4 ml of water is added and the tube is chilled in an ice bath. The crystals are removed by centrifugation or filtration, dissolved in the minimum amount of alcohol, and precipitated by addition of water, and then separated and converted to the picrate as outlined on page 298. However, if the melting point of the solid alkyl β-naphthyl ether is sharp, the conversion to the picrate is unnecessary.

Aryl Halides

The relatively slow rates of reaction of halogen atoms when they are attached to benzene nuclear carbon atoms render most of the reactions discussed in the preceding section inapplicable to aryl halides. Whenever it is feasible to prepare Grignard reagents from the aromatic halides, the anilide, p-toluidide, or naphthalide may be prepared. However, there are several typical aromatic reactions by which aryl halides may be derivatized.

Nitration should be the first reaction to be considered. Good results are obtained from about 50 mg of the halogen compound and in some cases the sample can be as low as 5–10 mg. Several (three to five) recrystallizations may be necessary for purification of the crude derivative.

Chlorosulfonylation of aromatic halides and subsequent ammonolysis to produce sulfonamides should be considered if the reaction proceeds without nuclear chlorination. Suitable derivatives may be prepared by other reactions:

1. Reactive bromo and iodo aromatic compounds may be converted into Grignard reagents and these reacted with isocyanates. Bromobenzene, for example, may be derivatized by converting it to phenylmagnesium bromide and then to the naphthalide.

2. Side chains, particularly methyl groups, are oxidized to the carboxylic stage. Thus the chloro, bromo, and iodo toluenes with one C may be readily oxidized to the corresponding halobenzoic acids.

3. A number of polycyclic compounds, as, for example, the α-chloro- and β-chloronaphthalenes, may be derivatized by preparation of the picrates.

4. Aryl halides in which the halogen atom is activated by nitro groups form derivatives with piperidine.

5. Aryl iodides react with chlorine (addition) to form dichlorides which may serve as derivatives.

The brief discussion above indicates that whenever nitration and chlorosulfonylation is not feasible, a judicious selection of the reaction to be employed for derivatization depends on the probable nature of the aromatic halide. A search of the literature will, in most cases, suggest the proper derivative to be prepared. For example, in the derivatization of 3,4-dichloro-1-nitrobenzene, one of the halogens may be ammonolyzed to form a chloronitroaniline (m.p. 104–105°). The ammonolysis requires heating of the aromatic halide with alcoholic ammonia at 210°—that is, under pressure. An easier method for derivatization of this compound, which contains activated halogen due to the presence of the nitro group, is to react it with morpholine. The halogen compound is refluxed for 2–3 hours with 2–3 moles of morpholine; the reaction mixture is treated with an excess of dilute hydrochloric acid, and the insoluble product is separated and crystallized from aqueous alcohol. The morpholine derivative of 3,4-dichloro-1-nitrobenzene melts at 127°. Other aromatic halides that give derivatives with morpholine are o-nitrochlorobenzene, p-nitrochlorobenzene, 2,4-dinitrochlorobenzene, and 3,5-dinitro-2-chlorobenzoic acid.

Nitro Derivatives of Aryl Halides

Discussion

For the preparation of mononitro derivatives the halide is dissolved or dispersed in concentrated sulfuric acid and then treated with an equal volume of concentrated nitric acid. The mixture is frequently shaken and kept at 45–55° for 5 minutes and then diluted with water.

An alternate method that is particularly applicable to the mononitration of micro quantities is to treat the nitro compound with specially prepared 100 per cent nitric acid. About 50 mg of the compound is mixed with 75–300 mg of the acid and kept at room temperature, or heated at 45–50°, for 5–15 minutes. The mixture is then diluted with water, and the product separated and crystallized. When two or three nitro groups are to be introduced, one of the following mixtures is used: (a) fuming nitric acid and concentrated sulfuric acid; (b) specially prepared 100 per cent nitric acid and concentrated sulfuric acid; (c) fuming sulfuric acid and nitric acid described under (a) or (b).

The selection of the proper procedure is of importance in obtaining nitro derivatives. For example, consider the nitration of chlorobenzene. A search in the literature discloses the following nitro derivatives: 2-nitro, m.p. 32°; 4-nitro, m.p. 83°; 3-nitro, m.p. 44°; and 2,4-dinitro, m.p. 52°. The desirable derivatives are 4-nitrochlorobenzene and 2,4-dinitrochlorobenzene. Since there are no precise directions in the literature for obtaining these derivatives starting with about 100 mg of chlorobenzene, the experimental condition and procedures can be easily determined by a series of test-tube experiments. In the following trials 50 mg of chlorobenzene was treated with the quantity (mg) of acid appearing outside the parentheses, while the numbers within the parentheses represent the melting point of the crude product before crystallization: (a) 150 conc. HNO_3 and 200 conc. H_2SO_4 for 15 minutes at 25° (oil); (b) 150 fuming HNO_3 for 15 minutes at 25° (oil); (c) 300 HNO_3 100 per cent for 5 minutes at 25° (81–82); (d) 150 fuming HNO_3 and 300 conc. H_2SO_4 for 15 minutes at 50–55° (30–34); (e) same as in (d), but heated at 60° for 30 minutes (40–44); (f) 150 fuming HNO_3 and 300 fuming H_2SO_4 for 30 minutes at 80–90° (48–50). From these results it is possible to select procedure (c) to obtain the mononitro derivative melting at 83° and procedure (e) or (f) to obtain the dinitro derivative melting at 52°. It is also clear from these results that several crystallizations would be required to remove the undesirable nitro derivatives from the crude nitration product. In most cases of nitration the formation of more than one nitro compound is the rule rather than the exception; therefore several crystallizations are necessary for purification of nitro derivatives.

General Method for Nitration

For the introduction of one nitro group in compounds that undergo nitration with ease, proceed as follows: place 200 mg of the compound to be nitrated in an 8-inch test tube; add first 2 ml of concentrated sulfuric acid; immerse the tube in a beaker containing cold water and add slowly 2 ml of concentrated nitric acid, shaking the tube from time to time so as to prevent a rise in temperature. Place the tube in a water bath at 50–55° and heat for 15–20 minutes. The tube is shaken frequently to ensure better contact of the halide with the nitrating mixture. After heating, the tube is cooled for a minute or two, and then its contents are poured into another tube, which contains 8–10 ml of cold water. The diluted mixture is returned into the original tube and cooled. The nitro compound that separates is filtered and washed twice with 3–4 ml of water; it is then transferred to the reaction tube and dissolved in the minimum amount of boiling methanol or ethanol. The hot solution is filtered, and water is added cautiously dropwise until a permanent cloudiness appears. The tube is heated until the solution is clear, and then it is cooled. The crystallization is repeated until the crystals from two successive crystallizations give a variation of 0.5–1° in the melting point.

For the introduction of two nitro groups or one nitro group in unreactive halides, the same procedure is used as described in the preceding paragraph, but fuming nitric acid is used in place of concentrated nitric acid. (**Caution:** *Care should be exercised in handling fuming nitric acid; its addition to the mixture of sulfuric acid and halide should be at such rate that no great amount of brown fumes appear.*) The mixture is then heated at 45–50°. If the melting point of the nitration product after one crystallization shows a difference of 10° or more from the value listed in the literature, the nitration is repeated and the temperature is raised first to 80°; if the result is not satisfactory, the nitration is repeated at 90–100°.

Example 11.2: Nitration of Chlorobenzene

Place in an 8-inch tube 2 ml of concentrated sulfuric acid and 200 mg of chlorobenzene. Cool and add slowly 2 ml of fuming nitric acid. Heat the tube in a bath at 90–100° for 30 minutes and mix the contents by shaking frequently. Cool and add the mixture to 10 ml of water and crystallize as described on page 000. Two to three crystallizations are required to obtain a product that melts at 52°. The yield is 100–120 mg.

Note: When fuming nitric acid is used, care should be taken if oxidizable groups are present. If a considerable amount of brown fumes are evolved during nitration, it is advisable to use a lower temperature.

For the preparation of 4-nitrochlorobenzene, mix in a 4-inch test tube 100 mg of the halide and 0.3 ml of 100 per cent nitric acid, and allow to stand at room temperature for 20 minutes. Add 2 ml of water and, after cooling, extract with 2 ml of ether. Draw the ether layer by means of a capillary pipet and place in another tube. Add one pellet of sodium hydroxide and shake gently for a few minutes. Evaporate the ether from a small dish and dissolve the residue in 2 ml of hot methanol. Filter with suction and add water dropwise until a permanent cloudiness results. Cool and filter the crystals. About 50–70 mg of 4-nitrochlorobenzene is obtained; the compound melts at 82–83°. For the preparation of 100 per cent nitric acid, see p. 468 of the larger work by the authors.

Sulfonamides of Aryl Halides

As indicated earlier, aryl halides undergo chlorosulfonylation by reaction with chlorosulfonic acid. The resulting sulfonchloride may then be converted by ammonolysis to a sulfonamide, according to Eq. (11.4). In actual

$$X\text{-}C_6H_5 \xrightarrow{2ClSO_2OH} X\text{-}C_6H_4\text{-}SO_2Cl \xrightarrow{2NH_3} X\text{-}C_6H_4\text{-}SO_2NH_2 \qquad (11.4)$$

practice the direction of the main reaction in a number of aryl halides is different from that indicated in the above equation. In a number of cases nuclear chlorination takes place. For example, p-diiodobenzene yields 1,2,4,5-tetrachlorobenzene. In general, aryl iodides do not yield good results. In the cases of fluorobenzene, iodobenzene, o-dichlorobenzene, and o-dibromobenzene the chief product is a sulfone instead of a sulfonchloride. Though the sulfones can be purified and employed as derivatives, more suitable compounds can be obtained by nitration procedures. The general method for the preparation of sulfonchlorides and their conversion to sulfonamides is described in the more extensive work of the authors (pp. 527-29).

References

Identification of Alkyl and Aryl Halides

Conversion of esters to sulfonamides: Huntress and Carten, *J. Am. Chem. Soc.* 62, 511 (1940).

Oxidation of thioethers to sulfones: Bost *et al., J. Am. Chem. Soc.* 54, 1985 (1932); 73, 1967 (1951).

Use of iodoso chlorides: Nicol and Sandin, *J. Am. Chem. Soc.* 67, 1307 (1945).

N-Alkyl-p-bromobenzenesulfon-p-anisides: Gillespie, *J. Am. Chem. Soc.* 56, 2740 (1934).

9-Alkylfluorene carboxylic acids from alkyl halides: Bavin, *Anal. chem.* 32, 554 (1960).

S-Alkyl isothiourea picrate for tertiary alkyl halides: Veibel and Lillelund, *Bull. soc. chim.* (5) 5, 1153 (1938).

S-Alkyl isothiourea picrates: Brown and Campbell, *J. Chem. Soc.*, 1699 (1937); Levy and Campbell, *J. Chem. Soc.*, 1442 (1939); Schotte, *Arkiv. Kemi* 5, 11 (1952); *Acta Chem. Scand.* 7, 1357 (1953).

S-Alkyl-2-mercapto-4,5-dihydroglyoxalinium salts and their picrates from alkyl halides using ethylenethiourea: Boyd and Meadow, *Anal. Chem.* 32, 551 (1960).

Alkyl mercuric halides: Hill, *J. Am. Chem. Soc.* 50, 167 (1928); Marvel, Gauerke, and Hill, *J. Am. Chem. Soc.* 47, 3009 (1925); Slotta and Jacobi, *J. prakt. Chem.* 120, 249 (1929).

Alkyl saccharins: Merritt, Levey, and Cutter, *J. Am. Chem. Soc.* 61, 15 (1939).

N-Alkyl tetrachlorophthalimides: Allen and Nicholls, *J. Am. Chem. Soc.* 56, 1409 (1934).

N-Alkyl-p-toluenesulfontoluides: Young, *J. Am. Chem. Soc.* 56, 2167, 2783 (1934); 57, 773 (1935).

Anilides, p-toluidides, and α-naphthalides: Gilman and Furry, *J. Am. Chem. Soc.* 50, 1214 (1928); Schwartz and Johnson, *J. Am. Chem. Soc.* 53, 1063 (1931); Underwood and Gale, *J. Am. Chem. Soc.* 56, 2117 (1934).

3,5-Dinitrobenzoates from alkyl halides, Furter, *Helv. Chim. Acta* 21, 872 (1938); Benfey *et al., J. Org. Chem.* 20, 1782 (1955).

Ethers of 2,4-dinitrothiophenol: Bost *et al., J. Am. Chem. Soc.* 73, 1967 (1951).

Ethers of p-hydrobenzoic acids: Lauer *et al., J. Am. Chem. Soc.* 61, 3050 (1939).

Ethers of p-hydroxydiphenylamine: Houston, *J. Am. Chem. Soc.* 71, 395 (1949).

Microscopic identification of aromatic halide derivatives: Dunbar and King, *Microchem J.* 3, 153 (1959).

6-Nitro-2-mercaptothiazole derivatives: Cutter *et al., J. Am. Chem. Soc.* **69**, 831 (1947); *Anal. Chem.* **25**, 198 (1953).

3-Nitrophthalimides: Sah and Ma, *Ber.* **65B**, 1930 (1932); *Science Repts., Natl. Tsing Hua Univ.* **2**, 147 (1933).

4-Nitrophthalimides and their use for saponification equivalent numbers: Billman and Cash, *J. Am. Chem. Soc.* **75**, 2499 (1953).

Picrates of alkyl β-naphthyl ethers: Baril and Megrdichian, *J. Am. Chem. Soc.* **58**, 1415 (1936); Dermer and Dermer, *J. Org. Chem.* **3**, 289 (1938).

Piperidyl derivatives of aromatic halogenonitro compounds: Seikel, *J. Am. Chem. Soc.* **62**, 750 (1940).

Triiodophenyl ethers: Drew and Sturtevant, *J. Am. Chem. Soc.* **61**, 2666 (1939).

Identification of Fluorine Compounds

Amides: Husted and Ahlbrecht, *J. Am. Chem. Soc.* **75**, 1605 (1953).

Analysis of fluorine-containing compounds: Simons, *Fluorine Chemistry*, Vol. II (N.Y.: Academic Press, 1954).

Detection of fluoride: Steiger, *J. Am. Chem. Soc.* **30**, 219 (1908).

2,4-Dinitrophenylhydrazones: Husted and Ahlbrecht, *J. Am. Chem. Soc.* **74**, 5422 (1952); Simons, Black, and Clark, *J. Am. Chem. Soc.* **75**, 5621 (1953).

Hydrates and amine adducts of aldehydes and ketones: Hauptschein and Braun, *J. Am. Chem. Soc.* **77**, 4930 (1955); Husted and Ahlbrecht, *J. Am. Chem. Soc.* **74**, 5422 (1952).

p-Toluenesulfonates: Tiers, Brown, and Reid, *J. Am. Chem. Soc.* **75**, 5978 (1953).

Chromatographic Procedures

Moynihan and O'Colla, *Chem. Ind. (London)* **1951**, 407.
Schmeiser and Jerchel, *Angew. Chem.* **65**, 366 (1953).
Winterringham *et al., Nature* **166**, 999 (1950).

HYDROCARBONS

Alkanes, Cycloalkanes, Alkenes, and Alkynes

Alkanes and cycloalkanes do not form derivatives suitable for characterization work despite the fact that they readily undergo a variety of reactions. A few cycloalkanes, such as cyclohexane and cyclopentane, can be oxidized by hot nitric acid to adipic and glutaric acids, respectively. Therefore, the identification of alkanes and cycloalkanes is based on physical constants: boiling point, refractive index, infrared spectra, and density. The determination of these constants leads to fairly accurate results with pure compounds; impurities, particularly those due to isomers, render characterization extremely difficult.

The characterization of alkenes, cycloalkenes, and alkynes through derivatization is beset with the same difficulties in spite of the large number of addition reactions which they undergo. Hence, to a very large extent, characterization is based on physical constants. However, as shown in

Table 11.2, there are a number of derivatives that are applicable to a limited number of hydrocarbons. For example, 2,4 dinitrobenzenesulfenyl

TABLE 11.2

Derivatives for the Identification of Hydrocarbons

Alkanes and Cycloalkanes	Nitrosates, nitrosites, and nitrosyl chlorides (for terpenes)
None	Dithiocyanates

Alkenes, Cycloalkenes, Alkynes, Dienes	Aromatic
Dinitrophenyl sulfides	*Nitro derivatives
Bromine addition products	*Trinitrofluorenone adducts
Mercuric salts	Acetamino derivatives
*Hydration or Friedel-Crafts acetylation to ketones and conversion to 2,4-dinitrophenylhydrazones or semicarbazones	Sulfonamides
	Picrates
	Aroyl benzoic acids
	2,4-Dinitrophenyl sulfides
	*Oxidation of side chains

* Recommended derivative.

chloride yields with symmetrical olefins a crystalline sulfide (adduct), as shown by Eq. (11.5). Another reaction for the characterization of a limited

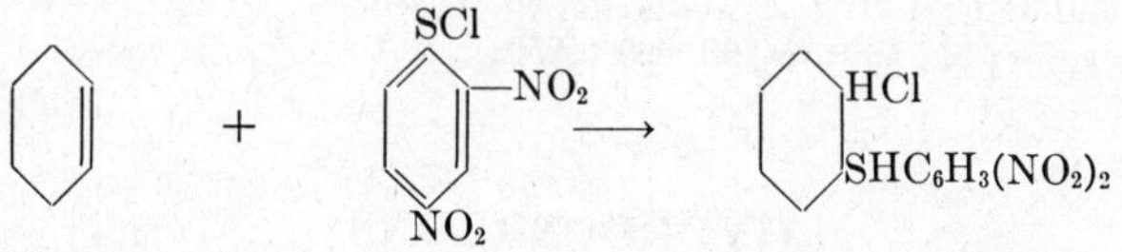

$$ \text{Cyclohexene} \quad + \quad \overset{SCl}{\underset{NO_2}{\bigcirc}}-NO_2 \quad \longrightarrow \quad \overset{HCl}{\underset{SHC_6H_3(NO_2)_2}{\bigcirc}} \tag{11.5} $$

Cyclohexene

2-Chlorocyclohexyl-2′,4′-dinitrophenyl
sulfide
(Adduct of cyclohexene)

number of alkenes is their reaction with acid chlorides and anhydrides in the presence of Friedel-Crafts catalysts to produce ketones, which are then readily derivatized by means of their 2,4-dinitrophenylhydrazones or semicarbazones, as shown by Eq. (11.6). The procedure has been applied to

$$ CH_3COCl + CH_3CH{=}CH_2 \xrightarrow{SnCl_4} CH_3CH{=}CHCOCH_3 + HCl \tag{11.6} $$

about twelve alkenes in quantities of about 1 g. Since the yields of the derivatives are in the range of 1.5 g and above, it is possible to apply this procedure to smaller quantities. The catalysts used for the Friedel-Crafts acetylation are stannic chloride and polyphosphoric acid.

A number of terpenes form crystalline addition products with addition of nitrosyl chloride, nitrogen tetroxide, and nitrogen trioxide, as shown by Eqs. (11.7–11.9). The nitrosochlorides are generally prepared by addition

$$RCH{=}CHR' + NOCl \longrightarrow RCH(NO)CHClR' \qquad (11.7)$$
Nitrosochloride

$$RCH{=}CHR' + N_2O_4 \longrightarrow RCH(NO)CH(ONO_2)R' \qquad (11.8)$$
Nitrosate

$$RCH{=}CHR' + N_2O_3 \longrightarrow RCH(NO)CH(ONO)R' \qquad (11.9)$$
Nitrosite

of concentrated hydrochloric acid to a mixture of the hydrocarbon and ethyl or n-butyl nitrite dissolved in acetic acid.

The preparations of D-limonene nitrosochloride and α-pinene nitrosochloride described in this section serve as examples of the application of this method to a limited number of terpenes.

Other reactions by which alkenes, dienes, and alkynes may form crystalline derivatives are: (a) addition of bromine; (b) addition of thiols, thiophenols, and thio acids; (c) addition of mercuric iodide by alkynes to give simple mercuric salts $(RC{\equiv}C)_2Hg$. The following examples illustrate the limited preparation of derivatives from cycloalkanes, cycloalkenes, and terpenes.

Oxidation of a Cycloalkane to a Dicarboxylic Acid

Discussion

A number of cycloalkanes are readily oxidized by hot nitric acid, undergoing rupture of the ring to form a mixture of dicarboxylic acids. Thus cyclohexane heated with nitric acid gives a mixture of adipic and glutaric acids and smaller amounts of succinic acid and nitro compounds. However, conditions can be chosen under which the oxidation gives mostly adipic acid and only small amounts of the other products. The method consists in heating nitric acid to boiling and adding the hydrocarbon 1 drop at a time and waiting until it has reacted *completely* before adding another drop. The reaction is very vigorous and the method is not recommended for the oxidation of more than 0.5 ml of the hydrocarbon added in portions of 0.05 ml. About 0.2–0.3 ml or less of the hydrocarbon yields a sufficient amount of the dicarboxylic acid for determination of its melting point after one or two crystallizations.

Example 11.3: Preparation of Adipic Acid from Cyclohexane

Place 2 ml of concentrated nitric acid in an 8-inch test tube and add 1 or 2 boiling stones. Clamp the tube to a stand in the hood and heat the acid to boiling. Reduce the flame of the microburner and, by means of a dropper, add cautiously 2 drops of cyclohexane; shake the tube and, when the vigorous reaction has subsided, repeat the addition. Add a total of 0.2 ml (8–10 drops) of the hydrocarbon over a period of 10 minutes. Boil gently for 1 minute and cool. Filter the crystals of adipic acid. Recrystallize from 2–3 ml of boiling water. The yield is 70–80 mg, melting at 152–153°.

Derivatives of Olefins and Terpenes

Example 11.4: Addition of Bromine to an Alkene: Styrene Dibromide

Dissolve 0.2 ml of styrene in 1 ml of dry carbon tetrachloride. Add 10 drops of bromine (use care in handling bromine) and then cool the tube. Add to the solid mass of crystals that separates out 5 ml of methanol, heat until the mixture dissolves, filter, and cool the filtrate. The yield of *styrene dibromide*, melting at 71–72°, is about 300 mg.

Example 11.5: Oxidation of a Cycloalkene to a Dicarboxylic Acid: Adipic Acid from Cyclohexene

In a 125-ml Erlenmeyer flask place 1.5 g of potassium permanganate, 25 ml of water, and 1 ml of 6N sodium hydroxide solution. Warm to effect solution of the permanganate; add 0.3 ml of cyclohexene and, after stoppering the flask with a solid rubber stopper, shake at intervals for 10–15 minutes or until the odor of cyclohexene has completely disappeared. Filter with suction and evaporate the filtrate to dryness; add 2 ml of 6N hydrochloric acid solution and 3 ml of water. Extract three times with 5 ml of ether. Evaporate the ether and crystallize the residue from hot water. The yield is 50–60 mg of adipic acid, melting at 149–151°.

Example 11.6: D-Limonene Nitrosochloride

About 0.2 ml of D-limonene (practical grade) is mixed in a 3- or 4-inch tube with 0.3 ml of n-butyl nitrite; a fine glass rod serves as a stirrer. The mixture is cooled to about −10° in an ice–salt mixture. Then, dropwise and with constant stirring, 0.5 ml of a mixture of equal parts of glacial acetic acid and hydrochloric acid of specific gravity 1.155 is added. The latter is prepared by the addition of 1 ml of water to 5 ml of concentrated acid of specific gravity 1.19. The addition of acid to the hydrocarbon should take about 2–3 minutes and then the mixture is allowed to stand in the ice bath for about 1 hour. The crystals are filtered and immediately crystallized from a mixture of 0.5 ml chloroform and 1.5 ml methanol. About 250 mg of crystals of the nitrosochloride is obtained, melting at 104–105°.

Note: The same procedure is used for the preparation of α-pinene nitrosochloride. The quantities used are: 0.5 ml of α-pinene, 0.5 ml of n-butyl nitrite, and 1.2 ml of glacial acetic acid, and a mixture (1:1) of 1.3 ml hydrochloric acid (sp. gr. 1.155) and glacial acetic acid. After one or two crystallizations, about 100–125 mg of crystals of the pure nitrosochloride is obtained, melting at 108–109°.

Aromatic Hydrocarbons

The aromatic hydrocarbons can be derivatized by: (a) nitration; (b) nitration followed by reduction of the nitro compound and acetylation of the amine to form the acetamino derivative; (c) chlorosulfonylation, followed by ammonolysis of the sulfonyl chloride to the sulfonamide; (d) condensation with phthalic anhydride in the presence of aluminum chloride to form the o-aroyl benzoic acids; (e) addition products with picric acid and 2,4,7-trinitrofluorenone; (f) oxidation of the side chains to carboxylic acids (this is applicable to toluene and the other alkyl benzenes); and (g) reaction of a number of aromatic hydrocarbons with 2,4-dinitrobenzenesulfenyl chloride in the presence of catalysts whereby aryl 2,4-dinitrophenyl sulfides are formed.

The first two methods should be tried first; in nitration the formation of dinitro and trinitro derivatives is preferable, as these are less likely to be contaminated by isomers than the mononitro compounds. The monoalkyl benzenes do not yield solid nitro derivatives readily and can be derivatized by formation of the mononitro compound and conversion to the 4-acetamino derivative; their nitration can be carried out with 10–20 mg of the hydrocarbon. A few monoalkyl benzenes, for example, the monobutylbenzenes, may be readily characterized by formation of the aryl 2,4-dinitrophenyl sulfides; at least 200 mg of the hydrocarbon is required. The picrates of the hydrocarbons form very readily, but they are not useful in most cases, with the exception of anthracene and naphthalene, because they are not stable enough to be isolated in the pure form. The 2,4,7-trinitrofluorenone addition products are stable and are therefore useful for the characterization of polynuclear hydrocarbons. For the conversion of aromatic hydrocarbons to sulfonamides at least 250 mg of the hydrocarbon is required, and for the preparation of aroyl benzoic acids, at least 400 mg.

Nitration of Aromatic Hydrocarbons

Discussion

Micro nitration of aromatic hydrocarbons in most cases even with milligram quantities of material gives fairly good yields because of the low solubility and solid nature of most nitro compounds. The usual nitrating agents may be used with slight modifications; however, particular care must be taken to select conditions which will give the minimum amount of undesirable nitro products.

The optimum conditions can best be ascertained by several trials in which the nitrating mixture, temperature, and time of contact between reactants are varied. For the characterization of alkyl benzenes it is possible to prepare the mononitro compounds and then to convert them without separation to the 4-acetamino derivatives.

The common nitrating agents are: (*a*) ordinary nitric acid (density 1.40–1.41), containing about 58 per cent acid, mixed either with concentrated or fuming sulfuric acid; (*b*) fuming nitric acid, which usually contains about 88–90 per cent acid; (*c*) ordinary or fuming nitric acid in acetic acid and acetic anhydride; and (*d*) alkali nitrates and sulfuric acid. In addition, the following have also been used: organic nitrates, such as benzoyl and acetyl nitrates; mixtures of nitric, acetic, and sulfuric acids; mixtures of nitric and sulfuric acids in chloroform, acetone, or ethers; and nitrogen oxides.

When difficulties are encountered because of migration of labile groups during nitration, it is advisable to work with quantities larger than 500 mg. Although it is possible to vary the amount of the acid or the solvent, the effect of nitric acid alone should be tried first in such cases. An experimental approach to determine the optimum conditions of nitration has been described in the larger work of the authors (pp. 465–68).

Example 11.7: Nitration of Toluene

Place in an 8-inch test tube 1.5 ml of concentrated sulfuric acid and 0.25 ml of toluene. Cool and add slowly 1.5 ml of fuming nitric acid. Heat in a water bath for 15 minutes. Remove the tube from the bath every few minutes and shake in order to mix the two layers. Cool and then add 7–8 ml of cold water or the same amount of ice. Filter the solid and wash twice with 2–3 ml of water. Dissolve in 4 ml of hot methanol, filter with suction, and add 1–2 drops of water to the filtrate. Cool for 10 minutes, filter the solid, and repeat the crystallization. The yield is 110–140 mg of 2,4-dinitrotoluene, melting at 69–70°.

Note: As stated under the discussion of aromatic hydrocarbons, the procedure outlined above will not be successful for the nitration of the three xylenes. It is recommended that the entire discussion on nitration given in the more extensive work of the authors be read in order to determine the optimum conditions. For example, the best conditions for dinitration of mesitylene were found to be as follows: a mixture of 500 mg of mesitylene, 4 ml of $CHCl_3$, and 2 ml of H_2SO_4 in a tube is cooled to −5°, and 0.4 ml of fuming nitric acid is added dropwise keeping the temperature below 0°. After 20 minutes the $CHCl_3$ layer is separated and washed successively with 1-ml portions of H_2O, 5 per cent Na_2CO_3, and H_2O. The solvent is evaporated and the crude crystallized from ethanol-water. About 450 mg of crystals is obtained, melting at 86–88°.

Example 11.8: Nitration of m-Xylene

Mix in an 8-inch test tube 3 ml of concentrated sulfuric and 1.5 ml of fuming nitric acid. Add 0.25 ml of *m*-xylene. Stopper the tube with a solid rubber stopper and shake for 2–3 minutes; at first heat is generated by the reaction, but the tube should not be immersed in cold water. Let it stand for 2 minutes and then heat in a water bath at 70–80° for 15 minutes. Add 8 ml of cold water and allow the oil that first separates out to crystallize. Filter the solid, wash with water, and recrystallize from 5–6 ml of methanol; repeat crystallization twice, using slightly less solvent for the second and third crystallizations. The yield is 100–110 mg of 2,4,6-trinitro-m-xylene, melting at 181–182°.

Note: For the preparation of the mononitro derivatives of the alkyl benzenes, see the original article by Ipatieff and Schmerling.

Example 11.9: Nitration of p-Xylene

Mix 2 ml of concentrated sulfuric acid and 2 ml of fuming nitric acid in an 8-inch test tube. Add 0.2 ml of p-xylene and immerse for 30 minutes in a water bath at 90–95°, shaking the tube frequently. Cool and add 20 ml of water. Filter the solid and recrystallize twice from 7–8 ml of methanol. After the hot alcoholic solution has been filtered, about 1 ml of water is added with shaking. The filtrate should be cooled for 10 minutes before the crystals are removed. The yield is 100–120 mg of 2,3,5-trinitro-p-xylene, melting at 136°.

Picrates and 2,4,7-Trinitrofluorenone Adducts of Aromatic Hydrocarbons

Discussion

A large number of hydrocarbons react with picric acid and other trinitro compounds to form addition products; however, only 2,4,7-trinitrofluorenone will be discussed, since it gives adducts whose melting points do not alter with purification. The melting points of 650 aromatic hydrocarbon adducts have been reported since the initial introduction of this reagent by Orchin and collaborators in 1946.

Example 11.10: Preparation of the 2,4,7-Trinitrofluorenone (T.N.F.) Adduct of Anthracene

Dissolve 100 mg of T.N.F. (2,4,7-trinitrofluorenone) in a mixture of 10 ml of absolute methanol or ethanol and 2 ml of benzene. Boil for a few seconds and add a solution of 60 mg of anthracene in 3.5 ml of methanol and 1.5 ml of benzene. Heat for 30 seconds and cool. Filter the red flocculent crystals, wash with 1 ml of methanol, and dry. The yield is 50–60 mg melting at 192–193°. The complex may be recrystallized from absolute alcohol or alcohol–benzene.

Oxidation of Side Chains

Aromatic hydrocarbons with side chains are oxidized by either alkaline permanganate or chromic acid solution to aryl carboxylic acids, which in many cases can be used as derivatives. The general methods for the oxidation of side chains are described on pages 326–27. It is obvious, however, that the number of hydrocarbons that can be directly derivatized by oxidation of the side chains to yield either monocarboxylic or dicarboxylic acids is small. Toluene and all monoalkyl benzenes yield benzoic acid, however. The three isomeric xylenes give different dicarboxylic acids, and mesitylene a tricarboxylic acid.

References

Addition compounds of dienes: Diels and Adler, *Ber.* **62**, 2081 (1939), 2337 (1939).

Addition of mercaptans to unsaturated hydrocarbons: Jones and Reid, *J. Am. Chem. Soc.* **60**, 2452 (1938).

Addition of nitriles to unsaturated hydrocarbons: Ritter *et al.*, *J. Am. Chem. Soc.* **70**, 4045, 4048 (1948); **71**, 4128, 4130 (1949).

Addition products of aromatic hydrocarbons with picric acid: Barel and Hauber, *J. Am. Chem. Soc.* **53**, 1087 (1931).

Characterization of acetylene by hydration to carbonyls: Sharefkin and Boghosian, *Anal. Chem.* **33**, 640 (1961).

Characterization of alkenes by epoxidation: Sharefkin and Swertz, *Anal. Chem.* **33**, 635 (1961).

Characterization of alkenes by Friedel-Crafts acetylation to ketones and derivatization of the latter to 2,4-dinitrophenylhydrazones or semicarbazones: Sharefkin and Sulzberg, *Anal. Chem.* **32**, 993 (1960).

Characterization of aromatic hydrocarbons with 2,4-dinitrobenzenesulfenyl chloride: Buess and Kharasch, *J. Am. Chem. Soc.* **72**, 3529 (1950).

Characterization of olefins with 2,4-dinitrobenzesulfenyl chloride: Grani, *J. Am. Chem. Soc.* **71**, 3883 (1949); Kharasch *et al.*, *J. Am. Chem. Soc.* **71**, 2724 (1949); **74**, 3422 (1952); **75**, 1081 (1953).

Chlorination (semimicro) of hydrocarbons using chlorine: Cheronis, *J. Chem. Educ.* **20**, 611 (1943).

Chlorination of hydrocarbons using sulfuryl chloride: Cutter and Brown, *J. Chem. Educ.* **21**, 443 (1944).

Conversion of aromatic hydrocarbons to sulfonamides: Huntress and Autenrieth, *J. Am. Chem. Soc.* **63**, 3446 (1941).

Derivatives of saturated hydrocarbons with desocholic acid: Huntress and Phillips, *J. Am. Chem. Soc.* **71**, 458 (1949).

Hydration of alkyl acetylenes: Thomas, Campbell, and Hennion, *J. Am. Chem. Soc.* **60**, 718 (1938).

Hydration of disubstituted acetylenes: Johnson, Swartz, and Jacobs, *J. Am. Chem. Soc.* **60**, 1883 (1938).

Identification of alkyl benzenes by means of the acetamino derivatives: Ipatieff and Schmerling, *J. Am. Chem. Soc.* **59**, 1056 (1937); **60**, 1476 (1938); **65**, 2470 (1943).

Identification of aromatic hydrocarbons: Levy and Campbell, *J. Chem. Soc.*, 141–42 (1939).

Identification of aromatic hydrocarbons by means of substituted *o*-aroyl benzoic acids: Lewenz, *J. Am. Chem. Soc.* **75**, 4087 (1953); Underwood and Walsh, *J. Am. Chem. Soc.* **57**, 940 (1935).

Identification of monosubstituted acetylenes by mercuric salts: Johnson and McEwen, *J. Am. Chem. Soc.* **48**, 469 (1926).

Identification of olefins as dithiocyanates: Dermer and Dysinger, *J. Am. Chem. Soc.* **61**, 750 (1939).

Nitrosochlorides and nitrolamines of unsaturated hydrocarbons: Perrot, *Compt. rend.* **203**, 329 (1936).

Reaction of aliphatic olefins with thiophenol: Ipatieff, Pines, and Friedman, *J. Am. Chem. Soc.* **60**, 2731 (1938).

Reaction of thiol compounds with olefins: Ipatieff and Friedman, *J. Am. Chem. Soc.* **61**, 70 (1939).

Semimicro preparation of nitrosochlorides and bromides of terpenes: Commanday and Cheronis, *Microchem. J.* **4**, 201 (1960).

Separation of normal from branched paraffins by urea: Zimmerschield *et al.,* *J. Am. Chem. Soc.* **71**, 2947 (1949).

2,4,7-Trinitrofluorenone adducts of polynuclear hydrocarbons: Beaton and Tucker, *J. Chem. Soc.,* 3870 (1952); Bergmann and Orchin, *J. Am. Chem. Soc.* **71**, 1917 (1949); Campbell and Kidd, *J. Chem. Soc.,* 2155 (1954); Cason and Philips, *J. Org. Chem.* **17**, 298 (1952); Chu *et al., J. Phys. Chem.* **57**, 504 (1953); Descamps and Martin, *Bull. soc. chim. Belges* **60**, 223 (1952); Gross and Lankelma, *J. Am. Chem. Soc.* **73**, 3429 (1952); Herran *et al., J. Org. Chem.* **16**, 899 (1951); Hofer and Peebles, *Anal. Chem.* **24**, 822 (1952); Howkins *et al., J. Chem. Soc.,* 3280 (1950); Huisgen and Sorge, *Ann.* **566**, 162 (1950); King and King, *J. Chem. Soc.,* 1375 (1954); Klemm and Sprague, *J. Org. Chem.* **19**, 1464 (1954); Kloetzel and Metel, *J. Am. Chem. Soc.* **72**, 4786 (1950); Lambert and Martin, *Bull. soc. chim. Belges* **61**, 224 (1953); Laskowski, Grabar, and McCrone, *Anal. Chem.* **25**, 1400 (1953); Laskowski and McCrone, *Anal. Chem.* **26**, 1497 (1954); **30**, 542 (1958); Newman and Hart, *J. Am. Chem. Soc.* **69**, 298 (1947); Newman and Kosak, *J. Org. Chem.* **14**, 375 (1949); Newman and Wheatley, *J. Am. Chem. Soc.* **70**, 1915 (1948); Newman and Whitehouse, *J. Am. Chem. Soc.* **71**, 3664 (1949); Orchin *et al., J. Am. Chem. Soc.* **68**, 1727 (1946); **69**, 505, 1225 (1947); **70**, 1745 (1948); **71**, 3002 (1949); **73**, 436, 1877 (1951); *J. Org. Chem.* **18**, 609 (1953); Price and Halpern, *J. Am. Chem. Soc.* **70**, 1915 (1948); Riegel *et al., J. Am. Chem. Soc.* **70**, 1073 (1948); Riegel, Siegel, and Kritchensky, *J. Am. Chem. Soc.* **70**, 2950 (1948); Soffer and Stewart, *J. Am. Chem. Soc.* **74**, 567 (1952); Stubbs and Tucker, *J. Am. Chem. Soc.* **3288** (1950); 231 (1954); Takemura, Cameron, and Newman, *J. Am. Chem. Soc.* **75**, 3280 (1953); Tucker, *J. Chem. Soc.,* 2182 (1949); Tucker, Forrest, and Whalley, *J. Chem. Soc.,* 3194 (1949); Tucker and Whalley, *J. Chem. Soc.,* 632, 3213 (1949); Woolfolk, Orchin, and Storch, *Fuel* **26**, 78 (1947); Laskowski and McCrone *Anal. Chem.* **30**, 542 (1958).

Physical Methods

Doss, *Physical Constants of the Principal Hydrocarbons* (N.Y.: The Texas Co., 1939).

Egloff, G., *Physical Constants of Hydrocarbons* (N.Y.: Reinhold, 1939).

Esafoy, *J. App. Chem. (U.S.S.R.)* **14**, 140 (1941), translated in *Foreign Petroleum Tech.* **9**, 344 (1941). Detection and determination of diene hydrocarbons with a conjugated system of double bonds.

Forziati *et al., J. Research Natl. Bur. Standards* **36**, 129 (1946). Purifications and properties of 29 paraffin, 4 alkylcyclopentane, 10 alkylcyclohexane, and 8 alkylbenzene hydrocarbons.

Ferris, *Handbook of Hydrocarbons* (N. Y.: Acadmenic Press, 1955).

Gibbon *et al., J. Am. Chem. Soc.* **68,** 1130 (1946). Purification and physical constants of aromatic hydrocarbons.

Gilman-Miller, *Applications of Infrared and Ultraviolet Spectra to Organic Chemistry* (N.Y.: Wiley, 1953).

Gooding, Adams, and Rall, *Ind. Eng. Chem., Anal. Ed.* **18,** 2 (1946). Determination of aromatics, naphthenes, and paraffins by refractometer methods.

Grosse and Linn, *J. Am. Chem. Soc.* **61,** 151 (1939). Refraction data on propane hydrocarbons.

International Critical Tables (N.Y.: McGraw-Hill, 1932).

Jacob, *Chim. & ind.* (*Paris*), Special No., 341 (Sept., 1926). Identification of hydrocarbons by magnetic rotatory power.

Javelle, *Chim. & ind.* (*Paris*), 264 (April, 1928). Identification of hydrocarbons by magnetic rotatory power.

Mousseron and Winternitz, *Bull. soc. chim.* **12,** 70 (1945). Constants of hydrocarbons of the cyclohexadiene series.

Moutte, *Chim. & ind.* (*Paris*), Special No., 202 (April, 1928). Analysis of hydrocarbons by means of their refractive dispersions.

Randall, H. M., R. G. Fowler, N. Fuson, and J. R. Dangl, *Infrared Determination of Organic Structure* (N.Y.: Van Nostrand, 1949).

Miscellaneous methods: Identification of hydrocarbons by gas–liquid partition chromatography, Harvey and Chalkley, *Fuel* **34,** 191 (1955); James and Phillips, *J. Chem. Soc.* **1953** (1900); Lichtenfels, Fleck, and Burow, *Anal. Chem.* **27,** 1510 (1955); Phillips, *Discussions Faraday Soc.* **7,** 241 (1949); Ray, *J. Appl. Chem.* **4,** 21, 82 (1954).

KETONES

The derivatives listed in Table 10.4 (page 257) for the characterization of aldehydes are also suitable for the characterization of ketones. Of the following derivatives—2,4-dinitrophenylhydrazones, *p*-nitrophenylhydrazones, phenylhydrazones, semicarbazones, thiosemicarbazones, dimethones, and oximes—only the *dimethones cannot* be prepared from the ketones. For all the rest the same general procedures described on pages 258–64 also apply to the preparation of derivatives from ketones. The modifications that have to be introduced into the general method are essentially an increase in the time of heating the reaction mixture and an increase in the amount of solvent employed for the crystallization of the derivative. For example, in the preparation of piperonal phenylhydrazone the reaction mixture (carbonyl compound, phenylhydrazine, and acetic acid) is heated for 3–4 minutes (Example 10.13), while in the preparation of the same derivative of benzophenone (Example 11.11) the reaction mixture is heated for 15 minutes. In general, the carbonyl group of the ketones reacts at a slower rate than that of an aldehyde. Similarly, the amount of solvent employed for the crystallization of the same quantity of phenylhydrazone from a ketone is much greater than that employed for an aldehyde of the same molecular type.

Oximes

Oximes are suitable derivatives for a number of ketones. Eq. (11.10) represents the formation of an oxime from a ketone. In the preparation of

$$(CH_3)_2C{=}O \quad + \quad H_2NOH \quad \longrightarrow \quad (CH_3)_2C{=}NOH \quad + \quad H_2O \qquad (11.10)$$

Acetone   Hydroxylamine   Acetoxime

oximes it is advisable to use 500 mg of the compound since the yields are relatively poor. Furthermore, there is always the danger of rearrangement of the oximes to N-substituted amides.

General Method for Preparation

Place in an 8-inch test tube 500 mg of the carbonyl compound, 500 mg of hydroxylamine hydrochloride, 3 ml of pyridine, and 3 ml of absolute alcohol. Arrange for reflux and boil the mixture gently for 2 hours on a steam bath. Pour the mixture into an evaporating dish and remove the solvent by a current of air under a hood. Scrape the residue by means of a microspatula and grind with 3 ml of cold water; filter and recrystallize the oxime from methanol, or methanol–water mixture.

Note: In the preparation of some oximes, the pyridine may be replaced with 4 ml of $1N$ sodium hydroxide solution, and absolute alcohol with ordinary methanol. The mixture is heated for 10–20 minutes and cooled in an ice–salt mixture.

When there is a danger of rearrangement the mixture is not refluxed but warmed and then allowed to stand for 24 hours. If on dilution with water and cooling the derivative does not separate out, the mixture is heated for a longer period of time and then cooled again. The crystals that separate out are crystallized from the minimum amount of aqueous methanol.

Substituted Hydrazones, Semicarbones, and Other Derivatives

The description of the general methods for substituted hydrazones, semicarbazones, and other derivatives in Chapter 10 (pages 258–64) applies as well for ketones. A few brief examples will be given here.

Example 11.11: Preparation of the Phenylhydrazone of an Aryl Ketone: Benzophenone Phenylhydrazone

Place in an 8-inch tube 150 mg of benzophenone and 5 ml of methanol. Heat for a few seconds to effect solution; then add 0.1 ml of phenylhydrazine and 1 drop of glacial acetic acid. Then heat to boiling point under reflux for about 15 minutes. Add dropwise 1.5 ml of water until a permanent clouding results. Cool, filter the crystals, and wash with 1 ml of water containing 1 drop of acetic acid. Crystallize the product from 10 ml of methanol. It may require two crystallizations for complete purification. The yield of phenylhydrazone is about 90–100 mg, melting at 136–137°.

Note: A number of ketones, such as acetophenone and cyclohexanone, react anomalously with phenylhydrazine. For example, following the same method as outlined, acetophenone yields a product that shows an initial melting point of 98°, which is 7° below the melting point of the pure derivative. Even after crystallization, the product undergoes decomposition when dried in air.

Example 11.12: Preparation of Acetone 2,4-Dinitrophenylhydrazone

Proceed by the same general method as described in Example 10.14 (page 260), and use 0.2 ml of acetone. The yield is about 50 mg, melting at 126°. If it is desired to recrystallize the derivative, use 3–4 ml of methanol and add water to the hot alcohol solution until a cloudiness results.

Example 11.13: Preparation of the Semicarbazone of a Ketone: Butanone Semicarbazone

Place in an 8-inch test tube 200 mg of semicarbazide hydrochloride, 300 mg sodium acetate, and 2 ml of water. Warm for a few seconds over a small flame to effect solution. Add 0.2 ml of butanone with a pipet or a dropper. Stopper with a cork provided with a reflux condenser, place the tube in a beaker containing water at 70–75°, and heat at this temperature for 10 minutes. Allow the tube to remain in the water bath for 10 additional minutes. Filter and wash crystals with 5 drops of cold water. Keep about 5 mg of the crystals and recrystallize the main portion from 1.5–2 ml of water. The yield is 60–70 mg, melting at 135–136°.

Note: For acetone, it is advisable to heat at 50° for about 1 hour; otherwise the yield is poor. For the higher aliphatic ketones, use 2 ml of methanol and 2 ml of water as a solvent for the reaction mixture. Since the solubilities of the semicarbazones decrease with increase in the complexity of the molecule, good results are obtained by using 100 mg of the carbonyl compound. Thus, 2-heptanone (methyl-*n*-pentyl ketone) and cyclohexanone give about 100 mg of pure derivative from 0.1 ml of the compound. Aromatic ketones react more slowly, and a longer period of heating the reaction mixture is recommended.

Example 11.14: Chromatographic Separation and Detection of Small Quantities of Ketones

The same method is employed as described for aldehydes (page 264). A mixture of an aliphatic ketone (acetone or butanone) and an aromatic ketone (acetophenone or benzophenone) of the same concentration as described for aldehydes is used with the same volume of reagent. When developed, the spots of the aliphatic ketones will generally be found on the upper part of the chromatogram; those of the aromatic ketones about the middle.

References

Action of hypochlorite on ketones: Hurd and Thomas, *J. Am. Chem. Soc.* **55**, 1646 (1933).

Improved hydroxylamine method for the determination of aldehydes and ketones: Bryant and Smith, *J. Am. Chem. Soc.* **57**, 57 (1935).

Optical properties of derivatives of ketones: Grammaticakis, *Bull. soc. chim.* **8**, 28 (1941); **7**, 527 (1940).

Reaction of acetophenone derivatives with sodium hypochlorite: Van Arendonk and Cupery, *J. Am. Chem. Soc.* **53**, 3184 (1931).

Oximes: Bachmann and Barton, *J. Org. Chem.* **3**, 307 (1938); Bachmann and Boatner, *J. Am. Chem. Soc.* **58**, 2099 (1936); Buck and Ide, *J. Am. Chem. Soc.* **53**, 1541 (1931).

Salts of oximes: Grammaticakis, *Compt. rend.* **224**, 1568 (1947).

Substituted hydantoins: Henze and Speer, *J. Am. Chem. Soc.* **64**, 2502 (1942).

Chromatography of metal β-diketone chelates: Berg and Strassner, *Anal. Chem.* **27**, 1131 (1955).

For other references on chromatographic identification of ketones, see aldehyde references (pages 266–67).

NITROGEN FUNCTIONS

Besides the amines (page 271) the nitrogen functions include the nitriles; nitro and nitroso compounds; azo, azoxy, and hydrazo compounds; isocyanates; and isocyanides. Of these only the derivatization of nitriles by hydrolysis and reduction, and the derivatization of nitro compounds will be discussed. For derivatives of the other groups of compounds containing the nitrogen function, the student is referred to the more extensive work of the authors.

Nitriles

Table 11.3 gives a summary of the suitable derivatives for the characterization of nitriles.

TABLE 11.3
Derivatives for the Identification of Nitriles

*Carboxylic acids	α-Iminomercaptoacetic acid hydrochlorides
Amides	Hydrazides
Amines (substituted thioureas)	Ketones

* Recommended derivative.

Hydrolysis of Nitriles to Carboxylic Acids

The following two examples illustrate the general methods for acid hydrolysis and alkaline hydrolysis of nitriles.

Example 11.15: Acid Hydrolysis of Acetonitrile

Place in an 8-inch distilling tube 4 ml of phosphoric acid (85 per cent), 2 ml of sulfuric acid (75 per cent), and 0.5 ml of acetonitrile. Add 2 boiling stones and attach to the tube a reflux condenser; boil gently for 1 hour. Add 3 ml of water and distill until 3.5 ml of distillate have been collected in an 8-inch tube, which serves as a receiver. Add 1 drop of phenolphthalein and sufficient sodium hydroxide solution (10 per cent) to develop a pink color; then make the solution just acid to the phenolphthalein test with 2–3 drops of dilute hydrochloric acid. Add 200 mg of *p*-nitrobenzyl bromide and 12–15 ml of methanol so that, when the solution is barely refluxing, it is homogeneous. Reflux for 2 hours. Cool for about 30 minutes and filter the crystalline mass that separates out. Recrystallize from 5 ml of alcohol (see page 244). The yield is 70–90 mg.

Note: When the carboxylic acid is not volatile with steam and boils above 200°, it is separated from the phosphoric–sulfuric acid mixture by extraction; the reaction mixture is first diluted with 3 ml of water, cooled, and partially neutralized with 6*N* sodium hydroxide solution. It is then extracted with three 5-ml portions of ether. If the acid is solid (from aryl cyanides), the ether is evaporated; otherwise it is used for the preparation of the *p*-toluidide or anilide. Addition of 100 mg of solid sodium chloride to sulfuric acid before boiling increases the rate of hydrolysis. Since hydrochloric acid is more effective for many nitriles than sulfuric acid[1] it can be employed for nitriles, which are resistant to hydrolysis.

An alternative method for the characterization of the carboxylic acid is to neutralize the hydrolytic mixture and prepare the *S*-benzylthiouronium salt directly.

Example 11.16: Alkaline Hydrolysis of Benzonitrile

Place 4 g of glycerol, 2 g of potassium hydroxide pellets, and 0.2 ml of benzonitrile in an 8-inch test tube. (**Caution:** *Care should be used in handling benzonitrile as it is a lachrymator.*) Attach a condenser to the tube and boil gently for 1 hour. Dilute with 1 ml of water, cool, and add 2 ml of ether. Shake gently and allow the immiscible layers to separate. Pour off the ether from the aqueous viscous layer. Cool the tube and make the solution just acid by slow addition of 6*N* hydrochloric acid solution. Extract three times with 4–5 ml portions of ether. Distill the ether or evaporate cautiously in a dish over the steam bath and crystallize the crude benzoic acid by dissolving it in the minimum amount of 90 per cent hot methanol and adding water to the hot solution until a permanent cloudiness results. The yield is 50–70 mg.

Note: The yield of the carboxylic acid from 200 mg of benzonitrile or phenylacetonitrile is usually about 150 mg, but the product melts 2° or more below the melting point of the pure compound. When this method is used with the lower aliphatic nitriles, it is advisable to begin with 400–500 mg of substance.

[1] Kilpatrick, *J. Am. Chem. Soc.* **69**, 40 (1947).

When the carboxylic acid is not a solid, the ether is transferred in a distilling tube; after removal of the solvent, the procedure described in Example 10.5 (page 240) is followed to convert the acid to the p-toluidide.

Reduction of Nitriles to Amines and Their Characterization by Preparation of Substituted Thioureas

Discussion

The reduction of the nitrile to a primary amine is represented by Eq. (11.11). The most common method for reduction is to employ sodium

$$RCN + 4(H) \longrightarrow RCH_2NH_2 \qquad (11.11)$$

in the presence of alcohol as the reducing agent. The reaction mixture is distilled to remove the alcohol, then rendered alkaline. The amine is distilled and derivatized by means of phenylisothiocyanate to yield a substituted thiourea.

The yield in the reduction of nitriles to amines is not very good since side reactions occur and hence it is necessary to use 1–1.5 g of the nitrile. When the quantity available is 500 mg or less, catalytic hydrogenation of the nitrile in 90 per cent ethanol containing 10 per cent ammonium carbamate should be tried.

General Method for Reduction of Nitriles

An 8-inch distilling tube provided with a reflux condenser is charged with 10 ml of absolute ethanol and 0.5 g of an aliphatic nitrile or 0.75 g of an aromatic nitrile. About 0.75 g of finely cut sodium is added at such a rate that the reaction proceeds vigorously under control. To add the sodium, momentarily lift the cork if a microcondenser is employed or insert it through the top of the condenser if a Liebig-type condenser is used.

After the reaction is complete (15 minutes) the mixture is cooled and 5 ml of concentrated hydrochloric acid are added dropwise with stirring of the contents of the tube. Test the reaction mixture and if it is not distinctly acid add a small amount more of hydrochloric acid. The tube is arranged for distillation, 2 boiling stones are added, and the mixture is distilled until 10 ml of distillate (ethanol) has been collected, which is set aside. The flask is cooled by an ice-salt mixture and 7 ml of 40 per cent sodium hydroxide is added in small amounts. The temperature of the mixture should not be allowed to rise above 30°. The tube is arranged for distillation; the receiving 8-inch tube is charged with 2 ml of water and 1 ml of 6N hydrochloric acid; the delivery tube is arranged so that it just dips into the dilute hydrochloric acid. The distillation of the amine–water mixture is continued until the volume of the distillate is 7–8 ml.

The distillate in the tube is cautiously neutralized by dropwise addition of sodium hydroxide solution to pH 8.0, and 0.25–0.30 ml (12–14 drops) of phenylisothiocyanate is added. The amine is converted to the substituted phenylthiourea according to the procedure described on pages 275–76.

References

Addition compounds of nitriles with mercaptoacetic acid: Condo, Hinkel, Fassero, and Shriner, *J. Am. Chem. Soc.* **59**, 230 (1937).

Alkyl (2,4,6-trihydroxyphenyl) ketones: Howells and Little, *J. Am. Chem. Soc.* **54**, 2451 (1932).

Alkyl phenyl ketones: Shriner and Turner, *J. Am. Chem. Soc.* **52**, 1267 (1930).

Conversion of nitriles to ketones and derivatization as semicarbazones: Plieninger and Werst, *Ber.* **88**, 1956 (1955).

Conversion of nitriles to methylene bisamides: Margat *et al., J. Am. Chem. Soc.* **73**, 1031 (1951).

Hydrolysis of nitriles with potassium hydroxide in diethylene glycol or glycerol: Hovira and Palfray, *Compt. rend.* **211**, 396 (1940).

Reduction to amines: Cutter and Taras, *Ind. Eng. Chem., Anal. Ed.* **13**, 830 (1941).

Reduction of nitriles by means of lithium aluminum hydride: Brown in Adams (ed.), *Organic Reactions,* Vol. VI (N.Y.: Wiley, 1951), pp. 469–509; Cheronis, N. D., *Micro and Semimicro Methods,* Vol. VI of Weissberger (ed.), *Technique of Organic Chemistry* (N.Y.: Interscience, 1957), pp. 252–56.

Nitro Compounds

The reactions by which the nitro compounds may be derivatized are: (*a*) reduction to amines, which are then converted to *N*-substituted benzamides, aryl sulfonamides, or substituted phenylthioureas; (*b*) further nitration, or introduction of other substituents; (*c*) oxidation of aromatic hydrocarbons having side chains to carboxylic acids. For example, *p*-nitrotoluene may be reduced to *p*-toluidine, nitrated to 2,4-dinitrotoluene, or oxidized to *p*-nitrobenzoic acid. In general, most aromatic mononitro compounds may be converted to dinitro or trinitro derivatives; in addition, other substituents already present may be altered; for example, the aryl nitro compound may be brominated or an alkyl side chain may be oxidized. The selection of the derivative should be the result of a judicious consideration of all the factors involved. The example of *p*-nitrotoluene may be further considered as an illustration; if *p*-nitrotoluene is reduced to *p*-toluidine, the melting point of the amine is low (45°) and must be acylated for identification. The second alternative is to oxidize *p*-nitrotoluene to *p*-nitrobenzoic acid; the disadvantage of this method lies in the high melting point of the acid. The third alternative, which is selected as a trial, is the preparation of 2,4-dinitrotoluene. The ease of nitration and the purification of the dinitro compound as compared with the preparation of other possible derivatives are the factors that suggest this selection.

The derivatization of dinitro, trinitro, and, in general, polynitro compounds must be considered individually for each compound. Addition compounds of polynitro derivatives often prove desirable derivatives. At least two nitro groups on each benzene ring are required for the formation of addition compounds. The relative position of the nitro groups and the nature of other substituents present in the ring also have an influence on the formation of the addition compound. Nitro groups ortho to each other and methyl groups situated between nitro groups appear to hinder addition compound formation. α-Naphthol is stated in the literature as having a greater tendency to form addition compounds than naphthalene; the latter, however, is the reagent recommended because it is easily available in the pure form.

Derivatization of Nitro Compounds by Reduction to Amines

Discussion

The reduction of nitro compounds to amines may be accomplished by: (a) metals, such as tin or zinc, in acid media; (b) catalytic hydrogenation; and (c) ions, such as S^{-2} and $S_2O_4^{-2}$.

Tin and hydrochloric acid are most commonly used for reduction in acid media. After the reduction of the nitro compound the solution is rendered alkaline and the amine is extracted with ether. If the product is a lower alkyl amine, the alkaline solution is distilled, and the distillate is collected in a small amount of dilute acid. The distillate is directly used for the preparation of derivatives by means of acylation. In some cases it is possible to reduce the nitro compound to the substituted hydroxylamine, which can be identified by its melting point or by derivatization.

Catalytic hydrogenation is advisable when the quantity of the nitro compound available is less than 100 mg. The general method described in the more extensive work of the authors is used.

Example 11.17: Reduction by Metals in Acid Media: Reduction of 1-Nitropropane to n-Propylamine

Place in an 8-inch test tube 200 mg of 1-nitropropane and 3 ml of 6N hydrochloric acid solution. Add 500 mg of tin in two portions over a period of 10 minutes, warming at first to start the reaction. Boil the mixture gently under reflux for 30 minutes or until the odor of the nitro compound has disappeared. Cool the mixture by immersion of the tube in running tap water and add slowly 6 ml of 6N sodium hydroxide. Transfer the contents of the tube to an 8-inch distilling tube; wash the vessel with 1–2 ml of water and unite the washings with the alkaline mixture. Add 2 boiling stones and distill the alkaline solution until 4 ml of distillate have been collected in a receiving tube containing 2 ml of 6N hydrochloric acid and

1 drop of aqueous methyl red or methyl orange solution. If the distillate becomes alkaline, a small amount of additional hydrochloric acid is added. Add to the distillate 0.4 ml of benzoyl chloride and then, while the tube is cooled in tap water, 8 ml of 6N sodium hydroxide solution. The tube is stoppered with a solid rubber stopper and shaken vigorously at intervals for 10 minutes. An oil separates out that, on cooling and shaking, solidifies. Then proceed to prepare the benzoyl derivative as directed in Example 10.21. The yield is 30–35 mg of the pure derivative, melting at 84°.

Note: If the amine boils much above 100°, it is best to extract the alkaline solution with ether. In such a case the acid solution that contains the amine salt is made alkaline, care being taken not to use a great excess of alkali; it is then extracted with three portions of 4–5 ml of ether (which is free from alcohol), and this extract is used directly for the preparation of the derivative.

Derivatization by Oxidation of Side Chains

Acid dichromate or alkaline permanganate solution may be used for the oxidation of alkyl side chains of aromatic nitro hydrocarbons to the carboxylic stage. Generally, permanganate oxidation is preferred for the more resistant side chains—that is, in the presence of nitro groups, or whenever extensive degradation by oxidation is necessary. The dichromate and permanganate procedures are illustrated by the following examples.

Example 11.18: Permanganate Oxidation of o-Nitrotoluene to o-Nitrobenzoic Acid

Place 1.5 g of solid potassium permanganate, 25 ml of water, 0.5 ml 8–10 drops) of 6N sodium hydroxide, and 2 boiling stones in an 8-inch tube arranged for heating under reflux. Lastly, add 400–500 mg of o-nitrotoluene and boil gently for 1 hour or longer until the purple color of the permanganate has disappeared. Cool the reaction mixture and carefully acidify with dilute sulfuric acid; then heat to boiling. If there is an appreciable amount of manganese dioxide present, add a small amount of solid sodium bisulfite. Cool and filter the acid; recrystallize from 4–5 ml of hot alcohol. Filter the hot alcoholic solution and add water dropwise until a permanent cloudiness results. Cool, and filter off the crystals. The yield is 300–400 mg.

Note: m- and p-Nitrotoluene are oxidized to the respective nitrobenzoic acids by the same method. Similarly, o-, m-, and p-xylenes are oxidized, respectively, to phthalic, isophthalic, and terephthalic acids. In the case of xylenes, since there are two side chains to be oxidized, the amount of all reagents except the hydrocarbon is doubled. A 125-ml Erlenmeyer flask is used for the boiling vessel, and the mixture is heated for 1.5–2 hours.

It is recommended that the mixture be boiled in the hood since frequent bumping occurs. If after heating 2 hours the permanganate color persists, cool and add bisulfite until the color has been discharged and the manganese dioxide has dissolved.

Example 11.19: Dichromate Oxidation of p-*Nitrotoluene to* p-*Nitrobenzoic Acid*

In an 8-inch tube dissolve 1 g of sodium dichromate in 3 ml of water and add 2 ml of concentrated sulfuric acid. Add 200–250 mg of *p*-nitrotoluene and 2 boiling stones. Boil for 20–30 minutes. Cool and add 2–3 ml of water; then filter. Wash three times with water. Recrystallize from 4–5 ml of hot methanol. Filter the hot methanol solution and add water to the filtrate until a permanent cloudiness results. Cool, and filter off the crystals. The yield is 180–230 mg.

Derivatization of Nitro Compounds by Further Nitration

The discussion of nitration on pages 305–07 and 313–15 should be reviewed. The derivatization of *o*- and *p*-nitrotoluenes by further nitration is based on the fact that both yield the same derivative, 2,4-dinitrotoluene.

Example 11.20: Nitration of o-*Nitrotoluene and* p-*Nitrotoluene to 2,4-Dinitrotoluene*

The general method on page 306 is employed. Use 200 mg of the mononitro compound. In the case of *p*-nitrotoluene, after one crystallization 60–70 mg of the pure dinitro compound (m.p. 70°) are obtained. In the case of *o*-nitrotoluene, two crystallizations are required, and the yield of the pure derivative is 40–50 mg.

A number of polynitro compounds with substituents other than nitro groups may be derivatized by alteration of their substituents. For example, 2,4-dinitrochlorobenzene may be converted by hydrolysis to 2,4-dinitrophenol; similarly, 2,4,6-trinitroanisole and the 2,4,6-trinitrophenetole may be hydrolyzed to give picric acid. Oxidation of methyl groups is feasible, although such oxidation should be done *with care even with semimicro quantities;* for example, 2,4,6-trinitrotoluene may be converted by oxidation to 2,4,6-trinitrobenzoic acid, but this is not recommended unless precautions are used and permission is obtained from the instructor.

The polynitro compounds may be derivatized by preparation of adducts with α-naphthol or naphthalene. To prepare such adducts with naphthalene, equimolecular amounts of the polynitro compound and the reagent are heated cautiously until a homogeneous melt is obtained. The melt is cooled, recrystallized from alcohol, dried rapidly, and the melting point determined.

References

Derivatives of nitroparaffins: Dermer and Hutcheson, *Proc. Oklahoma Acad. Sci.* **23**, 60 (1943); *C.A.* **38**, 2008 (1944).

Identification of nitro compounds by catalytic hydrogenation at atmospheric pressure: Cheronis and Koeck, *J. Chem. Educ.* **20,** 488 (1943); Cheronis and Levin, *J. Chem. Educ.* **21,** 603 (1944).

Identification of polynitro compounds as addition compounds: Asahina and Shinomiya, *J. Chem. Soc. Japan* **59,** 341 (1938); Dermer and Smith, *J. Am. Chem. Soc.* **61,** 748 (1939); Shinomiya, *Bull. Soc. Japan* **15,** 92 (1940).

Nitro compounds from halides (microscopic identification): Dunbar and King, *Microchem. J.* **3,** 143 (1959).

Reduction of nitro compounds by means of lithium aluminum hydride: Brown in Adams (ed.), *Organic Reactions,* Vol. VI (N.Y.: Wiley, 1951), pp. 469–509; Cheronis, *Micro and Semimicro Methods,* Vol. VI of Weissberger (ed.), *Techinque of Organic Chemistry* (N.Y.: Interscience, 1957), pp. 252–56.

PHENOLS

Table 11.4 lists *only the most important* derivatives for the characterization of phenols. For a complete list the reader is referred to Table 10.3

TABLE 11.4
Derivatives for the Identification of Phenols

*3,5-Dinitrobenzoates	Diphenyl and other urethans
*α-Naphthylurethans	Acetates, benzoates, *p*-nitrobenzoates, and
*Aryloxyacetic acids	other esters
*Bromo derivatives	Sulfonic acid esters
2,4-Dinitrophenyl ethers	Pseudosaccharin esters
Nitro derivatives	Picrates

* Recommended derivative.

(page 248), which lists the derivatives of alcohols, because in general most of the derivatives used for the characterization of alcohols may also be used for phenols. The selection of the most suitable derivative depends on the nature of the phenol. With monocylic phenols the 3,5-dinitrobenzoate, or *p*-nitrobenzoate, should be considered first if the melting point of the derivative is not above 200°. If the phenol is dicyclic or polycyclic and has a melting point of 100° or above, acetylation should be considered. Although the formation of urethans is slow, it should be considered wherever possible, for the reaction can be catalyzed. The preparation of aryl oxyacetic acid and 2,4-dinitrophenyl ethers is based on the greater reactivity of the phenolic function as compared to the alcoholic. The preparation of these derivatives is recommended if the quantity of material available is over 200 mg. Finally, a number of typical "aromatic" reactions, such as bromination and, to a lesser extent, nitration and oxidation, may at times be employed for the derivatization of phenols. In the case of nitrophenols, the corresponding amino compounds obtained by reduction should be considered because the formation of esters and urethans is slow and in some cases not feasible.

For example, p-nitrophenol can be readily derivatized, even in 10-mg quantities, by catalytic reduction to p-aminophenol.

3,5-Dinitrobenzoates

Example 11.21: Preparation of β-Naphthyl 3,5-Dinitrobenzoate

Place in a 6-inch tube 100 mg of $β$-naphthol, 110 mg of 3,5-dinitrobenzoyl chloride, 2 ml of pyridine, and 2 boiling stones. Arrange for reflux and boil gently for 1 hour. Cool, add 1 ml of 5 per cent sulfuric acid and 5 ml of water; shake well and filter. Replace the crystals, together with the filter paper, into the test tube, add 5 ml of 2 per cent sodium hydroxide, shake well to remove the 3,5-dinitrobenzoic acid, and filter; then wash twice with 2 ml of water. Suspend the crystals in 5 ml of methanol, heat almost to boiling, and filter. The crystals *remaining on the filter* are used for the melting-point determination. The yield is 80–100 mg, melting at 209–210°. A small crop of crystals may also be obtained from the filtered alcoholic solution.

Note: The solubility of the 3,5-dinitrobenzoates of naphthols in alcohol is small as compared with the like derivatives of phenols and cresols. For this reason the derivative is purified by removing soluble impurities, which in this case are mainly traces of dinitrobenzoic acid and unreacted naphthol. When the derivative separates out as an oil after washing with acid, the aqueous layer is poured out and 5–8 ml of ethyl ether or isopropyl ether is added, after which the ether solution is washed successively with water, 2 per cent sodium hydroxide solution, and finally with water. The ether is then evaporated and the residue is crystallized from methanol or ethanol.

General Method for Preparation of Benzoates and Acetates

Methods for the preparation of benzoates are described on pages 254 and 273. For the preparation of acetates the usual method with acetic anhydride is employed and a drop of sulfuric acid is added to catalyze the reaction. If this method does not yield good results 1–2 millimoles of the phenol in 5 ml of dry benzene is refluxed for about 1 hour with 1.1 millimoles of acetyl chloride and 100 mg of magnesium powder. The reaction mixture is diluted with 5 ml of ether and added to a mixture of 5 ml of 5 per cent sodium carbonate solution and 5 g of ice. The mixture is stirred well, transferred into a separatory funnel, and the aqueous layers removed. The benzene–ether solution is washed with water and then evaporated. The crude acetate is crystallized from aqueous methanol.

Urethans

General Method for Preparation

The general methods are the same as those for alcohols (page 252). If the phenol is relatively reactive, the isocyanate and the phenol are mixed in a perfectly dry vessel and heated for about 2–5 minutes. If the phenol is not reactive 1 ml of pyridine and a drop of 10 per cent solution of tri-methylamine in hexane or heptane are added and the mixture is heated for 20–30 minutes. The mixture is cooled and if the urethan does not separate out, 1 ml of 5 per cent sulfuric acid is added. The crude urethan is purified by crystallization from petroleum ether. Generally the introduction of proton-repelling substituents in Ar—OH produce a retardation in the for-mation of urethans. Thus phenol, *m*-cresol, and thymol yield α-naphthyl-urethans readily, but *p*-nitrophenol does so with difficulty, even in the presence of catalysts; picric acid does not react.

Example 11.22: Preparation of Thymol α-Naphthylurethan

Dry a test tube as directed in Example 10.10. While the tube is cooling, fit a cork (of a size equal to that with which the tube is stoppered) with a cal-cium chloride tube. Place rapidly in the dry test tube 200 mg of thymol and 0.25 ml of α-naphthyl isocyanate and stopper it with the cork holding the calcium chloride tube. Clamp the tube on a stand and heat it by means of a small direct flame so that the mixture boils gently for 2 minutes. Allow to cool for 3 minutes and then rub the mixture with a glass rod until it sets into a crystalline mass. Add 10 ml of petroleum ether, cool, and filter as described under the preparation of *n*-propylurethan in Example 10.10. Ex-tract the residue with another portion of 8–10 ml of boiling petroleum ether. Filter the crystals and wash them with 1 ml of petroleum ether. The yield is 120–140 mg of crystals, melting at 159°. On crystallizing from 7 ml of petroleum ether, 75–100 mg of pure urethan is obtained, melting at 160°.

Note: Nitrophenols do not yield urethans with ease. *p*-Nitrophenol, heated for 1 hour with the isocyanate, gives a product melting 30° below the temperature recorded in the literature as the melting point of the derivative. Naphthols react slowly, and therefore it is advisable to heat, under reflux, the naphthol and isocyanate dissolved in 5–10 ml of petroleum ether for 0.5–1 hour. Tertiary amines catalyze the reaction; a drop of 10 per cent solution of trimethyl, triethyl, or tributylamine in petroleum ether accelerates formation of urethans. The tertiary amine may be also used to induce crystallization in case the reaction mixture of phenol and isocyanate forms a viscous oil after heating.

Example 11.23: Preparation of p-tert-Butylphenol Phenylurethan

An 8-inch dry test tube is provided with a microcondenser and arranged for reflux. About 1 ml of a petroleum distillate (b.p. 160–180°), obtained by fractionating 10 ml of kerosene, is placed in the tube together with 100 mg each of *p-tert*-butylphenol and phenylisocyanate. Care is used in handling the isocyanate. Two boiling stones are added; the reaction mix-ture is gently refluxed for 2 hours and then cooled. The crystals that separate out are filtered and purified by crystallization from petroleum ether (b.p. 90–110°) as described in Example 10.10. The yield is 80–90 mg, melting at 148–149°.

Aryloxyacetic Acids

Phenols react with chloroacetic acid in presence of sodium hydroxide to yield aryloxyacetic acids, as represented by Eq. (11.12). The aryloxyacetic

$$C_6H_5ONa + ClCH_2COO^-Na^+ \longrightarrow C_6H_5OCH_2COO^-Na^+ + NaCl \qquad (11.12)$$

$$\text{Sodium chloroacetate} \qquad\qquad \text{Sodium phenoxyacetate}$$

acids are crystalline solids having well-defined melting points; in addition, the determination of their neutralization equivalents may be used as a confirmatory test. For semimicro work it is necessary to have available at least 200 mg of the phenol; otherwise the yield is not sufficient for beginners to handle.

General Method for Preparation

For the preparation of the derivative 0.5 ml of 50 per cent aqueous chloroacetic acid is added to 200 mg of the phenol and 1 ml of a 6N solution of sodium hydroxide in a small test tube. More water should be added if the phenol salt does not dissolve completely. The test tube is provided with a microcondenser and heated in a water bath at 90–100° for 1 hour. The solution is then cooled, two volumes of water added, and then acidified to Congo red with dilute hydrochloric acid. The derivative is extracted with two 4-ml portions of ether. The ether extract is washed with 2 ml of water and then extracted with 5 per cent sodium carbonate solution. The sodium carbonate extract is next acidified with dilute HCl to precipitate the aryloxyacetic acid. The acid is recrystallized from water. The melting points of about fifty derivatives of phenols have been reported in the literature.

Example 11.24: Preparation of an Aryloxyacetic Acid from m-Cresol

Place in an 8-inch test tube 200 mg of m-cresol, 1 ml of 6N sodium hydroxide solution, and 0.5 ml of a 50 per cent solution of chloroacetic acid. Provide the tube with a microcondenser arranged for reflux and heat in a water bath at 90–100° for about 1 hour. Cool and add 3 ml of water and 1 ml of 6N hydrochloric acid. Extract with two 4-ml portions of ether; wash the combined ethereal solutions first with 2 ml of water and then with 5 ml of 10 per cent sodium carbonate solution, which removes the aryloxyacetic acid. Transfer the sodium carbonate solution to a beaker and slowly add dilute hydrochloric acid until the solution is distinctly acid. Cool and filter the crystals. The yield is 45–55 mg of m-toloxyacetic acid.

Bromo Derivatives

Phenols react rapidly with bromine to give bromo-substituted phenols, which in many cases form useful derivatives. Thus, according to Eq. (11.13),

$$C_6H_5OH + 3Br_2 \longrightarrow C_6H_2Br_3OH + 3HBr \qquad (11.13)$$
$$\text{2,4,6-Tribromophenol}$$

phenol forms 2,4,6-tribromophenol, while *o*-cresol and *m*-cresol yield, respectively, dibromo and tribromo derivatives. The following example illustrates a general method for bromination of phenols. Since this reaction is relatively easy, the preparation of the bromo derivative should be considered.

Example 11.25: Preparation of Tribromophenol

Place in an 8-inch tube 0.8 g of potassium bromide and add 5 ml of water. Shake the tube until the salt dissolves; add *carefully* 0.5 g of bromine. Place in a 6-inch test tube 100 mg of phenol, 1 ml of methanol, and 1 ml of water. Add about 1.5 ml of the prepared bromine solution and shake the tube; continue the addition of bromine solution until the mixture retains a yellow color after shaking. Add 3–4 ml of water and shake vigorously. Filter the bromophenol and wash well with water. Dissolve the crystals in hot methanol and filter; add water dropwise to the methanol solution until a permanent cloudiness results. The yield is 180–200 mg of crystals, which melt at 95°.

References

Acetates: Chattaway, *J. Chem. Soc.*, **1931**, 2495.

Aryl oxyacetamides: Namstkin *et al.*, *Zhur. Anal. Khim.* **5**, 7 (1950); *C.A.* **44**, 4375 (1950).

Aryl oxyacetic acids: Hayes and Branch, *J. Am. Chem. Soc.* **65**, 1555 (1943); Koelsch, *J. Am. Chem. Soc.* **53**, 304 (1931).

Benzoates: Baumann, *Ber.* **19**, 3218 (1886); Schotten, *Ber.* **17**, 2544 (1884).

p-Bromo-, *p*-chloro-, *β*-naphthyl-, *p*-nitro-, and 3,5-dinitro-4-methylurethans: Sah *et al.*, *Rec. trav. chim.* **58**, 453, 582, 591, 595 (1939).

3,5-Dinitrobenzoates: Brown and Kremers, *J. Am. Pharm. Assoc.* **11**, 607 (1922); Phillips and Keenan, *J. Am. Chem. Soc.* **53**, 1924 (1931); Reichstein, *Helv. Chim. Acta* **9**, 799 (1926).

2,4-Dinitrophenyl ethers: Bost and Nicholson, *J. Am. Chem. Soc.* **57**, 2368 (1935).

3,5-Dinitrophenylurethans: Hoeke, *Rec. trav. chim.* **54**, 514 (1935); Sah and Ma, *J. Chinese Chem. Soc.* **2**, 229 (1934); Veibel and Lillelund, *Dansk Tidsskr. Farm.* **14**, 241 (1940); Veibel *et al.*, *Dansk Tidsskr. Farm.* **17**, 187 (1943).

Diphenylurethans: Herzog, *Ber.* **40**, 1831 (1907).

α-Naphthylurethans: French and Wirtel, *J. Am. Chem. Soc.* **48**, 1736 (1926); Sah *et al.*, *Rec. trav. chim.* **58**, 453, 582, 591, 595 (1939).

p-Nitrobenzoates: Adamsen and Kenner, *J. Chem. Soc.*, 287 (1935); Armstrong and Copenhaver, *J. Am. Chem. Soc.* **65**, 2252 (1943); Henstock, *J. Chem. Soc.*, **1933**,

216; King, *J. Am. Chem. Soc.* **61**, 2383 (1939); Meisenheimer and Schmidt, *Ann.* **475**, 157 (1929).

p-Nitrobenzyl ethers: Lyman and Reid, *J. Am. Chem. Soc.* **42**, 615 (1920); Reid, *J. Am. Chem. Soc.* **39**, 304 (1917).

p-Nitrophenylacetates: Ward and Jenkins, *J. Org. Chem.* **10**, 371 (1945).

Phenylurethans: McKinley *et al.*, *Ind. Eng. Chem.*, *Anal. Ed.* **16**, 304 (1944).

Picrates: Baril and Hauber, *J. Am. Chem. Soc.* **53**, 1087 (1931).

Pseudosaccharin ethers: Meadow and Reid, *J. Am. Chem. Soc.* **65**, 457 (1943).

Sulfonic acid esters: Hazlet, *J. Am. Chem. Soc.* **60**, 399 (1941); Sekera, *J. Am. Chem. Soc.* **55**, 421 (1933).

Use of substituted azides for urethans: Hoeke, *Rec. trav. chim.* **54**, 514 (1935); Sah *et al.*, *J. Chinese Chem. Soc.* **2**, 229 (1934); *Rec. trav. chim.* **58**, 453, 582, 591, 595, 1013 (1939); **59**, 238, 357 (1940); *Science Repts.*, *Natl. Tsing Hua Univ.* (*A*) **3**, 109 (1935); Veibel *et al.*, *Dansk Tidsskr. Farm.* **14**, 241 (1940).

SULFUR FUNCTIONS

Table 11.5 lists the derivatives that have been described in the literature for the characterization of *sulfonamides, sulfonchlorides, sulfonic acids, thioethers,* and *thiols.*

TABLE 11.5

Derivatives for the Identification of Sulfur Functions

Sulfonamides	*Phenylhydrazine and pyridine salts
	p-Nitrobenzylpyridinium salts
*Hydrolysis and characterization of the amine and sulfonic acid	Sulfonamides and sulfo-α-naphthylamides
*N-Xanthylsulfonamides	Chlorosulfoanilides and chloroarylsulfoanilides
N-Sulfonylphthalimides	Xanthydrol derivatives
Monoacetyl and diacetyl derivatives	
	Thioethers (Sulfides)
Sulfonchlorides	
	*Sulfones
*Sulfonamides	
Sulfonanilides	*Thiols (Mercaptans and Thiophenols)*
Sulfontoluidides	
	*2,4-Dinitrophenyl thioethers
Sulfonic Acids	*3,5-Dinitrophenyl thioesters
	Anthraquinone α-alkyl thioethers
*S-Benzylisothiouronium salts	Saccharin derivatives
p-Toluidine and other aryl amine salts	Nitrosylmercaptides

* Recommended derivative.

Sulfonamides and Sulfonchlorides

Sulfonchlorides are readily converted to sulfonamides, which are crystalline and serve as derivatives, as shown by Eq. (11.14). Sulfonamides are

$$RSO_2Cl + 2NH_3 \longrightarrow RSO_2NH_2 + NH_4Cl \tag{11.14}$$

generally characterized by hydrolysis, which gives a sulfonic acid and ammonia or an amine, according to Eqs. (11.15) and (11.16). The hydrol-

$$RSO_2NH_2 + HOH \xrightarrow{HCl} RSO_2OH + NH_4Cl \tag{11.15}$$

$$RSO_2NHR' + HOH \xrightarrow{HCl} RSO_2OH + R'NH_3Cl \tag{11.16}$$

ysis is effected by heating with 25 per cent hydrochloric acid, 80 per cent sulfuric acid, or a mixture of 85 per cent phosphoric acid and 80 per cent sulfuric. In the case of a substituted sulfonamide, the amine may be separated by making the solution alkaline and distilling, if the amine is volatile, or extracting with an appropriate solvent; thus it is possible to identify both the amine and the sulfonic acid.

Unsubstituted sulfonamides, RSO_2NH_2 may be reacted either with phthalyl chloride to give N-sulfonylphthalimides or with xanthydrol to form N-xanthylsulfonamides, as shown by Eqs. (11.17) and (11.18), respectively.

$$\tag{11.17}$$

N-Sulfonylphthalimide

$$\tag{11.18}$$

Xanthydrol   *N*-Xanthylsulfonamide

The preparation of N-xanthylsulfonamides may be applied successfully, using semimicro quantities; however, only about a dozen derivatives have been reported, and the method is not successful in benzenoid amides that contain branched alkyl groups on the ring. The alkylation of sulfonamides that have amino hydrogen has been used to prepare derivatives. Either alkyl halides, such as methyl iodide and ethyl bromide, or alkyl sulfates may be used for the alkylation. The chlorination and the preparation of N-acetyl and diacetyl derivatives should be considered for the characterization of small quantities of sulfonamides.

Example 11.26: Hydrolysis of a Sulfonamide

Place in a distilling tube 1 ml of concentrated sulfuric acid (sp.g. 1.84) and add cautiously 5 drops of water and 1 ml of 85 per cent phosphoric acid in the order given. Add 500–800 mg of the sulfonamide, place a thermometer in the tube, and heat gradually until the temperature reaches 160°. Keep the temperature at 155–165° for about 5–10 minutes or until the sulfonamide has passed into solution. Cool the dark viscous solution and add to it 6 ml of water. While the mixture is being cooled, add slowly

25–30 per cent sodium hydroxide solution until the solution is distinctly alkaline. In the case of an unsubstituted sulfonamide, ammonia will be liberated at this stage, and in the case of a substituted sulfonamide, an amine will be liberated.

If the amine resulting from the hydrolysis of the sulfonamide is volatile, the tube is arranged for distillation, and, after addition of boiling stones, the alkaline mixture is distilled until the volume is reduced to one half; after 8 ml of distillate has been collected, a drop is collected separately and tested with litmus or pH paper; if it is alkaline, the distillation is continued; otherwise it is discontinued. The distillate may be used directly for benzoylation as described in Example 10.21, or, if the free amine is desired, it may be extracted with ether; the extract, after drying with a few pellets of sodium hydroxide, is distilled to remove the solvent.

If the amine resulting from the hydrolysis of the sulfonamide has a low volatility, it may be extracted with ether directly from the cold alkaline solution. The ether extract is contaminated with hydrolytic decomposition products; therefore, after the evaporation of ether, the tarry mass is boiled with 3 ml of water, 1 ml of $6N$ hydrochloric acid solution, and a pinch of charcoal, and filtered. The filtrate is made alkaline, and the amine is extracted or derivatized. The residue remaining in the distilling tube is poured into an evaporating dish, treated with 100 mg of charcoal, evaporated to about 4–5 ml, and filtered while hot. The solution is carefully neutralized and then used for the preparation of the arylamine salt or benzylisothiouronium derivative (Example 11.29).

Example 11.27: Preparation of N-Xanthylsulfonamides

Place into an 8-inch tube 10 ml of glacial acetic acid, 200 mg of xanthydrol, and 200 mg of the sulfonamide. Insert a clean solid rubber stopper, and shake the mixture for 2–3 minutes. Filter the solid and crystallize from dioxane–water (3:1 mixture). One crystallization is usually sufficient.

Note: The following fourteen sulfonamides have been derivatized, using xanthydrol as a reagent (the number following the abbreviation of the sulfonamide is the melting point [uncorrected]): Benzenesulfonamide, 200–200.5; 2-Me-, 182–183.5; 4-Me-, 197–197.5; 4-Et-, 196; 4-n-Pr-, 200; 4-n-Bu-, 186; 4-n-Am-, 165; 3,4-di-Me-, 190; 2,4-di-Me-, 188; 2,5-di-Me-, 176; 2,4,6-tri-Me, 204; 4-NH₂-, 208; Saccharin-, 199; Sulfanilamide, 208.

Sulfonic Acids

The three most important derivatives for the characterization of sulfonic acids are: (a) the S-benzylisothiouronium derivatives prepared by reacting the alkali sulfonate with S-benzylisothiouronium chloride; (b) salts formed by the reaction of the sulfonic acid and an aryl amine such as aniline or one of the toluidines; (c) sulfonamides or sulfo-α-naphthylamides formed by conversion of the sulfonic acid to the sulfonchloride followed by reaction with ammonia or naphthylamine.

Other derivatives that have been proposed for the characterization of sulfonic acids are listed in Table 11.5.

S-Benzylisothiouronium Derivatives of Sulfonic Acids

Discussion

Equation (11.19) represents the formation of the S-benzylisothiouronium derivative. A number of sulfonic acid derivatives suitable for character-

$$C_6H_5CH_2SC(NH_2)_2Cl + RSO_2ONa \longrightarrow C_6H_5CH_2SC(NH_2)_2OSO_2R + NaCl \quad (11.19)$$

S-Benzylisothiouronium chloride   S-Benzylisothiouronium derivative

ization work are prepared by reacting the alkali sulfonate with S-benzylisothiouronium chloride. The reagent is easily prepared by refluxing an alcoholic solution of benzyl chloride and thiourea. A concentrated neutral solution of the sodium or potassium salt of the sulfonic acid to be derivatized is added with stirring to a slight excess of the reagent dissolved in water; the method is satisfactory for mono- and disulfonic acids if other functional groups are absent. The presence of hydroxy or amino groups is disadvantageous.

A number of important naphthalene-substituted sulfonic acids may be identified in micro quantities by microscopic examination of the benzoyl or S-benzylisothiouronium derivatives.

The preparation of the reagent and of the derivatives is described in the following examples.

Example 11.28: Preparation of S-Benzylisothiouronium Chloride

If S-benzylisothiouronium chloride is not available for the preparation of the derivatives, it may be prepared conveniently by heating for 20–30 minutes under reflux 2 g of benzyl chloride, 1.2 g of thiourea, and 3 ml of methanol. The pale yellow solution is cooled in an ice–water mixture and the mass of crystals is filtered by suction then washed twice with 1-ml portions of ethylacetate. The product is dried rapidly by pressing between filter papers and placed in a stoppered tube. The yield is 2.5–3.0 g.

Example 11.29: Preparation of S-Benzylisothiouronium Derivatives of Sulfonic Acids

Dissolve 200 mg of the sodium or potassium salt of the sulfonic acid in the minimum amount of water; in the case of the free sulfonic acid, dissolve 200 mg in dilute sodium hydroxide solution (0.5 ml of 10 per cent sodium hydroxide solution and 1 ml or more of water). Add a drop of phenolphthalein and neutralize the excess sodium hydroxide by addition of dilute hydrochloric acid solution. Prepare separately in a test tube a water solution of 250 mg of S-benzylisothiouronium chloride for each acidic group present in the sulfonic acid molecule of the 200-mg sample (for example, use 250 mg of the reagent if the acid taken is 1-naphthalene sulfonic acid, but use 500 mg of the reagent if the acid taken is naphthalene-2,7-disulfonic acid). Cool both solutions and mix by adding the sulfonic salt solution to the reagent slowly with shaking. If this procedure fails to give the derivative, dissolve the benzylisothiouronium chloride in sufficient hot alcohol to give a 15 per cent solution and add to this the sulfonic acid salt solution. The derivative is filtered, washed with water, and recrystallized by dissolving in the minimum amount of hot alcohol; add water dropwise until a permanent cloudiness results. Dry the crystals rapidly by pressing between filter paper or by placing in a vacuum desiccator. The derivatives often develop an offensive odor owing to the formation of benzylthiol (benzylmercaptan) by decomposition of the benzylisothiouronium chloride.

Aryl Amine Salts of Sulfonic Acids

Discussion

The formation of an aryl amine salt of the sulfonic acid is represented by Eq. (11.20). The aryl amines that have been proposed in the literature for

$$RSO_2ONa + [ArNH_2\overset{+}{H}]\overset{-}{C}l \longrightarrow [ArNH_2\overset{+}{H}]OS\overset{-}{O}_2R + NaCl \qquad (11.20)$$
Sodium sulfonate Aryl amine salt   Aryl amine sulfonate

the identification of sulfonic acids are aniline, o-toluidine, p-toluidine, pyridine, p-nitrobenzylpyridine and phenylhydrazine. The first three aryl amines are recommended for beginners. The salts are easily prepared by heating together an aqueous solution of the free acid or of the alkali salt, a slight excess of the amine, hydrochloric acid, and enough water to bring all of the material into solution at the boiling point. The salt separates out on cooling, and, after filtration, it is recrystallized from 1 per cent acetic acid to minimize hydrolysis. Aromatic aminosulfonic acids must first be acetylated in order to form the aryl amine salt. An alternative method is to remove the amino group; the amino sulfonic acid is diazotized, and the diazo group is replaced by chlorine through the Sandmeyer reaction; this last method is suitable for about 1 g of material and should not be tried with less than 500 mg.

Example 11.30: Preparation of Aryl Amine Salts of Sulfonic Acids

The following directions apply to the preparation of the sulfonates of aniline, o-toluidine, and p-toluidine.

Dissolve about 200 mg of the sodium salt of the sulfonic acid in water in an 8-inch tube; if the free sulfonic acid is available, use the same amount and dissolve it in the minimum amount of water or dilute sodium hydroxide. For the barium salt of the sulfonic acid, use 300 mg and boil it with 2 ml of water and 1 ml of 6N sulfuric acid; add a minute amount of charcoal or Filter-cell and filter the hot solution to remove the barium sulfate.

To the solution of the alkali sulfonate or free sulfonic acid, add 300 mg of the aryl amine (aniline, o-toluidine, or p-toluidine), 1–2 ml of 6N hydro-chloric acid, and enough water to bring all the material into solution at the boiling point. Add about 50–100 mg of charcoal, filter the hot solution, and cool. Filter the aryl amine sulfonate and recrystallize to constant melting point from 1 per cent acetic acid.

Note: The aryl amine salt should be thoroughly dried before the melting point is determined. When the salt melts above 180°, the sample may be dried by pressing the material on a filter paper and then filling the capillary; however, the capillary should be placed in the bath when the temperature is below 100° to ensure proper drying while the temperature of the bath rises.

In selecting the aryl amine it is suggested to the beginner to use either aniline or p-toluidine, since these are commonly available in the laboratory in a greater state of purity than o-toluidine.

Thioethers

The most convenient method for the preparation of derivatives of thio-ethers is to oxidize them to sulfones, according to Eq. (11.21). The general

$$
O_2N-\!\!\!\!\bigcirc\!\!\!\!-S-R \xrightarrow{(KMnO_4)} O_2N-\!\!\!\!\bigcirc\!\!\!\!-\overset{O}{\underset{O}{S}}-R \tag{11.21}
$$

2,4-Dinitrophenyl thioether   2,4-Dinitrophenyl sulfone

procedure is described in the two following examples.

Example 11.31: Oxidation of 2,4-Dinitrophenyl Thioethers to Sulfones

Dissolve 3 millimoles of the thioether in the minimum quantity of gla-cial acetic acid and treat it with 0.7 g of potassium permanganate dissolved in 25 ml of water. Add the permanganate solution in portions of 2–3 ml, shaking after each addition until the color is discharged. Continue the addition of permanganate until the color persists after shaking for several minutes. Remove the excess of permanganate by careful addition of sodium bisulfite solution; the sulfone precipitates at this point on cooling by addi-tion of 25–30 g of ice. Filter the solid and dry by pressing the solid between filter paper. Purify the sulfone by crystallization from methanol.

Example 11.32: Preparation of Methionine Sulfone

To prepare the sulfone dissolve 135 mg (1 millimole) of methionine in 1 ml of water and 1.25 ml of 1M perchloric acid. To this solution add 2 drops

(0.1 ml) of $0.5M$ ammonium molybdate and then 6 drops (0.3 ml) of 30 per cent hydrogen peroxide. The white precipitate that forms on the addition of the molybdate dissolves when the peroxide is added, and a yellow solution results. Immerse the tube containing the solution in a water bath at 20° for 2 hours; then add equal volumes of methanol and butylamine or amylamine to pH 9. Add about 25 ml of acetone and, after 10 minutes, wash the precipitated sulfone several times with acetone by decantation, then filter by suction, and wash with acetone and finally with ether. Dry by continuous suction and then place the powder in a watch glass and heat for 10 minutes at about 100°. The yield is 140–150 mg.

Thiols (Mercaptans and Thiophenols)

The derivatives for the characterization of open-chain thiols (mercaptans) and thiophenols are listed in Table 11.5 (page 333). The most suitable (asterisked) are the 2,4-dinitrophenyl ethers and their sulfones and the 3,5-dinitrobenzoyl thioesters.

Thioethers

Discussion

A 2,4-dinitrophenyl thioether may be formed by reacting the mercaptan with 2,4-dinitrochlorobenzene, as represented by Eq. (11.22). The reaction

$$\text{RSH} \xrightarrow{\text{NaOH}} \text{RSNa} + \text{Cl}\langle\!\!\!\bigcirc\!\!\!\rangle\text{NO}_2 \longrightarrow \text{NO}_2\langle\!\!\!\bigcirc\!\!\!\rangle\text{—S—R} + \text{NaCl} \qquad (11.22)$$

2,4-Dinitrochlorobenzene   2,4-Dinitrophenyl thioether

takes place with ease when a sodium mercaptide solution is added to an alcoholic solution of the aromatic nitrohalide. The thioether separates on cooling; for further identification the thioether may be oxidized to the corresponding sulfone. The preparation of these derivatives is convenient for the characterization of many mercaptans.

Another type of solid thioether may be prepared by reacting the thiol with sodium anthraquinone α-sulfonate, as shown by Eq. (11.23); the thio-

$$\alpha\text{-C}_{14}\text{H}_7\text{O}_2\text{SO}_3\text{Na} \xrightarrow{\text{RSH}} \alpha\text{-C}_{14}\text{H}_7\text{O}_2\text{SR} + \text{NaHSO}_3 \qquad (11.23)$$

Sodium anthraquinone   Anthraquinone
α-sulfonate α-alkyl thioether

ether may be oxidized to the corresponding sulfone; another anthraquinone derivative proposed for the same purpose is 1,5-butyl-anthraquinone-sulfone sodium sulfonate. One disadvantage of this method is that the reaction takes place slowly, requiring several hours of heating.

The general procedure for the preparation of 2,4-dinitrophenyl ethers is described in Example 11.33.

Example 11.33: Preparation of 2,4-Dinitrophenyl Ethers from Thiols

Place in an 8-inch tube 8 ml of methanol, 3 millimoles of the mercaptan, and 3 millimoles of sodium hydroxide (9–10 drops of $6N$ sodium hydroxide solution). Add the sodium mercaptide solution to a tube containing 600 mg of 2,4-dinitrochlorobenzene dissolved in 4 ml of methanol. Add a boiling stone and arrange for reflux. Boil the mixture gently for 5–10 minutes and filter the solution rapidly while hot. Cool for 10 minutes and filter the solid thioether. Recrystallize once or twice from methanol.

Note: If a red coloration results when the sodium hydroxide solution is added to the alcoholic solution of the mercaptan, a slight excess of the latter is used in order to remove the color caused by excess of alkali.

For further identification of the thioether, it may be converted by oxidation to a sulfone, as outlined in Example 11.31.

Thioesters

Discussion

Eqs. (11.24) and (11.25) represent the formation of thioesters by reacting the mercaptan with 3,5-dinitrobenzoyl chloride or 3-nitrophthalic an-

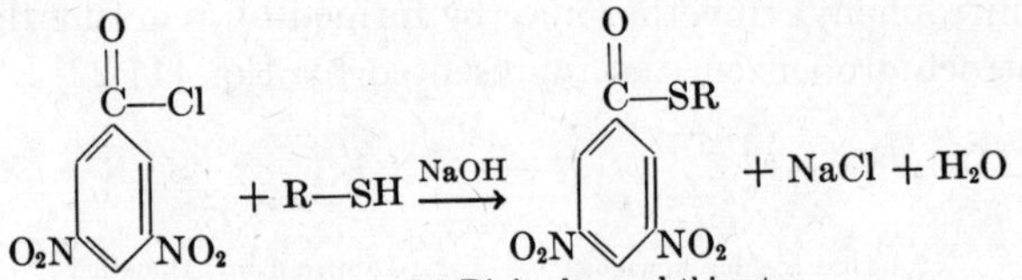

$$\text{(11.24)}$$

3,5-Dinitrobenzoyl thioester

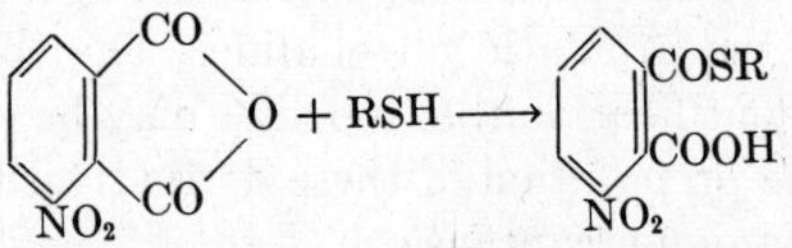

$$\text{(11.25)}$$

3-Nitrophthalic thioester

hydride. The preparation of the 3,5-dinitrobenzoyl thioester should be a second choice for the beginner. The method for the preparation of the esters is similar to that used for the alcohols. About 200 mg of the acid chloride or the anhydride is heated with 5–6 drops of the thiol until a uniform melt has been obtained; a few drops of pyridine may be added in the reaction of the thiol with the acid chloride to aid the removal of hydrogen chloride. After addition of water, the solid derivative is filtered and purified by crystallization.

The substituted thiocarbonic esters formed by reaction of thiols with azides may serve as derivatives. For example, *m*-nitrobenzazide yields with thiols *m*-nitrophenylurethans, according to Eq. (11.26). The general pro-

$$NO_2C_6H_4CON_3 + RSH \longrightarrow NO_2C_6H_4NHCOSR + N_2 \qquad (11.26)$$

m-Nitrobenzazide   Thiol   *m*-Nitrophenylurethan

cedure for the preparation of thioesters is given in Example 11.34.

Example 11.34: Preparation of 3,5-Dinitrothiobenzoates

Dry an 8-inch tube by heating it over a flame and then stopper it with a cork and allow to cool. Place 200 mg of 3,5-dinitrobenzoyl chloride and arrange for reflux. Add to the solid chloride 5–6 drops of the thiol and 1 drop of pyridine. Adjust the microburner so that the reaction mixture melts into a homogeneous mass and heat in this manner for about 10 minutes. If at this point a strong odor of the thiol persists, add 25–50 mg of the chloride and heat for an additional 5 minutes. Add 2 ml of water, cool, and stir by means of a glass rod until the oily mass solidifies. Filter with suction and wash with water. To remove the small amount of 3,5-dinitrobenzoic acid and to crystallize the derivative, follow the directions given in Example 10.9 (pages 251–52).

References

Sulfinic Acids

Aryl and alkyl mercuric chlorides from sulfinates: Coffey, *J. Chem. Soc.* **1926**, 637; Kharasch and Chalkey, *J. Am. Chem. Soc.* **43**, 607 (1921); Marvel *et al., J. Am. Chem. Soc.* **68**, 2735 (1946); Peters, *Ber.* **38**, 2567 (1905); Whitmore *et al., J. Am. Chem. Soc.* **45**, 1066 (1923).

1,2-Dialkylsulfonylethanes: Allen, *J. Org. Chem.* **7**, 23 (1942).

Sulfonamides

Acetyl derivatives: Baggesgaard-Rasmussen *et al., Dansk Tidsskr. Farm.* **31**, 53 (1957).

Microcharacterization: Chiarino *et al., Anales asoc. quim. arg.* **31**, 72, 233 (1943); *C.A.* **38**, 530 (1944); **39**, 255 (1945); Vonesch, *Anales farm. y bioquim. (Buenos Aires)* **14**, 81 (1943); *C.A.* **39**, 1430 (1945).

Paper chromatography: Bray *et al., Biochem. J.* **46**, 271 (1950); Longenecker, *Anal. Chem.* **21**, 1042 (1949); de Reeder, *Anal. Chim. Acta* **8**, 325 (1953); Robinson, *Nature* **168**, 512 (1951); San and Ultee, *Nature* **169**, 586 (1952); Steel, *Nature* **168**, 877 (1951).

N-Xanthylsulfonamides: Phillips and Frank, *J. Org. Chem.* **9**, 9 (1944).

Sulfonic Acids

Aniline salts of aromatic sulfonic acids: Dermer and Dermer, *J. Org. Chem.* **7**, 581 (1942).

Identification of anthraquinone sulfonic acids: Večeřa and Borecky, *Chem. Listy.* **51**, 974 (1957).

Identification of aromatic sulfonic acids containing an amino group: Allen *et al., J. Org. Chem.* **7**, 15 (1942); **10**, 1 (1945).

Identification of sulfobenzoic acid: Suter and Campaign, *J. Am. Chem. Soc.* **67**, 1860 (1945).

Microscopic identification of some important substituted naphthalene sulfonic acids: Chambers and Scherer, *Ind. Eng. Chem.* **16**, 1272 (1924); Garner, *J. Soc. Dyers Colour-*

ists **43**, 12 (1927); **52**, 302 (1936); Hann and Keenan, *J. Phys. Chem.* **31**, 1082 (1927); Whitmore and Gebhart, *Ind. Eng. Chem., Anal. Ed.* **10**, 654 (1938).

Micro identification of naphthalene sulfonic acid by means of benzylisothiourea: Garner, *J. Soc. Dyers Colourists* **52**, 302 (1936).

p-Nitrobenzylpyridinium salts of aromatic sulfonic acids: Huntress and Foote, *J. Am. Chem. Soc.* **64**, 1017 (1942).

N-Sulfonylphthalimides: Evans and Dehn, *J. Am. Chem. Soc.* **51**, 3651 (1929).

Phenylhydrazine salts of aliphatic sulfonic acids: Latimer and Bost, *J. Am. Chem. Soc.* **59**, 2501 (1937).

Pyridine salts of acetylated aminosulfonic acids: Chen and Gross, *J. Soc. Dyers Colourists* **59**, 144 (1943).

S-Benzylisothiouronium salts of sulfonic acid: Chambers and Scherer, *J. Ind. Eng. Chem.* **16**, 1272 (1924); Chambers and Watt, *J. Org. Chem.* **6**, 376 (1941); Campaign and Suter, *J. Am. Chem. Soc.* **64**, 3040 (1942); Donleavy, *J. Am. Chem. Soc.* **58**, 1005 (1936); Hann, *J. Am. Chem. Soc.* **57**, 2166 (1935); Večeřa and Boreky, *Chem. Listy* **51**, 974 (1957); Veibel, *J. Am. Chem. Soc.* **67**, 1867 (1945); Veibel and Lillelund, *Bull. soc. chim.* (*5*) **5**, 1153 (1939).

o- and *p*-Toluidine salts of aromatic sulfonic acids: Dermer and Dermer, *J. Org. Chem.* **7**, 581 (1942).

p-Toluidine salts of monoarylsulfates: Barton and Young, *J. Am. Chem. Soc.* **65**, 294 (1943).

p-Toluidine salts of sulfonic acids: Feiser, *J. Am. Chem. Soc.* **51**, 2463 (1929).

Xanthydrol derivatives of sulfonic acids: Phillips and Frank, *J. Org. Chem.* **9**, 9 (1944).

Thioethers and Thiols

Anthraquinone α-alkyl thioethers: Reid and Ellis, *J. Am. Chem. Soc.* **54**, 1687 (1932); Reid and Hoffman, *J. Am. Chem. Soc.* **45**, 1837 (1923); Reid, Mackall, and Miller, *J. Am. Chem. Soc.* **43**, 2104 (1921).

Derivatives with Chloramine-T: Večeřa and Petránek, *Chem. Listy* **50**, 240 (1956).

Derivatives with 2,4-dinitrobenzenesulfenyl chloride: Bohme and Stachel, *Z. Anal. Chem.* **154**, 27 (1957).

2,4-Dinitrophenyl derivatives: Vorohstov and Yacobson, *C.A.* **52**, 12784h.

Nitrosylmercaptide from reaction with nitrous acid: Rheinboldt, *Ber.* **59**, 1311 (1926); **60**, 184 (1927); Tasker and Jones, *J. Chem. Soc.* **95**, 1917 (1909); Vorlander and Mittag, *Ber.* **52**, 422 (1919).

Use of 3,5-dinitrobenzoyl chloride and 3-nitrophthalic anhydride: Wertheim, *J. Am. Chem. Soc.* **61**, 3660 (1939).

Use of 2,4-dinitrochlorobenzene for thioethers: Bost, Turner, and Norton, *J. Am. Chem. Soc.* **54**, 1985 (1932); **55**, 4956 (1933); Conn, *J. Am. Chem. Soc.* **55**, 4956 (1933).

Use of saccharin chloride: Meadow and Cavagnol, *J. Org. Chem.* **17**, 488 (1952).

Chromatographic Procedures

Schmeiser and Jerchel, *Angew. Chem.* **65**, 366 (1953). Paper chromatography of sulfur compounds.

Tables of Organic Compounds with Their Constants and Derivatives

A more extensive listing of compounds may be found in the larger text by these authors, and the tables of the larger text contain data on the constants of many additional types of derivatives.

Compounds melting above 20° are classed as solids and are arranged according to increasing *melting points*. Compounds melting below 20° are arranged according to increasing *boiling points*. Compounds melting between 20° and 30° are usually listed in both the solid and liquid sections of a table. All temperatures are degrees Centigrade; density is given at 20°/4°, and refractive index as n_D^{20}, unless otherwise indicated.

An asterisk following the melting point or boiling point of a derivative indicates that the value is corrected. The degree sign is usually omitted after the figures denoting *m.p.* or *b.p.* (or *m* or *b*). An exclamation point indicates that several values were found and the one so designated was chosen as the most reliable. Brackets [ ] indicate that the derivative has been prepared only indirectly.

When a derivative exists in two or more forms, each having a different melting point, both values are given, separated by a comma; thus 118,145 indicates that the particular derivative exists in two forms, one that melts at 118° and another at 145°. On the other hand, there are many derivatives for which several melting points are listed in the literature; in some cases, besides the value selected for the tables in this book, one or two additional values have been listed in parentheses after the selected value. For a more elaborate discussion of this topic, see pages 41-47.

The reader should consult pages 225-28 with reference to the selection of the derivative to be prepared.

In case the available data indicate a particular compound but the prepared derivative does not give the melting point listed in the appropriate table, the reader should consult pages 47-52.

For easier reference and also for identification in the index, numbers appear in sequence (1, 2, 3, etc.) to the left of the names of the compounds listed in the left column of each table.

The melting points of hundreds of compounds may be found in these tables even though they are not listed under the specific class name and hence the names of these compounds are not included in the index. For example, there are no "tables" of such compounds as semicarbazones or phenylhydrazones, but the melting points of many of the common oximes and phenylhydrazones may be found by consulting the appropriate columns of derivatives in the tables for aldehydes and ketones.

Commercial products often melt over a range of at least two degrees. The catalogs of commercial suppliers of organic chemicals usually indicate the purity of their products by giving the ranges over which the chemicals boil or melt.

Abbreviations

As far as possible, abbreviations conform to those used in the *Chemical Abstracts*. Thus, for example, MeOH stands for methanol, EtOH for ethanol, Ac₂O for acetic anhydride, Et₂O for ether, and so forth. Listed below are a few abbreviations used in the tables that may be less familiar.

N (after the boiling point, or melting point): Indicates a note to be found at the end of the group of compounds, in which this particular compound appears. The reference number for the note will be found in the column headed "Note."	*r.h.*: rapid heating *s.h.*: slow heating *s.t.*: sealed tube *T.S.*: test solution *v.s.*: volatile with steam

In the case of addition compounds or salts with HCl and the like, the base or compound to be identified is indicated by its initial letter (capital) followed by a center dot to separate it from the rest of the addition compound, a subscript being added when two or more molecules of base are involved: thus, P₂·H₂SO₄ under physostigmine denotes the physostigmine sulfate with two molecules of the base.

Literature References

A few references to specific articles that give data on derivatives are included in the Notes to these tables. General references, which the reader may consult for further data on derivatives and their constants, are listed on pages 222-24. Selected references to current literature appear the the ends of Chapters 10-11.

TABLE 1

Acetals

	Name of compound	B.p.	Products of hydrolysis	
			Alcohol	Aldehyde
1	Methylal	42–3	Methanol	Methanol
2	Dimethylacetal (1,1-Dimethoxyethane)	64	Methanol	Ethanal
3	2-Methyl-1,3-dioxolane	82	Ethanediol	Ethanal
4	Ethylal	88	Ethanol	Methanal
5	Acetal	104	Ethanol	Ethanal
6	1,3-Dioxane	106	1,3-Propanediol	Methanal
7	2-Methyl-1,3-dioxane	110	1,2-Ethanediol	Ethanal
8	Isopropylacetal	122	2-Propanol	Ethanal
9	Ethylpropylal (1,1-Diethoxypropane)	124	Ethanol	Propanal
10	Ethyl acral (1,1-Diethoxy-2-propene)	126	Ethanol	2-Propenal
11	Propylal (Di-n-propoxymethane)	137	Propanol	Methanal
12	Ethylbutylal (1,1-Diethoxybutane)	143	Ethanol	Butanal
13	Propylacetal (1,1-Di-n-propoxyethane)	147	Propanol	Ethanal
14	1,1-Diethoxy-2-chloroethane	157	Ethanol	Chloroethanal
15	Isobutylal (1,1-Diisobutoxymethane)	164	2-Methylpropanol	Methanal
16	1,1-Diethoxy-2-bromoethane	170	Ethanol	Bromoethanal
17	sec-Butylacetal (1,1-Di-sec-butoxyethane)	171	2-Butanol	Ethanal
18	Isobutylacetal	176	2-Methylpropanol	Ethanal
19	Butylal	181	Butanol	Methanal
20	1,1-Diethoxy-2,2-dichloroethane	184	Ethanol	Dichloroethanal
21	Butylacetal	186	Butanol	Ethanal
22	Benzaldehyde dimethylacetal	199	Methanol	Benzaldehyde
23	Amylal	219	Pentanol	Methanal
24	Amylacetal	222	Pentanol	Ethanal
25	Benzaldehyde diethylacetal	222	Ethanol	Benzaldehyde
26	Hexylal	255	Hexanol	Methanal

TABLE 2

Acid Anhydrides

No.	Name of compound	Note	B.p.	M.p.	Acid		M.p. of recommended derivatives		
					B.p.	M.p.	Amide	Anilide	p-Toluidide
1	Trifluoroacetic		39		72		75	88	
2	Acetic		140		118	16	82	114	147
3	Propionic		167		141		81	106	126
4	Isobutyric		182		154.4		130	105	107
5	Pivalic		190		164	35	154	129	120
6	Maleic		198			130	181	187(di),173-5(mono)	142(di)
7	Butyric		198		162.5		115	96	75
8	γ-Butyrolactone	1	206						
9	Citraconic		214			92d	185-7d(di)	175(di)	107
10	Isovaleric		215		176		136		
11	Dichloroacetic		216d		194		98s	118	153
12	Valeric		218		186		106	63	74
13	Trichloroacetic		223		197	57-8	141	97(94)	113
14	Crotonic		248		189	72	161(158)	118(115)	132
15	Caproic		254-7(245)		205		100	95(92)	75(73)
16	Heptanoic		258	17	223		96	70(65)	81
17	Caprylic		280-5		239		110(106)	57(55)	70
18	Oleic			22	216/5 mm	16	76	41	43
19	Capric		285	24	268-70	31	108(98)	70	78
20	Undecanoic			37	284	30	103(99)	71	80
21	o-Toluic			39		104-5	143	[125]	
22	Bromoacetic			41-2	208	50	91	131	144
23	Lauric			42	299	44	110(102)	78	87
24	Benzoic		360	42		122	130	163	158
25	Chloroacetic			46	189	63	121	134	162
26	Tridecanoic			50	312	44	100	80	88
27	Maleic		200	52-4		130	181	187(di)	142(di)
28	Myristic			54	202/16 mm	54	107(103)	84	93
29	Glutaric			56	200/20 mm	97	175-6(di)	224	218
30	Suberic (dimer)			56-7		144(141)	127(mono),217(di)	128(mono),186(di)	218(di)
31	Palmitic			64	222/16 mm	63	106-7	90	98
32	Margaric			67	231/16 mm	61	108		
33	Itaconic	2		67-8		165	192(di)	190(185)	
34	Sebacic (dimer)			68	243/15 mm	133s	210(di),170(mono)	201(di),122(mono)	201
35	Stearic			70		70	109	95	102

No.	Name							
36	m-Toluic		71		113	94	126	118
37	Phenylacetic		72		76.5	156	118	136
38	Arachidic		77.5	204/1 mm	77	108-9	92	96
39	o-Chlorobenzoic		79		142(140)	142	114(118)	131
40	m-Chlorobenzoic		95		158(155)	134	122	
41	p-Toluic		95		179-80	160	145	160
42	Anisic		99		184-6	167(163)	169-71	186
43	α-Naphthoic		106		162	202	163	
44	p-Ethoxybenzoic		108		198	202	170	
45	3,5-Dinitrobenzoic		109		204-5	183	234	
46	4-Nitrophthalic		119		165	200d	192	172(mono)
47	Succinic (131/10 mm)	261	119-20	235d	186-8	157(mono),260(di)	148(mono),230(di)	180(mono),255(di)
48	Nicotinic		123		237-8s	128	85	150
49	Phthalic	295	131.6		206d	220(di),149(mono)	253-5(170)	201(150)
50	o-Nitrobenzoic		135		146	176	155	
51	Cinnamic		136		133	148(142)	151(153)	168
52	α-Naphthoic		146		162	202	163	
53	2,4-Dinitrobenzoic		160		183	203		
54	m-Nitrobenzoic		160		140	143	154	162
55	3-Nitrophthalic		162		218	201d(di)	234(di)	226(di)
56	1,2-Naphthalic		169		175d	265d		
57	p-Nitrobenzoic		189		241	201(198)	211(204)	204(192)
58	p-Chlorobenzoic		194		240	179(170)	194	
59	Diphenic		217		229	212(di)	230	
60	p-Bromobenzoic		218		251	189	197	
61	D-Camphoric		221		188	177(mono),193(di)	204(mono),226(di)	α-212-4,β-190-6
62	Tetrachlorophthalic		256		250d			
63	2,3-Naphthalic		266(246)		241d			
64	1,8-Naphthalic[3]		274					
65	Tetrabromophthalic		275		266			
66	Tetraiodophthalic		329-31		327			

Notes on Acid Anhydrides

1. On ammonolysis with concentrated ammonia gives amide, m.p. 99(87).
2. Not v.s. (differentiation from citraconic). Anilide obtained only by boiling excess of amine with acid.
3. Naphthalimide, m.p. 300, is formed by heating compound with excess aqueous NH3; derivative is purified by boiling with NaCO3 solution. The compound boiled with aniline forms N-phenylnaphthalimide, m.p. 202.

TABLE 3

Acid Halides

Name of compound	M.p.	B.p.	M.p. of recommended derivatives	
			Amide	Anilide
		Fluorides		
1 Acetyl fluoride		20.5	82	114
2 Propionyl fluoride		44–6	81	106
3 Fluoroacetyl fluoride		50.5–51.0	108	
4 Trichloroacetyl fluoride		66–8	141	97(94)
5 Butyryl fluoride		67	115	96
6 Chloroacetyl fluoride		73–5	120	137(134)
		Chlorides		
7 Acetyl chloride		51–2	82	114
8 Oxalyl chloride	−12	64	419d	246
9 Fluoroacetyl chloride		71.5–73	108	105
10 3-Acrylyl chloride		76	85	106
11 Propionyl chloride		80	81	105
12 Isobutyryl chloride		92	130	105
13 Butyryl chloride		101–2	115	96
14 Pivalyl chloride		105–6	154	129
15 Chloroacetyl chloride		105–6	120	137(134)
16 Dichloroacetyl chloride		108	98s	118
17 Methoxyacetyl chloride		113	97	58
18 Isovaleryl chloride.		115	136	109
19 Trichloroacetyl chloride		118	141	97(94)
20 Valeryl chloride		126	106	63
21 Crotonyl chloride		126	161(158)	118(115)
22 Isocaproyl chloride		147	121	112(110)
23 Caproyl chloride		153	100	95
24 Heptanoyl chloride (Oenanthoyl)		175	96	70
25 Succinyl chloride	20	190d	157(mono),273(di)	148(mono),230(di)
26 Caprylyl chloride		196	110(106)	57(55)
27 Benzoyl chloride		197	130	163
28 Diethylmalonyl dichloride		197	146(mono),224(di)	
29 Phenylacetyl chloride		210	156	118
30 Pelargonyl chloride (Nonanoyl)		215.3	99	57
31 Glutaryl chloride		218	175–6(di)	224
32 p-Chlorobenzoyl chloride	16	222	179(170)	194
33 m-Chlorobenzoyl chloride		225	134	122
34 Decanoyl chloride		232	108(98)	70

No.	Compound				
35	o-Chlorobenzoyl chloride		238	142	114(118)
36	m-Bromobenzoyl chloride		243(239)		136(146)
37	o-Methoxybenzoyl chloride		254	129	131
38	m-Nitrobenzoyl chloride		278	143	154
39	Phthaloyl chloride	15-6	281	149(mono),220(di)	170(mono),253-5(di)
40	Lauroyl chloride	-17	145/18 mm	110(102)	78
41	Myristoyl chloride	1-3	174/16 mm	107(103)	84
42	Palmitoyl chloride	11-12	194/17 mm	106-7	90
43	Salicylyl chloride	19-20	92/15 mm	142	136
44	o-Nitrobenzoyl chloride	20	148/9 mm	176	155
45	Stearoyl chloride	23	202-3/6 mm	109	94
46	p-Anisoyl chloride	24	145/14 mm	162	169
47	m-Nitrobenzoyl chloride	35	278	143	154
48	Cinnamoyl chloride	36	257.5	149	153
49	p-Bromobenzoyl chloride	42	245-7d	189	197
50	2,4-Dinitrobenzoyl chloride	46		203	198
51	p-Nitrophenylacetyl chloride	48		198	
52	Phenacyl chloride (Other: Oxime, 89.0)	54.5-56.0	150-2/15 mm		
53	3,5-Dinitrobenzoyl chloride	68-9(74)		183	234
54	p-Nitrobenzoyl chloride	75		201(198)	211(204)
55	3-Nitrophthaloyl chloride	77		201d(di)	234(di)
56	Picryl chloride (See Table 23A)	83			
57	Diphenylcarbamyl chloride	86			
	Bromides				
58	Acetyl bromide		81	82	114
59	Propionyl bromide		103	81	106
60	Chloroacetyl bromide		127	120	137(134)
61	Butyryl bromide		128	115	96
62	Bromoacetyl chloride		134	91	131
63	Isovaleryl bromide		138-40	136	
64	Bromoacetyl bromide		149	91	131
65	α-Bromopropionyl bromide		153	123	99(110)
66	Benzoyl bromide		218	130	163
67	Phenacyl bromide (Other: Oxime, 89)	51			
68	p-Iodobenzoyl bromide	55		217	210
69	3,5-Dinitrobenzoyl bromide	60		183	234
70	p-Nitrobenzoyl bromide	64		201(198)	211(204)
	Iodides				
71	Acetyl iodide		108	82	114
72	Propionyl iodide		127	81	106
73	Butyryl iodide		146-8	115	96

TABLE 4
Amino Acids[a]

Name of compound	M.p.	Melting point of derivatives			
		Recommended			Others
		p-Toluene-sulfonyl	Phenyl-urea	Dinitro-phenyl	Benzoyl
1 L-(+)-Valine	93-6(315.t.)	147			127
2 N-Phenylglycine	127				63
3 L-Ornithine	140N				240(mono),189(di)
4 Anthranilic acid (o-Aminobenzoic) (See Table 9B)	147(145)	217			182
5 D-(+)-Valine	156-7(315)	147	147		
6 m-Aminobenzoic acid (See Table 9B)	174				248
7 L-Canavanine (C5H12O3N4) ([α]$_D^{20}$, +7.90)	184				86d(tri)
8 p-Aminobenzoic acid (See Table 9B)	188		300		278
9 Betaine (Trimethylglycine)	193				
10 DL-Glutamic acid	199(227)	117			153(155-7)
11 β-Alanine	200(196)		168(174)		120
12 L-Isoserine	200d				107-9*
13 DL-Proline (monohyd., m.p. 191)	203(205)				
14 L-Arginine (monohyd., m.p. 191) ([α]$_D^{20}$ +11.37)	207d		170	252	298(mono),235(di)
15 L-Glutamic acid	211-3N	131			104
16 Sarcosine (N-Methyl glycine)	212-3(210)		102		99(di)
17 L-Canaline (C4H10O3N2)	214d				
18 β-Hydroxyvaline	218d		182		153(mono)
19 L-(-)-Proline	222d		170N	137	156(mono)
20 L-Citrulline (C6H13O3N3)	222(226)				
21 D- or L-Lysine	224-5d	130-3	184		149-50(di)N,235(mono)
22 β-L-Asparagine (β-Aspartamide)	227N,(234-5)(243d)		164		189
23 DL-Threonine (α-Amino-β-hydroxybutyric acid)	227-9N	175	177-8	152	145N
24 L-Serine	228d				
25 Glycine	228-32N,(245d)(262d)	147(150)	197*(163)	195	187.5
26 β-Hydroxynorvaline (C5H11O3N)	230-1d		156		170-1(mono)
27 DL-Thyroxine	231-3N				210-5d
28 3-Aminosalicylic acid	235d				189
29 DL-Allothreonine	237-9N				175-6
30 DL-Arginine	238				230(di,anh.),176(di,hyd.)
31 DL-Serine (α-Amino-β-hydroxypropionic acid)	246d,N	213	169	199	149-50

No.	Amino acid					
32	DL-Isoserine (β-Aminolactic acid)	248d,N		184		151*
33	p-Hydroxyphenylglycine	248(200d)				117
34	D- or L-Threonine	251-2				147-8N
35	L(—)Cystine ($C_6H_{12}O_4N_2S_2$)	260d	204-5*	160	109	147-8,181(di)N
36	L(—)Glycylglycine (Diglycine)	260-2	178	176		208
37	DL-Phenylalanine	264d,N(273)	134-5	182	186	188*
38	DL-Aspartic acid	270d(280)	140	162		119(hyd.),176-7(anh.)
39	D- or L-Aspartic acid	270-1	153	175	196	185*
40	L(—)Hydroxyproline	274N	176			100(mono),92(di)
41	DL-Tryptophane	275-82(293)		151		145
42	L-(+)-Alloisoleucine	278d,N				
43	DL-Methionine	281(272)				
44	L(—)Methionine	283d,N	105			117
45	5-Aminosalicylic acid	283(280d)				252
46	D- or L-Phenylalanine	283r.h.(320)N	164-5(161)	181*		146
47	L-(+)Isoleucine	283d,N	130-2	121		117
48	L-Histidine	288N(253)	202-4d	166	175	230d(mono)
49	L-Tryptophane	290N	176	120	166	183
50	DL-Isoleucine	292N	141*			118
51	L-(+)-α-Aminobutyric acid	292d(303s.t.)(285)		165		121
52	DL-Leucine	293-5d,s.t.(332)	124*	115		137-41
53	D- or L-Leucine	293-5r.h.(337)	139	174d,N		105-7(anh.)
54	DL-Alanine	293-5	133(92-4)	175d(168)		166*
55	D- or L-Alanine	297d	124			151
56	DL-Norleucine	297-300(275s.t.)(332)	110	164		132
57	DL-Valine	298d,s.t.(282)(292)				53
58	L-(+)- or D-(—)- Norleucine (s. 275)	301				
59	Creatine	303(292)		117		
60	DL-Norvaline	303s.t.		170		147
61	DL-α-Aminobutyric acid	304(307)				
62	Creatinine	305d(260)				
63	L-(+)- or D-(—)-Norvaline	307d	N-188*,114(di)	104,194		64(97 anh.)
64	L-(—)-Tyrosine	314-8d,r.h.,290-5d,s.h.	224-6			N-166-7,211-2(di) N-197
65	DL-Tyrosine	340r.h.,290-5s.h.				166(mono),85(di)
66	Djenkolic acid ($C_7H_{14}O_4N_2S_2$) [α]$_D^{20}$ -25	300-50d(250)				
67	DL-Ornithine (cryst.)		188(mono)	192		4-N-285-8,188(di)
68	DL-Lysine (solid)			196		249(mono)N,146(di)
69	DL-Norvaline	303s.t.,N		117		
70	L-Cysteine ($C_3H_7O_2NS$)(oxidized to L-Cystine, m 260 q.v.)	none				

[a] Melting points of amino acids are strictly decomposition points, which vary widely depending on the rate of heating and the temperature at which the crystals are dried. See discussion page 229ff. For optical activities of amino acids, see Schmidt, *The Chemistry of Amino Acids and Proteins* (Baltimore: Charles C. Thomas, 1938), p. 580. Other derivatives are given in the notes as numbered.

352 ACIDS: 4. AMINO ACIDS

TABLE 4 (*continued*)

Notes on Amino Acids

Number in parenthesis refers to the compound as listed in Table 4.

(1): Acetyl, 156; *N*-Benzenesulfonyl, 153; 3,5-dinitrobenzoyl, 157-8; Phenylhydantoin 131-3; Carbobenzoxy, 64-65.

(2): Acetyl, 194; Formyl, 125; β-Naphthalenesulfonyl, 189.

(3): No melting point is given in any of the standard works. D-Ornithine is listed as a syrup. The m.p. of 140 listed was obtained from Vickery and Cook, *J. Biol. Chem.* **94**, 398 (1931). Picrate, 208; Picrolonate, 220-1(mono), 235-6(di); β-Naphthalenesulfonyl, 189; Flavinate, 234-5; Hydrochloric salt, 233; α,δ-Dicarbobenzoxy, 112-4.

(4): Acetyl, 185; Formyl, 169; Picrate, 104; 3,5-Dinitrobenzoyl, 278; *N*-Benzenesulfonyl, 214.

(5): Acetyl, 156; Formyl, 156.

(6): Acetyl, 250; 3,5-Dinitrobenzoyl, 270.

(7): Picrate, 163-4; Flavinate, 212.

(8): Acetyl, 252; Formyl, 268; 3,5-Dinitrobenzoyl, 290.

(9): Formyl, 183.

(10): Acetyl, 187.5; Formyl, 182; Picrolonate, 184d; α-Naphthylurea, 236; Phenylhydantoin, 165*; *N*-Chloroacetyl, 123; Hydrochloride salt, 202(193).

(11): Carbobenzoxy, 106; 3,5-Dinitrobenzoyl, 202.5; α-Naphthylurea, 231-3; Amide, 41.

(12): *N*-Chloroacetyl, 143.

(13): Amide, 93; Anilide, 170; 3,5-Dinitrobenzoyl, 217; Phenylhydantoin, 118*; Picrate, 135-7.

(14): The m.p. of D-Arginine, 207, is often confused with the m.p. of DL-Arginine, 238. For the picrolonate of D-Arginine see Heyl, *J. Am. Chem. Soc.* **41**, 681 (1919). 3,5-Dinitrobenzoyl, 150; Flavinate, 258-60 (mono), 220(di); Monocarbobenzoxy, 175; β-Naphthalenesulfonyl, 87-9; Picrate, 217(mono), 190d(di)

(15): L-Glutamic acid melts at 197-8d. Its HCl salt melts at 202 on slow heating but at 213 on rapid heating. Acetyl, 199; *N*-Benzenesulfonyl, 129-32; *N*-Chloroacetyl, 143; Hydrochloride salt 202 and 213; *p*-Toluenesulfonyl, 131.

(16): Acetyl, 135; 3,5-Dinitrobenzoyl, 153.5; Hydrochloride salt 168-70.

(17): Flavinate, 211d; Hydrochloride salt, 166d; Picrate, 192-3d.

(18): β-Naphthalenesulfonyl, 261; Phenylhydantoin, 125; Phenylurethan, 162; 2-Naphthalenesulfonamide, 261.

(19): The L-Proline phenylurea derivative may precipitate as a resin, and should be converted to the hydantoin, which melts at 118 and 144. The reineckate derivative has m.p. 199d. β-Naphthalenesulfonyl, 138(anh.), 134(hyd.); Phenylhydantoin, 144; Picrate, 154.

(20): Flavinate, 218d; Hydrochloride salt, 185; Picrate, 206.

(21): The dibenzoyl derivative of DL-lysine melts at 145-6. The DL-lysine also forms a monohydrochloride, m.p. 235-6, and a dihydrochloride, m.p. 188-90. 5-Benzoyl, 235(mono); Favinate, 213d; Hydrochloride salt, 235-6(mono), 193(di); Phenylhydantoin, 183-4*; Picrate, 266d; Picrolonate, 246-52.

(22): Heated rapidly, the compound melts at 234-5; heated slowly (sealed tube) it melts at 226-7. The D- and L-forms have the same m.p. Amide, 131; *N*-Chloroacetyl, 148-9; 3,5-Dinitrobenzoyl, 196-199; α-Naphthylurea, 199; Picrate, 180d.

(23): DL-Allothreonine melts at 237-9. D- and L-Threonine melt at 251-2. The dibenzoyl derivative melts at 174, the monobenzoyl at 176, and the eutectic of the two at 145. Phenylhydantoin, 164-5.

(24): The methyl ester of L-serine forms a hydrochloride, m.p. 167. Carbobenzoxy, 121.

(25): Glycine turns brown above 220 and decomposes at 262. A commercial source lists the m.p. at 245. Glycine picrate contains 2 mol. of glycine; it softens at 199-200 and decomposes at 202. The earlier figure of 190 is stated to be erroneous. The barium salt crystallizes even from dilute solutions and is suitable for separation from other amino acids. Acetyl, 206; Anilide, 62; 3,5-Dinitrobenzoyl, 179; Formyl, 153-4; Hydrochloride salt, 185; β-Naphthalenesulfonyl, 159*; α-Naphthylurea, 191; Picrate, 190; Picrolonate, 214-5d; *p*-Toluidide, 107.

(26): Phenylhydantoin, 154-5.

(27): Thyroxine melts at 250 when heated 10°/minute and at 230-5 when heated 3°/minute; iodine is evolved in all cases. *N*-Chloroacetyl, 201-2d; Methyl ester, 156.

(28): Acetyl, 215; *N*-Benzenesulfonyl, 194.

(29): See note under (23).

(30): Nitrate, 230; Picrate, 200-1(mono), 196(di); Picrolonate, 248 and 231.

(31): L-Serine melts at 228d. *N*-Chloroacetyl, 122-3; 3,5-Dinitrobenzoyl, 95; Ethyl ester, 256d; Methyl ester hydro-

TABLE 4 (*continued*)

chloride, 114; β-Naphthalenesulfonyl, 214*
and 220; α-Naphthylurea, 192; Phenylhydantoin, 168-9; Phenylurethan, 159;
Picrolonate, 265d.

(32): D- and L-Isoserine melt at 199-201.
Phenylurethan, 183-4.

(33): Acetyl; 203(mono), 174-5(di); Amide,
135-6; Methyl ester 97-8.

(34): See note under (23).

(35): The phenylurea gives hydantoin, m.p.
117. The dibenzoyl derivative of DL-
cystine melts at 170. L-Cystine diethyl
ester dihydrochloride melts at 177; DL-
Cystine diethyl ester dihydrochloride
melts at 185. Mesocystine melts at
200-18d. 3,5-Dinitrobenzoyl, 180d; β-
Naphthalenesulfonyl, 226-30; Phenylhydantoin, 117.

(36): Acetyl, 187-9; N-Chloroacetyl, 178-
80; 3,5-Dinitrobenzoyl, 210; β-Naphthalenesulfonyl, 180-2; α-Naphthylurea, 217.

(37): Variable values are given in the literature for the decomposition point of
the pure compound and of the picrolonate.
Amide, 138-44; N-chloroacetyl, 130-1;
3,5-Dinitrobenzoyl, 93; Formyl, 168-9;
α-Naphthylurea, 143-4; Phenylhydantoin,
173-4; Picrate, 173; Picrolonate, 238N,
also 212 and 182.

(38): The DL-α-ethyl ester melts at 165 and
the β-ethyl ester melts at 200d. Picrolonate, 130.

(39): Amide, 131(di); N-Chloroacetyl, 142;
Dihydrazide, 135; β-Naphthylenesulfonyl,
153; α-Naphthylurea, 115d.

(40): 4-hydroxyproline exists in two isomeric forms. Natural 4-hydroxyproline
as $[\alpha]_D^{21}$-81 and forms a picrate that melts
at 188.
I. m.p. 274; $[\alpha]_D^{21}$ 75.2 in water.   m.p.
274 $[\alpha]_D^{26}$-74.6 D,L,m.p. 261
II. m.p. 237-41; $[\alpha]_D^{18}$ 58.6   m.p. 238-41;
$[\alpha]_D^{18}$ 58.1 D,L,m.p. 250
Phenylhydantoin, 123-4.

(41): N-Benzenesulfonyl, 185d; 3,5-Dinitro-
benzoyl, 240.

(42): L-(+)-Isoleucine melts at 283-4d and
D-(−)-isoleucine melts at 285-6d; their
$[\alpha]_D^{20}$ is 10.7. They give the same derivative. L-(+)-Alloisoleucine melts at 278d
and D-(−)-Alloisoleucine melts at 274-5d;
their $[\alpha]_D^{20}$ is 14. They give the same derivative. DL-Isoleucine and DL-Alloleu-
cine have been reported as melting at 275
in a sealed tube. N-Benzenesulfonyl, 147-
8; Formyl 126; α-Naphthylurea, 166-8.

(43): Acetyl, 114; Formyl, 99-100; Picrolonate, 179-80.

(44): D- and L-Methionine shrink and
darken above 278 and melt at 283 with
decomposition. Acetyl, 98-9; 3,5-Dinitro-

benzoyl, 94-5(hyd.) and 150(anh.); α-Naphthylurea, 188; p-Tolyurea, 157-8.

(45): Acetyl, 184(di) and 218(mono).

(46): The decomposition point is variously
given as 283, 275-80, and 320. N-Chloroacetyl, 126; 3,5-Dinitrobenzoyl, 93;
Formyl, 167; α-Naphthylurea, 155; Picrolonate, 208d.

(47): See note (42). N-Benzenesulfonyl,
147-8; N-Chloroacetyl, 72-4; Formyl,
156; α-Naphthylurea 178-9d.

(48): The melting point of L-histidine is
variously reported as 253, 272, 277, and
288. It forms a reineckate, m.p. 220d.
3,5-Dinitrobenzoyl, 189; Flavinate, 224-
6d; β-Naphthalenesulfonyl, 149-50; Picrate, 86; Picrolonate 232(mono) and
265d(di).

(49): The melting point is variously reported as 252, 278, 282, and 290. N-Benzenesulfonyl, 185d; Carbobenzoxy, 126,
N-Chloroacetyl, 159; 3,5-Dinitrobenzoyl,
233d; Formyl, 195-6; β-Naphthalenesulfonyl, 185; α-Naphthylurea, 159-60; Picrolonate, 203-4.

(50): See note (42). Formyl, 121; α-Naphthylurea, 178; Phthalyl, 120-1.

(51): N-Chloroacetyl, 119; Formyl, 126;
β-Naphthalenesulfonyl, 148; α-Naphthylurea, 195; Phenylhydantoin, 120-7d.

(52): Amide, 106-7; 3,5-Dinitrobenzoyl,
187; β-Naphthalenesulfonyl, 145-6; α-
Naphthylurea, 163; p-Nitrobenzl ester,
184-5; Picrolonate, 150r.h., 180s.h.;
Phthalyl, 140-1.

(53): 3,5-Dinitrobenzoyl, 187; Formyl,
141-4; β-Naphthalenesulfonyl, 68*;
Phenylhydantoin, 163.5; Phthalyl, 115-6;
Picrolonate, 150.

(54): The phenyl urea derivative has been
reported as melting at 150, 168, 174, and
190. Amide, 62; Acetyl, 137; 3,5-Dinitro-
benzoyl, 177; Naphthalenesulfonyl, 152-3*;
α-Naphthylurea, 198; p-Nitrobenzyl ester,
229; Phthalyl, 160-1; Picrolonate 216d.

(55): Amide, 72; Acetyl, 116; β-Naphthalenesulfonyl, 122-3; α-Naphthylurea, 202d and
198; Picrolonate, 217d(mono) and 145(di).

(56): N-Chloroacetyl, 104-7; Formyl,
114.5; Phthalyl, 111.5-112.5.

(57): D- or L-Valine melts at 315. Amide,
78-80; 3,5-Dinitrobenzoyl, 158; Formyl,
140-5; α-Naphthylurea, 204; Phenylhydantoin, 125; Phthalyl, 101.5-102.0; Picrolonate, 150r.h. and 250s.h.

(58): Formyl, 115-6; β-Naphthalenesulfonyl,
149.

(59): Acetyl, 165(di); Picrate, 218-20.

(60): Hydrochloride salt, 188d; Ethyl ester,
65; Formyl, 132; Phenylhydantoin, 103;
Phthalyl, 103.4.

TABLE 4 (*continued*)

(61): *N*-Chloroacetyl, 130; α-Naphthylurea, 194; Phenylhydantoin, 126; Phthalyl, 95.5-96.5.

(62): Picrate, 220.

(63): Acetyl, 137; *N*-Chloroacetyl, 107.

(64): The di-α-naphthalenesulfonyl derivative of L-form gives at 100-2 a viscous oil. It is very slightly soluble in hot water and very soluble in hot alcohol. DL-Tyrosine quickly heated decomposes at 340 but, in a preheated bath, at 295. Forms both di- and mono-derivatives. See Mcherney and Swann, *J. Am. Chem. Soc.* **59**, 1117. The m.p. of L-Tyrosine has been confused in some works with the m.p. of the DL-form. The decomposition point of the L-form with rapid heating is 314-8, and with slow heating 290-5 (Fischer, *Ber.* **32**, 3641). Amide, 153-4; Acetyl, *N*-148 and 172(di); *N*-Chloroacetyl, 155-6; Formyl, *N*-171-4*d* and 147(di); β-

Naphthalenesulfonyl, 102*N*; α-Naphthylurea, 205-6(mono); Picrolonate 260*d* (mono).

(65): 3,5-Dinitrobenzoyl, 252-4(di); α-Naphthylurea, 205-6; Picrolonate, 260*d*.

(66): Dihydantoin, 200.

(67): See note (1). Hydrochloride salt, 215*d* (mono); α-Naphthylurea, 221*d*(mono) and 236(di); Oxalate salt, 218; Picrolonate, 203(di).

(68): The m.p. of the 1-*N*-Benzoyl derivative is reported at 235-49; the benzoyl derivative is reported at 254-68. The 1-*N*-Benzoyl-5-*N*-*p*-toluenesulfonyl, m.p. 140; 5-*N*-benzoyl-1-*N*-*p*-toluenesulfonyl, m.p. 199. Hydrochloride salt, 235-6*d* (mono) and 188-90(di); Picrate, 225*d*(mono).

(69): The D- and L-acids sinter at 307. Ethyl ester, 65; Formyl, 132*; Phenylhydantoin, 103.

(70): Hydrochloride salt, 175-8*d*.

TABLE 5A

Carboxylic Acids—Liquid[a]

| Name of Compound | Note | B.p. | Melting point of derivatives | | | |
| | | | Recommended | | Others | |
			p-Toluidide	Anilide	2-Naphthyl-amide	p-Nitrobenzyl ester
1 Trifluoroacetic		71		88		
2 Thioacetic		93		76		
3 Formic (m.p. 8.4)	1	101	130	50	129	31
4 Acetic (m.p. 16)		118	147*	114.2*	132	78
5 Difluoroacetic		134-5		52		
6 Acrylic (m.p. 13)	2	140	141	105		
7 Propionic	3	141	126	106		31
8 Propiolic (m.p. 18)	4	144d		87		
9 Isobutyric	5	154.4	107	105		
10 Butyric		162.5	75	96	125	35
11 Vinylacetic (3-Butenoic)	6	163(169)		58		
12 Fluoroacetic (m.p. 31-2)		164				
13 Pyruvic (m.p. 13)	7	165d	109(130)	[104]s		
14 Isocrotonic (m.p. 15)(cis)	8	169	132	102		
15 Isovaleric	9	176	107	110	138.5	
16 DL-2-Methylbutanoic	10	176	93	112		
17 3,3-Dimethylbutanoic		184	134	132		
18 DL-α-Chloropropionic		186	124	92		
19 n-Valeric (Pentanoic)	11	186	74	63	112	
20 2,2-Dimethylbutanoic		187(190)	83	92		
21 DL-2,3-Dimethylbutanoic		192	113	78		
22 Dichloroacetic		194	153	118		
23 2-Ethylbutanoic (Diethylacetic)		195	116	127		
24 DL-2-Methylpentanoic		196	81	95		
25 DL-3-Methylpentanoic		197	75	87		
26 4-Methylpentanoic (Isocaproic)		199	63	112		
27 Methoxyacetic	12	204		58		
28 Caproic (Hexanoic)	13	205	75	95	107	
29 2-Ethylpentanoic		209	129	94		

(continued)

TABLE 5A (continued)

Name of Compound	Note	B.p.	Recommended		2-Naphthyl-amide	Others
			p-Toluidide	Anilide		p-Nitrobenzyl ester
30 2-Methylhexanoic		210	85	98		
31 α-Bromobutyric		217d	92	98		49
32 4-Methylhexanoic		218		77		
33 Heptanoic (Heptoic)		223	81	70	101	
34 Caprylic (m.p. 16)(Octanoic)		239	70	57	103	
35 Pelargonic (m.p. 12.3)(Nonanoic)		255	84	57	103	
36 Oleic (m.p. 16)		216/5 mm	43	41	169	
37 D,L-Lactic (m.p. 17)		226/10 mm	107	59	137.5	

a The m.p. of the amides of most of these acids may be found by referring to the Index of Tables and to Table 8.

TABLE 5B[a]

Carboxylic Acids—Solid

Name of Compound	Note	M.p.	Recommended		2-Naphthyl-amide	Others
			p-Toluidide	Anilide		p-Nitrobenzyl ester
1 Undecylenic (b.p. 275)		24.5	68	67		
2 α-Bromopropionic (b.p. 205d)		26	125	99		
3 Undecanoic (b.p. 284)(Undecylic)		30(29)	80	71		
4 Hexahydrobenzoic (b.p. 233)		30		146		
5 Fluoroacetic (b.p. 167-9)(amide,108)		31-2				
6 Capric (b.p. 268-70)(Decanoic)		31	78	70	104	
7 Levulinic (b.p. 246)	14	33-5	109	102		
8 Erucic (b.p. 252-4/12 mm)		34(30)	58	65	87	61
9 Pivalic (b.p. 164)(Trimethylacetic)		35	[120]	129		
10 α-Methylhydrocinnamic (b.p. 272)		36.6	130			
11 Lauric (b.p. 299)		43	87	78	106	
12 Tridecanoic (b.p. 312)		44	88	80		
13 α-Bromoisovaleric (b.p. 230d)		44	124	116		

No.	Name	M.P.	p‑Toluidide	Anilide	Amide	p‑Bromophenacyl ester
14	Elaidic (b.p. 234/15 mm)(amide,94)	44‑5(51)				36
15	Hydrocinnamic (b.p. 280/754 mm) [15]	48	135		135	88
16	Dibromoacetic (b.p. 232‑5d)(amide,156)	48				
17	α‑Bromoisobutyric (b.p. 198‑200)	48‑9				
18	Bromoacetic (b.p. 208)	50	93	98		
19	Pentadecanoic (b.p. 212/16 mm)	52	91	83		
20	Myristic (b.p. 202/16 mm)	54	93	131	134	80
21	Trichoroacetic (b.p. 197)	57‑8	113	78	108	49
22	Margaric (b.p. 231/16 mm)	61	98	84		42
23	β‑Bromopropionic	62.5	162	97(94)		
24	Palmitic (b.p. 222/16 mm)	63	100	90		67
25	Chloroacetic (b.p. 189) [16]	63	102	137	174	65
26	Cyanoacetic	66		198	109	
27	D‑Chaulmoogric (b.p. 222/10 mm) [17]	68.5		89		
28	Stearic (Octadecanoic)	70‑1		95	109	80
29	α‑Crotonic (*trans*)(b.p. 189w.s.)	72	132	118	117‑8	
30	Phenylacetic (b.p. 265)	76.5	136	118	157.5	107
31	Arachidic (b.p. 204/1 mm) [18]	77(75)		92	96‑8	100
32	α‑Hydroxyisobutyric (b.p. 212)	79	96	136	112	
33	Glycolic [19]	80	143	97	159	71
34	Iodoacetic	83	133	143‑4		
35	Dibenzylacetic	89	175	155		
36	o‑Benzoylbenzoic (anh. m 128) [20]	90(93)	170(mono)	195		69
37	Citraconic [21]	92d		175.5(di)		
38	Glutaric (b.p. 200/20 mm)	97(98)	218	224		
39	Phenoxyacetic	99		99		102
40	Citric (monohyd.)(anh. m 153) [22]	100	189(tri)	199(tri)(192)		113
41	o‑Methoxybenzoic	100‑1	[62]	131	112	87.2(mono)
42	L‑Malic (Hydroxysuccinic)	100‑1	206‑7(di)	197(di)		124.5(di); 204(di)
43	Oxalic (dihydrate) [23]	101	268(di)	254(di)		91
44	o‑Toluic [24]	104‑5	144	125	138	44
45	Pimelic (b.p. 223/15 mm) [25]	105	206	109(mono)155(di)		75
46	Azelaic (b.p. 237/15 mm) [26]	107	201	108(mono)		87
47	Ethylmalonic [27]	111	118	150		
48	m‑Toluic	111‑3	164	126	117	
49	Pyrotartaric [28]	115	115	200(di)		
50	α‑Phenoxypropionic	115‑6		117(118‑9)		
51	Benzylmalonic	117d		217(di)		119(di)
52	DL‑Mandelic (D m 133; L m 134)	118	172	152		124
53	Benzoic (*v.s.*) [29]	122.36*	158	163	189	89
54	Picric (See Table 29B)	122.5				

(continued)

TABLE 5B (continued)

| | | | Melting point of derivatives | | | |
| | | | Recommended | | Others | |
Name of Compound	Note	M.p.	p-Toluidide	Anilide	2-Naphthyl-amide	p-Nitrobenzyl ester
55 Trichlorolactic		124		164		
56 Diethylmalonic		125		141		91(di)
57 2,4-Dimethylbenzoic (anh.)		127		195		
58 o-Benzoylbenzoic (anh.)		128				100
59 Tribromoacetic (amide, 122)		131		140		
60 2,5-Dimethylbenzoic		132		155		
61 α-Naphthylacetic		133				
62 Furoic (Pyromucic)	30	133-4	107.5	123.5		133.5
63 Sebacic (b.p. 243/15 mm)		133s	201	201(di)		73.5(di)
64 Cinnamic (trans)	31	133	168	153		117
65 Maleic	32	134	142(di)	187(di)		91
66 Malonic (Propanedioic)	33	135	253(di)	230(di)		86
67 Acetylsalicylic	34	135		136		90.5
68 Acetone dicarboxylic	35	135d		155(di)		
69 Glutaconic (cis and trans)	36	136-8		228(di)		
70 Phenylpropiolic	37	137	142	126		
71 Methylmalonic		138d	145d(mono)228(di)(214)	184		83
72 m-Nitrobenzoic		140	162	154		141
73 meso-Tartaric [amide, 187(di)]		140				93
74 β-Naphthylacetic (amide, 200)		142				
75 o-Chlorobenzoic		142	131	118		106
76 Suberic (Octanedioic)		142	218(di)	128(mono)186(di)		85(di)
77 o-Nitrobenzoic		146		155		112
78 o-Chlorophenoxyacetic		146		121		
79 o-Aminobenzoic (Anthranilic) (See Tables 4 and 9B)		147(145)	151	131		205
80 Oxanilic		148-9		154(di)		
81 Diglycolic		148(142)	148(mono)[173]	118(mono)152(di)		
82 Diphenylacetic		148		180	191-2	
83 o-Bromobenzoic		150		141		110
84 Benzilic	39	150	[190]	175		99.5
85 Citric (anh.) (monohyd. m 100)	22	153	189(tri)	199(tri)(192)		
86 p-Nitrophenylacetic		153	210	198(212)		102
87 Adipic (b.p. 216/15 mm)(Hexanedioic)		153	241	151-3(mono)241(di)		106

No.	Acid	Ref.	M.P.			
88	m-Bromobenzoic		155		136	105
89	m-Chlorobenzoic		158		122	107
90	p-Chlorophenoxyacetic		158	156	125	98
91	Salicylic (v.s.)	40	158.3*	188.9*	136	
92	α-Naphthoic		162		163	111
93	o-Iodobenzoic		162		141	
94	4-Nitrophthalic		165	172(mono)	192	91
95	Itaconic	41	165		190	
96	3,4-Dimethylbenzoic (amide, 133)		166(164)		104	
97	3,5-Dimethylbenzoic (amide, 133)		166			
98	D,L-Phenylsuccinic	42	167		222(di)	
99	D-Tartaric	38	169-71	180d(mono)264d(di)		163
100	m-Aminobenzoic (See Tables 4 and 9B)		174	140		201
101	p-Toluic (b.p. 275)	43	178	160(165)	145(148)	104.5
102	4-Chloro-3-nitrobenzoic acid		181-2		131	
103	p-Fluorobenzoic (amide, 154)		182.6			
104	2,4-Dinitrobenzoic (amide, 203)	44	183			142
105	p-Anisic		184-6	186	169-71	132
106	β-Naphthoic		185.5	192	171	
107	Acetylanthranilic		185		167	
108	p-Aminobenzoic (See Table 4)		186			
109	Hippuric		187		208	136
110	3-Nitroanisic		187		163	
111	m-Iodobenzoic (amide, 186)		187			121
112	Succinic (b.p. 235d)(Butanedioic)	45	188	180(mono)255(di)	148(mono)230(di)	88
113	D-Camphoric (L m 187)	46	188	α-212-4, β-190-4	204(mono)226(di)	65.5
114	Chlorofumaric		191-2	186		138.5(di)
115	Dimethylmalonic [amide, 269(di)]		193s			83.6
116	p-Ethoxybenzoic		198	170		110
117	DL-Glutamic (L m 211-3)(See Table 4)	47	199D			
118	Protocatechuic		200d		166	188
119	Fumaric (m. 287 s.t.)	57	200s		314	151
120	m-Hydroxybenzoic	48	200	163	157(155)	106-8
121	DL-Tartaric (monohyd.)(anh. m 206)	38	204		236	148
122	m-Nitrocinnamic (trans)(amide, 196)		205			174
123	p-Coumaric (cis m 138)		206			152
124	o-Phthalic	49	208d	201(di)	253(di)	155
125	3,5-Dinitrobenzoic		208-9(203)	145-7	234	157

(continued)

TABLE 5B (*continued*)

	Name of Compound	Note	M.p.	Recommended		Others	
				p-Toluidide	Anilide	2-Naphyl-amide	p-Nitrobenzyl ester
126	o-Coumaric (*trans*)	50	208d				152.5
127	Oxamic		210		148-9		
128	l-Glutamic (See Table 4)		211-3				
129	Mucic	51	214d				310
130	p-Hydroxybenzoic	52	215	204	198		182(192)
131	β-Resorcylic (2,4-Dihydroxybenzoic)	52	216d		126-7		189
132	Piperic		216				145
133	3-Nitrophthalic		218	226(di)	234(di)		189
134	2-Hydroxy-3-naphthoic		223*	222	244(249)		
135	4-Hydroxy-2-naphthoic		225-6*	206			
136	2,2'-Diphenic		229		230		187
137	5-Nitrosalicylic		229-30		224		
138	Nicotinic	53	237-8s	150	85,132N		
139	o-Nitrocinnamic (*trans*) (*cis* m 146-7)		240				
140	Mesaconic	59	240.5	212(di)	186(di)		132
141	p-Nitrobenzoic		241	204	211		134
142	p-Chlorobenzoic		242		194		129
143	p-Bromobenzoic		254(258)		197		141
144	Gallic	54	254d		207		39
145	p-Iodobenzoic		270		210		141
146	p-Nitrocinnamic (*trans*)(*cis* m 143)	55	285				186
147	Muconic (*trans-trans* m 296-8, *cis-cis* m 195)	56	289d				
148	Terephthalic	58	300s		334-7(di)		263
149	Isophthalic	60	348s				202

^{a}The m.p. of the amides of most of these acids may be found by referring to the Index of Tables and Table 8.

Notes on Carboxylic Acids

1. Decomposed by concentrated H_2SO_4; readily reduced to H_2CO by Mg and HCl.
2. Polymerizes, especially when warmed.
3. Salted out by $CaCl_2$.
4. Na amalgam reduces it to propionic.
5. Odor of rancid butter.
6. Odor of butyric acid.
7. Violet-blue color with nitroprusside.
8. Sharp odor.
9. Offensive odor, like decayed cheese.
10. Odor similar to preceding, but weaker.
11. Same odor as isovaleric. The m.p. of

the p-phenylphenacyl ester obtained by regular methods of purification is 63.5, but purified by chromatographic methods the compound melts at 69°.

12. Viscous, oily.

13. Oily; odor unpleasant; v.s.

14. Often deliquesces to a liquid at ordinary temperatures; forms semicarbazone, m.p. 192.

15. Treated with $AlCl_3$ it ring-closes to 1-indanone, m.p. 42. [Bull. soc. chim. (4), 41, 942 (1927)].

16. Heilbron lists three isomers: α, m.p. 61.3; β, 56.2; γ, 52.5.

17. Upon solidifying, the crystals grow upward and branch out, if pure; if impure, the surface remains flat. This behavior is also true of α-hydnocarpic acid, $C_{16}H_{22}O_2$.

18. Sublimes. Warmed with dilute H_2SO_4 and MnO_2, has odor of benzaldehyde.

19. Long heating at 100 gives the anhydride, m.p. 128-30.

20. Steam distillation gives the anhydride.

21. b.p. 302-4.

22. Monoanilide melts at 164; dianilide at 179. Aqueous solution by boiling yields the anhydrous acid, m.p. 153. Floats on CCl_4 (differentiation from tartaric acid). Warmed with acetic anhydride and pyridine it exhibits a carmine coloration.

23. Melts at 101 when heated rapidly; sublimes (anh.) between 150 and 160; anh. acid 189.5. Treated with acetic anhydride, it decomposes to CO_2 and CO.

24. b.p. 259/751 mm.

25. Sublimes but not v.s.

26. Boils above 360 with a light decomposition.

27. b.p. 263; s; v.s.

28. Forms two monoanilides: one (from ethyl acetate) melts at 159; the other (from chloroform) at 123.

29. b.p. 249. m.p. as determined by Bureau of Standards: 122.36.

30. b.p. 230-2. Gives the pyrrole test.

31. b.p. 300.

32. If pure, melts at 137; ordinarily contains 3% fumaric acid and melts at 130; heated to 160, it converts to anhydride, m.p. 60 (b.p. 196); monoanilide, m.p. 198(187); mono amide, m.p. about 173 but begins decomposition at about 153.

33. About 100mg boiled in t.t. with 3 ml acetic anhydride 3 minutes and diluted with 3 ml acetic acid gives a yellow-red solution with green-yellow fluorescence (differentiation from pyromucic).

34. Heated rapidly, m.p. 135.

35. Decomposed by hot water, acids, or alkalies to acetone + CO_2. Violet color with $FeCl_3$ (enol). Forms oxime, m.p. 53-4.

36. Labile (cis) form melts at 136 but changes on melting into the stable (trans) form, m.p. 138; mono anilide, m.p. 167. With acetic anhydride at 40 it is dehydrated, 6-hydroxy-α-pyrone, m.p. 88.

37. Sublimes, melts under water at 80.

38. Warmed with acetic anhydride and pyridine gives an emerald color. So do racemic acid and D-tartaric acids.

39. 1 mg dissolved in 3 drops H_2SO_4 immediately gives intense orange-red coloration, soon red-violet at edges.

40. Sublimes; v.s. Treated with H_2SO_4 and MeOH and warmed gives the characteristic odor of oil of wintergreen (methyl salicylate).

41. Not v.s. (differentiation from citraconic). Anilide obtained only by boiling excess of amine with acid.

42. D- and L-forms melt at 173-4; anhydride, m.p. 84. DL-acid, m.p. 167; α-amide, 159; β-amide, 145; anhydride, 54; imide, 90; α-anilide, 175; β-anilide, 171; α-p-toluidide, 175; β-p-toluidide, 169.

43. Sublimes; v.s.

44. b.p. 275-80.

45. Ammonium salt distilled with Zn dust gives the pyrrole test.

46. L-acid melts at 187, DL at 202(208); D-acid forms two p-toluidides: α, m.p. 212-4, and β, 190-6; mono anilide melts at 204; mono amide melts at 177.

47. The D-acid melts at 224-5; the L-acid melts at 213 (rapid heat).

48. Sublimes; v.s. Tastes faintly sweet.

49. Melts at 191 in sealed tube and at 230 by rapid heating; converts to anhydride on heating; mono-p-toluidide melts at 150 (slow) and 160-5 (rapid); mono anilide melts at 170; mono amide, 149; imide, 233-5.

50. Sublimes, but not v.s.

51. Melts at 233-4 if rapidly heated. Gives pyrrole test.

52. Titrated with bromothymol blue as indicator.

53. The anilide melts at 265 recrystalized from water, but at 132 from a benzene-ligroin mixture. See also Table 37.

54. Melting point may vary, often 222-40d. Aqueous solution treated with a few drops of KCN solution gives red color, disappearing except at surface, and reappearing on shaking.

55. Cis form melts at 143.

56. Melts at 289 (slow heat) and at 306 (rapid heat). The trans-trans form melts at 296-8 and the cis-cis at 195.

57. m.p. 300-2 (rapid heat), 286-7 (sealed tube). Sublimes above 200 and at 230 dehydrates to maleic anhydride.

58. Sublimes without melting at about 300. Ba salt ($+4H_2O$) very insoluble (differentiation from isophthalic).

59. Two mono amides: α, m.p. 222; β, m.p. 174. Sublimes, but not v.s.

60. Sublimes below m.p. with formation of anhydride. Ba salt ($+6H_2O$) very soluble (differentiation from terephthalic).

TABLE 6A

Alcohols—Liquid

#	Name of compound	Note	M.p.	B.p.	Melting point of derivatives		
					Recommended		Other
					α-Naphthyl-urethan	3,5-Dinitro-benzoate	p-Nitro-benzoate
1	Methyl			64.65	124	108*	96
2	Ethyl			78.32	79	93	57
3	Isopropyl			82.4	106	123	110(108)
4	tert-Butyl (See Table 6B)		25.5	82.5			
5	Allyl			97.1	108	49	28
6	n-Propyl			97.15	105	74	35
7	DL-sec-Butyl	1		99.5	97	76	26
8	tert-Amyl			102.35	72	116	85
9	Isobutyl			108.1	104	87	69
10	DL-3-Methyl-2-butanol	2		114	109	76	
11	3-Pentanol (Diethyl carbinol)			116.1	95	101	17
12	n-Butyl			117.7	71	64(62.5)	36
13	2,3-Dimethyl-2-butanol			118	101	111	82
14	DL-2-Pentanol (sec-Amyl)	3		119.85	74.5(76)	62	17(24-5)
15	3,3-Dimethyl-2-butanol			120.4		107	
16	2,2-Dimethyl-3-butanol			120-1	128		98.5
17	2-Methyl-2-pentanol			121	104	72	69.5
18	3-Methyl-3-pentanol			123	83.5	96.5	67
19	2-Methoxyethanol ("Methyl cellosolve")	4		124.5	113		51
20	1-Chloro-2-propanol			127		77	
21	2-Methyl-3-pentanol			127.5		85	
22	2-Methylbutanol			128.9	82	70	
23	2-Chloroethanol			131	101	95	
24	4-Methyl-2-pentanol			132	88	65	26
25	Isoamyl (3-Methylbutanol)			132	68	61	21
26	D,L-2-Chloro-1-propanol	5		133-4		76	
27	3-Methyl-2-pentanol			134	72(84)	43.5	
28	2-Ethoxyethanol ("Ethyl cellosolve")	4		135	67	75	
29	3-Hexanol			136	72	77	
30	2,2-Dimethylbutanol			136.7	81	51	
31	n-Amyl (1-Pentanol)			138*	68	46.4	11

No.	Compound		m.p.	B.P.			
32	...anol						
33	2,4-Dimethyl-3-pentanol			138-9	60.5	38.5	155
34	Cyclopentanol	6		140	95(99)	115	62
35	2,3-Dimethylbutanol			140.8	118	51.5	
36	2-Methylpentanol			145	76	50.5	
37	2-Ethylbutanol			148	60	51.5	
38	Trichloroethanol	7	19	151	120	142.3	71
39	3-Methylpentanol			152-3	58,40-1(DL)	38	
40	4-Methylpentanol			151-2	58-9	72(70)	
41	4-Heptanol (Dipropyl carbinol)			156(154)	80	64	35
42	n-Hexyl (1-Hexanol)			157.5	59(62)	58.4	+5
43	DL-2-Heptanol			158.7	54	49	
44	Trimethylene chlorohydrin			161d	76	77	
45	Cyclohexanol (See Table 6B)	8	25	161.1	129	113	50
46	DL-4-Methylhexanol	9		165	50		
47	2-Methylcyclohexanol (cis or β)			165.5	154-5	99	56
48	Diacetone alcohol (See Table 21A)			166	155	55	48
49	2-Methylcyclohexanol (trans or α)	10	21	166.5	45-7	115	65
50	DL-3-Methyl hexanol			168-9	130	81	
51	2-Aminoethyl alcohol (See Table 9A)			171			
52	Furfuryl	11		172	128-9	91-2	76
53	2,6-Dimethyl-4-heptanol	6		172-3	160	134	118
54	3-Methylcyclohexanol (cis or β)			173/745 mm	122(118)	140	65
55	4-Methylcyclohexanol (cis or β)			173-4/750 mm (52/2 mm)	115	97-8	94
56	4-Methylcyclohexanol (trans or α)			173-5/745 mm	73	47	67
57	3-Methylcyclohexanol (trans or α)			174-5	62	83-4	58
58	1,3-Dichloro-2-propanol			176	90	32	10
59	Trimethylene bromohydrin			176d	63-4		46-8
60	Heptyl (1-Heptanol)	11		176.8	93	61	28
61	Tetrahydrofurfuryl			177-8/743 mm	61	169	37-8
62	DL-2-Octanol			179	67		
63	2,3-Dichloropropanol			182	176		
64	2-Ethylhexanol			184.6	55.5		
65	DL-1,2-Propanediol	12		187.4	53	42.8*	127
66	Diethylene glycol monomethyl ether ("Methyl carbitol")	13		194			92
67	n-Octyl (1-Octanol)		-16	195	67	61	12
68	Glycol (Ethylene glycol)			197.85	176	169	140
69	2-Nonamol			198.2	55.5		
70	L-Linalöl			199	53	42.8*	70

(continued)

TABLE 6A (*continued*)

| Name of compound | Note | M.p. | B.p. | Melting point of derivatives | | |
| | | | | Recommended | | Other |
				α-Naphthyl-urethan	3,5-Dinitro-benzoate	p-Nitro-benzoate
71 DL-α-Methylbenzyl		20	202	106	95	43
72 Benzyl			205.5	134	113	85
73 DL-1,3-Butanediol	14		207.5	184	178	
74 1,3-Propanediol			210-2	164(di)		119
75 DL-2-Decanol			211	69	44	
76 n-Nonyl (1-Nonanol)	15		215(213.5)	65.5	52	10
77 Triethylene glycol dimethyl ether (See Table 13A)			216			
78 m-Methylbenzyl (m-Tolylcarbinol)			217	116		
79 α-Ethylbenzyl (α-Phenylpropyl)			219	102		59-60
80 Phenethyl (β-Phenylethyl)			219.8	119	108	62*
81 Geraniol			230	48	63	35
82 1,4-Butanediol		19.5	230(235)	199(bis)		175(di)
83 n-Decyl (1-Decanol)			231	73	57	30
84 3-Phenylpropanol			237.4(235)		92	47
85 Pentamethylene glycol			238	147(di)		104.5(di)
86 Undecyl		15.85	243	73	55	
87 Diethylene glycol	16		244.5	149	151*	
88 o-Methoxybenzyl			247	136		
89 Dodecyl (Lauryl)		24(26)	259	80	60	45(42)
90 Glycerol	17		290d	192(tri)		188(tri)

TABLE 6B

Alcohols—Solid

No.	Name of compound	Note	M.p.	B.p.	Recommended α-Naphthyl-urethan	Recommended 3,5-Dinitro-benzoate	Other p-Nitro-benzoate
1	α-Methylbenzyl (α-Phenylethyl) (See Table 6A)		20	202			
2	2-Methylcyclohexanol (trans or α) (See Table 6A)		21	167.4			
3	Dodecyl (Lauryl) (See Table 6A)		24	259			
4	Cyclohexanol		25	161.1	129	113	50
5	tert-Butyl		25.45	82.5	101	142	116
6	m-Nitrobenzyl (See Table 27)		27	175–80/3 mm			
7	Diethanolamine (See Table 9B)		28	217–18/150 mm			
8	Tridecanol		30.63	155–6/15 mm	114	121	37.4*
9	Cinnamyl		33	257	152(147)	79	78
10	α-Terpineol (D, m.p. 36.9; L, m.p. 37)	18	35	221	149	104	97
11	DL-Fenchyl	19	38–9	201.5	82	67	109
12	Tetradecanol (Myristyl)	20	39	170–3/20 mm			51.2*
13	Pinacol (Tetramethylethylene)	21	43	173			
14	L-Menthol	22	44N	216	119(126)	153	62
15	α-Pentadecanol		44		72		45.8*
16	Cetyl (Hexadecanol)		49	190/15 mm	82	66	58.4*
17	2,2-Dimethylpropanol (Neopentyl)		52–3	113	100		
18	α-Heptadecanol		54	310	88.5		53.8*
19	Piperonyl	23	58			77	64
20	Octadecanol (Stearyl)		59.5	210.5/15 mm	89	117–8	
21	p-Methylbenzyl(p-Tolylcarbinol)		59–60	217 v.s.			58.9*
22	Nonadecanol		62				132
23	Benzhydrol (Diphenylcarbinol)	24	68	180/20 mm	136(139)	141	
24	Erythritol	25	72	270			
25	o-Nitrobenzyl (See Table 23)		74				
26	Myricyl (10-Nonadecanol)		85		54		
27	o-Hydroxybenzyl (Saligenin) (See Table 25)	26	86–7	118–20/11 mm			128.6
28	Phenacyl (See Table 21B)	27	86N				
29	D-Sorbitol (anh.)		89–93				
30	p-Nitrobenzyl (See Table 23)	28	93	185/12 mm	140	140	123
31	DL-Benzoin (See Table 21B)		137	344	176		185(190–3)
32	Cholesterol (anh.) (L-Cholesterol)		148.5	360d	202		
33	Ergosterol (anh.)	29	165				
34	D-Borneol		208(205)	212	132(127)	154–5	137(153)

TABLE 6 (*continued*)

Notes on Alcohols

1. D-form; $[\alpha]_D^{20}$ is +13.87.

2. The m.p. of derivatives are for the DL-form of the alcohol; for example the hydrogen phthalate of the D- or L-form melts at 34. D-form b.p. 110-2, $[\alpha]_D^{20}$ is +5.34 in ethanol.

3. The m.p. of derivatives are for the DL-form of the alcohol; for example, the hydrogen phthalate of the D- or L-form melts at 34. α-naphthylurethanes, α, m.p. 88-91; β, m.p. 71-3.

4. The lower monoalkyl ethers of ethylene glycol are commonly known as "cellosolves." The esters resulting from the reaction of the hydroxyl group are, as a rule, liquids and unsuitable for derivatives. For urethane derivatives of cellosolves, reference is made to *J. Am. Chem. Soc.* **62**, 3136 (1940), and for phenyl phenacyl esters to *J. Am. Chem. Soc.* **62**, 1635 (1940).

5. Density, 1.103; refractive index, 1.436. 2-Chloro-1-propanol heated with alkalis yields propylene oxide.

6. Camphor-like odor.

7. The melting points for the D- and L-α-naphthylurethan are: D, 38-40; L, 97-8.

8. Camphor-like odor. Oxidation with chromic acid gives cyclohexanone; with nitric acid, adipic acid.

9. Odor of amyl alcohol.

10. The hexahydro-o-cresol obtained from o-cresol consists of a mixture of two stereoisomers, each of which can be resolved into two optically active forms.

The derivatives listed are for the DL-, *cis*, or β-form; likewise those listed for the *trans* isomer (m 167.4) are for the DL-form.

11. Furfuryl alcohol instantly reduced permanganate in cold water and decolorizes bromine water, whereas tetrahydrofurfuryl does not.

12. Viscous liquid; tastes sweet. A few drops distilled with anh. $ZnCl_2$ gives propionaldehyde. Monostearate, m.p. 59.5; distearate, m.p. 72.3.

13. The 3-nitrophthalate monohydrate melts at 87-90, and the anhydrous at 91.4-92.2.

14. The b.p. of the D,L-compound is also reported as 204. The phenylurethane of the D-form melts at 115-6 and the di-phenyl-urethane of the L-isomer melts at 127-8.

15. Other values for the m.p. of the phenyl-urethan are 69, 62-4.

16. Somewhat viscous and slightly sweet.

17. Tribenzoate 71-72, also 75-6; tri-p-toluenesulfonate, m 103.

18. The D and L isomers melt at 37 and yield phenyl urethanes, m.p. 110.

19. The D-α isomer melts at 45 and boils at 201-2. The phenyl urethane m.p. is 145. The L-α isomer melts at 47 and gives the following derivatives: phenylurethane, 82; acid pthalate, 146; p-nitrobenzoate, 109. The L-β isomer melts at 3-4 and yields an acid phthalate, m.p. 153, and a p-nitrobenzoate, m.p. 83.

20. p-Bromobenzenesulfonyl derivative, m.p. 51.5; p-methoxyphenylurethane m.p. 83; 3,4-dimethoxyphenylurethane, m.p. 79.5.

21. Pinacol hydrate over NaOH loses water after several days, forming eventually the anhydrous compound. When boiled with dilute H_2SO_4 it gives pinacolone, having a strong peppermint odor; heated with B_2O_3, it gives a good yield of pinacolone. The diacetate melts at 65.

22. L-Menthol (strong peppermint odor) exists in four allotropic forms with melting points of 44(42.5), 35, 33, and 31. The acid phthalate exists in two forms, one stable, m.p. 122, and one labile, m.p. 110. The benzoate melts at 55. D-Menthol melts at 38-40. The DL-form melts at 34 and yields a phenylure-thane, m.p. 104, and an acid phthalate, m.p. 130. The hydrogen phthalate derivative, m.p. 110, slowly changes in contact with the mother liquor to the stable form, m.p. 122.

23. Benzoate, m.p. 66; phenylurethan, m.p. 102.5.

24. Easily oxidized by chromic acid mixture to benzophenone.

25. Tetraacetate, m.p. 53.

26. Benzoate, m.p. 118.5; m-nitrobenzoate, m.p. 104-5; o-nitrobenzoate, m.p. 124.5.

27. The m.p. is also reported at 112; hexa-acetate, m.p. 99; hexabenzoate, m.p. 216.

28. Acetate, m.p. 78; benzoate, m.p. 94-5.

29. Acetate, m.p. 176; benzoate, m.p. 168.

Aldehydes—Liquid

	Name of compound	Note	B.p.	Recommended		Others	
				Semi-carbazone	2,4-Dinitro-phenyl-hydrazone	Phenyl-hydrazone	Methone
1	Formaldehyde (Methanal)	1	−21	169	166	145	191(189)
2	Acetaldehyde (Ethanal)	2	20.2	162	148,168	57(63),99	140
3	Perfluoro-n-butyraldehyde		29		107		
4	Propionaldehyde (Propanal)	3	47.5–49.0	89,154	155	(oil)	155
5	Glyoxal (m.p. 15)		50	270	328	180	228
6	Acrolein		52.4	171	165	52(Pyrazoline)	192
7	Isobutyraldehyde		64	125–6	187	(oil)	154
8	n-Butyraldehyde (Butanal)		74.7	106(96)	123		135(142)
9	Pivaldehyde (Trimethylacetaldehyde)		75	190	209		
10	Isovaleraldehyde		92.5	107	123	(oil)	154
11	2-Methyl-1-butanal (α-Methyl-n-butyraldehyde)		92–3	103	120		
12	Chloral		98	90d	131		104.5
13	n-Pentanal (Valeraldehyde)		103		107(98)		
14	Crotonaldehyde		104	199	190	56	186
15	Diethyl acetaldehyde (α-Ethyl-n-butyraldehyde)		117	99	130		102
16	n-Hexanal (Caproaldehyde)		131	106	104(107)		108.5
17	Tetrahydrofurfural		144–5/740 mm	166	204		123
18	n-Heptanal		153	109	108		135(103)
19	Furfural	4	161.7(90/65 mm)	202	230,212–4	97	160
20	Hexahydrobenzaldehyde		162	173			
21	n-Octanal (Caprylaldehyde)		171	101	106		90
22	Bromal		174				
23	Benzaldehyde	5	179	222N	237	158,154–5	193
24	5-Methylfurfural		187	N		147–8	
25	n-Nonanal (Pelargonaldehyde)		190	100	100		86
26	Salicyladelhyde (See Table 25)	6	197	231	248,252N	142	208N
27	m-Tolualdehyde		199	224(204)		91(84)	
28	o-Tolualdehyde		200(197)	212	194	106	
29	p-Tolualdehyde		204	234(215)	234	114	
30	D-Citronellal		207	84(91–2)	78		77-9

(continued)

Melting point of derivatives

TABLE 7A (continued)

	Name of compound	Note	B.p.	Recommended		Others	
				Semi-carbazone	2,4-Dinitro-phenyl-hydrazone	Phenyl-hydrazone	Methone
31	n-Decanal (Capraldehyde)		207-9	102	104		91.7
32	o-Chlorobenzaldehyde (m.p. 11)	7	213-4(208)	229-30,146	213.6(209)	86	205d
33	m-Chlorobenzaldehyde (m.p. 18)	8	216-4(208)	228	256(248)	134	
34	Hydrocinnamaldehyde		224	127	149		
35	Citral (Geranial)	9	228(118/20 mm)	164N	110		
36	Citral (Neral)	9	228(103/12 mm)	171N	96		
37	m-Methoxybenzaldehyde (m.p. 3-4)		230	233d		76	
38	m-Bromobenzaldehyde		234	205		141	
39	Cumaldehyde	10	236	211	244-5(241)	129	170-1
40	Anisaldehyde (m.p. 20) (p-Methoxybenzaldehyde)	11	248	210	253-4d	120-1	145
41	Cinnamaldehyde	12	252	215(208)	255d	168	213

TABLE 7B

Aldehydes—Solid

	Name of compound	Note	M.p.	Recommended		Others	
				Semi-carbazone	2,4-Dinitro-phenyl-hydrazone	Phenyl-hydrazone	Methone
1	Palmitaldehyde		34	108-9			
2	α-Naphthaldehyde (b.p. 292)		34	221		80	
3	Phenylacetaldehyde (b.p. 195)		34	156(153)	121	63(58)(mono),101-2(di)	165
4	Piperonal (Heliotropin) (b.p. 263)	13	37	234	266d	102-3(106)	177-8(193N)
5	o-Iodobenzaldehyde		37	206	253	79	
6	o-Methoxybenzaldehyde (b.p. 236)		38	215	101		
7	Stearaldehyde		38(55)	108-9			
8	o-Aminobenzaldehyde		40	247		221	
9	o-Nitrobenzaldehyde	14	41	256	250d(265)	156	

No.	Compound						
10	Veratraldehyde (b.p. 285)		44(58)	177	264-5	121	173
11	Lauraldehyde		44-5	106(103)	106		
12	p-Chlorobenzaldehyde (b.p. 214.5-216.5)	15	47	230(233)	265	127	
13	Chloralhydrate (b.p. 96)		53				
14	2,3-Dimethoxybenzaldehyde		54	231		138	
15	Phthalaldehyde		56			191(di)	
16	m-Iodobenzaldehyde		57	226		155	
17	m-Nitrobenzaldehyde		58	246	293d	120(124)	
18	2-Naphthaldehyde		60	245	270	206(217-8)	
19	p-Bromobenzaldehyde	16	67	228	128	113	
20	p-Aminobenzaldehyde		72	173		156(175)	
21	p-Dimethylaminobenzaldehyde		74	222	325	148	
22	p-Iodobenzaldehyde		78	224		121	
23	Vanillin (See Table 25)		81	230	271d	105	196-8*
24	Phenylglyoxal		91	217d		152(di)	
25	m-Hydroxybenzaldehyde (See Table 25)		104(108)	198	260d	131,147	
26	p-Nitrobenzaldehyde	17	106	221(211)	320	159	
27	2,3-Dihydroxybenzaldehyde		108	226d		167	
28	p-Hydroxybenzaldehyde (See Table 25)	18	116	224(280d)N	280(260)N	184(178)	190*(184)
29	Terephthalaldehyde		116			278d(di)	
30	2,4-Dihydroxybenzaldehyde (β-Resorcylic aldehyde)		136	260d	286	158	
31	DL-Glyceraldehyde (dimer)		142(138.5)	160d	166-7*	175-6d,121-8	
32	3,4-Dihydroxybenzaldehyde (Protocatechualdehyde)		153-4d	230d	275d	145d	
33	3,5-Dihydroxybenzaldehyde (α-Resorcylic aldehyde)		156-7	223-4			

Notes on Aldehydes

1. Commercial "formalin" is a 40% aqueous solution of formaldehyde. It generally contains 10-15% methanol to prevent polymerization.

2. Acetaldehyde yields iodoform by the usual technique. This fact distinguishes it from formaldehyde or propionaldehyde.

3. The semicarbazone melts at 154 if recrystallized from water; at 89 if recrystallized from benzene-ligroin. The 2,4-dinitrophenylhydrazone also is reported as red crystals melting at 150 and orange crystals melting at 148.

4. Furfural 2,4-dinitrophenylhydrazone exists as red crystals, m.p. 230 corrected, and as yellow crystals, m.p. 212-4. A mixture of the forms melts below 200. The α-oxime, from petroleum ether, melts at 75-6; the β-oxime, from alcohol, melts at 91-2.

5. The semicarbazone melts at 233-5 when heated rapidly. The α-oxime (stable) melts at 35; a β-oxime, needles from ether, melts at 130. The p-nitrophenylhydrazone has been reported in other forms melting at 262 and 234-6.

6. Salicylaldehydedimethone crystallizes directly as the anhydride from 70% alcohol, m.p. 208 cor. Salicylaldehyde gives the phenol test with ferric chloride. The 2,4-dinitrophenylhydrazone melts at 248 (from absolute alcohol) and at 252 from benzene. The dimethone forms the

TABLE 7 (*continued*)

anhydride when crystallized from 70% alcohol, m.p. 208.

7. α-oxime, m.p. 76; β-oxime, m.p. 101-3. Semicarbazone (1) leaflets, m.p. 229-30; (2) yellow prisms, m.p. 146.

8. α-(*anti*)oxime, m.p. 70-1; β(*syn*)oxime, m.p. 118 r.h. The m.p. of the 2,4-dinitrophenylhydrazone from xylene is 248.

9. Ordinary citral in the presence of NaC₂H₃O₂ with semicarbazone hydrochloride yields a mixture of the semicarbazones, m.p. 132; in the absence of NaC₂H₃O₂ only the geranial isomer precipitates, m.p. 164.

10. α-oxime (from alcohol), m.p. 52(61); β-oxime, m.p. 112.

11. β-(*syn*) oxime (needles from benzene), m.p. 133. α-(*anti*)oxime is more soluble than the β-form and exists as two forms: (1) leaflets, m.p. 45, and (2) needles, m.p. 65 (soluble in benzene).

12. *Syn*-oxime, m.p. 138.5, forms acetyl derivative, m.p. 69-70. *Anti*-oxime, m.p. 64-5, forms acetyl derivative, m.p. 35.

13. Piperonaldimethone was reported to melt at 193, but a later report gives 177-8. The anhydride has m.p. 220 corrected. Piperonal is readily reduced to piperonyl alcohol, m.p. 52. *Syn*-oxime (from methanol), m.p. 146, → acetyl derivative,

m.p. 99. *Anti*-oxime (from water), m.p. 112, → acetyl derivative, m.p. 86.

14. *Anti*-oxime, needles, m.p. 102. *Syn*-oxime (from benzene), m.p. 154.

15. α-oxime, m.p. 110. β-oxime, m.p. 146. *o*-nitrophenylhydrazone, m.p. 203-4.

16. *Syn*-oxime, m.p. 157. *Anti*-oxime, m.p. 111.

17. *Syn*-oxime, m.p. 182-4. *Anti*-oxime, m.p. 133(128-9).

18. *p*-Hydroxybenzaldehyde-2,4-dinitrophenylhydrazone crystallizes from water as red crystals, m.p. 260, and from acetic acid as purple crystals, dec. 280.

TABLE 8

Amides and Ureas

Note: The identification of amides and substituted amides is usually made by hydrolyzing the compound and identifying the products. Direct derivatives for some amides are given in the Notes (Hg salts, amide oxalates, N-xanthylamides, and N-acylphthalimides). Several of these amides are also found in Table 37 on physiologically active compounds.

Name of compound	Note	B.p.		Name of compound	Note	M.p.
Liquids			47	Tridecananilide		80
			48	Propionamide	1,2	81
1 N,N-Dimethylformamide		153	49	Acetamide	1,2	82
2 N,N-Diethylformamide		176	50	Aceto-N-methyl-p-toluidide		83
3 Formamide		195d		(N-Methyl-p-acetotoluidide)		
			51	α-Bromoisobutyranilide		83
Solids		M.p.	52	Myristanilide		84
4 N,N-Dimethylbenzamide		41	53	Nicotinanilide		85
5 Oleanilide		41	54	Acrylamide		85
6 Ethyl urethane		49	55	Allylurea		85
7 Malonamide		50(mono)	56	Acetoacetanilide		85
		170(di)	57	α-Undecylenamide		87
8 Formanilide		50	58	Propiolanilide		87
9 N-Propylacetanilide		50	59	DL-3-Methylpentananilide		87
10 Methyl urethane		52	60	o-Chloroacetanilide		88
11 Difluoracetamide		52	61	n-Butyl oxamate		88
12 N-Phenylurethane		53	62	D-Chaulmoogranilide		89
13 N-Benzyl acetamide		53-4	63	Palmitanilide (Hexadecanoic)		90
14 N-Ethylacetanilide		54	64	Bromoacetamide		91
15 Butyl urethane		54	65	Isopropyl urethane		92
16 Acetoacetamide		54	66	o-Nitroacetanilide		92
17 Isobutyl urethane		55	67	Arachidanilide		92
18 N-Methyl-o-acetotoluidide		56	68	DL-α-Chloropropionanilide		92
19 Caprylanilide		57(55)	69	2,2-Dimethylbutanilide		92
20 Pelargonanilide		57	70	o-Nitroacetanilide		92
21 DL-Lactanilide		59	71	4-Methyl-2-nitroacetanilide		94
22 Propyl urethane		60	72	2-Ethylpentananilide		94
23 n-Valeranilide (Pentatoic)		63	73	m-Toluamide	2	94(97)
24 Isoamyl urethane		64	74	N-Methyl-N-α-naphthyl-		
25 Aceto-m-toluide		66		acetamide		94
26 Aceto-N-methyl-m-toluidide		66	75	n-Butyranilide		95
27 Ethyl oxanilate		66-7	76	Stearanilide (Octadecanoic)		95
28 Heptananilide		70(65)	77	Iodoacetamide		95
29 Capranilide		70	78	Azela-amide		95(mono)
30 Undecananilide		71	79	DL-2-Methylpentananilide		95
31 Oenanthoanilide		71	80	Caproanilide (Hexanoic)		95(92)
32 2-Methylhexanamide		72	81	Trichloroacetamide		96
33 Vinylacetamide		73	82	Heptamide	1	96
34 N,N-Diphenylformamide		73	83	Semicarbazide	3	96
35 Oleamide		76	84	Glycollanilide		97
36 DL-Lactamide		76	85	Trichloroacetanilide		97(94)
37 Tiglamide		76	86	Methoxyacetamide		97
38 Thioacetanilide		76	87	Hydrocinnamanilide		98(96)
39 4-Methylhexananilide		77	88	Dichloroacetamide		98s
40 Tiglanilide		77	89	α-Hydroxyisobutyramide		98
41 DL-2,3-Dimethylbutanilide		78	90	2-Methylhexananilide		98
42 Pentadecananilide		78	91	4-Methylhexanamide		98
43 Lauranilide		78	92	Octanamide		99
44 N,N-Diethylcarbanilide		79	93	α-Bromopropionanilide		99(110)
45 DL-2-Methylpentanamide		80	94	Phenoxyacetanilide		99
46 DL-α-Chloropropionamide		80	95	Pelargonamide	1	99

(*continued*)

TABLE 8 (*continued*)

	Name of compound	Note	M.p.		Name of compound	Note	M.p.
96	Caproamide		100	152	N-Ethyl-p-nitro acetanilide		118
97	Tridecanamide		100	153	Cyanoacetamide	1	120
98	Phenylpropiolamide		100(109)	154	Glycollamide		120
99	N,N-Diphenylacetamide		101	155	Isocaproamide	1	120-1
100	Phenoxyacetamide		101	156	Chloroacetamide	1	121
101	Methyl urea	4	101	157	o-Chlorophenoxyacetanilide		121
102	N-Methylacetanilide		102	158	Tribromoacetamide		122
103	Pentadecanamide		102	159	Sebacanilide		122(mono)
104	Levulinanilide		102	160	N,N'-Dimethylcarbanilide		122
105	Isocrotonanilide		102	161	m-Chlorobenzanilide		122
106	Isocrotonamide		102	162	α-Bromopropionamide		123
107	Undecanamide		103(99)	163	Furanilide		123.5
108	3,4-Dimethylbenzanilide		104	164	Pyruvamide		124
109	Pyruvanilide		104s	165	3-Methylpentanamide		125
110	2-Ethylpentanamide		105(103)	166	p-Chlorophenoxyacetanilide		125
111	Isobutyranilide		105	167	Benzo-m-toluidide		125
112	Hydrocinnamamide		105	168	o-Toluanilide		[125]
113	Acrylanilide		105	169	Acetyl-β-phenylhydrazine		125-6
114	Propionanilide		106	170	Adipanilide		125-30
115	Palmitamide		106				(mono)
116	D-Chaulmoogramide		106	171	n-Butylethyl barbituric acid		125
117	sym-Dimethylurea		106	172	Ethyl-n-hexyl barbituric acid		126
118	n-Valeramide		106	173	Succinimide	1	126
119	Fluoracetamide		107	174	m-Toluanilide		126
120	Myristamide		107(103)	175	Angelanilide		126
121	Margaramide		108	176	β-Resorcylanilide		126-7
	(Heptadecanoic)				(2,4-Dihydroxybenzoic)		
122	Capramide (Decanoic)		108(98)	177	2-Ethylbutananilide		127
123	Thioacetamide		108	178	Suberamide (Octanedioic)		127(mono)
124	Fluoracetamide		108	179	p-Methoxyacetanilide		127
125	Levulinamide		108d	180	Angelamide		127-8
126	Azela-anilide		108(mono)	181	Dibenzylacetamide		128
127	Arachidamide		108-9	182	Nicotinamide		128
128	Anthranilamide		109	183	Phenylpropiolanilide		128(126)
129	Pimelanilide		109(mono)	184	Suberanilide (Octanedioic)		128(mono)
130	Stearamide	1	109	185	Isovaler-p-bromanilide		128
131	α-Methylhydrocinnamamide		109	186	o-Anisamide		128
132	Isovaleranilide		110	187	Pivalanilide		129
133	Caprylamide (Octanoic)		110(104)	188	Piperine		129
134	DL-2-Methylbutananilide		110	189	o-Methoxybenzamide		129
135	Lauramide		110(102)	190	Isobutyramide	1	129
136	β-Bromopropionamide		111	191	3,4-Dimethylbenzamide		130
137	m-Aminobenzamide		111	192	Benzamide	1,2	130
138	2-Ethylbutanamide		112(107)	193	Anthranilanilide		131
139	4-Methylpentananilide		112(110)	194	Bromoacetanilide		131
	(Isocaproic)			195	o-Methoxybenzanilide		131
140	DL-2-Methylbutanamide		112	196	DL-2,3-Dimethylbutanamide		132
141	Aceto-o-toluidide		112	197	2,2-Dimethylbutanamide		132(103)
142	o-Aminobenzamide		114	198	3,3-Dimethylbutananilide		132
143	Acetanilide		114*	199	3,3-Dimethylbutanamide		132
144	Ethyl oxamate		114	200	Malonanilide		132(mono)
145	o-Chlorobenzanilide		114(118)	201	Nicotinanilide	12	132N(85)
146	n-Butyramide	1,2	115	202	Urea (Carbamide)	5	132.8*
147	Methacrylamide		116	203	DL-Mandelamide		132
148	α-Bromoisovaleranilide		116	204	Mesitylenamide		133
149	Phenylacetanilide		118		(3,5-Dimethylbenzoic)		
150	α-Crotonanilide (*trans*)		118(115)	205	α-Bromoisovaleramide		133
151	Dichloroacetanilide		118	206	p-Chlorophenoxyacetamide		133

TABLE 8 (*continued*)

	Name of compound	Note	M.p.		Name of compound	Note	M.p.
207	2,4-Dimethylacetanilide		133	263	N-Phenylsuccinimide		156
208	N-β-Naphthylacetamide		134	264	Dibromoacetamide		156
209	Chloroacetanilide		134	265	Phenylacetamide		156
210	m-Chlorobenzamide	2	134	266	L-Malamide		156.5-158.0
211	Phenacetin		134	267	Succinamide		157(mono)
212	Isovaleramide	1	135(137)	268	m-Hydroxybenzanilide		157(155)
213	Acetylsalicylanilide		136	269	DL-Phenylsuccinamide (α)		158
214	m-Bromobenzanilide		136		(β m 145)		
215	Salicylanilide		136	270	Benzo-p-toluidide		158
216	α-Hydroxyisobutyranilide		136	271	N-α-Naphthylacetamide		159
217	m-Aminobenzanilide		140	272	Nitrourea		159(150)
218	β-Phenylalanine amide		140	273	p-Toluamide	1,2	160
219	Trichloroacetamide		141	274	N-α-Naphthylbenzamide		161
220	o-Iodobenzanilide		141	275	Crotonamide		161(158)
221	o-Bromobenzanilide		141	276	N-β-Naphthylbenzamide		162
222	o-Toluamide	1,2	142	277	p-Aminoacetanilide		162
223	m-Tolylurea		142	278	p-Hydroxybenzamide (hyd.)		162
224	o-Chlorobenzamide	2	142	279	α-Naphthanilide		163
225	Salicylamide	2	142	280	Benzanilide		163
226	Furamide		143	281	Trichloroacetanilide		164
227	m-Nitrobenzamide		143	282	o-Benzoylbenzamide		165
228	Iodoacetanilide		143-4	283	L-Glutamamide		165
229	Benzo-o-toluidide		144	284	Protocatechuanilide		166
230	p-Toluanilide		145(148)	285	p-Bromoacetanilide		167
231	α-Bromoisovalerylurea		145	286	Acetylanthranilanilide		167
232	α-Triphenylguanidine		145	287	p-Anisamide		167(163)
233	DL-Phenylsuccinamide (β)		145	288	Diphenylacetamide		168
	(α m 158)			289	Maleamide		168-70
234	Diethylmalonamide		146(mono)				(190)
235	Hexahydrobenzanilide		146(131)	290	α-Benzoyl-β-phenylhydrazine		168
236	Phenylurea		147	291	p-Hydroxyacetanilide		169
237	Diphenylguanidine		147	292	p-Anisanilide		169-71
238	Succinanilide		148(mono)	293	m-Hydroxybenzamide		170
239	Cinnamamide		148(142)	294	Alloxan	6	170d
240	α-Bromoisobutyramide		148	295	Sebacamide		170(mono)
241	4-Methyl-3-nitroacetanilide		148	296	Malonamide		170(di)
242	Benzylurea		149	297	DL-Phenylsuccinanilide		β-170
243	Phthalamide		149(mono)				(mono)
	(Phthalamic acid)						(α-175
244	Ethylmalonanilide		150				(mono))
245	o-Chlorophenoxyacetamide		150	298	p-Ethoxybenzanilide		170
246	Adipanilide		151-3	299	β-Naphthanilide		171
			(mono)	300	Azela-amide		172(di)
247	Cinnamanilide		151(153)	301	Diallylbarbituric acid		173
248	DL-Mandelanilide		152	302	p-Phenetylurea (Dulcin)		173
249	Aceto-p-toluidide		153	303	Maleanilide		173-5
250	N-Methyl-p-nitroacetanilide		153				(mono)
251	Ethylisoamylbarbituric acid		154	304	Ethylphenyl barbituric acid		174
252	p-Fluorobenzamide		154	305	N-Bromosuccinimide		174
253	Pivalamide		154	306	Pimelamide		175(di)
254	m-Nitrobenzanilide		154	307	Citraconanilide		175(di)
255	Benzilamide		155	308	DL-Phenylsuccinanilide (α)		175(mono)
256	Pimelanilide		155(di)		(β m 170(mono))		
257	o-Bromobenzamide	2	155	309	p-Hydroxyphenylacetamide		175
258	Dibenzylacetanilide		155	310	Glutaramide		175-6(di)
259	o-Nitrobenzanilide		155	311	o-(p-Toluyl) benzamide		175-6
260	m-Bromobenzamide		155	312	o-Nitrobenzamide		176
261	α-Naphthylacetanilide		155	313	Acetylanthranilamide		177
262	m-Nitroacetanilide		155	314	D-Camphoramide		177(mono)

(*continued*)

TABLE 8 (*continued*)

Name of compound	Note	M.p.	Name of compound	Note	M.p.
315 Mesaconamide		177(di)	369 4-Nitrophthalamide		200*d*
316 *p*-Chloroacetanilide		179	370 Methylsuccinamide		200(di)
317 2,4-Dimethylbenzamide		179	371 β-Naphthylacetamide		200
318 *p*-Chlorobenzamide		179	372 Isatin		200-1
319 *p*-Cyanobenzanilide		179	373 3-Nitrophthalamide		201*d*(di)
320 *o*-Nitrocinnamamide		179(170)	374 Sebacanilide		201(di)
321 D-Tartaranilide		180*d*	375 *p*-Nitrobenzamide	1	201
		(mono)	376 Ethylisopropylbarbituric acid		201
322 Diphenylacetanilide		180	377 *p*-Ethoxybenzamide		202
323 α-Acetyl-β-methylurea		180	378 α-Naphthoamide		202
324 α-Naphthylacetamide		181	379 2,4-Dinitrobenzamide		203
325 Maleamide		181(di)	380 D-Camphoranilide		204(mono)
326 Dimethylurea (unsym.)		182	381 p-Nitrocinnamamide		204
327 Thiourea		182	382 N-Phenylphthalimide		205
328 *p*-Tolylurea		183	383 Gallanilide		207
329 Hippuramide		183	384 Dicyanodiamide		207
330 3,5-Dinitrobenzamide		183	385 Hippuranilide		208
331 *p*-Iodoacetanilide		184	386 *o*-Coumaramide		209*d*
332 *o*-Iodobenzamide		184	387 Sebacamide		210(di)
333 *o*-Nitrocinnamamide		185	388 *p*-Iodobenzanilide		210
334 *N,N'*-Diacetyl-*o*-phenylene			389 Citramide		210-5*d*
diamine		185			(tri)
335 Citraconamide		185-7(di)	390 DL-Phenylsuccinamide		211(di)
336 *m*-Iodobenzamide		186	391 *p*-Nitrobenzanilide		211(204)
337 Mesaconanilide		186(di)	392 Protocatechucamide		212
338 Hexahydrobenzamide		186	393 Diphenamide		212(di)
339 Suberanilide		186(di)	394 *p*-Nitrophenylacetanilide		213
340 Hippuric acid		187	395 Ethylmalonamide		214(di)
341 Maleanilide		187(di)	396 *p*-Nitroacetanilide		216
342 *meso*-Tartaramide		187(di)	397 Suberamide		217(di)
343 Diethylbarbituric acid		188	398 Methylmalonamide		217(206)
344 *as*-Diphenylurea		189	399 *p*-Iodobenzamide		217
345 Aconitanilide		189(di)	400 Benzylmalonanilide		217(di)
346 *p*-Bromobenzamide	2	189	401 4-Hydroxy-2-naphthamide		217-8
347 Picramide (See Table 9B)		190	402 Glutar-*p*-toluidide		218
348 Itaconanilide		190(185)	403 Acetourea		218
349 *N,N'*-Diacetyl-*m*-phenylene			404 *s*-Di-*m*-tolylurea		218
diamine		191	405 2-Hydroxy-3-naphthamide		218
350 Mucamide		192*d*	406 Hydantoin		218
		(mono)	407 Phthalamide	10	220(di)
351 β-Naphthamide		192(195)	408 Mucamide		220(di)
352 Itaconamide		192(di)	409 Adipamide		220(di)
353 4-Nitrophthalanilide		192	410 Saccharin	7	220
354 Biuret		192*d*	411 DL-Phenylsuccinanilide		222(di)
355 *o*-Tolylurea		192	412 β-Resorcylamide		222
356 D-Camphoramide		193(di)	413 *p*-Cyanobenzamide		223
357 *p*-Chlorobenzanilide		194	414 Diethylmalonamide		224(di)
358 *p*-Coumaramide		194	415 Glutaranilide		224
359 *o*-Benzoylbenzanilide		195	416 5-Nitrosalicylanilide		224
360 *m*-Nitrocinnamamide		196	417 5-Nitrosalicylamide		225
361 D-Tartaranilide		196*d*	418 Pyrotartaramide		225(di)
362 2-Methyl-4-nitroacetanilide		196	419 Benzylmalonamide		225(di)
363 *p*-Bromobenzanilide		197	420 Malonanilide		225(di)
364 L-Malanilide(Hydroxysuccinic)		197	421 DL-Tartaramide		226
365 *p*-Nitrophenylacetamide		198	422 D-Camphoranilide		226(di)
366 *p*-Hydroxybenzanilide		198	423 Succinamide		230(di)
367 Cyanoacetanilide		198	424 Diphenanilide		230
368 Citranilide (monohyd.)		199(tri)	425 Nitroguanidine		230*d*
		(192)	426 Benzo-2,4,6-tribromoanilide		232

TABLE 8 (*continued*)

	Name of compound	Note	M.p.		Name of compound	Note	M.p.
427	Phthalimide	1	233.5*	443	Succinamide		260d(di)
			(238)	444	Theophylline		264
428	Caffeine	8	234	445	D-Tartaranilide		264d(di)
429	3,5-Dinitrobenzanilide		234	446	Fumaramide		266d(di)
430	3-Nitrophthalanilide		234(di)	447	s-Di-p-tolylurea		268
431	Terephthalanilide		234-7	448	Dimethylmalonamide		269(di)
432	DL-Tartaranilide		236	449	Isophthalamide		280
433	Carbanilide(s-Diphenylurea)		238(240)	450	Creatinine		292d
434	Muconamide		240(di)	451	Mucanilide		310
435	p-Hydroxy-N-methyl-			452	N,N'-Diaceto-p-phenylene-		
	acetanilide		240		diamine		310
436	Adipanilide		241(di)	453	Fumaranilide		314
437	2-Hydroxy-3-naphthanilide		244(249)	454	Creatine		315d
438	Gallamide		245	455	Theobromine		337s
439	Barbituric acid		245	456	Xanthine		360
440	s-Di-o-tolylurea		250	457	Uric acid	9	400d
441	Phthalanilide		253-5	458	Oxamide	11	419d(di)
442	Oxanilide		254(di)				

Notes on Amides and Ureas

1. *J. Am. Chem. Soc.* **65**, 1355 (1943). Xanthyl (primary amides). The following gives the amide and the melting point of the derivative: Acetamide, 238-40; Benzamide, 222.5-223.5; n-butyramide, 185-7; chloroacetamide, 208-9; cyanoacetamide, 222-3; n-heptamide, 154-5; isobutyramide, 210-11; isocaproamide, 159-60; isovaleramide, 182-3; p-nitrobenzamide, 231-3; palmitamide, 140-2; pelargonamide, 147.5-148.5; phenylacetamide, 194-5; phthalimide, 176-7; propionamide, 210-11; stearamide, 139-41; succinimide, 245-7; o-toluamide, 199-200.5; p-toluamide, 224-5.

2. *J. Am. Chem. Soc.* **64**, 1738 (1942). Mercuric oxide derivatives. The following gives the amide and the melting point of the mercury derivative: acetamide, 196-7; o-anisamide, 241; benzamide, 222; m-bromobenzamide, 235; p-bromobenzamide, 266; butyramide, 222-4; m-chlorobenzamide, 245; p-chlorobenzamide, 258;

propionamide, 201; o-toluamide, 196; m-toluamide, 200; p-toluamide, 260.

3. Semicarbazide is easily derivatized by use of carbonyl compounds, such as acetone.
4. It forms a picrate, m.p. 127d.
5. It forms a picrate, m.p. 148. Treated in very dilute solutions of acetic acid (50%) with xanthydrol dissolved in alcohol forms derivative, m.p. 265 (dixanthylurea).
6. On oxidation gives alloxantin, m.p. hydrate 201-3.
7. Forms xanthyl derivative, m.p. 199.
8. Aqueous soln. of compound forms with sat'd. solution of $HgCl_2$ derivative, m.p. 246 (corrected).
9. On oxidation with dilute nitric acid gives alloxan, m.p. 170d.
10. Monoamide, mp. m.p. 149.
11. M.p. determined in sealed tube.
12. Anilide crystallizes from benzene-ligroin, m.p. 132; from water as a di-hydrate, m.p. 85.

TABLE 9A

Amines (Primary and Secondary) — Liquid

| | Name of compound | Note | B.p. | Melting point of derivatives | | |
| | | | | Recommended | | Other |
				Benzene-sulfonamide	Benzamide	Phenyl-thiourea
1	Methylamine		-6	30	80	113
2	Dimethylamine		7	47	41	135
3	Ethylamine		19(16.5)	58	71	106(135)
4	Isopropylamine		33	26	100	101
5	tert-Butylamine		46		134	120
6	n-Propylamine		49	36	84	63
7	Methylisopropylamine		50		(oil)	120
8	Diethylamine		56	42	42	34
9	Allylamine		58(56)	39		98
10	sec-Butylamine		63	70	76	101
11	Isobutylamine		69	53	57	82
12	n-Butylamine		77		42	65
13	Di-isopropylamine		84(86)	94		102
14	Isoamylamine		96			69
15	n-Amylamine		104			101
16	Piperidine		106	93-4	48	69
17	Di-n-propylamine		109	51		102
18	Ethylenediamine	1	116	168(di)	244(di)(249)	
19	DL-1,2-Diaminopropane		119-20		192(di)	
20	p-Aminoazobenzene		126		211	
21	n-Hexylamine		129	96	40	77
22	Morpholine		130	118	75	136
23	Cyclohexylamine		134	89	149	148
24	1,3-Diaminopropane		136	96		
25	Diisobutylamine		139	55(57)	147(di)	
26	n-Heptylamine		155			113
27	1,4-Diaminobutane (m.p. 27) (See Table 9B)		159			75
28	Di-n-butylamine		159			
29	2-Aminoethyl alcohol (Ethanolamine)	2	171			86
30	1,5-Diaminopentane		178-80	119	135(di)	148
31	Aniline	3	183-4(184.4)	112	163	154
32	Benzylamine		184-5	88	105	156

No.	Name	B.P.				Note
33	α-Phenylethylamine (α-Aminoethylbenzene)	185(187)		120	72	
34	Diisoamylamine	187–8		185		
35	p-Fluoroaniline	188	79	63	87	
36	N-Methylaniline	196	69	116	135	
37	β-Phenylethylamine (β-Aminoethylbenzene)	198			136	
38	o-Toluidine	200	124	146		
39	m-Toluidine	203	95		89	
40	N-Ethylaniline	205	(oil)		92(104)	
41	Diamylamine	205				
42	o-Chloroaniline	207(209)	129	125	156	
43	N-Methyl-o-toluidine	208		60		
44	N-Methyl-p-toluidine	210		99		
45	2,5-Dimethylaniline (m.p. 15)	215	138	66	148	
46	2,6-Dimethylaniline	215		53	204	
47	o-Ethylaniline	216(210)		140		
48	p-Ethylaniline	216(214)		168	152	
49	2,4-Dimethylaniline	217	130	147		
50	N-Ethyl-p-toluidine	217		151		
51	o-Amino-N,N-dimethylaniline	220	72	192	153	
52	3,5-Dimethylaniline	220		39		
53	N-Ethyl-m-toluidine	221		136	104	
54	N-n-Propylaniline	222	54	72		
55	2,3-Dimethylaniline	222		189		
56	2-Chloro-4-methylaniline	223	110	137	136	
57	o-Anisidine (m.p. 5–6)	225 v.s.	89	60		
58	Cumidine (p-Isopropylaniline)	225	132	162		
59	α-Methyl-α-phenylhydrazine	227(131/35 mm)	137	153	193	
60	2,4,6-Trimethylaniline (Mesidine)	229(232–3)		204	137	
61	o-Phenetidine	229	102	104	124(116)	
62	m-Chloroaniline	230(236)	121	120		4
63	Tetrahydroisoquinoline	233		129		
64	2-Bromo-4-methylaniline (m.p. 26) (See Table 9B)	240	154			
65	2-Amino-p-cymene	241		102		
66	N-Butylaniline	241.6		56		
67	Phenylhydrazine (m.p. 19)	243	148	168	172	
68	p-Phenetidine (m.p. 3–4)	248(254)	143	173	136	
69	m-Phenetidine	248		103	138	
70	Tetrahydroquinoline (m.p. 20)	250	67	75		5

(continued)

TABLE 9A (*continued*)

	Name of compound	Note	B.p.	Melting point of derivatives: Recommended: Benzene-sulfonamide	Melting point of derivatives: Recommended: Benzamide	Other: Phenyl-thiourea
71	*o*-Aminoacetophenone (m.p. 20) (See Table 21A)		250-2*d*		98	
72	*m*-Bromoaniline (m.p. 18)		251		120(136)	143
73	3-Bromo-2-methylaniline		253-5		176-7	
74	Methyl anthranilate (m.p. 24-5)		260*d*(255)	107	100	
75	*p*-Amino-diethylaniline		261		172	
76	Ethyl anthranilate (m.p. 13)		265*d*	92	98	
77	Ethyl-*m*-aminobenzoate		294		148	
78	*N*-Methyl-1-naphthylamine		294(296)		121	
79	Dibenzylamine		300	68	112	
80	*N*-Methyl-2-naphthylamine	6	315	107	84	

TABLE 9B

Amines (Primary and Secondary)—Solid

	Name of compound	Note	M.p.	Melting point of derivatives: Recommended: Benzene-sulfonamide	Melting point of derivatives: Recommended: Benzamide	Other: Phenyl-thiourea
1	Methyl anthranilate (b.p. 260*d*) (See Table 9A)		24-5			
2	3-Bromo-4-methylaniline (b.p. 254-7)		25-6	117	132	
3	2-Bromo-4-methylaniline (b.p. 240)		26		149	154
4	1,4-Diaminobutane (b.p. 159)		27		177(di)	
5	2,2′-Iminodiethanol (b.p. 270)		28	130	99	
6	4-Chloro-2-methylaniline (b.p. 241)		29	125	142	
7	3-Aminobiphenyl	7	30			
8	*o*-Bromoaniline (b.p. 250)		32		116	146(161)
9	*as*-Diphenylhydrazine	8	34		192	

No.	Amine		M.P.			
10	p-Amino-N-methylaniline (b.p. 258)		36		165	
11	N-Benzylaniline (b.p. 298)		37	119	107	103
12	N-Methyltribroaniline		39	101		
13	2-Chloro-4,6-dimethylaniline		40		148	
14	p-Amino-N,N-dimethylaniline (b.p. 262)		41		228	
15	2-Amino-6-methylpyridine (b.p. 209)		41		90	
16	1,6-Diamino hexane		42	154(di)	155(di)	141
17	4-(Methylamino)-2-nitrotoluene		45(57)		278	
18	2,4'-Diaminobiphenyl		45			
19	p-Toluidine (b.p. 200)		45(43)		158	
20	3,4-Dimethylaniline (b.p. 224)		48	120		
21	2-Aminobiphenyl (b.p. 299)		49	118	102	
22	2-Bromo-4,6-dimethylaniline		49-50		186	
23	1-Naphthylamine		50	167	160	165
24	2,5-Dichloroaniline		50		120	166
25	Indole (b.p. 253)	9	52		68	
26	4-Aminobiphenyl		53	254	230	
27	Diphenylamine	10	53-4	124	180N	152
28	p-Anisidine (b.p. 240)		58	95	154(157)	157(171)
29	4-Bromo-2-methylaniline (b.p. 240)		59		115	
30	1-Chloro-2-naphthylamine		59(60)		98-9	
31	2-Aminopyridine (b.p. 204)		60(57.5)			
32	o-Iodoaniline		61(58)	165(di)	139	
33	p-Iodoaniline		62		222	153
34	3-Chloro-1-naphthylamine		62		162	
35	N-Phenyl-1-naphthylamine		62		152	
36	m-Phenylenediamine	11	63	194	240(di)	
37	p-Amino-N-methylacetanilide		63			
38	2,4-Dichloroaniline		63	128	117	
39	3-Aminopyridine (b.p. 250-2)		64		119	
40	2,5-Diaminotoluene (b.p. 273-4)		64	147(mono)(2N)	307	
41	p-Tolylhydrazine (b.p. 240-4d)	12	65		146N	
42	p-Bromoaniline (b.p. 245)		66	134	204	148
43	1,8-Diaminonaphthalene		66.5		311-2(di)	
44	m-Nitro-N-methylaniline		68		105	
45	Pseudocumidine		68	83	167	
46	2,4-Dimethyl-6-nitroaniline		70	136	185	
47	o-Nitroaniline v.s.		71	104	110(98)	
48	p-Chloroaniline		72	122	192	152
49	4-Nitromesidine		75	163	169	
50	4-Aminodiphenylamine (anhyd.)		75		203	

(continued)

TABLE 9B (*continued*)

| | | | Melting point of derivatives | | Other |
| | | | Recommended | | |
Name of compound	Note	M.p.	Benzene-sulfonamide	Benzamide	Phenyl-thiourea
51 4-Methyl-3-nitroaniline		78	160	172	171
52 2,4,6-Trichloroaniline	13	78		174	
53 2,4-Diaminophenol		78–80		253(di)	
54 N-p-Tolyl-1-naphthylamine		79		140	
55 Di-p-tolylamine (b.p. 330)		79		125	171
56 2,4-Dibromoaniline		79		134	
57 2-Aminophenylamine		79–80		136	
58 4-Chloro-3-methylaniline		83		119.5	
59 2-Amino-4-chloroanisole		84		78	
60 3,4-Diaminotoluene (b.p. 265)		89–90	178–9(di)	263–4(di)	
61 Ethyl-p-aminobenzoate	14	90		148	
62 2,4-Diaminochlorobenzene		91		178(di)	
63 2-Methyl-3-nitroaniline (6-Nitro-o-toluidine)		92		168	
64 2,6-Dibromo-4-chloroaniline		93		194	
65 2,4-Dibromo-6-chloroaniline		95		192	
66 Semicarbazide		96		225	
67 p-Nitro-N-ethylaniline		96		98	
68 2,4-Di-iodoaniline		96		181	
69 2-Methyl-6-nitroaniline		97		167	
70 8-Nitro-1-naphthylamine		97	194		
71 1,2-Diaminonaphthalene	15	98	215d,N	291(di)	
72 2-Amino-4-methylpyridine		98		114	
73 m-Aminoacetophenone (See Table 21B)		99			
74 2,4-Diaminotoluene	16	99	192(di)	224(di)	
75 o-Phenylenediamine		102	185	301(di)	
76 N-p-Tolyl-2-naphthylamine		103		139	
77 Piperazine (b.p. 140)	17	104	282(di)	196(di)75(mono)	
78 4-Aminoacetophenone (b.p. 294) (See Table 21B)		106	128	205	
79 p-Bromophenylhydrazine (Other: Acetophenone, 112)		106			
80 2-Methyl-5-nitroaniline (4-Nitro-o-toluidine)		107	172	186	

No.	Name		M.P.			
81	N-Phenyl-2-naphthylamine		108		148(136)	129
82	2-Naphthylamine		112	102	162	
83	4-Amino-3-nitrophenetole		113	72		
84	m-Nitroaniline		114	136	155(158)	160
85	4-Bromo-8-nitro-1-naphthylamine	23	116			
86	4-Methyl-2-nitroaniline		117(115)	102(99)	148(143)	
87	2-Amino-4-nitroanisole		118		160-1	
88	5-Nitro-1-naphthylamine		119	183		
89	1,4-Diaminonaphthylene	17	120		280(di)	
90	2,4,6-Tribromoaniline	13	122(119)		198	
91	m-Hydroxyaniline (m-Aminophenol)	18	122		174N	156
92	2,4-Dimethyl-5-nitroaniline		123	149	200	
93	4-Aminobenzophenone		124		152	
94	4-Aminoazobenzene		125(127)		211	
95	1-Nitro-2-naphthylamine		126	156	168	
96	Benzidine		127	232(di)	352(di)203-5(mono)	
97	4-Amino-3-nitroanisole		129		140	
98	o-Tolidine	19	129(123)			
99	2-Methyl-4-nitroaniline (5-Nitro-o-toluidine)		130	158	265(di)N	
100	o-Dianisidine		137-8		236(di)	
101	2-Amino-5-nitroanisole		139-40	181	149-50	
102	p-Phenylenediamine (b.p. 267)		142	247(di)	300(di)128(mono)	
103	5-Nitro-2-naphthylamine		144		182	
104	2-Nitro-1-naphthylamine		144		175	
105	2,5-Dimethyl-4-nitroaniline		144-5	162		
106	o-Aminobenzoic acid (See Tables 4 and 5B)		147	214	182	
107	p-Nitroaniline		147	139	199	
108	p-Nitro-N-methylaniline		152	121	112	
109	p-Nitrophenylhydrazine		157d		193	
110	2,7-Diaminonaphthalene		166		267(di)	
111	4-Methyl-2,6-dinitroaniline		168(172)		186	
112	Picramic acid		169		300	
113	o-Hydroxyaniline (o-Aminophenol)	20	174	141	167N	146
114	m-Aminobenzoic acid (See Tables 4 and 5B)		174			
115	5-Amino-o-cresol		175		194(ON-Di)	
116	2,4-Dinitroaniline		180		202(220)	
117	p-Hydroxyaniline (p-Aminophenol)	21	184s	125	216-7N	150

(continued)

TABLE 9B (continued)

Name of compound	Note	M.p.	Melting point of derivatives		Other
			Recommended		Phenyl-thiourea
			Benzene-sulfonamide	Benzamide	
118 p-Aminobenzoic acid (See Tables 4 and 5B)		188	212	278	
119 2,4,6-Trinitroaniline	22	190	211	196	
120 1-Amino-4-nitronaphthalene		195	158(173)	224	
121 2,4-Dinitrophenylhydrazine		199-200		206-7	
122 1,8-Dinitro-2-naphthylamine		226		352(di)	
123 -4,4'-Diaminostilbene (trans)		231		252	
124 2,4-Dinitro-1-naphthylamine		242		98	
125 Carbazole		246		228	
126 2-Aminoanthraquinone		305-8(302)	271		

Notes on Primary and Secondary Amines

1. Hydrate with 1 H_2O, m.p. 10.
2. 1-Naphthyl urea, m.p. 186; picrate, m.p. 160.
3. Hydrochloride, m.p. 198.
4. 1,2,3,4-Tetrahydroisoquinoline boils at 233; the 5,6,7,8- isomer boils at 218.
5. 1,2,3,4-Tetrahydroquinoline boils at 250; the 5,6,7,8- isomer boils at 222 and forms a picrate, m.p. 157.
6. With HNO_2 yields N-nitroso derivative, m.p. 88.
7. Acetamide, m.p. 148.
8. Acetamide, m.p. 184.
9. N-nitroso derivative, m.p. 171.
10. Yields N-nitroso derivative readily, m.p. 67. The benzoyl derivative has one modification that melts at 107-8; this form melts, resolidifies, is transformed into the stable modification between 130 and 140, and then melts at 180.
11. The di-substituted derivatives are obtained with difficulty.
12. The 1-N-benzamide melts at 68-70; the 2-N-benzamide melts at 146.
13. The acetyl derivative is easily prepared.
14. The 4-p-toluenesulfonamide melts at 140. The 3-N-acetamide melts at 95 and the 4-N-acetamide melts at 131-2; 3-N-benzamide, m.p. 158, and 4-N-benzamide, m.p. 193-4; 3-N-benzenesulfonamide, m.p. 134-5 and 4-N-benzenesulfonamide, m.p. 146-7.
15. 1-N-benzenesulfonamide.
16. Both of the mono-benzenesulfonamides (2-N- and 4-N-) melt at 138. 2-N-acetamide, m.p. 140, and 4-N-acetamide, m.p. 161.5; 4-N-benzamide, m.p. 142; 4-N-p-toluenesulfonamide, m.p. 160.
17. 1-N-benzamide, m.p. 186; 1-N-p-toluenesulfonamide, m.p. 187-8.
18. Diacetyl derivative, m.p. 101; N-benzoyl derivative, m.p. 174; O-benzoyl derivative, m.p. 153.
19. Mono-acetamide-H_2O, m.p. 103; mono-benzamide, m.p. 198-200.
20. The N-benzoyl derivative melts at 167; the O-benzoyl derivative melts at 185.
21. The N-benzoyl derivative is reported with m.p. of 216-7, 205, and 227; the O-benzoyl derivative melts at 235; the O-N-dibenzoyl derivative is reported at 234. Mono-acetyl derivative, m.p. 168; diacetyl, 150d.
22. Forms with naphthalene in acetic acid, addition compound, m.p. 168.
23. Acetamide, m.p. 202

TABLE 10A

Amines (Tertiary)—Liquid

No.	Name of compound	Note	B.p.	Picrate	Quaternary methyl iodide
1	Trimethylamine	8	3	216(225)	230d
2	Triethylamine	8	89	173	117
3	Pyridine	8	116	167	230
4	2-Methylpyridine (α-Picoline)	8	129	169	230
5	β-Dimethylaminoethyl alcohol		135	96-7	
6	2,6-Dimethylpyridine		143	168	233
7	3-Methylpyridine (β-Picoline)	1	143	150	92
8	4-Methylpyridine (γ-Picoline)	8	143.1	167	152
9	3-Chloropyridine		149	135	
10	Tri-n-propylamine		156.5	116	207-8
11	2,4-Dimethylpyridine (Lutidine)	8	159	183	113
12	2,5-Dimethylpyridine		160	169	
13	2,3-Dimethylpyridine		164	188	
14	3-Bromopyridine		170		165
15	D-Conine		170(167)	75	
16	2,4,6-Trimethylpyridine (2,4,6-Collidine)		172	156	
17	N,N-Dimethylbenzylamine		181	93	179
18	N,N-Dimethyl-o-toluidine		185	122	210
19	N,N-Dimethylaniline (m.p. 1.5-2.5)		193(196)	163	228d (220)
20	3-Ethyl-4-methylpyridine		195-6	148-50	
21	N-Ethyl-N-methylaniline		201(209)	134-5(129)	125
22	N,N-Diethyl-o-toluidine		206(210)	180	224d
23	N,N-Dimethyl-o-chloroaniline		207 (85-7/10 mm)	132	152
24	N,N-Dimethyl-p-toluidine		210	129	219
25	Tributylamine	8	211(216.5) (93-5/15 mm)	106	180
26	N,N-Dimethyl-m-toluidine		212	131	177
27	N,N-Diethylaniline		218(216) (93-5/10 mm)	142	102
28	N,N-Diethyl-p-toluidine		229	110	184
29	Quinoline (m.p. −15.6)	2,8	239	203	133N
30	Isoquinoline (m.p. 26.5) (See Table 10B)	3	243		
31	DL-Nicotine		243	218	219
32	N,N-Dipropylaniline		245	261	156
33	Triisoamylamine		245 (111-3/10 mm)	125	
34	Quinaldine (2-Methylquinoline)	8	247	191(195)	195
35	8-Methylquinoline		248	200	
36	6-Methylquinoline		258	229	219(216)
37	3-Bromo-N,N-dimethylaniline		259	135	
38	Lepidine (4-Methylquinoline)		263	210-1	173-4
39	2,4-Dimethylquinoline		264	194	264(252)
40	2-Phenylpyridine		268-9	175	
41	N,N-Di-n-butylaniline		271	125	
42	N,N-Dimethyl-1-naphthylamine		273	145	
43	6-Bromoquinoline (m.p. 19)		278	217	273
44	6-Methoxyquinoline (m.p. 20(28))		286	300d	236
45	N-Benzyl-N-methylaniline		306	127	164

TABLE 10B
Amines (Tertiary)—Solid

Name of compound	Note	M.p.	Picrate	Quaternary methyl iodide
1 6-Bromoquinoline (b.p. 278) (See Table 10A)		19		
2 6-Methoxyquinoline (b.p. 286) (See Table 10A)		20(28)		
3 Isoquinoline (b.p. 243)	3	26.5	222	159
4 N-Ethyl-N-benzylaniline		34	120(112)	
5 2-Chloroquinoline (b.p. 267)		38	122	
6 7-Methylquinoline (b.p. 252)		39	237	
7 6-Chloroquinoline (b.p. 262)		41		248
8 N,N-Dimethyl-2-naphthylamine (b.p. 305)		47	206	
9 2-Bromoquinoline		49		210
10 8-Methoxyquinoline (b.p. 283)		50	143	160
11 2,6-Dimethylquinoline		60	178(186-7)	237
12 m-Nitro-N,N-dimethylaniline		60	119	205
13 α,α'-Dipyridyl		69	159	
14 N,N-Dibenzylaniline		72(70)	131d	135
15 p-Dimethylaminobenzaldehyde (See Table 7B)		74		
16 8-Hydroxyquinoline		75	204	143d
17 p-Dinitrosodimethylaniline		85	140	
18 Tribenzylamine	5,8	91	190	184
19 4,4'-Tetramethyldiaminodiphenylmethane (b.p. 390)		91	185(mono) 178(di)	
20 Acridine — Other: Acid oxalate, 162		111	208	224
21 3,5-Dibromopyridine (b.p. 222)		112		274
22 Antipyrine		113	188	
23 p-Dimethylaminoazobenzene		117		174
24 Methyleneaminoacetonitrile (b.p. 210)	6	129	127	
25 6-Nitroquinoline		154		245
26 6-Nitroquinaldine		164(174)		214
27 4,4'-Bis(dimethylamino)-benzophenone (See Table 21B)		174	156	105
28 6-Hydroxyquinoline		193	236	
29 Hexamethylenetetramine — Other: Dibromide, 198-200	7	(280s)	179d(157)	190(204)

Notes on Tertiary Amines

1. Oxidizes to nicotinic acid, m.p. 228.
2. Yields ethiodide, m.p. 159.
3. Yields ethiodide, m.p. 148.
4. B · HCl, m.p. 177.
5. Ethiodide, m.p. 190.
6. Acid hydrolysis gives glycine.
7. Reacts with dilute acids to yield formaldehyde and the ammonium salt of the acid used.
8. β-Resorcylic acid readily forms salts with many tertiary amines. The following are the melting points (corrected) of these salts for certain amines [Anal. Chem. 23, 1042 (1951)].

   triethyl, 120
   tri-n-butyl, 121
   tribenzyl, 141
   pyridine, 145
   γ-picoline, 125
   2,4-lutidine, 143
   quinoline, 128
   quinaldine, 145
   α-picoline, 141

TABLE 11
Carbohydrates[a]

Name of compound	Note	M.p.	$[\alpha]_D^{20}$	Derivatives[b] Azoates M.p.	$[\alpha]_{5438, CHCl_3}^{25}$
1 β-Melibiose dihydrate ($C_{12}H_{22}O_{11}$ + $2H_2O$)		82–85	+129.5	280	172
2 2-Deoxy-D-ribose		90	+2.88 → 2.13		
3 D-Ribose		95	−23.7(−21.5)		
4 β-Maltose hydrate ($C_{12}H_{22}O_{11}$ + H_2O)	(a)	102.5	+136	275	+2
5 β-D-Fructose		102–4(A)	−92.0	125	−440
6 Glucosamine		105–10	+48(+44)		
7 α-D-Lyxose		106–7(101)	+5.5 → −14.0		
8 Raffinose	(b)	119	+123.1	145	+146
9 β-L-Rhamnose	(c)	122–6(Ac)	+9.1		
10 D-Mannose	(d)	133	+14.5	157	+244
11 D-Xylose	(e)	143*	+19.1(W4½)		
12 α-L-Fucose		145	−75.9		
13 α-D-Glucose (anh.)		146	+52.7	266	+223
14 β-D-Glucose		148–50	+18.7 → 52.7		
15 Melezitose dihydrate		155	+88.8	130(sinters)	+188
16 β-D-Arabinose		158–9	−175 → −105		
17 β-D-Galacturonic acid		160	+55.3		
18 β-L-Arabinose		160	+104.5	262	+755
19 DL-Galactose		163(144)			
20 L-Sorbose		165(159–60)	−43.7		
21 α-D-Galactose		165.5*	+81.7	276	+436
22 β-D-Glucuronic acid		165*	+36.3		
23 β-Inulin			−40		
24 Sucrose	(f)	185	+66.5	125	+35
25 L-Ascorbic acid		190(187)	+49		
26 α-Lactose monohydrate		202	+52.3		
27 α,α-Trehalose		203	+197.1	134.5	+210
28 α-Lactose (anh.)		223	+53.6	288	+320
29 β-Lactose (anh.)		252	+35.2		
30 β-Cellobiose		225	+189	273	105
31 Starch (sol.)		dec.			
32 Glycogen		240d			

(continued)

TABLE 11 (continued)

ᵃSpecial symbols used in this table:

(A) = ethyl alcohol as solvent	(P-A) = 50-50 pyridine-alcohol as solvent
(W) = water as solvent	(M) = methyl alcohol as solvent
(P) = pyridine as solvent	(Aa) = acetic acid as solvent

(Ac) = acetone as solvent
(Chl) = chloroform

Where a letter followed by a number appears within parentheses, the letter stands for the solvent and the number signifies the number of hours required to attain equilibrium in mutarotation. Thus, for the phenylhydrazone of D-Mannose, +33.8 (P 56) means that pyridine is used as a solvent and that equilibrium is attained in 56 hours. The initial value, not given in the table, is $[\alpha]_D = +26.3$. This value changes to −6.3 in 9 hours. After 56 hours it again becomes positive and attains a state of equilibrium with a value of $[\alpha]_D = +33.8$; which is the recorded value.

Where the solvent (for rotation or crystallization) is water, the (W) is omitted, unless admixed with alcohol or where for some special reason it is deemed best to specify it.

ᵇFor other derivatives, see Notes under number of compound.

Notes on Carbohydrates

(a) According to the literature maltose, when carefully heated to 160 (in vacuo) loses water and is resolved into the anhydride maltosan ($C_{12}H_{20}O_{10}$), a brown amorphous powder, which does not regenerate maltose when boiled with water and melts at 145-50. β-Maltose hydrate is stated to melt at 160-5 and at 102-3; the anhydrous form at 108. Beilstein (XXI, 386) gives the following melting points for hydrate: 110-25d, 106-12, 115 115-25, 125-30; for the anhydrous form it gives no melting point but only the temp. of dehydration.

(b) Raffinose crystallizes with five molecules of water and is freed therefrom by very slow heating.

(c) α-L-Rhamnose crystallizes with one molecule of water, and by cautious heating for several days is converted into anhydrous β-L-rhamnose, which can be recrystallized from acetone.

(d) D-Mannose does not occur in nature as such, but only in the form of the polymerized anhydride mannan, from which it may be obtained by hydrolysis. It can be made by oxidation of D-mannitol.

(e) D-Xylose, formerly called L-xylose, is best identified microscopically by conversion into "Cd xylonobromide," $Cd(C_5H_9O_6)_2 \cdot CdBr_2 \cdot 2H_2O$, by means of Br_2 and $CdCO_3$ (Bertrand, Bull. soc. chim. (3) 5, 556; 7, 501; 15, 592; cf. Widtsoe, Ber. 33, 136); or as the brucine salt of D-xylonic acid (Neuberg, Ber. 35, 1470-3).

(f) Sucrose, when crystallized from methyl alcohol, m 169-70, from ethyl alcohol m 179-80, from ethyl alcohol and water m 184-5. [Heldermann, Z. physik. Chem. 130, 396 (1927) and Pictet and Vogel, Helv. Chim. Acta 11, 436 (1928)].

Notes on Other Derivatives of Carbohydrates

Other derivatives with their melting points and specific rotation are given as follows: The number corresponds to the number of the compound on Table 11. Following the number is the name of the derivative, the melting point, and the specific rotation in parenthesis: Thus, (3): p-bromophenylhydrazone 170* (−5.7A) signifies the p-bromophenylhydrazone of D-ribose which has a melting point of 170 and a specific rotation $[\alpha]_D^{20°} = -5.7$ in ethanol as a solvent.

(1): Phenylosazone, 176-8 (+43.2P); p-Bromophenylosazone, 181-2; Octaacetyl, 177.5 (+102.5); Oxime, 186; 2-Phenylhydrazone, 145, also 160.

(2): Benzylphenylhydrazone, 127-9.

(3): Phenylosazone, 160 and 164; p-Bromophenylhydrazone, 170* (−5.7A); p-Bromophenylosazone, 180-5; Dimethylmethane dimethyldihydrazone, 141-2.

(4): Phenylosazone, 208; Semicarbazone, 213d (+80W); β-Naphthylhydrazone, 176 (+10.6M).

(5): p-Nitrophenylhydrazone, 180-1 and 176 (+16P-A); o-Nitrophenylhydrazone, 156-7* (+31P-A); Phenylosazone, 210; Pentaacetyl, α-form, 70 (+34.7chl.), β-form, 108-9 (−120.5chl.)

(6): Phenylosazone, 210; N-Acetyl, 190; Oxime, 127; Phenylurea, 210; Semicarbazone, 165.

(7): Phenylosazone, 164; p-Nitrophenylhydrazone, 172; p-Bromophenylhydrazone, 162 and 156-7; Benzylphenylhydrazone, 116 and 128 (+26.4A).

(8): α-Methylphenylhydrazone, 190*; Tritrityl, 130.

(9): p-Nitrophenylhydrazone, 190-1 (+21.4); p-Bromophenylhydrazone, 222; α-Methylphenylhydrazone, 124; Phenylosazone, 222 and 185 (+94P).

(10): p-Bromophenylhydrazone, 208; Methylphenylhydrazone, 181 (+8.6M); Nitrophenylhydrazone, 195A and 203 (+56.0P-A); Phenylhydrazone, 199 (+33.8P56); Phenylosazone (Glucosazone), 210.

(11): β-Naphthylhydrazone, 124*; m-Nitrophenylhydrazone, 163A; Phenylosazone, 164* (A-W); Phenylosotriazole, 89*; Benzylphenylhydrazone, 95 (−33A); Methylphenylhydrazone, 108.

(12): p-Bromophenylhydrazone, 181-4 (A-W); p-Nitrophenylhydrazone, 210-1; Phenylhydrazone, 170; p-Toluene sulfonylhydrazone, 174 (−17.0P); Methylphenylhydrazone, 174 (α_D^{19} 17.0P); Oxime, 188-9.

(13): β-Naphthylhydrazone, 178* (+11A); p-Nitrophenylhydrazone, 190-[196] (+21.5P-M); α-Penta-acetate, 132; Phenylosazone, 210[211W] (−1.5P-A); Phenylosotriazole, 195-6* (−81.6P); p-Bromophenylhydrazone, 164-6 (−43.6 → +18.9P); 2,4-Dinitrophenylosazone, 256-7; p-Nitrophenylosazone, 257.

(14): Phenylosazone, 210; 2,4-Dinitrophenylosazone, 256-7; p-Nitrophenylosazone, 257; β-Penta-acetate, 132.

(15): Hendeca methyl ether, $[\alpha]_D$ + 114(CH₃OH).

(16): Phenylosazone, 162-3 and 160; Benzylphenylhydrazone, 173; Oxime, 136.

(17): p-Bromophenylhydrazone, 151d (+11.5M); Brucine salt, 180; Phenylhydrazone, 140.

(18): Benzylphenylhydrazone, 174* (−11.5M); Cinchonine salt, 178 (+139W) Diphenylhydrazone, 204-5* (+14.9P); Methylphenylhydrazone, 165*; β-Naphthylphenylhydrazone, 177(A); p-Nitrophenylhydrazone, 186(W); Oxime, 139 (+13.3W); Phenylhydrazone, 153 (+7.95P); Phenylosazone, 166; Semicarbazone, 190 (+23.8W); p-Bromophenylhydrazone, 155; Tetra-acetyl, 94-6 (α-form ether), 86 (β-form W).

(19): Phenylosazone, 206; Methylphenylhydrazone, 183; Phenylhydrazone, 158-60.

(20): p-Bromophenylhydrazone, 181; Phenylosotriazole, 159*; Phenylosazone 156 and 168; o-Nitrophenylosazone, 211-2; Penta-acetyl, 97 (+2.9chl.).

(21): α-Methylphenylhydrazone, 190-1* (inactive); Mucic Acid, 214d; p-Nitrophenylhydrazone, 197W* (+45.6); α-Penta-acetate, 95; β-Penta-acetate, 142; Phenylosazone, 182-4 and 201 (+0.34PA-24); Phenylosotriazole, 111*; o-Tolylhydrazone 176* (95%A); m-Tolylhydrazone 154*; p-Bromophenylhydrazone, 168; Benzylphenylhydrazone,

157; Diphenylhydrazone, 157.

(22): Cinchonine salt, 204* (+139.9W); p-Nitrophenylhydrazone, 225; Semicarbazone, 188; Thiosemicarbazone, 223.

(23): Triacetate, 150-60 (−45.5Ac).

(24): Octa-acetate, 75 and 69 (+59.6chl.); Trityl, 127.9 (+44.3).

(25): p-Nitrophenylhydrazone, 262(di); p-Bromophenylhydrazone, 170(di); Diphenylhydrazone, 187; Di-2,4-dinitrophenylhydrazone, 282.

(26): Mucic acid, 214d; p-Nitrophenylosazone, 258; Octa-acetate, 100; Phenylosazone, 210 and 200 (+7.9M9); Phenylosotriazole, 181*.

(27): Octa-acetate, 100-2* (+162chl.); Hexa-acetate, 93-96 ($[\alpha]_D^{19}$ +158.3chl.); Octa-Nitrate, 124.

(28): Phenylosazone, 200 and 210-2.

(29): p-Nitrophenylosazone, 258d; 2-Naphthylhydrazone, 203.

(30): Octa-Acetate (α) 229.5 (+42.0chl.); Octa-Acetate (β), 192 and 202 (−14.5chl.); Oxime, 123-5; Phenylhydrazone, 90; Phenylosazone, 200d(−6.5P-A); Phenylosotriazole, 165*; Semicarbazone, 183-5 (−5.2W); Thiosemicarbazone, 170d.

(32): Dibenzoate (+179.8chl.); Triacetate, (+170chl.).

TABLE 12A

Esters—Liquid[a]

	Name of compound	B.p.	$D_4^{20\,b}$	$n_D^{20\,b}$
1	Methyl formate	31.5	0.97421	1.344
2	Ethyl formate	54.2	0.92247	1.35975
3	Methyl acetate	57.1	0.9274	1.36170
4	Ethyl trifluoroacetate	62	1.1952	$1.3093\text{-}n^{15}$
5	Isopropyl formate	68(71)	0.8728	1.368
6	Butyl nitrate	75	0.911D	
7	Methyl chloroformate	75	1.2231	1.38675
8	Ethyl acetate	77.15	0.90055	1.372
9	Methyl propionate	79.9	0.9151	1.3779
10	Methyl acrylate	80.3	$0.961\text{-}D^{19.2}$	1.3984
11	n-Propyl formate	81	0.918	1.37789
12	Allyl formate	83	0.946	
13	Ethyl nitrate	87	1.106	1.3853
14	Isopropyl acetate	91(88)	0.872	1.377
15	Methyl isobutyrate	92.6	0.8906	1.3840
16	Ethyl chloroformate	93	1.13519	1.3974
17	sec-Butyl formate	97	0.884	1.384
18	tert-Butyl acetate	97.8	0.867	1.386
19	Isobutyl formate	98	0.88535	1.38568
20	Isoamyl nitrite	99	$0.880\text{-}D_4^{15}$	$1.38708\text{-}n_D^{21}$
21	Ethyl propionate	99.1	0.8889	1.3853
22	n-Propyl acetate	101	0.8834	1.38468
23	Methyl butyrate	102.3	0.8982	1.3879
24	Allyl acetate	104	0.9276	1.40488
25	n-Butyl formate	106.6	0.8885	1.38940
26	Ethyl isobutyrate	110	0.86930	1.3903
27	sec-Butyl acetate	112	0.872	1.3865
28	n-Propyl chloroformate	115	1.0901	1.40350
29	Methyl isovalerate	116.7	0.8808	1.3900
30	Isobutyl acetate	117.2	0.8747	1.39008
31	Ethyl butyrate	121.6	0.87917	1.40002
32	n-Propyl propionate	122	0.8809	1.39325
33	Isoamyl formate	123	0.8773	1.39756
34	n-Butyl acetate	126.1	0.881	1.3947
35	Diethyl carbonate	126.8	0.9752	1.3852
36	Methyl valerate	127.7	0.885	1.397
37	Isopropyl n-butyrate	128	$0.8787\text{-}D^0$	
38	Isobutyl chloroformate	130	$1.053\text{-}D^{15}$	$1.40711\text{-}n^{17.9}$He
39	Methyl (mono) chloroacetate	130(132)	1.238	1.4221
40	n-Amyl formate	132	0.885	1.39916
41	Ethyl isovalerate	134.7	0.86565	1.4009
42	Isobutyl propionate	137	$0.8876\text{-}D_4^0$	1.3975
43	Methyl pyruvate	138	1.154	
44	Ethyl crotonate	138	0.91752	1.42524
45	Isoamyl acetate	142	0.8674	1.40034
46	n-Propyl-n-butyrate	143	0.872	1.4005
47	Ethylene glycol monomethyl ether acetate ("Methylcellosolve acetate")	144	1.088	
48	DL-Methyl lactate	144.8	1.0931	1.4144
49	Ethyl (mono)chloroacetate	145	1.158	1.42274

[a] Many esters not listed in this table may be found as "derivatives" in Tables 5A, 5B, 6A, and 6B; sulfonate and sulfinate esters are not included and for references on these consult the References of Chapter 11.

[b] A small percentage of impurity in the compound may markedly affect the values given.

TABLE 12A (*continued*)

	Name of compound	B.p.	$D_4^{20\ b}$	$n_D^{20\ b}$
50	β-Chloroethyl acetate	145	1.178	1.4234
51	Ethyl valerate	145.5	0.8739	1.40094
52	*n*-Butyl propionate	146.8	0.895	1.401
53	Ethyl α-chloropropionate	146	1.087	
54	Benzyl chloroacetate	147	$1.2223\text{-}D_4^4$	$1.5246\text{-}n_D^{18}$
55	*n*-Amyl acetate	148.8	0.8756	1.4031
56	Isobutyl isobutyrate	149(147)	$0.8749\text{-}D_4^0$	1.3999
57	Methyl caproate	151.2	0.88464	1.405
58	Ethyl lactate	154.5	1.030	1.410
59	Ethylene glycol monoethyl ether acetate ("Cellosolve acetate")	156	0.976	
60	Ethyl dichloroacetate	158	1.2821	1.43860
61	Ethyl (mono)bromoacetate	159	1.506	1.451
62	Ethyl glycolate	160	1.082	
63	Isoamyl propionate	160	0.870	1.4065
64	*n*-Butyl *n*-butyrate	165(167)	0.869	1.406
65	*n*-Propyl *n*-valerate	167(166)	0.870	1.4065
66	Ethyl caproate	167.9	0.8710	1.40727
67	Ethyl trichloroacetate	168	1.380	1.450
68	Methyl acetoacetate	170	1.077	1.41964
69	Methyl enanthate (Methyl heptoate)	173.8	0.88011	1.412
70	Cyclohexyl acetate	175	0.970	1.442
71	*n*-Butyl chloroacetate	175	$1.081\text{-}D_4^{15}$	
72	Methyl methylacetoacetate	177	1.030	1.418
73	Isoamyl *n*-butyrate	178	0.864	1.411
74	Ethyl β-bromopropionate	179	1.425	
75	Ethyl acetoacetate	181	1.025	1.41976
76	Ethyl methylacetoacetate	181	1.006	1.419
77	Dimethyl malonate	181.5	1.1539	1.41398
78	*n*-Amyl *n*-butyrate	185.0	0.866	1.412
79	Diethyl oxalate	185.4	1.0785	1.41043
80	*n*-Propyl *n*-caproate	186	0.867	1.417
81	Butyl valerate	186	0.885	
82	Dimethyl sulfate	188	$1.3348\text{-}D^{15}$	1.3874
83	Butyl lactate	188	$0.984\text{-}D_{20}^{20}$	1.4216
84	Ethyl *n*-heptoate	189	0.888	1.413
85	Diisopropyl oxalate	189	1.00097	1.4100
86	Isoamyl isovalerate	190	0.870	$1.41300\text{-}n_D^{18.7}$
87	Ethylene glycol diacetate	190.2	1.1040	1.4150
88	Tetrahydrofurfuryl acetate	194	1.061	1.438
89	Methyl caprylate	194.6	0.878	1.417
90	Dimethyl succinate	196.0	1.1192	1.41965
91	Phenyl acetate	196.7	1.078	1.503
92	Diethyl malonate	199.3	1.05513	1.41618
93	Methyl benzoate	199.5	1.089	1.5164
94	Methyl cyanoacetate	200	$1.0962\text{-}D_4^{25}$	
95	Benzyl formate	203	1.080	1.51537
96	γ-Butyrolacetone	203-4	$1.1441\text{-}D^0$	
97	Ethyl levulinate	205.8	1.01114	1.42288
98	*n*-Amyl valerate	207.4	$0.8825\text{-}D_4^0$	$1.4181\text{-}n_D^{15}$
99	*o*-Tolyl acetate (*o*-Cresyl acetate)	208	1.048	
100	Ethyl caprylate	208.5	0.8667	1.41775
101	Trimethylene glycol diacetate	210	1.069	
102	Phenyl propionate (m.p. 20)	211	1.050	
103	Ethylene glycol dipropionate	212	$1.045\text{-}D_4^{25}$	

(*continued*)

TABLE 12A (*continued*)

	Name of compound	B.p.	$D_4^{20\ b}$	$n_D^{20\ b}$
104	*m*-Tolyl acetate (*m*-Cresyl acetate)	212	1.049	1.4978
105	*p*-Tolyl acetate (*p*-Cresyl acetate)	212.5	1.051	1.500
106	*n*-Propyl oxalate	213	1.038	1.416
107	Ethyl benzoate	213.2	1.0465	1.506
108	Methyl *o*-toluate	215	1.068	
109	Methyl *m*-toluate	215	1.061	
110	Benzyl acetate	217	1.055	1.5200
111	Diethyl succinate	217.7	1.0398	1.41975
112	Diethylene glycol monoethyl ether acetate	218	1.013	
113	Isopropyl benzoate	218	1.023	
114	Diethyl fumarate	218.4	1.052	1.44103
115	Methyl phenylacetate	220	1.068	1.507
116	*n*-Propyl levulinate	221	0.9895	
117	Diethyl maleate	222.7	1.066	1.44156
118	Methyl salicylate (See Table 29A)	223	1.184	1.5369
119	Methyl caprate	226	0.873	1.426
120	Phenyl *n*-butyrate	227	1.023	
121	L-Menthyl acetate	227	0.9185	
122	Ethyl phenylacetate	227.5	1.0333	1.500
123	*n*-Propyl benzoate	230	1.023	1.500
124	Ethyl salicylate (See Table 25)	234	1.1396	1.52542
125	Diethyl glutarate	234(237)	1.02229	1.42395
126	Isopropyl salicylate	237	$1.095\text{-}D_4^{19}$	
127	Benzyl *n*-butyrate	238	$1.033\text{-}D_4^{16}$	
128	*n*-Propyl salicylate	239	$1.098\text{-}D_4^{15}$	
129	Isobutyl benzoate	241	0.999	
130	Dimethyl L-malate	242	1.2334	1.4425
131	Geranyl acetate	242	$0.9174\text{-}D_4^{15}$	1.4660
132	*n*-Butyl oxalate	243	1.010	
133	Thymyl acetate	245	1.009	
134	Diethyl adipate	245	1.0090	1.42765
135	Ethyl caprate	245	0.8650	1.42575
136	*n*-Propyl succinate	246	1.006	1.425
137	Diethylene glycol monobutyl ether acetate	246	0.983	
138	*n*-Butyl benzoate	249	1.000	1.497
139	Diethyl L-malate	253	1.1290	1.4362
140	*n*-Butyl phenylacetate	254	0.994	1.489
141	Glyceryl triacetate	258	$1.161\text{-}D_4^{15}$	
142	Isoamyl oxalate	262	0.961	1.427
143	Isoamyl benzoate	262	1.004	1.495
144	Isobutyl succinate	265	0.974	1.427
145	Methyl laurate	268	0.870	1.432
146	Ethyl laurate	269	0.862	1.4321
147	Ethyl anisate	269	1.1038	1.5254
148	Ethyl cinnamate	271	1.0490	1.55982
149	Isoamyl salicylate	277	$1.065\text{-}D_4^{15}$	
150	Resorcinol diacetate	278	1.179	
151	Diethyl D-tartrate	280	1.2028	1.44677
152	Dimethyl phthalate	283.8	1.191	1.5138
153	Diethyl phthalate	289.5(298)	1.1175	1.5019
154	Diethyl sebacate	307	0.9631	1.43657
155	Benzyl benzoate (m.p. 21)	324*	1.1224	1.5681
156	Dibutyl phthalate	340.7	$1.047\text{-}D_{20}^{20}$	1.4900
157	Tricresyl phosphate	400*d*	$1.197\text{-}D_4^{25}$	1.5568
		(275-80/20 mm)		

TABLE 12B

Esters—Solid[a]

	Name of compound	M.p.		Name of compound	M.p.
1	Phenyl propionate (b.p. 211)	20	28	Dimethyl 3-nitrophthalate	69
2	Ethyl palmitate (b.p. 185)	24	29	Phenyl phthalate	70
3	Methyl anthranilate	24.4	30	2-Naphthyl acetate	71
	(o-Amino methylbenzoate)		31	Glyceryl tristearate	71
4	D-Bornyl acetate (b p. 221)	29	32	p-Cresyl benzoate	71
5	Ethyl o-nitrobenzoate	30	33	Glyceryl tribenzoate	72
6	Methyl m-bromobenzoate	32	34	Ethylene glycol dibenzoate	73
7	Ethyl stearate	33	35	Methyl m-nitrobenzoate	78
8	Methyl cinnamate (b.p. 261)	36	36	Benzyl oxalate	80
9	Methyl sebacate (b.p. 288d)	38	37	Methyl p-bromobenzoate	81
10	Methyl stearate	38.8	38	Dimethyl DL-tartrate (b.p. 282)	90(87)
11	Benzyl cinnamate	39	39	2-Naphthyl salicylate	95.5
12	Benzyl phthalate	42		(See Table 25)	
13	Phenyl salicylate (See Table 25)	42	40	Methyl p-nitrobenzoate	96
14	Benzyl succinate	42	41	Ethyl 3,5-dinitrosalicylate	99
15	Methyl p-chlorobenzoate	44	42	Ethyl 5-nitrosalicylate	102
16	Methyl anisate (b.p. 255)	45	43	Diphenyl adipate	106
17	Diethyl 3-nitrophthalate	46	44	2-Naphthyl benzoate	107
18	Ethyl m-nitrobenzoate (b.p. 296)	47	45	Methyl 3,5-dinitrobenzoate	108
19	1-Naphthyl acetate	49	46	Ethyl p-hydroxybenzoate	116
20	Triphenyl phosphate	49	47	Resorcinol dibenzoate	117
21	Dimethyl oxalate	54	48	Ethyl 3-nitrosalicylate	118
22	m-Cresyl benzoate	54	49	Methyl 5-nitrosalicylate	119
23	Ethyl-p-nitrobenzoate	56	50	Hydroquinone diacetate	124
24	Ethyl trichloroacetate	66	51	Methyl 3,5-dinitrosalicylate	127
	[b.p. 162(167)]		52	Methyl p-hydroxybenzoate	131
25	Ethyl oxanilate	66-7	53	Methyl 3-nitrosalicylate	132
26	Coumarin (b.p. 290)	67	54	Methyl terephthalate	141
27	Phenyl benzoate (b.p. 299)	69	55	Pyrogallol triacetate	164-5

[a] Many esters not listed in this table may be found as "derivatives" in Tables 5A, 5B, 6A, and 6B; sulfonate and sulfinate esters are not included, and for references on these consult the References of Chapter 11.

TABLE 13A

Ethers—Liquid[a]

	Name of compound	Note	B.p.	D_4^{20}	n_D^{20}
1	Methyl ethyl ether		10	$0.7252\text{-}D_4^0$	
2	Ethylene oxide (Epoxyethane)	1	10.7	$0.882\text{-}D_{10}^{10}$	$1.3614\text{-}n^4$
3	Furan		31.27	0.9366	1.42157
4	Ethyl ether	2	34.60	0.71425	1.3526
5	1,2-Epoxypropane (Propylene oxide)		35	0.830	1.466
6	Ethyl vinyl ether		36	0.760	
7	Chloromethyl methyl ether		59	1.015	1.397
8	2,3-Epoxybutane	3	N		
9	1,2-Epoxybutane (α-Butylene oxide)		61–2	0.837^{17}_6	1.385^{17}_n
10	2-Methylfuran (Sylvan)		64	0.913	1.434
11	Tetrahydrofuran		65	0.889	1.407
12	Allyl ethyl ether		65	0.765	1.388
13	Isopropyl ether		67.5	0.726	1.3688
14	n-Butyl methyl ether		70	0.774	1.374
15	Chloromethyl ethyl ether		80(83d)	1.014	1.404
16	Ethylene glycol dimethyl ether		85	0.867	1.37965
17	Dihydropyran		86	0.923	1.440
18	Tetrahydropyran		88	0.881	1.421
19	n-Propyl ether		90.1	0.74698	1.3885
20	n-Butyl ethyl ether		92.3	0.7506	1.3820
21	α-Chloroethyl ether		98	0.966	1.405
22	n-Amyl methyl ether		99	0.761	1.387
23	1,4-Dioxane	4	101.4	1.03361	1.4232
24	α,α'-Dichloromethyl ether		105	$1{,}328\text{-}D_4^{15}$	1.436
25	β-Chloroethyl ether		107	0.989	1.411
26	α-α'-Dichloroethyl ether		116	$1.138\text{-}D_4^{12}$	1.423
27	n-Amyl ethyl ether		118	0.762	1.393
28	Ethylene glycol diethyl ether		121	0.841	
29	sec-Butyl ether		121	0.760	1.396
30	Cyclopentyl ethyl ether		122	0.853	1.423
31	Isobutyl ether		123	0.762	1.397
32	n-Hexyl methyl ether		126	0.772	1.397
33	Cyclohexyl methyl ether		134	0.875	1.435
34	Ethylene glycol monoethyl ether (See Table 6A)		134.8	0.9297	1.40797
35	n-Hexyl ethyl ether		142	0.772	1.401
36	n-Butyl ether		142.4	0.76829	1.400
37	Cyclohexyl ethyl ether		149	0.864	1.435
38	Anisole	5	153.8	0.99393	1.52211
39	Diethylene glycol dimethyl ether		162.0	$0.9440\text{-}D_{20}^{20}$	1.4090
40	Benzyl methyl ether		171*	0.9649	1.5008
41	o-Cresyl methyl ether ("Methyl anisole")		171	0.9666	1.505
42	Ethylene glycol mono-n-butyl ether (See Table 6A)	6	171/743 mm	0.9188	1.4177
43	Phenetole		172	0.9666	1.5080
44	Isoamyl ether		172.5	0.778	1.409
45	p-Cresyl methyl ether		173	0.970	1.512
46	m-Cresyl methyl ether		173	0.972	1.513
47	1,8-Cineole ("Eucalyptol")		176	$0.9267\text{-}D_{20}^{20}$	1.45839
48	β-β'-Dichloroethyl ether		178	1.220	1.457
49	o-Cresyl ethyl ether		184	0.953	1.505
50	Benzyl ethyl ether		184-6*	0.9478	1.4958
51	n-Amyl ether		187.5	0.78298	1.416
52	Diethylene glycol diethyl ether		188	0.906	1.411
53	Phenyl n-propyl ether		188	0.949	1.501
54	p-Cresyl ethyl ether		191	0.949	1.505
55	m-Cresyl ethyl ether		191	0.949	1.506

TABLE 13A (*continued*)

No.	Name of compound	Note	B.p.	D_4^{20}	n_D^{20}
56	Diethylene glycol monomethyl ether (See Table 6A)		194	$1.035\text{-}D_{20}^{20}$	1.4244
57	m-Chloroanisole		194	1.191	1.545
58	o-Chloroanisole		195		
59	Diethylene glycol monoethyl ether (See Table 6A)		196	$1.023\text{-}D_{20}^{20}$	1.4298
60	p-Chloroanisole		200	$1.1851\text{-}D_4^{12.8}$	
61	Guaiacol		205		
62	n-Butyl phenyl ether		206		
63	Veratrole (m.p. 22.5)		207	1.080	
64	o-Chlorophenetole (m.p. 17)		208	1.134	1.530
65	m-Bromoanisole		210		
66	p-Chlorophenetole (m.p. 21)		212	$1.1231\text{-}D_{20}^{20}$	$1.5227\text{-}n_D^{19}$
67	γ,γ'-Dichloropropyl ether		215	1.140	
68	p-Bromoanisole		215	$1.494\text{-}D_4^{9}$	
69	Methyl thymyl ether		216	$0.954\text{-}D_4^{0}$	
70	Triethylene glycol dimethyl ether		216	$0.9871\text{-}D_{20}^{20}$	1.4233
71	Resorcinol dimethyl ether		217*	1.050	1.4233
72	o-Bromophenetole		218		
73	o-Bromoanisole		218		
74	o-Cresyl butyl ether		223		
75	n-Hexyl ether		229		
76	Ethyleneglycol bis-2-chloroethyl ether		230		
77	Diethylene glycol mono-n-butyl ether		231	$0.9954\text{-}D_{20}^{20}$	
78	Safrole		233 v.s.	1.100	1.5383
79	p-Bromophenetole		233(229)		
80	Anethole (m.p. 22)		235	0.98	1.558
81	Resorcinol diethyl ether		235		
82	o-Iodoanisole		240	1.800	
83	Eugenol methyl ether		244	1.0336	1.5360
84	Ethylene glycol monophenyl ether		245	$1.1094\text{-}D_{20}^{20}$	
85	o-Iodophenetole		246	1.800	
86	Isosafrole		248	1.122	1.5782
87	Phenyl ether (m.p. 28)		252(258)	1.073	
88	n-Heptyl ether		263(260)	0.805	1.427
89	Isoeugenol methyl ether		264	1.0528	1.5692
90	Tetraethylene glycol dimethyl ether		139/12 mm 266(275)	1.009	1.432
91	o-Nitrophenetole		267		
92	Methyl 1-naphthyl ether		271*	1.09159	1.6940
93	o-Nitroanisole		272	1.254	1.562
94	Ethyl 1-naphthyl ether		280.5*	1.074	1.5973
95	n-Octyl ether		288(291.7)	0.806	1.433
96	Benzyl ether		300d	1.0428	

a For methods for preparing derivatives of ethers and data for derivatives, see the larger text by these authors.

TABLE 13B

Ethers—Solid[a]

No.	Name of compound	Note	M.p.	B.p.	No.	Name of compound	Note	M.p.	B.p.
1	p-Chlorophenetole (See Table 13A)		21	212	19	2,4,6-Trichlorophenetole		44	246
2	Anethole (See Table 13A)		22	235	20	Hydroquinone monomethyl ether		52.5	244/754.2 mm
3	Veratrole (See Table 13A)		22.5	207	21	Dihexadecyl ether (dicetyl)		54	dec.
4	n-Amyl 2-naphthyl ether		24.5	327.5*	22	p-Nitroanisole		54	259
5	p-Iodophenetole		27	252	23	Hydroquinone dimethyl ether		56	213
6	Isoamyl-2-naphthyl ether		28	231*	24	p-Nitrophenetole		60	283
7	Phenyl ether		28	252(258)	25	2,4,6-Trichloroanisole		60	
8	o-Methoxybiphenyl		29	274	26	Trioxane	7	61-3	
9	β-Bromoethyl phenyl ether		32		27	Hydroquinone diethyl ether		72	
10	Isobutyl 2-naphthyl ether		33	304	28	2,4,6-Tribromophenetole		72	
11	m-Nitrophenetole		34	284	29	Methyl 2-naphthyl ether		73*	273
12	sec-Butyl 2-naphthyl ether		34	298.5*	30	2,4,5-Tribromophenetole		73	
13	n-Butyl 2-naphthyl ether		35.5	309*	31	Benzyl 1-naphthyl ether		77	200/12 mm
14	Ethyl 2-naphthyl ether		36	282*	32	2,4,6-Tribromoanisole		87	
15	m-Nitroanisole		39	258	33	p-Methoxybiphenyl		90	
16	Isopropyl 2-naphthyl ether		40	285	34	2,4-Dinitroanisole		95	
17	n-Propyl 2-naphthyl ether		40	297	35	Benzyl 2-naphthyl ether		101.5*	dec.
18	Catechol diethyl ether		43	217	36	3,5-Dinitroanisole		105-6	
					37	Hydroquinone dibenzyl ether		127	

[a] For derivatives of ethers, see the larger text by these authors.

Notes on Ethers

1. Forms a polymer in aqueous alkali, m.p. 56.

2. Furan may be detected by applying it to a pine splinter that has been moistened with hydrochloric acid; a green color develops.

3. Trans-2,3-epoxybutane, b.p. 53-4/741 mm; D_4^{25} = .8010; cis-2,3-epoxybutane, b.p. 58-9/745 mm; D_4^{25} = .8226. The normal crude mixture is 65% trans and 35% cis.

4. 1,4-Dioxane is soluble in water, with which it forms a constant-boiling mixture at 82.8 which contains 48 mole % of dioxane. Dioxane forms addition products with bromine, m.p. 65-6, and with iodine, m.p. 84-5.

5. Anisole forms a dinitro derivative that exists in two forms, m.p. 87 and 95.5. The 2,4-dibromoanisole melts at 61.

6. Compound boils at 170-6/743 mm. It also forms the xanthate with KOH and CS_2 [Ind. Eng. Chem., Anal. Ed. 7, 128 (1935)].

7. The picrate is unstable on exposure to air. With 1,3,5-trinitrobenzene addition compound melts at 86.5.

TABLE 14

Alkyl and Cycloalkyl Halides

Name of compound	Note	B.p.	D_4^{20}	n_D^{20}	Melting point of derivatives	
					S-Alkyl-isothiourea picrate	Picrate of β-Naphthyl ethers
Chlorides						
1 Methyl		−24			224	117
2 Vinyl		−14			104	
3 Ethyl		13	.917-D_6^6		188	104
4 Isopropyl		36.5	.859	1.378	196(148)	92
5 Allyl		44.6	.940	1.416	155	99
6 n-Propyl		46.4	.889	1.388	181(176)	81(75)
7 tert-Butyl		50.7	.846	1.386	160(188)	85
8 sec-Butyl		68	.874	1.397	190(166)	84
9 Isobutyl		68.9	.881	1.398	174	
10 Neopentyl		85	.879			
11 Isoamyl		100	.872	1.409	179(173)	94
12 n-Amyl		106	.882	1.412	154	67
13 n-Hexyl		133	.878	1.420	157	
14 Cyclohexyl		143	.989	1.462		
15 n-Heptyl		159	.877	1.426	142	123
16 Benzyl	1	179.4	1.100	1.539	188	
17 Octyl		180(184)	.875	1.431	134	84
18 Phenethyl (β-Phenylethyl)		190				
19 n-Nonyl		202	.870	1.434	131	
20 Benzal		207				
21 p-Chlorobenzyl (m.p. 24–27)	2	214(222)			194	
22 n-Decyl		223	.868	1.437	137	
23 n-Undecyl		241	.868	1.440	139	
24 Cetyl (Hexadecyl)		289d			155(137)	
25 p-Bromobenzyl	3	36 m.p.			219	
26 p-Nitrobenzyl		71 m.p.				
Bromides						
27 Methyl		3.46			224	117
28 Ethyl		38.4	1.460	1.425	188	104
29 Propenyl (1-bromopropene)		60	1.4133	1.452		
30 Isopropyl		60	1.314	1.425	196(148)	92
31 Allyl		71	1.400	1.470	155	99
32 n-Propyl		71	1.355	1.435	181(177)	75(81)

(continued)

TABLE 14 (continued)

	Name of compound	Note	B.p.	D_4^{20}	n_D^{20}	Melting point of derivatives	
						S-Alkyl-isothiourea picrate	Picrate of β-Naphthyl ethers
33	tert-Butyl		73.2	1.211	1.435	160(188)	84
34	Isobutyl		91	1.253	1.437	174(167)	85
35	sec-Butyl		91.2	1.256	1.440	190(166)	67
36	n-Butyl		101.6	1.274	1.440	180(177)	
37	Neopentyl		109	1.225			
38	Isoamyl		120.6	1.213	1.442	179(173)	94
39	n-Amyl		129	1.219	1.445	154	67
40	Cyclopentyl		137	1.387	1.489		
41	n-Hexyl		155(157)	1.175	1.448	157	
42	n-Heptyl		180	1.140	1.451	142	
43	Benzyl		198	1.438		188	123
44	Octyl		201.5	1.112	1.453	134	
45	Phenethyl (β-Phenylethyl)		218	1.359	1.556		84
46	n-Nonyl		220	1.090	1.454	131	
47	Cetyl (m.p. 14)		201/19 mm	1.001	1.462	155(137)	
48	p-Chlorobenzyl		62 m.p.			194	
49	p-Bromobenzyl		62 m.p.			219	
	Iodides						
50	Methyl		43	2.282	1.532	224	117
51	Ethyl		72.4	1.940	1.514	188	104
52	Ethylene		82			270	
53	Isopropyl		89.8	1.703	1.499	196(148)	92
54	n-Propyl		102.5	1.743	1.505	181(176)	75(81)
55	Allyl		103	1.777	1.578	155	99
56	tert-Butyl		103.3(98)			160(188)	85
57	sec-Butyl		120	1.592	1.499	190(166)	84
58	Isobutyl		120.4	1.602	1.496	174(167)	67
59	n-Butyl		130	1.616	1.499	180(177)	94
60	Isoamyl		148	1.503	1.493	179(173)	67
61	n-Amyl		155	1.512	1.496	154	
62	n-Hexyl		179	1.437	1.493	157	
63	n-Heptyl		204	1.373	1.490	142	
64	Nonyl		219.5			131	
65	Octyl		225.5	1.330	1.489	134	
66	Cetyl (n-Hexadecyl)		22 m.p.			155(137)	
67	Benzyl		24 m.p.			188	123

Notes on Alkyl and Cycloalkyl Halides

1. Forms quaternary salt with dimethylaniline, m.p. 110.

2. On oxidation gives *p*-chlorobenzoic acid, m.p. 242.

3. On oxidation gives *p*-bromobenzoic acid, m.p. 251.

TABLE 15A

Aryl Halides— Liquids[a]

Name of compound	Note	B.p.	D_4^{20}	n_D^{20}	Recommended derivative Nitro	
					Position	M.p.
Fluorobenzene		87.4	1.024	1.466		
o-Fluorotoluene		114	1.004	1.4704		
m-Fluorotoluene		116	0.997	1.4691		
p-Fluorotoluene		117	0.998	1.469		
Chlorobenzene		131.8	1.107	1.525	2,4	52
Bromobenzene		156.2	1.494	1.560	2,4	70(75)
o-Chlorotoluene		159.3	1.082	1.524	3,5	63
m-Chlorotoluene		162.3	1.072	1.521	4,6	91
p-Chlorotoluene (m.p. +7)		162.4	1.071	1.521	2	38
m-Dichlorobenzene		173	1.288	1.546	4,6	103
1-Chloro-2-ethylbenzene		178.4	1.0569	1.5218	4,5	110
o-Dichlorobenzene		179	1.305	1.552	3,5	82
o-Bromotoluene		181.8	1.425	1.555	4,6	103
m-Bromotoluene		183.7	1.410	1.551		
1-Chloro-3-ethylbenzene		183.8	1.0529	1.5199	2	47
1-Chloro-4-ethylbenzene		184.4	1.0455	1.5175	4	171
p-Bromotoluene (m.p. 26)		184.5	1.390			
Iodobenzene		188.6	1.831	1.620		
1-Chloro-3,4-dimethylbenzene		191	1.069-D[15]	1.5168		
1-Chloro-2-isopropylbenzene		191	1.0341	1.580		
o-Bromochlorobenzene		195	1.646			
o-Chloroanisole		195(199)	1.248	1.5480		
1,3-Bromochlorobenzene		196	1.6365-D[15]	1.5773-n$_D^{17}$		
1-Chloro-4-isopropylbenzene (*p*-Chlorocumene)		198.3	1.0208	1.5117		
2,6-Dichlorotoluene		199	1.269	1.551	3	53
1-Bromo-2-ethylbenzene		199.3	1.3549			
p-Chloroanisole		200	1.1851-D[12.8]		2	95
2,4-Dichlorotoluene		200*	1.249	1.549	3,5	104
m-Iodotoluene		204	1.698		4,6	108
1-Bromo-4-ethylbenzene		205.1	1.3423	1.5448		
3,4-Dichloro-1-methylbenzene (3,4-Dichlorotoluene)		208.9	1.2526	1.5471	6	64

(continued)

TABLE 15A (continued)

	Name of compound	Note	B.p.	D_4^{20}	n_D^{20}	Recommended derivative Nitro	
						Position	M.p.
32	1-Bromo-2-isopropylbenzene		210.2	1.3020	1.5408		
33	o-Iodotoluene	1	211	1.698	1.594	6	103
34	1-Fluoronaphthalene		214	1.134	1.606	4	61
35	m-Dibromobenzene		219	1.952	1.5361		
36	p-Bromoisopropylbenzene		219	1.2854	1.609	4,5	114
37	o-Dibromobenzene		219.3	1.956	1.5761		
38	2,4-Dichlorobenzyl chloride		248	1.4068	1.665		
39	o-Bromoiodobenzene		257	2.262			
40	1-Chloronaphthalene		259.3	1.191	1.633	4,5	180
41	1-Bromonaphthalene		281.2	1.484	1.658	4	85

TABLE 15B

Aryl Halides—Solid[a]

	Name of compound	Note	M.p.	Recommended derivative Nitro	
				Position	M.p.
1	p-Bromotoluene (b.p. 184)		28.5	4	56
2	4-Iodotoluene (b.p. 211)	1	35	2	54
3	1,2,3-Trichlorobenzene (b.p. 218.5)		52	1,8	175
4	1,4-Dichlorobenzene (b.p. 173)		53		
5	2-Chloronaphthalene (b.p. 265)		56(61)	2	68
6	2-Bromonaphthalene (b.p. 281)		59	2	72
7	1,3,5-Trichlorobenzene (b.p. 208.4)		63		
8	1,4-Bromochlorobenzene (b.p. 196.9)		67	2,5	84
9	p-Nitrobenzyl chloride		72		
10	1,4-Dibromobenzene (b.p. 219)		89	2,5	171
11	4-Nitrobenzyl bromide		99		
12	1,4-Di-iodobenzene (b.p. 289)		129		
13	4,4'-Dichlorobiphenyl		149		
14	4,4'-Dibromobiphenyl		164		
15	Hexachlorobenzene (b.p. 309)		229s(226)		

[a] Alkyl-substituted aryl halides may be oxidized to the corresponding carboxylic acids as derivatives. For other derivatives, see the larger text by the authors.

Note on Aryl Halides

1. Oxidizes with nitric acid if heated at 200° for 3 hours; resistant to permanganate oxidation.

TABLE 16

Dihalides and Polyhalides (Nonaromatic)

	Name of compound	Note	M.p.	B.p.	D_4^{20}	n_D^{20}	Melting point of derivatives
1	Dichlorofluoromethane			8.9			
2	Trichlorofluoromethane			23.7			
3	1,2-Bromofluoroethylene			36	1.693		
4	Dichloromethane	1		40.7	1.336	1.4237	2-Naphthyl ether, 133
5	1,2-Dichloroethylene (trans)			48	1.2569	1.452	
6	Perfluoro-n-hexane			57.1	1.6995	$1.2515\text{-}n_D^{22}$	
7	1,1-Dichloroethane			57.4	1.175	1.4164	Di-1-naphthyl ether, 117
8	Perfluoro-2-methylpentane			57.7	1.7326	$1.2564\text{-}n_D^{22}$	
9	1,2-Dichloroethylene (cis)			60	1.282	$1.4428\text{-}n_D^{25}$	
10	Chloroform	2		61.3	1.489	1.446	
11	m-Nitrobenzal chloride			65			m-Nitrobenzoic acid, 140
12	2,2-Dichloropropane			70	1.093	1.4117	
13	1,1,1-Trichloroethane			74.1	1.349	1.4380	
14	Carbon tetrachloride			76.8	1.595	1.4630	
15	p-Nitrobenzal bromide			82			p-Nitrobenzoic acid, 241
16	1,2-Dichloroethane (sym)			84.1	1.256	1.4443	
17	1,1,2-Trichloroethylene			87	1.464	1.4773	
18	1,2-Dichloropropane			96.4	1.155	1.4388	
19	Dibromomethane	1		98.6	2.496	1.538	
20	Benzotrifluoride			102.2			
21	Bromo-trichloromethane			104.7			
22	1,1-Dichloro-2-methyl propane			106d			
23	1-Bromo-2-chloroethane			106.7	1.689	1.491	
24	1,2-Dichloro-2-methyl propane			108			
25	1,1-Dibromoethane			112	2.055	1.5128	
26	Trichloronitromethane			112			
27	1,1,2-Trichloroethane			114	1.443	1.4707	
28	2,3-Dichlorobutane			116			
29	1,3-Dichloro-1-butene			120-3	1.130	1.464	
30	1,1,2,2-Tetrachloroethylene			121	1.623	1.5055	
31	1,2-Dichlorobutane			123.5			
32	1,3-Dichloropropane			125(122)	1.188	1.452	
33	1,1,1,2-Tetrachloroethane			130.5			
34	1,2-Dibromoethane (m.p. +10)			131.5	2.179	1.5379	
35	1,2-Dibromopropene			132	2.008		
36	1,2-Dibromopropane			141.6	1.933	1.5203	
37	1-Bromo-3-chloropropane			143-4	1.594	1.4861	
38	s-Tetrachloroethane			146	1.600	1.4942	
39	1,2-Dibromo-2-methyl propane			149	1.783	1.512	
40	1,2-Dibromo-1-butene			150	$1.887\text{-}D^0$		
41	1,1,1,2-Tetrachloropropane			150	1.473	1.4867	
42	Bromoform (m.p. +8)			150.5	2.890	1.598	
43	2,3-Dibromobutane			157	1.792	1.515	
44	1,2,3-Trichloropropane			158	1.417	1.4585	
45	Pentachloroethane			161	1.681	1.504	
46	1,2-Dibromobutane			166.3	1.820		

(continued)

TABLE 16 (*continued*)

	Name of compound	Note	M.p.	B.p.	D_4^{20}	n_D^{20}	Melting point of derivatives
47	1,3-Dibromopropane			167.5	1.982	1.523	
48	1,3-Dibromobutane			174	$1.820\text{-}D^0$	1.507	
49	1,3-Dibromo-2-methyl-propane			174.6	$1.821\text{-}D_0^{20}$		
50	Di-iodomethane			181	3.325	1.7425	
51	1,1,2-Tribromoethane			188.9	2.6211	1.5933	
52	1,4-Dibromobutane			197.5	1.826	1.519	
53	o-Chlorobenzyl chloride			213-4			o-Chlorobenzoic acid, 142
54	Benzal chloride	3		214	$1.295\text{-}D^{16}$	1.5515	Benzoic acid, 122
55	m-Chlorobenzyl chloride			216	1.269		m-Chlorobenzoic acid, 158
56	1,2,3-Tribromopropane			220	2.402	1.582	
57	Benzotrichloride			220.8	$1.374\text{-}D_{15}^{17.5}$		Benzoic acid, 122
58	1,5-Dibromopentane	1		221	1.702	1.513	
59	sym-Tetrabromoethane			243.5	2.967	1.638	
60	m-Bromobenzyl chloride		23				m-Bromobenzoic acid, 155
61	p-Chlorobenzyl chloride		29				p-Chlorobenzoic acid, 240
62	o-Bromobenzyl bromide		31				o-Bromobenzoic acid, 150
63	m-Bromobenzyl bromide		41				m-Bromobenzoic acid, 155
64	1,7-Dibromoheptane	1	41.7	263	1.525^{15}	1.514^{15}	
65	p-Bromobenzyl chloride		50				p-Bromobenzoic acid, 251
66	p-Bromobenzyl bromide		63				p-Bromobenzoic acid, 251
67	1,2-Di-iodoethane		81		2.132^{10}		
68	Carbon tetrabromide		92	189.5			
69	1,1,1-Trichloro-2,2-bis(p-Chlorophenyl)-ethane (D.D.T.)	4	108				
70	1,2,3,4-Tetrabromobutane		115.8	282.7			
71	Iodoform	5	119				
72	α-Benzene hexachloride		157	288			
73	Hexachloroethane		187s	185			
74	β-Benzene hexachloride		310s				

Notes on Dihalides and Polyhalides

1. The following are derivatized as S-alkyl *bis* (thiourea picrates): dichloro- and dibromo-methane, m.p. 267d; 1,5-dibromopentane, m.p. 247d; 1,7-dibromoheptane, m.p. 208; 1,8-dibromo-octane, m.p. 214; 1,9-dibromononane, m.p. 193.
2. Gives carbylamine test with aniline or any RNH_2.
3. Compound (200 mg) heated to 50° in 1 ml conc. H_2SO_4 for 5 minutes, diluted, neu-tralized, and treated with semicarbazide, as directed on page 262, gives benzaldehyde semicarbazone.
4. Impure product melts as low as 90° due to the presence of isomers. Pure commercial product melts at 108. For ident. see *J. Am. Chem. Soc.*, **66**, 2129.
5. Characteristic odor; reduces with $Na_2As_2O_4$ to CH_2I_2; forms compound with quinoline, m.p. 65d.

TABLE 17

Alkanes and Cycloalkanes[a,b]

	Name of compound	B.p.	D_4^{20}	n_D^{20}
1	Neopentane (2,2-Dimethylpropane)	9.5	0.596	1.3513
2	Isopentane (2-Methylbutane)	28	0.620	1.3580
3	Pentane	36	0.626	1.3577
4	Cyclopentane	49.3	0.746	1.4068
5	2,2-Dimethylbutane	49.7	0.649	1.3689
6	2,3-Dimethylbutane	58	0.662	1.3750
7	2-Methylpentane	60.3	0.653	1.3716
8	3-Methylpentane	63.3	0.664	1.3764
9	Hexane	68.3	0.659	1.3749
10	Methylcyclopentane	72	0.749	1.4100
11	2,4-Dimethylpentane	80.5	0.673	1.3823
12	Cyclohexane (m.p. 6.5)	80.7	0.778	1.4264
13	2,3-Dimethylpentane	89.9	0.695	1.3920
14	Heptane	98.4	0.684	1.3877
15	2,2,4-Trimethylpentane	99.2	0.692	1.3916
16	Methylcyclohexane	100.9	0.769	1.4231
17	Octane	125.7	0.703	1.3975
18	1,2-Dimethylcyclohexane(cis)	129.7	0.796	1.4360
19	Ethylcyclohexane	131.7	0.788	1.4332
20	Nonane	150.8	0.718	1.4056
21	2,7-Dimethyloctane	160	0.724	1.4080
22	p-Menthane	169	0.796	1.4369
23	Decane	174	0.730	1.4114
24	Decahydronaphthalene(trans)	187.2	0.870	1.4695
25	Decahydronaphthalene(cis) (Decalin)	195.7	0.896	1.4810
26	Undecane	196	0.702	1.4190
27	Dodecane	217	0.749	1.4216

[a] For additional information, see Wendland, "A System of Characterization of Pure Hydrocarbons," *J. Chem. Educ.* **23**, 3-15 (1946).

[b] The table in the authors' larger text lists 146 alkanes and cycloalkanes.

TABLE 18

Alkenes, Alkynes, Cycloalkenes, and Dienes[a]

	Name of compound	Note	B.p.	D_4^{20}	n_D^{20}
1	2-Methyl propene		−6.9	$.6266^{-6.6}$	1.3467
2	1-Butene		−6.3	$.6255^{-6.5}$	1.3465
3	1,3-Butadiene	1	−4.54	$.650^{-6}$	1.4292^{-25}
4	2-Butene (cis)	2	3.73	$.6303^1$	
5	3-Methyl-1-butene (pol)		20	$.6320-_{25}^{15}$	1.3643
6	2-Butyne	16	27.2	.688	1.3893
7	1-Pentene (Amylene)		30.1	.6410	1.3710
8	Isopropene (2-Methyl-1,3-butadiene) (pol)	3	34.1	.6809	1.4219
9	trans-2-Pentene		35.85	.6486	1.3790
10	cis-2-Pentene		37.0	.6562	1.3822
11	2-Methyl-2-butene		38.5	.6620	1.3878
12	1-Pentyne		39.7	.6945	1.3847
13	1,3-Cyclopentadiene (pol)	4	41.0	.7983	1.4461
14	1,3-Pentadiene (Piperylene)	5,16	42.3	.6803	1.4309
15	Cyclopentene		44.2	.7736	1.4225
16	1,5-Hexadiene		59	.690	1.4034
17	2,3-Dimethyl-1,3-butadiene (pol)	16	68.9	.7263	1.4390
18	1-Hexyne		71	.719	1.3990
19	2,3-Dimethyl-2-butene (Tetramethylethylene)	16	73	.7081	1.4115
20	2,4-Hexadiene (Dipropenyl)	7,16	81	.7152	1.4493
21	Cyclohexene	8	83	.8088	1.4465
22	1-Heptyne		100	.7338	1.4084
23	2,4,4-Trimethyl-1-pentene ("Di-isobutylene")		101.2	.7151	1.4082
24	Δ'-Tetrahydrotoluene		107	$.8145_8^{18}$	1.4503^{18}
25	Cycloheptene (Suberene)	9	115*	.8228	1.4580
26	Ethynylbenzene (Phenylacetylene)		141.7	.9281	1.5485
27	Styrene (pol) (Vinyl benzene)	16	145.2	.9056	1.5470
28	α-Pinene	10,16	156	.8600	1.4560
29	Allylbenzene		157	.8912	1.5042^{25}
30	Myrcene	11	166-7	.7982	1.4706
31	Dipentene (DL-Limonene)	16	175-6	.8402	1.4727
32	Sylvestrene	12,16	175-8	.8479	1.4760
33	D-Limonene	13,16	176.7	.8446	1.4739
34	β-Methyl styrene (Propenylbenzene)		176-7	.935	1.5903^{16}
35	Indene (pol)	14,16	181	.9915	1.5764
36	1,1-Diphenylethylene		277		
37	1,4-Dihydronaphthalene (m.p. 24)		212	.998	1.5740
38	L-Camphene (m.p. 51.3)		159-60	.8555	1.46207
39	1,4-Diphenylbutadiene (m.p. 70)				
40	Stilbene (m.p. 125) (trans)	15	306s	$.970_{13}^{125}$	
41	1,4-Diphenylbutadiene (trans) (m.p. 148)				

[a] For additional information, see Wendland, "A System of Characterization of Pure Hydrocarbons," J. Chem. Educ. **23**, 3-15 (1946).

Notes on Alkenes, Alkynes, Cycloalkenes, and Dienes

1. 39 and 119 bromo(tetra). α-Naphthoquinone, m.p. 105-6 (*Ann.* 460, 98).
2. *trans*, b.p. 0.96.
3. Thiocyanate, m.p. 77; maleic anh. → tetrahydro-4-methylphthalic anh.,

m.p. 64.
4. Dimer, m.p. 32 (b.p. 170); + benzoquinone, m.p. 76; maleic anh. → cis-3,6-endomethylene-Δ⁴-tetrahydrophthalic anh., m.p. 164.
5. Maleic anh. → 3-methyl-1,2,3,6-tetrahy-

TABLE 18 (*continued*)

drophthalic anh., m.p. 61-2.

6. 1,3-Cyclohexadiene with bromine in CHCl₃ or *n*-C₆H₁₄ forms a dibromo compound with m.p. 68, but isomerizes rapidly to 1,4-dibromo-2-cyclohexene, m.p. 108, which adds no more bromine; but the isomer melting at 68, when treated with bromine, yields two tetrabromo derivatives, one form melting at 87-9 and the other at 155-6. Maleic anh. → 3,6-endoethylene-1,2,3,6-tetrahydrophthalic anh., m.p. 147; α-Naphthoquinone, m.p. 135.

7. Maleic anh. → 2,5-dimethyl-1,2,5,6-tetrahydrophthalic anh., m.p. 95.

8. Nitrosochloride, m.p. 153*d*; 1,2-dithiocyanate, m.p. 58; KMnO₄ → adipic acid, m.p. 154.

9. Nitrosochloride, m.p. 118; oxidation with HNO₃ gives pimelic acid, m.p. 105.

10. Oxidation with Hg(OAc)₂ → sobrerol, m.p. 150, and 8-hydroxycarvotanacetone (semicarbazone, m.p. 176); oxidation with KMnO₄ at 30° in acid solution gives pinonic acid, m.p. 103-5 (semicarbazone, m.p. 205). Nitrolbenzylamine, m.p. 123;

nitrolpiperidine, m.p. 119; nitrosochloride, m.p. 109; nitroso, m.p. 132.

11. Maleic anhydride addition, m.p. 33-4; dihydromyrcene tetrabromide, m.p. 96.

12. Dihydrochloride, m.p. 72; nitrosochloride, m.p. 107; nitrolbenzylamine, m.p. 71.

13. Dihydrochloride, m.p. 50; nitrosochloride, m.p. 100-4; nitrolbenzylamine, m.p. 92; nitroso, m.p. 72.

14. Bromohydrin, m.p. 128-9; picrate, m.p. 98 (very explosive); nitrosochloride, m.p. 150; nitrolamine, m.p. 157.

15. Na + C₂H₅OH → bibenzyl, m.p. 52; nitrosite, m.p. 196*d*; (α-α′-dinitrobenzyl, m.p. 236); picrate, m.p. 94-5; with 1,3,5-C₆H₃(NO₂)₃ gives compound, m.p. 120(101). Bromo: α-di, m.p. 237; β-di, m.p. 110.

16. The following bromo derivatives apply to the compounds in Table 18, the numbers for which are in parentheses: (6) 243(tetra); (14) 114(tetra); (17) 138(tetra); (19) 121 (di); (20) 185(tetra); (27) 74(di); (28) 169(di); (31) 124(tetra); (32) 135(tetra); (33) 104(tetra); (35) 32(di).

TABLE 19A

Aromatic Hydrocarbons—Liquid[a]

	Name of compound	Note	B.p.	D_4^{20}	n_D^{20}	Melting point of derivatives — Nitro Position	Melting point of derivatives — Nitro M.p.	Melting point of derivatives — Picrate[b]
1	Benzene	1	80.1	.8790	1.5011	1,3	89	84u
2	Toluene	1	110.6	.8670	1.4969	2,4	70	88u
3	Ethylbenzene		136.2	.8670	1.4959	2,4,6	37	96u
4	p-Xylene	2	138.3	.8611	1.4958	2,3,5	139	90u
5	m-Xylene	2	139.1	.8642	1.4972	2,4,6	183	91u
6	o-Xylene	2	144.4	.8802	1.5054	4,5	118(71)	88u
7	Isopropylbenzene (Cumene)		152.4	.8618	1.4915	2,4,6	109	
8	n-Propylbenzene		159.2	.8620	1.4920			103u
9	1,3,5-Trimethylbenzene (Mesitylene)		164.7	.8652	1.4994	2,4; 2,4,6	86; 235	97u
10	tert-Butylbenzene		169.1	.8665	1.4926	2,4	62	
11	1,2,4-Trimethylbenzene (Pseudocumene)		169.2	.8758	1.5049	3,5,6	185	97u
12	Isobutylbenzene		172.8	.8532	1.4865			
13	sec-Butylbenzene		173.3	.8621	1.4901			
14	m-Isopropyltoluene (m-Cymene)		175.2	.8610	1.4930			90.5
15	1,2,3-Trimethylbenzene		176.2	.8944	1.5139			
16	p-Isopropyltoluene (p-Cymene)		177.1	.8573	1.4909	2,6	54	
17	o-Isopropyltoluene (o-Cymene)		178.3	.8766	1.5006			
18	m-Diethylbenzene		181.1	.8641	1.4953	2,4,6	62	
19	Indene		182	.857				98
20	n-Butylbenzene		183.3	.8601	1.4898			
21	1,2,3,5-Tetramethylbenzene		197.9	.8899	1.5125	4,6	181	
22	1,2,3,4-Tetramethylbenzene (Isodurene)		205.0	.9053	1.5201	5,6	176 (157)	92-5
23	n-Pentylbenzene		205.4	.8585	1.4878	5,7	95	
24	Tetrahydronaphthalene (Tetralin)		206	.971		2,4,6	108	
25	1,3,5-Triethylbenzene		218					
26	n-Hexylbenzene		226.1	.8575	1.4864			
27	Cyclohexylbenzene		235-6	.9502	1.5329			
28	1-Methylnaphthalene	4	244.8	1.0200	1.6174	4	71	142
29	n-Heptylbenzene		245.5	.8567	1.4854			
30	1,1-Diphenylethane		268-70	1.003	1.5761			

TABLE 19B

Aromatic Hydrocarbons—Solid

Name of compound	Note	M.p.	B.p.	T.N.F.	Melting point of derivative — Nitro — Position	M.p.
1 Diphenylmethane	3	25.3			2,4,2',4'	172
2 2-Methylnaphthalene	4	34.4	241	125-6	1	81
3 Pentamethylbenzene	4	51			6	154
4 Bibenzyl	4	53			4,4'	180
					2,2',4,4'	169
5 Biphenyl (s, v.s.)		69.2			4,4'	237
					2,2',4,4'	165
6 1,2,4,5-Tetramethylbenzene (Durene)		79			3,6	205
7 Naphthalene (s, v.s.)	4	80.25	218	153-4	1	61(57)
8 1,3-Diphenylbenzene		89	363			
9 Triphenylmethane		92	390		4,4',4"	206
10 Retene	4	95.2s		175-6		
11 Acenaphthene	4	96.2s			5	101
12 Phenanthrene	4	96.3		197		
13 Fluorene	4	116	293-5	179	2(2,7)	156(199)
14 Pyrene		148	385	242-3		
15 1,2-Benzanthracene		159-60		160		
16 1,1'-Binaphthyl		160				
17 Hexamethylbenzene	4	165				
18 2,2'-Binaphthyl		188	358-62			
19 sym-Tetraphenylethane		211				
20 1,4-Diphenylbenzene		213				
21 Anthracene	4,5	216s		194		
22 Chrysene (C$_{22}$H$_{18}$)	4,6	251	448	248-9		
23 Picene (C$_{22}$H$_{14}$)		364	520	257-8		

Notes on Aromatic Hydrocarbons

1. Toluene is oxidized readily to benzoic acid, m.p. 121.7.
2. The three xylenes are oxidized by KMnO$_4$ to dicarboxylic acids: o-xylene to phthalic; m-xylene to isophthalic and p-xylene to terephthalic acids.
3. Oxidized by HNO$_3$ to benzophenone.
4. Melting points of molecular compounds with trinitrobenzene: 1-methyl naphthalene, 154; 2-methyl naphthalene, 123; pentamethylbenzene, 121; bibenzyl, 102; naphthalene, 153; retene, 139; acetnaphthalene, 168; phenanthrene, 145; fluorene, 105; hexamethylbenzene, 174; anthracene, 164; chrysene, 186.
5. Oxidizes to anthraquinone, bromination in acetic acid gives dibromo derivative, m.p. 122.
6. Dibromo derivative by bromination in acetic acid, m.p. 275.

TABLE 20A

Ketones—Liquid

	Name of compound	Note	B.p.	Recommended			Others	Oxime
				Semi-carbazone	2,4-Dinitro-phenyl-hydrazone	Phenyl-hydra-zone		
1	Trifluoroacetone		21		139			59
2	Acetone		56.11	190(187)	126	42		
3	Hexa fluoroacetyl acetone	1	63–5					b.p. 152
4	2-Butanone (Ethyl methyl)		80	146	114	(oil)		
5	3-Buten-2-one (Methyl vinyl)		81	141				
6	Biacetyl		88	278–9(di)	314–5*(di)	243(di)	76(mono),245–6*(di)	(oil)
7	2-Methyl-3-butanone		94.3	113–4	120	(oil)		
8	3-Pentanone (Diethyl)		102	138–9	156	(oil)		b.p. 165
9	2-Pentanone (Methyl propyl)		102.3	112(106)	143–4	(oil)		b.p. 167
10	Pinacolone		106*	157–8	125	(oil)		75(79)
11	4-Methyl-2-pentanone		116.8	132(134)	95			b.p. 176
12	Chloroacetone		119	150(164d)	125			
13	1,1-Dichloroacetone		120					
14	3-Hexanone		124	163	130			34
15	2,4-Dimethyl-3-pentanone		124	160*	88(94–8)			
16	2,3-Hexanedione		128					175(di)
17	2-Hexanone (Butyl methyl)		128	125*(122)	110	(oil)		49
18	Mesityl oxide	2	130	164N(133)	200(203)	142		48–9
19	Cyclopentanone, v.s.	3	130.7	210(203)	146*N	55(50)		56.5
20	1-Bromo-2-propanone		136	135d				
21	2-Methyl cyclopentanone		139	184				36
22	2,4-Pentanedione (Acetylacetone)		139	122(mono) 209(di)				149(di)
23	4-Heptanone (Dipropyl), v.s.		144	132	75			b.p. 193
24	Acetoin (3-Hydroxy-2-butanone)		145(148)	185(202)	318			
25	1-Hydroxy-2-propanone, v.s. (Acetol)		146	196	128.5*	103		71
26	2-Heptanone (Amyl-methyl)		151.2	123	89	207		
27	Cyclohexanone		156	166–7	162(160)	81(77)		91
28	3,5-Dimethyl-4-heptanone (Di-sec-butyl)		162 (170–3)	83–4				
29	2-Methylcyclohexanone		165	197d	137*			43

Melting point of derivatives

No.	Name		B.p.				
30	4-Hydroxy-4-methyl-2-pentanone (Diacetone alcohol)		166(164)		202-3		58
31	Isovalerone (2,6-Dimethyl-4-heptanone) (Di-isobutyl)		169	122(126)	66(92)		
32	Methylacetoacetate		170	152	155	94	
33	3-Methyl cyclohexanone		170	179	134	110	
34	4-Methyl cyclohexanone		171	203	58		39
35	2-Octanone (Hexyl methyl)		173	122-3*			
36	Methyl methylacetoacetate		177	138	140		
37	Methyl cyclohexyl ketone		180	177	148		60
38	Cycloheptanone		181	163	93		23
39	Ethyl acetoacetate		181	133(129d)			
40	Ethyl methylacetoacetate		181	86			
41	5-Nonanone (Dibutyl)		186-7	90			
42	Butyroin		180-90				
43	2,5-Hexanedione (Acetonylacetone)		194	185d(mono) 224(di)	99 257(di)	120	137(di)
44	D-Fenchone, v.s.	4	195-6 (193)	184(172)	140		165(α) 123(β)
45	Methyl levulinate		196	143	142	96	
46	Phorone (m.p. 28) (See Table 20B)		198				
47	Ethyl ethylacetoacetate		198	154			
48	Acetophenone (m.p. 20), v.s.	5	202(199)	198-9*(203)	238-40N(250)	105-6	60
49	β-Thujone		202	174	114	104	55
50	Ethyl levulinate		206	148	102		59
51	l-Menthone		209(207)	189(187)	146		
52	Methyl 2-thienyl ketone		213-4	190-1		53	81
53	o-Methylacetophenone		214	203-5	159	96	61
54	Isophorone		215	199.5d(191)			79.5(76)
55	o-Hydroxyacetophenone (m.p. 28) (See Table 20B)		215	209-10		68	
56	Benzyl methyl (m.p. 27) (See Table 20B)		216				
57	Propiophenone (m.p. 20)		218	182(174)	191	147	54
58	m-Methylacetophenone		220	198	207		55
59	Phenyl n-propyl ketone		218-21	188			50
60	Isopropyl phenyl ketone (Isobutyrophenone)		222	181	163	73	94
61	Pulegone		224(221-2)	174(175-6)	142		119
62	Isovalerophenone		225	210	142		72(64.5)

(continued)

TABLE 20A (*continued*)

	Name of compound	Note	B.p.	Melting point of derivatives: Recommended: Semi-carbazone	Recommended: 2,4-Dinitro-phenyl-hydrazone	Others: Phenyl-hydra-zone	Others: Oxime
63	1-Phenyl-2-butanone		226	135(146)			
64	*p*-Methylacetophenone (m.p. 28) (See Table 20B)		226				
65	*m*-Chloroacetophenone		228	232			88
66	*o*-Chloroacetophenone		229	160	206		113
67	D-Carvone	6	230	162-3*N*,142-3	191	109-10	72-3,56-7
68	*n*-Butyrophenone		230(218-21)	191(188)	190		50
69	*p*-Chloroacetophenone (m.p. 20)		232(236)	202-4(160;146)	231	114	95
70	Methyl β-phenylethyl ketone		235	142			87
71	*m*-Methoxyacetophenone		240	196			
72	*n*-Valerophenone		242	166	166	162	52
73	*o*-Methoxyacetophenone		245	183		114	83
74	2,4,5-Trimethylacetophenone		246-7	204			85-6
75	*o*-Aminoacetophenone (m.p. 20) (See Table 9A)		250-2*d*	290*d*		108	109
76	Ethyl benzoylacetate		265	125			
77	Dypnone		340-5	151			134(syn)78(anti)

TABLE 20B

Ketones—Solid

	Name of compound	Note	M.p.	Melting point of derivatives: Recommended: Semi-carbazone	Recommended: 2,4-Dinitro-phenyl-hydrazone	Others: Phenyl-hydra-zone	Others: Oxime
1	Acetophenone (b.p. 202) (See Table 20A)		20				
2	*p*-Chloroacetophenone (b.p. 232) (See Table 20A)		20				

No.	Compound	Ref.	M.P.				
3	Propiophenone (b.p. 128) (See Table 20A)		20				
4	o-Aminoacetophenone (b.p. 250-2d) (See Table 20A)		20				
5	Benzyl methyl ketone (b.p. 216)		27	198	156	87	68-70
6	p-Methylacetophenone (b.p. 226)		28	205	258	96(121)	88
7	Phorone (b.p. 198)		28	221	112	110	48
8	o-Hydroxyacetophenone (b.p. 215)		28	210	100	121 (128-9)	118
9	n-Dihexyl ketone (b.p. 255)		33				
10	1,3-Diphenyl-2-propanone (b.p. 330) (Dibenzyl)		34	145-6 (125-6)	220	146	125
11	α-Naphthyl methyl ketone (b.p. 302)		34	229	227(223)	142	139
12	p-Chloropropiophenone		36	175-6		159(157)	62-3
13	p-Methoxyacetophenone (b.p. 258)		38	198	258	130-1N	87
14	Benzalacetone (b.p. 262, v.s.)		42	187-8			117
15	1-Indanone, v.s.	7	42	233(239)			146
16	Benzophenone (b.p. 360)	8	48	167	238-9	137-8	144
17	Phenacyl bromide		50	146	237*(230)	126	89.5 and 97.0
18	p-Bromoacetophenone		51	208	199	177	128
19	Phenyl p-tolyl ketone		55	121	262	120	153-4
20	β-Naphthyl methyl ketone		56	236	245*	116	145
21	Benzalacetophenone	9	57-8	168N	212	132,90	140,68(116,75)
22	Phenacyl chloride		59	156(149)	204*	130	89
23	Desoxybenzoin (b.p. 320)		60	148	180	106	98
24	p-Methoxybenzophenone	10	62	178-9	269(di)	128(135)	146-7(137-8), 115-6
25	p-Toluquinone, v.s.		69	179		126	134-5d
26	1,2-Tricosanone (Laurone)		69.5				84
27	Dihydroxyacetone	11	72N		277-8		
28	Dibenzoylmethane		78	205		151-2	
29	p-Chlorobenzophenone		78		185	112	163
30	m-Nitroacetophenone		81	257	228		132
31	p-Bromobenzophenone		82		230		169
32	Fluorenone, v.s.		83	146	283-4		195-6*
33	Phenacyl alcohol (See Table 6B)		86				70
34	Benzil		95	174-5(182)(mono), 243-4(di)	189	134(mono),225(di), (235)	137-8(mono), 237(di)
35	m-Hydroxyacetophenone		96	195			
36	p-Chlorophenacyl bromide	12	96-8	196d			
37	m-Aminoacetophenone (See Table 9B)		99				

(continued)

TABLE 20B (*continued*)

Name of compound	Note	M.p.	Melting point of derivatives			
			Recommended			Others
			Semi-carbazone	2,4-Dinitro-phenyl hydrazone	Phenyl-hydra-zone	Oxime
38 p-Aminoacetophenone (b.p. 294) (See Table 9B)		106	250	263		148
39 p-Bromophenacyl bromide	13	108-9	199	261	151	115
40 p-Hydroxyacetophenone		109	187-90	180	153	145
41 Dibenzalacetone		112			152	142-4
42 Quinone (Benzoquinone), *v.s.*	14	116	243(di)	231*(di)	90	240(di)
43 Acenaphthenone		121				175
44 p-Phenylphenacyl bromide	15	124-5				
45 1,4-Naphthoquinone, *v.s.*	15	125	247(mono)	278(mono)	205-6d(mono)	198(mono)
46 p-Phenylphenacyl chloride	15	126-7				
47 Furoin		135	194	217	81	161
48 p-Hydroxybenzophenone (See Table 25B)	17	135		242.4*	144	152 and 81N
49 Benzoin (See Table 6B)	16	137(133)	205-6d	245(234)	158-9(α)106(β)	99(syn)(β)151-2(anti)
50 1,2-Naphthoquinone	18	145-7d(120)(115-20d)	184d		138	169(di)162-4(mono)
51 Methone (5,5-Dimethyl-dihydroresorcinol)		148-9				115(mono)
52 Furil		165		215	184	100(di)
53 Quinhydrone	19	171			152	161
54 Gallacetophenone		173	225		174-5	162-3
55 4,4'-Bis(dimethylamino)-benzophenone (See Table 10B)		174				233
56 D(+)-Camphor (b.p. 209)		179	247-8d*(238)	177	233	118-9
57 Camphorquinone, *v.s.*	20	198	236d;147N			170
58 9,10-Phenanthraquinone		207s	220d	312d	165	202d r.h.(di)158(mono)
59 Anthraquinone	21	286*N			183*	224
60 Alizarin	22	290*N				
61 Chloranil		290s.t.			220	

Notes on Ketones

1. Calcium chelate green crystals, m.p. 113-5.
2. Mesityl oxide semicarbazone exists as two isomers: α, m.p. 164; β, (from benzene) m.p. 133-4.
3. Crystallized from acetic acid, m.p. 146; from ethanol, m.p. 142.
4. D- or L-Fenchone forms two oximes: α+, m.p. 165; β+, m.p. 123. DL-Fenchone forms two oximes: α+, m.p. 159; β, m.p. 129.
5. Acetophenone-2,4-dinitrophenylhydrazone is reported to melt at 238-40 (crystallized from alcohol) and at 249-50 (crystallized from acetic acid).
6. D-Carvone forms two semicarbazones: (1) m.p. 162-3; (2) m.p. 141-2. The semicarbazone of the L-form melts at 162-3 and that of the DL-form at 154-6. D-Carvone forms two oximes: α (leaflets) from ethanol, m.p. 72-3v.s.; β, m.p. 56-7 (levo-rotatory in ethanol). L-Carvoxime has two forms: α-(-), m.p. 72v.s.; β-(+), m.p. 57-8. DL-Carvoxime melts at 93-4.
7. m.p. 134-5 if extracted with hot methanol and recrystallized.

8. p-Nitrophenylhydrazone: yellow, m.p. 154-5; red, m.p. 144.
9. Semicarbazone: α (colorless), m.p. 168; β (yellow), m.p. 170; γ (colorless); m.p. 178-9. Chalcone dimerizes to produce four dimers, m.p. 124, 178, 195, and 225-6.
10. Oxime: α-isomer, m.p. 146-7 (137-8 older ref.); β-isomer, m.p. 115-6.
11. Dihydroxy acetone (m.p. 72) dimerizes, m.p. 78-81. Dibenzoate (m.p. 120-5).
12. Acetate ester, m.p. 72.
13. Acetate ester, m.p. 86.
14. Chlorine-like odor, liberates I_2 from KI solution (acidified). Sublimes readily. Equimolecular quantities with hydroquinone gives quinhydrone, m.p. 171.
15. Acetate ester, m.p. 111.
16. Phenylhydrazone: α (needles), m.p. 158-9; β (prisms), m.p. 106. Oxime: α (syn) oxime, m.p. 99; β (anti) oxime, m.p. 151.2.
17. p-Hydroxybenzophenone oxime exists as two isomers, m.p. 81 and 152. The lower-melting isomer changes into the other by continued heating at its melting point.

18. The 1-p-Nitrophenylhydrazone derivative, m.p. 250-1 (246); the 2-p-nitrophenylhydrazone derivative, m.p. 236. The two isomeric mono-oximes are tautomeric with the corresponding nitroso phenols: (1) 2-nitroso-1-naphthol (2-oxime), m.p. 162-4; (2) 1-nitroso-2-naphthol (1-oxime), m.p. 109.
19. Oxidizes to quinone, m.p. 116, readily; and reduces readily to hydroquinone, m.p. 171.
20. The 3-semicarbazone has two forms: α (from ethanol), m.p. 236d; β (from benzene-petroleum ether), m.p. 147. The 3-p-bromophenylhydrazone (from acetic acid) melts at 215-6.
21. Several m.p. listed in literature: 273, 284-6, 285; b.p. 382. Compound boiled with NaOH solution and Zn dust gives deep red filtrate.
22. Sublimes above 110 (differentiation from flavopurpurin and anthrapurpurin, which sublime at 160 and 170, respectively), alizarin diacetate, from acetic anhydride and H_2SO_4 in cold water, precipitated on dilution and recrystallized from alcohol, m.p. 182.

TABLE 21

Azo-, Hydrazo-, and Azoxy- Compounds

	Parent compound	Azo-	M.p.
1	Toluene	o-Azo-	55
2	Toluene	m-Azo-	55
3	Benzene	Azo-	68
4	Phenetole	o-Azo-	131
5	Toluene	p-Azo-	144
6	Biphenyl	o-Azo-	145
7	Phenetole	p-Azo-	160
8	Naphthalene	1,1'-Azo-	190
9	Naphthalene	2,2'-Azo-	208

	Parent compound	Azoxy-	M.p.
1	Toluene	o-Azoxy-	59
2	Toluene	m-Azoxy-	39
3	Benzene	Azoxy-	36
4	Phenetole	o-Azoxy-	102
5	Toluene	p-Azoxy-	70
6	Biphenyl	o-Azoxy-	158
7	Phenetole	p-Azoxy-	138
8	Naphthalene	1,1'-Azoxy-	127
9	Naphthalene	2,2'-Azoxy-	168

	Parent compound	Hydrazo-	M.p.
1	Toluene	o-Hydrazo-	165
2	Toluene	m-Hydrazo-	38
3	Benzene	Hydrazo-	130
4	Phenetole	o-Hydrazo-	89
5	Toluene	p-Hydrazo-	134
6	Biphenyl	o-Hydrazo-	182
7	Phenetole	p-Hydrazo-	86
8	Naphthalene	1,1'-Hydrazo-	153
9	Naphthalene	2,2'-Hydrazo-	140

TABLE 22A

Nitriles—Liquid

	Name of compound	B.p.	Carboxylic acid formed by hydrolysis
1	Acrylonitrile	78(74)	b.p. 140
2	Acetonitrile	81.6	b.p. 118
3	Propionitrile	97.3	b.p. 141
4	Isobutyronitrile	104	b.p. 155
5	Butyronitrile	117.4	b.p. 162.5
6	Allyl cyanide (β-Butenonitrile)	119	b.p. 163(169)
7	α-Hydroxyisobutyronitrile	120	m.p. 79
8	Methoxyacetonitrile	120	b.p. 203-4
9	Chloroacetonitrile	127	m.p. 63

TABLE 22A

Nitriles—Liquid

Name of compound	B.p.	Carboxylic acid formed by hydrolysis
10 Isovaleronitrile	130	b.p. 176
11 Valeronitrile	141	b.p. 186
12 Furonitrile	147	m.p. 133-4
13 Isocapronitrile	155	b.p. 199
14 *n*-Capronitrile	165	b.p. 205
15 α-Mandelonitrile (m.p. 29)	170*d*	m.p. 133
16 β-Chloropropionitrile	178	m.p. 41
17 Lactonitrile	182-4	m.p. 25-6
18 Glycolonitrile	183*d*	m.p. 80
19 Heptanonitrile	183	b.p. 223
20 Benzonitrile	190.2	m.p. 122
21 γ-Chlorobutyronitrile	196-7	b.p. 196
22 Caprylonitrile	199	b.p. 239
23 *o*-Tolunitrile	205	m.p. 104
24 Ethylcyanacetate	207	
25 *m*-Tolunitrile	212	m.p. 113(110)
26 *p*-Tolunitrile (m.p. 27)	217	m.p. 179-80
27 Malonitrile (m.p. 30)	218-9	m.p. 135.6
28 β-Hydroxypropionitrile	220	
29 Phenylacetonitrile	234	m.p. 76.5
30 γ-Hydroxybutyronitrile	240	b.p. 204
31 Decanonitrile	245	m.p. 31
32 Cinnamonitrile (m.p. 20)	255-6	m.p. 133
33 Dodecanonitrile	280	m.p. 44
34 Glutaronitrile	286	m.p. 97
35 Adiponitrile (m.p. 0-1)(di)	295	m.p. 153

TABLE 22B

Nitriles—Solid

Name of compound	M.p.	Carboxylic acid formed by hydrolysis
1 Cinnamonitrile (b.p. 255-6)	20	m.p. 133
2 *p*-Tolunitrile (b.p. 217)	27	m.p. 179-80
3 Malonitrile (b.p. 218-9)(di)	31	m.p. 135
4 α-Naphthonitrile	36	m.p. 162
5 *m*-Bromobenzonitrile (b.p. 225)	38	m.p. 155
6 *m*-Chlorobenzonitrile, *v.s.*	41	m.p. 158(155)
7 *o*-Chlorobenzonitrile (b.p. 232), *v.s.*	47	m.p. 142(140)
8 *o*-Bromobenzonitrile (b.p. 251-3/754 mm), *v.s.*	53	m.p. 150
9 *o*-Iodobenzonitrile	55	m.p. 162
10 Succinonitrile	57	m.p. 186-8
11 *p*-Methoxybenzonitrile (b.p. 256-7)	61-2	m.p. 184-6
12 β-Naphthonitrile	66	m.p. 184
13 Cyanoacetic acid	66	m.p. 135-6
14 *p*-Chlorobenzonitrile	92(96)	m.p. 240
15 *o*-Nitrobenzonitrile	110	m.p. 146
16 *p*-Bromobenzonitrile	112	m.p. 251
17 *p*-Nitrophenylacetonitrile	116	m.p. 153
18 *m*-Nitrobenzonitrile	118	m.p. 140
19 Methylaminoacetonitrile	129	m.p. 232*d*
20 *p*-Nitrobenzonitrile	147	m.p. 241

TABLE 23A

Nitro Compounds—Liquid

	Name of compound	B.p.	Melting point of derivatives		
			Recommended		
			Benzene-sulfon-amide[a]	Benz-amide[a]	Others[b]
1	Nitromethane	101	30	80	
2	Chloropicrin	113			
3	Nitroethane	114	58	71	
4	2-Nitropropane	120.3			
5	Tetranitromethane (m.p. 13)	126			
6	1-Nitropropane	131.6	36	84	
7	2-Nitrobutane	140	70	76	
8	2-Methyl-1-nitropropane	140.5	53	57	
9	2-Methyl-2-nitrobutane	150			
10	2-Nitropentane	152-4			
11	1-Nitrobutane	153			
12	3-Methyl-1-nitrobutane	164			Acetamide, b.p. 230
13	1-Nitropentane	173			
14	1-Nitrohexane	193	96	40	
15	Nitrobenzene	210.85*	112	160	m-Dinitrobenzene, 90
16	o-Nitrotoluene	221.7	124	143	2,4-Dinitrotoluene, 70
17	o-Nitroethylbenzene	224		147	
18	2-Nitro-m-xylene (m.p. 13) (2,6-Dimethylnitrobenzene)	226			2,4,6-Trinitro-m-xylene, 182
19	Phenylnitromethane	226d	88	105	
20	m-Nitrotoluene (m.p. 16)	232.6	95	125	m-Nitrobenzoic acid, 140
21	2-Nitro-p-xylene (2,5-Dimethylnitrobenzene)	239/739 mm	138	140	2,3,5-Trinitro-p-xylene, 139
22	p-Nitroethylbenzene	241		151	2,4,6-Trinitroethylbenzene, 37
23	4-Nitro-m-xylene (m.p. 2)	244/744 mm	129	192	2,4,6-Trinitro-m-xylene, 182
24	3-Nitro-o-xylene (m.p. 15) (2,3-Dimethylnitrobenzene)	250			3,4-Dinitro-o-xylene, 82
25	2-Nitro-p-cymene	264		102	2,6-Dinitrocymene, 54
26	o-Nitroanisole (m.p. 10)	265	89		2,4,6-Trinitroanisole, 68
27	o-Nitrophenetole (m.p. 5-6)	268	102		2,4-Dinitrophenetole, 86

[a] The substituted amides refer to the derivatives of the amines formed by the reduction of the nitro compounds.

[b] For derivatives for the nitroparaffins, consult Table 9A for the primary amines that are formed by reducing the nitro compounds.

TABLE 23B

Nitro Compounds—Solid

	Name of compound	M.p.	Benzene-sulfon-amide[a]	Benz-amide[a]	Others
			Melting point of derivatives		
			Recommended		
1	2,4-Dinitrofluorobenzene	25-7	185		
2	m-Nitrobenzyl alcohol	27			m-Nitrobenzoic acid, 140
3	4-Nitro-o-xylene (3,4-Dimethylnitrobenzene)	30			3,4-Dinitro-o-xylene, 82 Acetamide, 99
4	o-Chloronitrobenzene	32	129	99	2,4-Dinitrochlorobenzene, 52(50)
5	2,4-Dichloronitrobenzene (b.p. 258)	33	128	115	
6	m-Nitrophenetole	34		103	
7	2-Nitrobiphenyl	37(33)		102	
8	m-Iodonitrobenzene	38			m-Iodoaniline, 33
9	m-Nitroanisole (b.p. 258)	38			3,5-Dinitroanisole, 106
10	o-Bromonitrobenzene	43		116	2,4-Dinitrobromobenzene, 72
11	p-Nitrobenzal chloride	43			p-Nitrobenzoic acid, 241
12	Nitromesitylene	44		204	Dinitromesitylene, 86
13	m-Chloronitrobenzene	45		118	
14	o-Iodonitrobenzene	49		139	o-Iodoaniline, 56
15	o-Nitrobenzyl chloride	49			o-Nitrobenzoic acid, 146
16	p-Nitrotoluene (b.p. 238.34*)	51.65*	120	158	2,4-Dinitrotoluene, 70
17	2,4-Dinitrochlorobenzene	52		178	2,4-Dinitrophenol, 114
18	p-Nitroanisole	52.5	95	154	2,4-Dinitroanisole, 89
19	2,5-Dichloronitrobenzene	54		120	
20	m-Bromonitrobenzene	56		136(120)	3,4-Dinitrobromobenzene, 59
21	1-Nitronaphthalene	58	167	160	
22	β-Nitrostyrene	58			
23	p-Nitrophenetole	58	143	173	2,4-Dinitrophenetole, 86
24	3-Nitrobiphenyl	61(59)			3-Aminobiphenyl, 30
25	m-Nitrobenzal chloride	65			m-Nitrobenzaldehyde, 58
26	2,6-Dinitrotoluene	66			2,4,6-Trinitrotoluene, 80(82)
27	2,4,6-Trinitroanisole	68			Picric acid, 122.5 Naphthalene adduct, 69-70
28	2,4-Dinitrotoluene	70			Naphthalene adduct, 60 Acetyl-2,4-diamino-toluene, 224
29	p-Nitrobenzyl chloride	72			p-Toluidine, 45 p-Nitrobenzoic acid, 241
30	2,4-Dinitrobromobenzene	72			2,4-Dinitrophenol, 114
31	o-Nitrobenzyl alcohol	74			o-Nitrobenzoic acid, 146
32	3,5-Dimethylnitrobenzene (b.p. 273)	75			Acetamide, 144
33	2-Nitronaphthalene	78	136	162	
34	2,4,6-Trinitrophenetole	78			Picric acid, 122.5 Naphthalene adduct, 39
35	2,4,6-Trinitrotoluene	80.1(82)			2,4,6-Trinitrobenzoic acid, 230
36	p-Nitrobenzal bromide	82			p-Nitrobenzoic acid, 241
37	Picryl chloride	83	211		Picric acid, 122.5
38	p-Chloronitrobenzene	84	121	192	p-Nitrophenol, 114
39	2-Nitroresorcinol	85			Dimethyl ether, 131
40	2,5-Dibromonitrobenzene	85			Acetamide, 171-2
41	4-Bromo-1-nitronaphthalene	85			4-Bromo-1-naphthyl-amine, 102

(continued)

TABLE 23B (*continued*)

	Name of compound	M.p.	Melting point of derivatives		
			Recommended		
			Benzene-sulfon-amide[a]	Benz-amide[a]	Others
42	2,4-Dinitrophenetole	86			2,4,6-Trinitrophenetole, 78
43	4-Chloro-1-nitronaphthalene	87(85)			4-Chloro-1-naphthyl-amine, 98
44	8-Nitroquinoline	88-9		98	8-Aminoquinoline, 70
45	m-Dinitrobenzene, v.s.	90	194	240	m-Nitroaniline, 114
					Naphthalene adduct, 52
46	3,5-Dinitrotoluene	92			3,5-Dinitrobenzoic acid, 204-5
47	p-Nitrobenzyl alcohol	93			
	(See Table 6B)				
48	8-Chloro-1-nitronaphthalene	94			8-Chloro-1-napthyl-amine, 96
49	2,4-Dinitroanisole	95			2,4-Dinitrophenol, 114
					Naphthalene adduct, 50
50	4-Chloro-2-nitroanisole	98			4-Chloro-o-anisidine, 84
51	p-Nitrobenzyl bromide	99			Esters of acids
	(See Table 15B)				
52	8-Bromo-1-nitronaphthalene	99-100			8-Bromo-1-naphthyl-amine, 90
53	5-Chloro-1-nitronaphthalene	111			5-Chloro-1-naphthyl-amine, 85
54	4-Nitrobiphenyl	114		230	Acetamide, 171
55	m-Nitrobenzonitrile	117-8			m-Nitrobenzoic acid, 140
56	o-Dinitrobenzene	118	186	301	o-Nitroaniline, 71
57	1,3,5-Trinitrobenzene	122			Compd. with anthracene, 164
58	5-Bromo-1-nitronaphthalene	122.5			5-Bromo-1-naphthyl-amine, 69
59	Trinitrophenol (Picric acid)	122.5			Acetate, 76
60	2,6-Dinitro-p-xylene	123			6-Nitro-p-2-xylidine, 98
61	2,4,6-Tribromonitrobenzene	125		198	2,4,6-Tribromoaniline, 122
62	p-Bromonitrobenzene	126	136	204	p-Nitrophenol, 114
63	2,2'-Dinitrobiphenyl	128(124)			2,2'-Diaminobiphenyl, 81
64	4-Chloro-1,5-dinitronaphthalene	138			4,8-Dinitro-1-naphthol, 235d
65	4-Bromo-1,5-dinitronaphthalene	143			4,8-Dinitro-1-naphthol, 235d
66	p-Nitrobenzonitrile	145-6			p-Nitrobenzoic acid, 241
67	4-Chloro-1,3-dinitronaphthalene	146.5			2,4-Dinitro-1-naphthol, 140
68	6-Nitroquinoline	149-50		169	6-Aminoquinoline, 114
69	Picramic acid (See Table 9B)	169		230(220)	
70	1,8-Dinitronaphthalene	170			1,3,8-Trinitronaphthalene, 218
71	4-Bromo-1,8-dinitronaphthalene	170			4,5-Dinitro-1-naphthol, 230d
72	p-Iodonitrobenzene	173		222	p-Iodoaniline, 62
73	p-Dinitrobenzene	173	247	300	p-Nitrophenol, 114
74	4-Chloro-1,8-dinitronaphthalene	180			4,5-Dinitro-1-naphthol, 230d
75	4,4'-Dinitrodiphenylmethane	183			4,4'-Dinitrobenzophenone, 189
76	3,3'-Dinitrobiphenyl	200			3,3'-Diaminobiphenyl, 94
77	1,5-Dinitronaphthalene	214			1,4,5-Trinitronaphthalene, 154
78	4,4'-Dinitrobiphenyl	240			Benzidine, 128

[a] The substituted amides refer to the derivatives of the amines formed by reduction of the nitro compounds.

[b] For derivatives for the nitroparaffins, consult Table 9A for the primary amines that are formed by reducing the nitro compounds.

TABLE 24

Nitroso Compounds[a]

	Name of compound	M.p.		Name of compound	M.p.
1	1-Nitroso-2,4-dimethylbenzene	41	14	1-Nitroso-2,5-dimethylbenzene	101
2	1-Nitroso,3,4-dimethylbenzene	45	15	5-Nitroso-o-anisidine	107
3	p-Nitrosotoluene	48	16	1-Nitroso-2-naphthol	109
4	N-Nitrosoacetanilide	51d	17	p-Nitroso-N-methylaniline	118
5	m-Nitrosotoluene	53	18	p-Nitrosophenol	125
6	N-Nitrosodiphenylamine	66	19	5-Nitroso-o-cresol	135
7	Nitrosobenzene	68	20	p-Nitrosodiphenylamine	144
8	o-Nitrosotoluene	72	21	1-Nitroso-2,6-dimethylbenzene	145
9	p-Nitroso-N-ethylaniline	78	22	2-Nitroso-1-naphthol	162d
10	p-Nitroso-N,N-diethylaniline	84(86)	23	Nitrosothymol	162
11	p-Nitroso-N,N-dimethylaniline	87(84)	24	p-Nitrosoaniline	174
12	1-Nitroso-2,3-dimethylbenzene	91	25	4-Nitroso-1-naphthol	194(198)
13	1-Nitrosonaphthalene	98	26	1-Nitrosoanthraquinone	224

[a] Nitroso compounds may be derivatized by reduction to the corresponding amine.

TABLE 25A

Phenols—Liquid[a]

| | Name of compound | Note | M.p. | B.p. | Melting point of derivatives | | | |
| | | | | | Recommended | | Others | |
					α-Naphthyl-urethan	3,5-Dinitro-benzoate	(Mono, di, tri) Bromo derivative	Aryloxy-acetic acid
1	o-Chlorophenol		7	175.6	120	143	48-9(mono),76(di)	145
2	Phenol (See Table 25B)		41.8	182				
3	o-Cresol (See Table 25B)		31	191-2				
4	o-Bromophenol		5	195	129		95	143
5	3-Chloro-p-cresol			196				108
6	Salicylaldehyde (See Table 7A)			197				132
7	p-Cresol (See Table 25B)		36	202				
8	o-Ethylphenol			202(207)		108		141
9	m-Cresol		12	203	128	165.4*	84(tri)	103
10	2,4-Dimethylphenol (See Table 25B)		27	211.5*				
11	o-Hydroxyacetophenone (See Table 20B)		28	215				
12	m-Ethylphenol			217				77
13	Methylsalicylate	1		224				
14	Ethylsalicylate	2		234				
15	p-Isobutylphenol			236				125
16	Carvacrol		1	237.8	116	83(77)	46	151
17	m-Methoxyphenol			243	129		104(tri)	111-3
18	p-Butylphenol		22	248				81
19	p-Amylphenol		23	248-53				90
20	Eugenol		16-9	254.8	122	130.8*	118(tetra)	81,100
21	Isoeugenol			267.5	150	158.4*		94,116

[a] For phenyl isocyanate derivatives of certain alkylated phenols, see Ind. Eng. Chem., Anal. Ed. 16, 304 (1944).

TABLE 25B

Phenols—Solid[a]

| Name of compound | Note | M.p. | B.p. | Melting point of derivatives | | | |
| | | | | Recommended | | Others | |
				α-Naphthyl-urethan	3,5-Dinitro-benzoate	(Mono, di, tri) Bromo derivative	Aryloxy-acetic acid
1 p-Butylphenol (See Table 25A)			248				
2 p-Amylphenol (See Table 25A)			248-53				
3 2,4-Dimethylphenol		27	211.5*	135	164.4		141
4 o-Methoxyphenol (Guaiacol), v.s.		28(32)	205	118	141.2*	116(tri)	116
5 o-Cresol		31	191-2	142	138.4*	56(di)	152
6 m-Bromophenol		32	236	108			108
7 2-Bromo-4-chlorophenol		33-4	123/10 mm				139-40*
8 m-Chlorophenol		33	214	158	156		110
9 p-Cresol		36	202	146	188.6*	108(tetra), 49(di)	135
10 2,4-Dibromophenol		36	238-9				153
11 m-Iodophenol		40					115
12 Phenol	3	41.8	182	133	145.8*	95(tri)	99
13 Phenyl salicylate (Salol)		42	173/12 mm		183	95(tri)	
14 p-Chlorophenol	4	43	217	166	186		156
15 2,4-Dichlorophenol	5	43	209			68	135
16 o-Iodophenol		43					135
17 o-Nitrophenol, v.s.	6	45		113	155	117(di)	158
18 p-Ethylphenol		47	219	128	133		97
19 2,6-Dimethylphenol		49	203	176.5	158.8*	79	139.5
20 Thymol		49.7	233	160	103.2	55	149
21 p-Methoxyphenol		56	244				110-12
22 2,3-Dichlorophenol, v.s.		56-7	206	141-2	181.6*	90(di)	
23 3,4-Dimethylphenol, v.s.		62.5	225	169	191	171(tri)	162.5
24 p-Bromophenol	7	64				95(tri)	157
25 4-Homopyrocatechol		65	252				58(di)
26 p-Chloro-m-cresol, v.s.		66	235	153-4			
27 4-n-Hexylresorcinol	8	67.5	335d				
28 3,5-Dichlorophenol, v.s.		68	233			189(tri)	
29 2,4,5-Trichlorophenol		68		188			157
30 2,4,6-Trichlorophenol		68	245		136		182
31 3,5-Dimethylphenol, v.s.		68(64)	220		195.4*	166(tri)	111(81)

(continued)

Table 25B (continued)

| Name of compound | Note | M.p. | B.p. | Melting point of derivatives | | | |
| | | | | Recommended | | Others | |
				α-Naphthyl urethan	3,5-Dinitro-benzoate	(Mono, di, tri) Bromo derivative	Aryloxy-acetic acid
32 2,4,6-Trimethylphenol		69				158(di)	142
33 Mesitol		70	220			158(di)	139.5
34 2,4,5-Trimethylphenol		71	232			35	132
35 2,5-Dimethylphenol		74.5	212	173	137.2*	178(tri)	118
36 2,3-Dimethylphenol		75					187
37 Vanillin (See Table 7B)		81					187
38 Saligenin (See Table 6B)		86–7					120
39 1-Naphthol		94	280	152	217.4*	105(2,4-di)	193.5
40 p-Iodophenol		93–4					156
41 2,4,6-Tribromophenol		95		153	174	120(tetra)	200
42 2-Naphthylsalicylate	9	95.5					
43 m-Hydroxyacetophenone (See Table 20B)		96					
44 m-Nitrophenol		97		167	159	91(di)	156
45 p-tert-Butylphenol		100	237	110		50	86
46 m-Hydroxybenzaldehyde (See Table 7B)		104					148
47 Pyrocatechol		105	245.6	175	152	193(tetra)	
48 1,2-Naphthalenediol		108					104–6(di)
49 Orcinol		108		160	190	104(tri)	217
50 p-Hydroxyacetophenone (See Table 20B)		109					
51 2,4,6-Trinitro-m-cresol	10	109–10					
52 Resorcinol	11	110	280.8*(275.9)		201	112(4,6-di)	195
53 Bromohydroquinone		110				186(di)	
54 p-Nitrophenol	6	114		150–1	186	142(2,6-di)	187
55 2,4-Dinitrophenol	12	114				118(6-Br)	
56 p-Hydroxybenzaldehyde (See Table 7B)		116–7					198
57 Phloroglucinol (anh.)		117				151(tri)	
58 m-Hydroxyaniline (See Table 9B)		122					
59 Trinitrophenol (Picric acid)	13	122					
60 2-Naphthol		123	286	157	210.2*	84	154
61 Toluhydroquinone		124					153
62 p-Cyclohexylphenol		132			168*		
63 Pyrogallol		133			205(tri)		
64 p-Hydroxybenzophenone (See Table 20B)		135				158(di)	198
65 1,8-Naphthalenediol		142		220			

No.	Compound					
66	2,4-Dinitroresorcinol	147.8				155
67	3-Nitrosalicylic acid (See Table 5B)	148-9				
68	p-Hydroxypropionphenone (See Table 20B)	148				191
69	Salicylic acid (See Table 5B)	158.3				
70	2,4,6-Triiodophenol	159		181		
71	Picramic acid (See Table 9B)	169				
72	Hydroquinone	171(172)	286	317	186(di)	250
73	3,5-Dinitrosalicylic acid (See Table 5B)	173-4				
74	o-Aminophenol (See Table 9B)	174				
75	p-Aminophenol (See Table 9B)	184s				
76	2,7-Naphthalenediol	190				149
77	Pentachlorophenol	190.2				196
78	2-Hydroxyquinoline (anh.) (See Table 10B)	199				
79	m-Hydroxybenzoic acid (See Table 5B)	200				206
80	p-Hydroxybenzoic acid (See Table 5B)	215				278
81	2,4-Dihydroxybenzoic acid (See Table 5B)	216d				
82	Phloroglucinol (dihydrate)	218		162		
83	5-Nitrosalicylic acid (See Table 5B)	229-30			151(tri)	
84	4,4'-Dihydroxybiphenyl	274-5				274

^{a} For phenyl isocyanate derivatives of certain alkylated phenols, see *Ind. Eng. Chem., Anal. Ed.* **16**, 304 (1944).

Notes on Phenols

1. Shaken with cold alkaline solution of $(CH_3CO)_2O$ gives a derivative, m.p. 52. Nitrated with fuming HNO_3 and H_2SO_4 at 0° gives methyl 3,5-dinitrosalicylate, m.p. 126-7.

2. Nitrated with fuming HNO_3 and H_2SO_4 gives ethyl 3,5-dinitrosalicylate, m.p. 92-3.

3. Compound crystallizes in three modifications: (1) m.p. 42; (2) m.p. 38.8; (3) m.p. 28.5 (*Z. Physik. Chem.* **29**, 71). For the m.p. of the phenylurethan see *Ber.* **40**, 1834, and *C.A.* **26**, 5556.

4. Br_2 in $CCl_4 \rightarrow$ 2-bromo-4-chlorophenol, m.p. 33-4. $2Br_2$ in HAc $\rightarrow$ 2,6-dibromo-4-chlorophenol, m.p. 90.

5. The β-naphthyluretane, m.p. 166.

6. Nitration with fuming HNO_3 (and concentrated H_2SO_4) gives picric acid, m.p. 122. p-Nitrophenol can be readily derivatized by reduction.

7. Nitration by means of fuming HNO_3 in acetic acid yields dinitro derivative, m.p. 76.

8. "Attempts to prepare the benzoate. p-nitrobenzoate, or 3,5-dinitrobenzoate derivatives gave only non-crystallizable or tarry products."

9. Refluxed with equivalent amounts of $(CH_3CO)_2O$ and fused with CH_3COONa for 4 hours gives a derivative, m.p. 136.

10. Methyl ether, m.p. 94.

11. Two forms: labile, m.p. 108.0-108.5; stable, m.p. 110.

12. Nitration with fuming HNO_3 and H_2SO_4 gives picric acid, m.p. 122.

13. Slowly heated sublimes, quickly heated explodes; addition compounds with naphthalene, m.p. 149(151); nitron picrate, m.p. 210, detects 1:250,000 (Bush, *Ber.* **38**, 4056).

TABLE 26[a]

Sulfonamides and Related Compounds

	Name of compound	M.p. of amide	Acid chloride	Acid	Thiouronium salt
1	Ethane-	58	b.p. 178	-17	115
2	Methane-	90	b.p. 60/21	20	
3	Hexadecane-	97	54	54	
4	Benzyl-	105	92		
5	m-Toluene-	108	12		
6	2-Naphthol-3-	110	112		
7	2,6-Dimethylbenzene-	113(96)	39	98	
8	2-Phenyl-1-ethane-	122	33	91	
9	D-Camphor-10-	132	67	193	
10	DL-Camphor-8-	133-5	106		
11	D-Camphor-8-	137	138		
12	p-Toluene	137	69	92	181-2
13	2,4-Dimethylbenzene-	139(137)	34	62(2H$_2$O)	146
14	2,4,6-Trimethylbenzene-	142	56	77	
15	3,4-Dimethylbenzene-	144	52	64	208
16	p-Chlorobenzene-	144	53	93(68)	175
17	2,5-Dimethylbenzene-	148	24-6	86(2H$_2$O)	184
18	1-Naphthalene-	150	68	90	137
19	Benzene-	153	15	66	148
20	o-Toluene-	156	68(10)	57	170
21	m-Nitrobenzyl-	159d	100	74(H$_2$O)	
22	m-Aminobenzene-	165			185
23	p-Bromobenzene-	166(161)	76	88-90	215-216.5
24	m-Nitrobenzene-	167	64	48	146
25	2,5-Dimethyl-3-nitrobenzene-	172-3	61	200	
26	Phenol-4-	177		93-7	169
27	2,5-Dichlorobenzene-	181	38	112	170
28	2,4,5-Trimethylbenzene-	181	61	100	
29	4-Chloro-2,5-dimethylbenzene-	185	50	145	
30	2,5-Dimethyl-6-nitrobenzene-	192	110	70(85)	
31	o-Nitrobenzene-	193	69	140	
32	2,5-Dimethyl-4-nitrobenzene-	198	75	105	
33	1-Nitronaphthalene-2-	214	121	91(122)	
34	2-Naphthalene-	217(213)	76(79)		191
35	1,3-Benzene-di-	229	63		214.3
36	p-Sulfobenzoic acid	236(di)	57(di)	94(3H$_2$O)	213
37	4-Chloro-2-nitrobenzene-	237	75	82	
38	2-Naphthol-6-	237-9		129(H$_2$O)	217(207)
39	2,7-Naphthalene-di-	242	158(162)		212

(continuation of Table 26)

No.	Name of compound				Thiouronium salt
40	1,5-Anthraquinone-di-	246	265-70	310d	206
41	1,2-Benzene-di-	254	143		180
42	1-Naphthylamine-5-	260	197		211
43	2-Anthraquinone-	261	129		81
44	1,6-Naphthalene-di-	298	203		171
45	4,4'-Diphenyl-di-	300	225	72	
46	2,6-Naphthalene-di-	305	183		256
47	1,5-Naphthalene-di-	310(340)		245	257(251)
48	1,8-Anthraquinone-di-	340	223	294	

A much larger listing of sulfonamides, sulfonchlorides, and their derivatives are included in the Tables of the larger text by these authors.

TABLE 27[a,b]

Sulfonic Acids

No.	Name of compound	Thiouronium salt	No.	Name of compound	Thiouronium salt
1	1-Amino-8-naphthol-3,6-di-	312d	11	2-Naphthol-3,6-di-	233.2
2	Anthraquinone-1-	191	12	2-Naphthol-6,8-di-	228
3	Benzothiazole-2-	170.5-171.0	13	1-Naphthalene-8-	300d
4	1-Bromocamphor-1-	133.7	14	2-Naphthalene-4,8-di-	209-11d
5	m-Diethylaminobenzene-	182.4	15	2-Naphthalene-5,8-di-	276
6	1-Naphthol-2-	169.4	16	2-Naphthalene-6,8-di-	276d
7	1-Naphthol-4-	103.4	17	N-Phenyl-1-naphthalene-8-	182-9d
8	1-Naphthol-3,6-di-	217	18	o-Sulfobenzoc	206
9	1-Naphthol-4,8-di-	205.2	19	m-Sulfobenzoc (monosalt)	163
10	2-Naphthol-8-	218	20	Thymolsulfonic	212.4

[a] Additional sulfonic acids may be found in Table 26.
[b] Data for the amides and acid chlorides are not available for the acids of this table.

TABLE 28A

Thioethers (Sulfides)—Noncyclic[a]

	Name of compound	Sulfide (b.p.)	Sulfone (m.p.)
1	Methyl	38	109
2	Ethyl methyl	65-7	36
3	Vinyl	84	
4	Methyl isopropyl	84.7	
5	Ethyl	91	70-1
6	Methyl propyl	95.5	
7	Chloromethyl methyl	106	
8	Ethyl isopropyl	107.4	
9	Ethyl propyl	118.5	
10	Isopropyl	119	36
11	tert-Butyl ethyl	120.4	
12	Chloromethyl ethyl	128	
13	Propyl isopropyl	132	
14	sec-Butyl ethyl	133.6	
15	Bromoethyl methyl	134	
16	Isobutyl ethyl	134.2	
17	Propyl	142.8	30
18	Butyl ethyl	144.3	50
19	tert-Butyl	149	
20	Chloromethyl propyl	150	
21	Chloromethyl	156	71-2
22	Isobutyl	169(173)	17

	Name of compound	Sulfide (m.p.)	Sulfone (m.p.)
23	Butyl	188-9 b.p.	46(42)
24	Methyl phenyl	192 b.p.	88
25	Ethyl phenyl	204 b.p.	42
26	Isoamyl	215 b.p.	31
27	Bis(2-chloroethyl (m.p. 13)	215-7d b.p.	56
28	Phenyl	296(290) b.p.	128(125)
29	Decyl (205-7/4 mm)	22(17-23)	
30	Dodecyl (260-3/4 mm)	40.0-40.5(37.5-40.0)	
31	Benzyl phenyl	41	146
32	Bis(4-chlorobenzyl)	41	
33	Bis-(p-methoxyphenyl)	46	130
34	Tetradecyl	49-50	
35	Benzyl	50	151.7
36	p-Tolyl	57	158
37	Octyl	57	76
38	Hexadecyl (Cetyl)	61.3	103.4
39	Octadecyl (b. dec)	68-9	
40	Bis-(2-phenylethyl)	92.2	100.6
41	Bis-(p-bromophenyl)	112	172
42	Methyl-1-naphthyl	119	
43	Methyl-2-naphthyl	127	
44	Bis-(p-nitrobenzyl)	158-9	260.5

[a] See J. Am. Chem. Soc. 69, 2053 (1947), for the melting points of many substituted diaryl sulfides and their sulfones.

TABLE 28B

Thioethers (Sulfides)—Cyclic

	Name of compound	B.p.	D_4^{20}	n_D^{20}		Name of compound	B.p.	D_4^{20}	n_D^{20}
1	Thiophene	84.2	1.06485	1.52890	14	2-Isopropylthiophene	153.	0.9678	1.5038
2	2-Methylthiophene	112.6	1.0193	1.5203	15	2-Methylthiacyclohexane	153.0	0.9428	1.4905
3	3-Methylthiophene	115.4	1.02183	1.52042	16	3-Isopropylthiophene	157.	0.9733	1.5052
4	2-Methylthiacyclopentane	133.2	0.95552	1.4909	17	2-Methyl-3-ethylthiophene	157.		
5	3-Ethylthiophene	136.	0.9980	1.5146	18	3-Methylthiacyclohexane	158.0	0.9473	1.4922
6	2,5-Dimethylthiophene	136.7	0.9850	1.5129	19	2-Propylthiophene	158.5	0.9687	1.5049
7	3-Methylthiacyclopentane	138.7	0.9634	1.4924	20	4-Methylthiacyclohexane	158.6	0.9471	1.4923
8	2,4-Dimethylthiophene	140.7	0.9956	1.5104	21	5-Ethyl-2-methylthiophene	160.1	0.9661	1.5073
9	2,3-Dimethylthiophene	141.6	1.0021	1.5192	22	3-Propylthiophene	161	0.9716	1.5057
10	Thiacyclohexane (Pentathiamethylene)	141.8	0.9856	1.5067	23	2-Ethyl-3-methylthiophene	161	0.9815	1.5105
11	trans-2,5-Dimethylthiacyclopentane	142.0	0.9222	1.4799	24	2-Methyl-4-ethylthiophene	163	0.9742	1.5098
12	cis-2,5-Dimethylthiacyclopentane	142.3	0.9222	1.4799	25	2,3,5-Trimethylthiophene	164.5	0.9753	1.5112
13	3,5-Dimethylthiophene	145.	1.008	1.5212	26	2,3,4-Trimethylthiophene	172.7	0.995	1.5208

TABLE 29

Thiols (Mercaptans and Thiophenols)[a]

	Name of compound	B.p.	M.p.	Melting point of derivatives	
				Recommended	
				2,4-Dinitro-phenyl Thioether	3,5-Dinitro-thio-benzoate
1	Methyl	6		128*	
2	Ethyl	36		115	62
3	Isopropyl	56		94	84
4	n-Propyl	67		81	52
5	Isobutyl	88		76	64
6	Allyl	90		72	
7	n-Butyl	97		66	49
8	tert-Amyl	97			
9	Isoamyl	119-21		59	43
10	n-Amyl	126		80	40
11	Dimethylene di-	147		248	
12	n-Hexyl	151		74	
13	Cyclohexyl	159		148	
14	Thienyl	166		119	
15	Trimethylene di-	169		194	
16	Phenyl (Thiophenol)	169		121	149
17	n-Heptyl	176		82	53
18	o-Thiocresol	194	15	101	
19	Benzyl	194		130	120
20	m-Thiocresol	195		91	
21	p-Thiocresol	195	43	103	
22	n-Octyl	199		78	
23	α-Phenylethyl	199		89	
24	n-Nonyl	220		86	
25	p-Aminothiophenol		45		
26	Cetyl (Hexadecyl)		50.5	91(96)	
27	β-Thionaphthol	162/20 mm	81	145	
28	Biphenyl		111	146	
29	α-Thionaphthol	161/20 mm		176	

[a] See *J. Am. Chem. Soc.* **65**, 1466 (1943) for data on α,ω-dimercaptans.

Appendix

SUGGESTED LIST OF APPARATUS FOR QUALITATIVE ORGANIC ANALYSIS

These suggestions are primarily for schools that offer instruction in qualitative organic analysis. In place of the equipment listed under "Macro Individual Desk Equipment" a *standard organic desk* may be assigned, supplemented by those asterisked items that are not a part of the desk. For semimicro work add to the standard organic desk *the asterisked items* listed under "Semimicro Individual Equipment."

Macro Individual Desk Equipment

*2 Beakers (50 ml)
2 Beakers (150 ml)
2 Beakers (400 ml)
*1 Crucible cover (No. 00)
1 Crucible tongs
5 Clamps, Bunsen (small)
1 Clamp, Bunsen (large)
6 Clamp holders
1 Condenser, short stem with rubber tubing
*1 Cylinder (10 ml)
1 Cylinder (50 ml)
*1 Desiccator (small)
1 File
*1 Filter flask (125 ml)
*1 Flask, distilling (25 ml)
1 Flask, distilling (125 ml)
1 Flask (200 ml, R.B.)
1 Flask (500 ml, F.B.)
*1 Flask, Erlenmeyer (50 ml)
1 Flask, Erlenmeyer (125 ml)
*1 Funnel, Hirsch (No. 000)
1 Funnel, Buchner (50 mm)

1 Funnel, separatory (60 ml)
1 Funnel, glass (7.5 cm), short stem
*4 Medicine droppers
*1 Medicine dropper, graduated
*1 Pipet, graduated (1 ml)
1 Ring iron (2.5 in.)
1 Ring iron (4 in.)
1 Spatula, steel
*6 Test tubes (10 × 75 mm)
*12 Test tubes (13 × 100 mm)
*6 Test tubes (15 × 150 mm or 16 × 125 mm)
1 Test tube brush
1 Test tube holder
1 Test tube support
1 Triangle, Chromel (2 in.)
1 Tube, drying
1 Tube, melting point
1 Thermometer (360°)
1 Watch glass (3 in.)
2 Watch glasses (4 in.)
1 Wing top
1 Wire gauge

Semimicro Individual Desk Equipment

1 Bath for heating (Sm)
*2 Beakers (25 ml)
2 Beakers (50 ml)
1 Beaker (150 ml)
1 Beaker (400 ml)
*1 Burner (Sm) with tubing
*1 Two-purpose clamp
12 Bottles, reagent (Sm, ½ oz. capacity),
   square with black screw cap holding
   dropper
6 Bottles for solids (½ oz. capacity)
   with screw cap for storing
   derivatives
3 Bunsen clamps (small)
4 Clamp holders
*2 Condensers (Sm) or 1-Liebig
   (160 mm), with adapter
1 Cork borer set (1-6 or 1-8)
1 Cylinder (10 ml)
*1 Disc perforated, beveled (20 mm)
*1 Distilling tube or flask (10 ml)
*1 Distilling tube or flask (25 ml)
1 Desiccator (small), tube or bottle
1 Evaporating dish (30 ml)
*1 pkg. Filter paper discs (25 mm)
1 File
2 Flasks, Erlenmeyer (25 ml)
1 Flask, Erlenmeyer (50 ml)
1 Flask, Erlenmeyer (125 ml)
*1 Funnel (4.5-5 cm diam.), sanded
*1 Funnel, separatory 10 ml or
   30 ml)

*4 Medicine droppers
*1 Medicine dropper, graduated
*1 Pipet, graduated (1 ml)
*1 pkg. (25) Paper drying discs (or
   porcelain)
25 Labels for reagent bottles
2 Rings, iron (2.5 in.)
1 ft. Rubber tubing (3 mm)
*8 ft. Rubber tubing (5 mm), for
   Sm condenser
2 ft. Rubber tubing for suction
*1 Stainless steel spatula
*6 Test tubes (10 × 75 mm)
6 Test tubes (13 × 100 mm)
6 Test tubes (15 × 150 mm or
   16 × 125 mm)
*3 Test tubes (100 × 25 mm)
*3 Test tubes (200 × 25 mm)
*1 Test tube (200 × 25 mm), with
   side arm
*3 Test tubes (3 ml) (tapered at
   the end, if possible
1 Test tube brush
1 Test tube holder
*1 Test tube support (organic to
   hold all tubes including
   200 × 25 mm)
1 Thermometer (360°)
1 Towel
2 Watch glasses (50 mm)
2 Watch glasses (75 mm)
1 Wire gauge

Permanent Laboratory Equipment for Loan or Side Bench

The *single asterisked* items are suggested for small-scale experimentation. The *double asterisked* items are optional if school budgets permit.

Balances, triple beam type or semi-
   quantitative with analytical weights
*Burner, micro (Alber), one for each bench
Burners, Tirrel, and wing top, one
   for each bench
Ceramic ink
**Chromatographic tube (Column double
   length)
*Chromatographic paper (Whatman
   No. 1, 21-24 cm for strips)
*Chromatographic jar for ascending or
   descending two-dimensional
   chromatography
*Claissen flasks (25 ml) for vacuum
   distillation

*Claissen flasks (25 ml), with Vigreaux
   fractionating column
*Column, fractionating (Sm), Heli-Grid
   packing
Corks
Densitometer (Fisher-Davidson type)
**Drying apparatus, Ma-Schenck or glass
   Abderhalden type
Drying oven, Electric or steam
*Filter stick
Glass rod, soft (6 mm)
*Glass rod, soft (4 mm)
*Glass tubing, soft (4 or 5 mm)
Glass tubing, soft (6 mm)
Glass tubing, soft (8 mm)

**Hot stage, Kofler, and microscope
  (50 *or* 100)
*Hydrogenator (Sm)
*Hot plate (micro)
Immersion heater
**Infrared spectrophotometer (Perkin
  Elmer No. 137B *or* Beckman IR No. 5)
Melting point tube, Thiele, modified
  Thiele, or any student apparatus
**Melting point apparatus, precision type
*Melting point apparatus, for m.p.
  (−50 to +40)
*Microapparatus for preparation of
  derivatives
*Microapparatus for preparation of
  Grignard reagents
*Microfiltration assembly (Ma)

*Microscope slides with depression
  and cover slips
**Microscope slides for m.p.
  determination
**Microsublimator
Polarimeter
**Pycnometer (1 ml) *or* Alber pipet
  pycnometer
Refractometer, Abbe *or* Fisher-Jelly
  type
**Refractometer, Nichols
Rubber stoppers
Suction pump with Manometer
  attachment
**Universal apparatus
**Universal micromanipulation stage
  (Schenck-Ma)

LIST OF REAGENTS AND CHEMICALS[1]

It is assumed that all laboratories are supplied with concentrated hydrochloric acid, nitric acid, sulfuric acid and ammonium hydroxide and with the usual $6N$ dilutions of these chemicals and sodium hydroxide, hence, they are not included in the listings. In many instances the same chemical is used in several different per cent concentrations in different parts of the text and where this is the case only the most concentrated solution is listed; for example, sodium hydroxide in aqueous solution is used as $6N$, 20 per cent, 10 per cent, 5 per cent, and 2.5 per cent solutions, but it is expected that the lower concentrations will be prepared as needed by diluting the 10 per cent stock solution. All of the chemicals that are recommended for any use in the text have been listed. A large percentage of the experiments can be made with a much less extensive list of chemicals since many of the listed chemicals are for the preparation of derivatives in cases where several different reagents can be effectively used.

For the preparation of derivatives only the derivatizing reagents necessary for the asterisked or recommended derivatives listed in Chapters 10 and 11 are included. It is important, before a student plans to prepare "other derivatives" than those which are asterisked, to check with the stockroom in order to determine whether the derivatizing reagent is available.

Special Reagents

Barfoed's reagent. Dissolve 16.6 g of crystallized copper acetate in 245 ml of water and add 2.4 ml of glacial acetic acid.

Benedict's reagent. Dissolve 4.3 g of finely pulverized copper sulfate in 25 ml of hot water; cool and dilute to 40 ml with water. Dissolve separately 43 g of sodium citrate and 25 g of anhydrous sodium carbonate (or an equivalent amount of the hydrate form) in 150 ml of water. Heat to effect solution; cool; then add the copper sulfate solution and dilute to 250 ml. Keep the solution in a cork-stoppered bottle.

Ceric nitrate reagent. Dissolve 90 g of ceric ammonium nitrate, $Ce(NH_4)_2(NO_3)_6$, in 225 ml of warm $2N$ nitric acid.

"Doctor" solution (sodium plumbite reagent). Dissolve 45 g of sodium hydroxide in 240 ml of water and then dissolve 7.5 g of litharge (PbO) in the hot caustic solution.

---

[1]This list of reagents and chemicals is intended for colleges and universities that offer a comprehensive course in Qualitative Organic Analysis. Many of the chemicals here listed are only used as alternate methods for detecting classes or for alternate methods for preparing derivatives.

TABLE A.1

Composition of Indicator Reagents for Davidson's System of Classification

Component		Solvent	Composition of reagent, ml			
			A-I	A-II	B-I	B-II
Alizarin yellow-R	0.1%	Methanol	25	–	–	–
Bromothymol blue	0.1%	Methanol	25	25	–	25
Bromocresol purple	0.1%	Methanol	–	37.5	–	37.5
Thymol blue	0.1%	Methanol[a]	–	25	–	25
Benzeneazodiphenylamine	0.1%	Acetic acid	–	–	25	–
Methylene blue	0.1%	Acetic acid	–	–	10	–
Potassium hydroxide	2M	Methanol	25	25	–	–
Hydrochloric acid	Conc.		–	–	4.5	4.5
Methanol			425	887.5	–	933
Pyridine			500	–	–	–
Acetic acid (glacial)			–	–	960.5	–

[a]Contains 0.3 ml of 2M KOH (in methanol) per 100 ml.

Lucas' reagent. Dissolve 320 g of anhydrous zinc chloride in 200 ml of concentrated hydrochloric acid.

Millon's reagent. Dissolve 100 g of mercury in 75 ml of concentrated nitric acid (do not heat externally); dilute the solution with 135 ml of water and stir concentrated nitric acid into the mixture as may be necessary to dissolve any precipitate.

Molisch's reagent. Dissolve 2.5 g of pure α-naphthol in 25 ml of ethanol or methanol. This reagent should have been made within a few days of the time it is used.

Schiff's fuchsin aldehyde reagent. Prepare 100 ml of a 0.1 per cent aqueous solution of p-rosaniline hydrochloride. Add 4 ml of a saturated aqueous solution of sodium bisulfite. Allow the mixture to stand for 1 hour and then add 2 ml of concentrated hydrochloric acid.

Another commonly used method of preparing this reagent is as follows: Dissolve 250 mg of p-rosaniline hydrochloride (fuchsin) in 50 ml of warm water. Cool and saturate the solution with sulfur dioxide until the pink color has disappeared. Add 250 mg of decolorizing carbon, shake, filter and dilute to 250 ml. If the stock solution develops a pink color, repeat the saturation with sulfur dioxide before using it.

Seliwanoff's reagent. Dissolve 125 mg of pure resorcinol in 250 ml of dilute hydrochloric acid (83 ml of concentrated acid and 167 ml of water).

General Reagents

All solutions are aqueous unless some other solvent is specified. In cases where the salt is usually obtained in the hydrated form, the actual weights of chemicals are given to make the desired concentration of the anhydrous compound. Solutions of reagents that are not stable for extended periods of time are not included in this list and should be prepared when needed. It is assumed that the less concentrated per cent solutions will be made as needed from the stock solutions listed here.

Barium chloride, 5 per cent.
Bromine, 2 per cent in carbon tetrachloride.
Ferric chloride, 10 per cent (16 g of the hexahydrate per 100 ml of solution).
Formaldehyde, 40 per cent (formalin).
Hydrochloric acid, 10 per cent; also 1.2N and 2N.
Hydrogen peroxide, 6 per cent.
Hydroxylamine hydrochloride, 1N in methanol (7 g in 100 ml); 1N in propylene glycol.
Iodine, 10 per cent in 20 per cent potassium iodide.

Lead acetate, 5 per cent; also, a saturated solution in ethanol.
Methone (dimethyldihydroresorcinol), 5 per cent in ethanol.
Morpholine, 20 per cent.
Nickel sulfate, 10 per cent.
Phenol, 4 per cent.
Potassium ferricyanide, 10 per cent.
Potassium fluoride, 30 per cent.
Potassium hydroxide, $5N$ in 80 per cent methanol (28 g dissolved in 20 ml of water, then diluted to 100 ml with methanol); also $2N$ in 80 per cent methanol (10 ml of the $5N$ diluted to 25 ml with 80 per cent methanol); also, $6N$ in water (33.6 g in 100 ml); also, 20 per cent in propylene glycol.
Potassium permanganate, 1 per cent.
Silver nitrate, 5 per cent; also, a saturated solution in ethanol.
Sodium acetate, 10 per cent (16.5 g of the trihydrate per 100 ml of solution).
Sodium bicarbonate, $1.5N$.
Sodium carbonate, 10 per cent.
Sodium hydroxide, 20 per cent; 10 per cent; also, $6N$ (24 g per 100 ml of solution) and $2.5N$.
Sodium hypochlorite, 5 per cent ("Clorox").
Sodium iodate, 4 per cent.
Sodium iodide, 2 per cent.
Sodium nitroprusside, 10 per cent.

Pure Chemicals

Acetaldehyde
Acetic acid, glacial
Acetic anhydride
Acetone
Acetyl chloride
Allyl alcohol
Aluminum chloride, anhydrous
4-Aminoantipyrine
Aminoguanidine
Ammonium sulfate
Anthrone
Aniline
Benzene
Benzenesulfonhydroxamic acid
Benzenesulfonyl chloride
Benzoyl chloride
Benzoyl peroxide
Benzylamine
Benzyl chloride
S-Benzylisothiouronium chloride
Bromine
p-Bromoaniline
N-Bromosuccinimide
Butanol
n-Butyl chloride
Butyl nitrite
Calcium chloride, anhydrous
Carbon disulfide
Carbon tetrachloride
Charcoal (also decolorizing carbon)
Chloracetic acid
Chloroform
Chlorosulfonic acid

Chromic oxide
Citric acid
Congo Red
Copper sulfate (hydrated and anhydrous)
Diethylene glycol
Dimethylaniline
Dimethylformamide
N,N-Dimethyl-α-naphthylamine
2,4-Dinitrobenzenesulfenyl chloride
3,5-Dinitrobenzoic acid
3,5-Dinitrobenzoyl chloride
2,4-Dinitrochlorobenzene
2,4-Dinitrofluorobenzene
2,4-Dinitrophenylhydrazine
Dioxane
Diphenylamine
N,N-diphenylbenzidine
Ethanol, 95 per cent
Ethanol, absolute
Ethylene glycol
Ethyl ether
Ferric ammonium sulfate
Ferric chloride
Ferrous ammonium sulfate
Ferrous sulfate
Hexane
2-Hydrazinobenzothiazole
Hydroxylamine hydrochloride
N-Iodosuccinimide
Isatin
Isopropyl alcohol
Lanthanum chloranilate
Mercuric oxide

432 APPENDIX

Methanol
Methone (dimethyldihydroresorcinol)
Methyl iodide
Methyl-p-toluenesulfonate
Molybdenum oxide
α-Naphthol
β-Naphthol
β-Naphthylamine
α-Naphthyl isocyanate
Nickel chloride
Ninhydrin
p-Nitrobenzenediazonium tetrafluoro-
    borate
p-Nitrobenzoyl chloride
p-Nitrobenzyl chloride
p-Nitrophenylhydrazine
p-Nitrophenylhydrazine hydrochloride
Phenol
Phenylhydrazine
Phenylhydrazine hydrochloride
2-Phenylindole
Phenyl isothiocyanate
Phlorglucinol
Phosphoric acid, 85 per cent
Phosphorous pentachloride
Phosphorous pentoxide
Phthalic anhydride
Picric acid
Potassium carbonate, anhydrous
Potassium dichromate
Potassium hydroxide
Potassium iodate
Potassium iodide

Potassium permanganate
Potassium thiocyanate
Propylene glycol
Pyridine
4-Pyridylpyridinium dichloride
Resorcinol
Semicarbazide hydrochloride
Soda-lime
Sodium
Sodium acetate, anhydrous
Sodium anthraquinone α-sulfonate
Sodium bisulfite
Sodium carbonate, anhydrous
Sodium dichromate
Sodium formate
Sodium hydroxide
Sodium nitrite
Sodium peroxide
Sodium sulfate, anhydrous
Sulfanilic acid
Sulfuric acid (fuming), 20 per cent SO_3
Tetraisopropyl titanate
Thionyl chloride
Thiourea
Tin
Titanium ammonium fluoride
Toluene
p-Toluenesulfonyl chloride
p-Toluidine
2,4,7-Trinitrofluorenone
Xanthydrol
Zinc, dust
Zinc chloride, anhydrous

COMMON DRYING AGENTS

Several common drying agents for organic compounds are listed in Table A.2. Desirable features to be considered in selecting a drying agent are: (1) it should not react or combine with the compound to be dried; (2) it should not catalyze polymerizations or condensation reactions.

TABLE A.2

Common Drying Agents for Organic Compounds

Anhydrous substance	Applicable to	Not applicable to	Drying power	Relative efficiency
Calcium chloride	Hydrocarbons, halides, ethers, esters	Hydroxy and amino compounds	High below 30°	Medium
Calcium sulfate ("Drierite")	All compounds	None	Low	Good
Magnesium sulfate	All compounds	None	High	Good
Potassium carbonate	Amines, alcohols, ketones	Acids	Medium	Medium
Phosphorus pentoxide	Halides, hydrocarbons, nitriles	Most compounds	High	Excellent
Sodium hydroxide	Amines, hydrazines, saturated hydrocarbons	Most compounds	High	Excellent
Sodium metal	Ethers, saturated hydrocarbons	Most compounds	High	Excellent

CLEANING SOLUTIONS

Chromic acid cleaning mixture. Dissolve 10 g of sodium dichromate in 10 ml of water in a 400 ml beaker. Add slowly with careful stirring 200 ml of concentrated sulfuric acid. The temperature will rise to nearly 80°. Allow the mixture to cool to about 40°, and place in a dry glass-stoppered bottle. Care should be exercised in making and handling this solution.

Trisodium phosphate solution. Glassware that does not contain tars may be cleaned with a 15 per cent solution of trisodium phosphate. A warm solution with the aid of an abrasive powder, such as pumice, is safer to handle and cleans as well as or better than chromic acid solutions.

THE NEUTRALIZATION EQUIVALENT

The neutralization equivalent of an acid or acidic compound is actually its equivalent weight and hence is of value in identifying the compound. If the substance is a liquid, weigh it in a dropping bottle and then transfer a few drops of the liquid to the titration flask and reweigh the dropping bottle. The weight of the sample is the difference in the two weights of the dropping bottle. If the substance is even reasonably volatile, it is best to have the solvent (water, alcohol, or a mixture of water and alcohol depending on the solubility of the substance) in the titration flask before adding the substance so as to retard its evaporation.

Transfer an accurately weighed sample (100-300 mg) of the acidic substance to a 125-250 ml flask and add 25-35 ml of water, alcohol, or a mixture of these solvents to dissolve the substance. Titrate the solution with a standardized solution of sodium hydroxide (approximately $0.1N$), using phenolphthalein as the indicator. Calculate the neutralization equivalent of the compound as follows:

$$N.E. = \frac{\text{wt. of sample in grams} \times 1000}{\text{ml of NaOH} \times N \text{ of the NaOH}}$$

This formula may be written as:

$$N.E. = \frac{\text{milligrams of sample}}{\text{milliequivalents of NaOH}}$$

THE SAPONIFICATION EQUIVALENT OF AN ESTER

The objective is to hydrolyze the ester and determine the amount of alkali that reacts with the acid produced. Saponification of an ester by aqueous sodium hydroxide is inhibited because most esters are sparingly soluble in the aqueous solution. Solutions of potassium hydroxide in diethylene glycol,[2] potassium hydroxide in ethylene glycol,[3] and potassium hydroxide in a mixture of glycol monoethyl ether[4] have been used. It has been found that a solution of potassium hydroxide in 80-90 per cent methanol, ethanol, propanol, or 2-propanol makes a satisfactory saponifying agent for esters except for those with very high weights or that for some reason are difficult to hydrolyze. In such cases, one of the glycols should be substituted for the lower alcohols. The alcoholic solution of potassium hydroxide tends to attack glass and to absorb carbon dioxide from the air during the refluxing of the mixture, hence it is wise to standardize the alkaline solution under conditions which are similar to those used for the saponification of the ester. Since many esters are at least moderately volatile it has been found advisable to have the alcoholic solution of potassium hydroxide in the flask before adding the weighed sample of the ester so as to dissolve the ester and decrease its loss by evaporation. Liquid esters should be *weighed by difference* from a dropping bottle.

[2] Redemann and Lucas, *Ind. Eng. Chem., Anal. Ed.* 9, 521 (1937).
[3] Maglio, *Chemist Analyst* 35, 29 (1946).
[4] Hahn, *Analytica Chim. Acta* 4, 577 (1950).

Dissolve approximately 3 g of potassium hydroxide in 60 ml of alcohol and allow any sediments to settle. Add exactly 25 ml of alcoholic solution of potassium hydroxide to each of two 125-250 ml flasks. To one of the flasks, add an accurately weighed sample (300-400 mg) of the ester and use the other flask as a "blank." Attach reflux condensers to both flasks and gently boil the solution for one hour. Cool the solutions and rinse each condenser with 10 ml of water, catching the washings in the solutions. Titrate each of the solutions with standardized hydrochloric acid (approximately 0.5N) using phenolphthalein as an indicator.

The difference in the volumes of hydrochloric acid required for the solution which contained the ester and that required for the "blank" represents the amount of potassium hydroxide which reacted with the ester. The number of milliliters of a solution multiplied by the normality of the solution equals the number of milliequivalents, hence the volume of the hydrochloric acid (the difference in the volumes required for the two titrations) multiplied by the normality of the hydrochloric acid equals the milliequivalents of potassium hydroxide required for the ester. Calculate the saponification equivalent of the ester as follows:

$$\text{S.E.} = \frac{\text{milligrams of ester}}{\text{milliequivalents of KOH}}$$

Methods have been developed[5] that eliminate the use of the "blank" in the above method by using double-indicator titrations instead. Details for this method may be found in the references. These are semimicro and micro methods.

Saponification Number

Industrially, the saponification number is more generally used than the saponification equivalent.[6] The saponification number is defined as the number of milligrams of potassium hydroxide required to saponify one gram of the oil or fat. It may be calculated from the data obtained from the determination of the saponification equivalent by the use of the following formula:

$$\text{S.N.} = \frac{\text{milliequivalents of KOH} \times 56.1}{\text{grams of ester}}$$

THE IODINE NUMBER OF A FAT OR OIL

The chemical literature records several fundamental methods and many modifications of these methods for the determination of the iodine number (centigrams of iodine absorbed per gram of the sample). Two procedures have been selected, one for a macro quantity, and one for a micro quantity. The iodine numbers, as determined by various methods, do not always agree.

Macro Method

The method used here is that of Wijs. More details of this method and also other methods are given in the *Official and Tentative Methods of the American Oil Chemists Society* and the *Official Methods of the Association of Official Agricultural Chemists*[7] publications.

---

[5]Rieman, *Ind. and Eng. Chem., Anal. Ed.* 15, 325 (1943); Rieman and Mercali, *Ind. and Eng. Chem., Anal. Ed.* 18, 144 (1946); Ketchum, *Ind. and Eng. Chem., Anal. Ed.* 18, 273, 460 (1946).

[6]For a direct method of determining the saponification number, see *Official Methods of Analysis* (8th ed., Association of Official Agricultural Chemists, P.O. Box 540, Benjamin Franklin Station, Washington, D. C., 1955).

[7]*Ibid.*

Add 100-400 mg of the sample to a 250-500 ml glass-stoppered flask containing 20 ml of carbon tetrachloride (or chloroform). Add 25 ml of the iodine solution (Wijs) which has been prepared by dissolving 13 g of pure iodine in a liter of pure acetic acid and then passing chlorine gas into the solution until the quantity of sodium thiosulfate required for titration is not quite doubled.[8] Moisten the stopper with 5 drops of a 15 per cent solution of potassium iodide. Let the flask stand in the dark for 30 minutes and then add 20 ml of 15 per cent potassium iodide and 100 ml of water. Titrate the solution with sodium thiosulfate (about $0.1N$) until the yellow color almost disappears and then add 3 drops of a 1 per cent starch suspension and continue the titration until the blue color entirely disappears. Near the end of the titration, shake the solution vigorously. Run a "blank" determination (without the fat or oil). The iodine number is the number of centigrams of iodine that reacts with one gram of fat or oil.[9]

$$I.N. = [(V \times N \text{ of } Na_2S_2O_3 \text{ for "blank"}) - (V \times N \text{ of } Na_2S_2O_3 \text{ for sample})]$$

$$\times 0.127 \times \frac{100}{\text{wt. of sample}}$$

Micro Method

The Kaufman method[10] is given here because it is recommended in the literature[11] and because it has proven satisfactory in the authors' laboratory. The reagents required are: (a) $0.1N$ bromine in methanol saturated with dry sodium bromide; (b) 10 per cent potassium iodide; (c) $0.05N$ sodium thiosulfate; (d) alcohol-free chloroform.

The fat or oil (10-30 mg) is dissolved in 2 ml of chloroform in a 150 ml iodine flask. Place 2 ml of chloroform in another flask and use it as a "blank" to be treated by the same procedures as the sample. Exactly 5 ml of the $0.1N$ bromine solution are added from a microburet. The mixture is shaken. The reaction is complete in 1-5 minutes. Add 3 ml of the potassium iodide solution and shake the mixture. The excess bromine reacts with the potassium iodide to liberate free iodine which is then titrated with the sodium thiosulfate solution. Since the number of milliequivalents of sodium thiosulfate is equal to the number of milliequivalents of excess bromine and since the milliequivalent weight of iodine may be considered as 0.127, the following equation may be used to calculate the iodine number.

$$I.N. = [(V \times N \text{ of } Na_2S_2O_3 \text{ for "blank"}) - (V \times N \text{ of } Na_2S_2O_3 \text{ for sample})]$$

$$\times 0.127 \times \frac{100}{\text{wt. of sample}}$$

CONSTRUCTION OF DROPPERS AND PIPETS WITH LONG CAPILLARY TIPS

Pipets with long capillary tips are indispensable for the separation of small volumes of immiscible liquids and the transfer of small volumes. Figure A.1 shows an ungraduated pipet with a long capillary tip; it can be obtained either with a sturdy tip so that it can be cleaned and used again, or (at a smaller cost) with a thin tip so that after use it

[8]An alternative method uses dichloramine T as the source of the chlorine. See *Analyst* **58**, 523 (1933); c.f. *C.A.* **27**, 5562 (1933).
[9]Scotti has modified the Wijs method to shorten the time required. See a. *Oil and Soap*, **16**, 236 (1939) [c.f. *C.A.* **34**, 901 (1940)]; b. *Olii minerali, grassi e saponi, colori e vernici* **18**, 96 (1938) [c.f. *C.A.* **34**, 429 (1940)].
[10]*Ber.* **70B**, 2554 (1937); [c.f. *C.A.* **32**, 1961 (1938)].
[11]*Fette u. Seifen* **43**, 155 (1936) [c.f. *C.A.* **31**, 2034 (1937)]; *Biochem. Z.* **296**, 174 and 180 (1938) [c.f. *C.A.* **33**, 420 (1939)]; *Oil and Soap* **16**, 69 (1939) [c.f. *C.A.* **33**, 4446 (1939)]; *Fette u. Seifen* **47**, 4 (1940) [c.f. *C.A.* **34**, 4291 (1940)]; *Woolen- u. Leinen- Ind.* **60**, 67 (1940) [c.f. *C.A.* **34**, 7635 (1940)]; *Inst. espan. oceanograf. Notas y resúmenes. Ser* **2**, No. 111, 5 (1942) [c.f. *C.A.* **37**, 5263 (1943)].

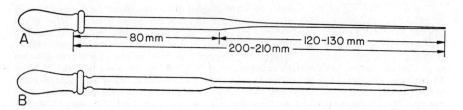

Figure A.1. Ungraduated pipet with long capillary tip. *A*. With thin tip (0.1 mm.). *B*. With sturdy tip.

must be discarded. However, the author recommends that the beginner prepare these pipets as directed, six or twelve at a time, and have them handy when needed.

To prepare a pipet with a long capillary tip, clean thoroughly with soap and water a piece of soft glass tubing 200 mm. in length and 6-8 mm. in diameter. Rinse well, first with tap water and then with distilled water. Allow the tubing to dry. Use either a good Bunsen flame (Figure A.2*a*), or, if not familiar with elementary manipulations, use a burner provided with a wing top. Grasp the ends of the glass tubing with both hands and rotate it between the thumb and index finger over the flame. When the glass has softened enough to bend easily, remove from the flame and draw gently and steadily lengthwise until the length has doubled. Hold in place until the glass has hardened and then lay it carefully on an asbestos mat. Cut the capillary at about the middle (Figure A.2*b*). Heat the wide end of the pipet until it is fire-polished. If the glass tubing used is of 6 mm. bore, the wide end should be flanged in order to form a tight fit with the rubber bulb. To flange the end, heat it in the flame until the tube has softened; then press firmly against an asbestos pad. Repeat this operation until a flange 7-8 mm. in diameter is formed (Figure A.2*c*). Pipet droppers of various sizes are made by a method similar to that described for the capillary pipet.

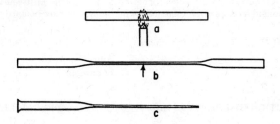

Figure A.2. Construction of dropper with long capillary tip.

SOLUTIONS AND DEVELOPING SOLVENTS FOR CHROMATOGRAPHIC PROCEDURES

Standard Solutions

Standard solutions of known compounds such as amino acids, sugars, derivatives of alcohols, and carbonyl and other organic compounds are employed in identification work by paper chromatography on guide strips alongside the paper strips containing the unknown. The solutions of known compounds are made by dissolving 10 mg of the compound in 10 ml of the appropriate solvent and then using 2-5 microliters of the solution for application on the guide strips. For sugars the solvent used is water. For amino acids and amines, the solvent is 80 per cent methanol or 10 per cent 2-propanol; a drop of 6M HCl is added if the amine or amino acid does not dissolve readily. For 3,5-dinitrobenzoates and 2,4-dinitrophenylhydrazones, a mixture of methanol and chloroform (1:1 by volume) is used as a solvent.

Solvents

The solvents given in Table 7.1, page 164, and described in greater detail in this section are for general use in identification work employing test tube or small tank ascending techniques.

In general, with an increase in the amount of water in the solvent mixture the rate of development and the R_f value increase. The addition of amines or ammonia to the solvent mixture causes an increase in the R_f values of basic substances. Nonpolar solvents tend to decrease the R_f values of polar compounds, but addition of small amounts of polar solvents can produce a desired increase. Thus a mixture containing 2,4-dinitrophenylhydrazine hydrochloride and the 2,4-dinitrophenylhydrazone of a carbonyl compound can be resolved by using a mixture of 90 per cent hexane, 5 per cent ether, and 5 per cent water. In a homologous series of compounds or derivatives the open chain compounds have higher R_f values than those containing cyclic and, particularly, aromatic rings. Introduction of hydroxyl and thiol groups into a structure tend to lower the R_f values. The solvent mixtures given below are prepared by mixing in an 8 inch tube, shaking for a few minutes, and then observing whether separation of a lower aqueous layer takes place. If a separation occurs, the upper layer is removed by means of a separatory funnel and stored in a bottle.

In many cases the water is not listed in the second column of Table 7.1 (page 164). However, in the following, the water is properly listed.

1-Butanol-acetic acid-water: 1-Butanol, 6 ml; acetic acid, 1.5 ml; and water, 2.5 ml.
Phenol-water: Phenol, 8 ml; water, 2 ml.
Phenol-ammonia: Phenol, 8 ml; water, 2 ml; ammonia (conc., 0.88), 1 drop.
2,4-Lutidine-water: 2,4-Lutidine, 6.9 ml; water, 3.1 ml.
tert-*Butanol-methylethyl ketone-water-diethylamine:* tert-Butanol, 3.8 ml; water, 3.8 ml; methylethyl ketone, 2.0 ml; diethylamine, 0.4 ml.
Ethyl acetate-pyridine-water: Ethyl acetate, 5.7 ml; pyridine, 2.4 ml; water, 1.9 ml.
tert-*Butanol-water:* tert-Butanol, 8 ml; water, 2 ml (a ratio of 7:3 may be employed if faster development is desired).
2-Propanol-water: 2-Propanol, 8 ml; water, 2 ml.
1-Butanol-pyridine-water: 1-Butanol, 4 ml; pyridine, 4 ml; water, 2 ml.
Ether-hexane-water: Ether, 0.5 ml; hexane (or low boiling petroleum ether), 9 ml; water, 0.5 ml.
1-Butanol-ammonia: 1-Butanol, 10 ml; ammonia (sp. g., 0.88), 1 drop.
Methanol-acetone-water: Methanol, 4 ml; acetone, 4 ml; water, 2 ml.

Index of Text

Aryl sulfenyl chloride, 274
Association, 80, 87, 197
Azeotropic mixtures, 199, 201
Azines, 185
Azo compounds, 107

B

Baeyer test, 112
Barbiturates, 269
Barfoed's solution, 135-136, 429
Bases, separation from mixtures, 199
Basicity, classification by, 95-101
  tests for, 101
Basicity constants, 86, 87, 98
Benedict's reagent, 135-136, 138
Benzaldehyde, derivatization of, 225,
  261, 262
  separation from acid, 197
Benzaldehyde dimethyl acetal, 228
Benzamides, 270, 271-273
Benzene, detection of, 108, 177-178
  nitro derivatives of, 145
  separation of, 196, 200, 206
  as solvent, 12
Benzenesulfonamides, in amine detection,
  130, 207, 231
  identification of, 335
  preparation from amines, 274-275
  preparation from others, 296, 297-298
Benzene sulfonhydroxamic acid, 137
Benzidine, 72, 87
Benzoates, 12, 329
Benzoic acid, 197
Benzonitrile, 322
Benzophenone phenylhydrazone, 319-320
p-Benzoquinone, 181-182
Benzo-p-toluidide, 240-241
Benzoyl chloride, reaction with alcohols,
  254
  reaction with amines, 271-273
  reaction with amino acids, 231-232
Benzoyl ureas, 202
Benzyl alcohol, 227
N-Benzylamides, 289-290
Benzylphenylhydrazones, 283-284
Benzyl ureas, 202
Boiling points, 52-54
Borates, 70
Bromcresol green, 164
Benzylamine, 87, 117-118
Benzyl benzoate, 198
S-Benzylisothiouronium chloride, 336-
  337
Benzyl ketones, 137

Bromination, of aryl ethers, 296, 297
  of phenols, 238, 332
Bromine, and alkene addition, 312
  detection of, 73-75, 180-181
  in unsaturation test, 111-112
p-Bromoanilides, 238, 241
Bromobenzene, 304
p-Bromobenzoic acid, 241-242
p-Bromophenacyl esters, 244
Bromophenols, 90
p-Bromophenylhydrazones, 281
N-Bromosuccinimide, in alcohol classification, 120
  in amine detection, 129, 133
p-Bromosulfonamides, 274, 275
o-Bromo-p-toluidides, 242
Bromphenol blue, 164
Buchner funnel, 20
Butanal dimethone, 263-264
Butanone, in paper chromatography, 167-
  168
  semicarbazone, 320
sec-Butylamine, 198
Butylbenzenes, 108
n-Butyl chloride, 302, 303
p-tert-Butylphenol, 330
n-Butyl phthalate, 291
n-Butyl sebacate, 198
n-Butyranilide, 241-242
Butyric acid, 90, 241

C

Calcium chloride, 24, 25-26
Calibration, of thermometers, 45
Camphor, in molecular weight determination, 62-65
Capillary pipet, 38
  construction of, 435-436
Capillary tubes, 42-43, 53, 54, 64, 65
n-Caproic acid, 177, 239
n-Caprylic acid, 239
Carbamates. See Urethans
Carbitols, 120
Carbohydrates, derivatization of, 280-287
  detection of, 127, 134-136, 287
  in paper chromatography, 163, 164
  separation from mixtures, 199, 206, 212
  specific rotation of, 286
Carbon, detection of, 69
Carbon disulfide, in alcohol detection,
  119-120
  in amine detection, 132-133
Carbon tetrachloride, 12, 194, 208
Carbonyl compounds, chemical detection
  of, 136-140, 198

Carbonyl compounds (*cont'd*)
    chromatographic detection of, 164, 167-
        168
    derivatization of, 257-267, 318-321
    infrared detection of, 173, 174-175, 177,
        181-182, 183-184
    separation from mixtures, 199, 206,
        208, 212
Carboxylic acids, acidity constants of, 89-
        90
    chemical detection of, 115-116, 117-118
    chromatographic detection of, 247
    derivatization of, 236-246
    formation from nitriles, 321-323
    infrared detection of, 176-177, 183
    separation from mixtures, 206
    solubility of, 12, 88
    See also Dicarboxylic acids
Catechol, 90
Cellosolves, 120
Ceric nitrate, 121, 429
Charcoal, 18
Chelation, 80, 87, 182, 197
Chlorine, detection of, 73-75, 180
Chloroacetic acid, 90, 331
Chloroanilines, 87
o-Chlorobenzaldehyde, 261
Chlorobenzene, 306, 307
p-Chlorobenzoic acid, 241-242
Chlorobutyric acids, 90
Chloroform, in amine detection, 132
    as solvent, 12, 194, 206, 208
Chloroform-aluminum chloride test, 107-
        108
Chloronaphthalenes, 305
Chlorophenols, 90
Chlorosulfonylation, of aryl halides, 304,
        307-308
    of ethers, 297-298
p-Chlorotoluene, 198
Chromatography. *See* Column chromatog-
        raphy; Gas-liquid chromatography;
        Paper chromatography; and under
        specific compound classes
Cinnamyl alcohol, 121
Citric acid, 133
Color, classification by, 107, 108
    in paper chromatography, 161, 164
Column chromatography, 157, 188, 208,
        209
Classification, by acid-base method, 95-
        101
    by color tests, 107, 108
    by solubility tests, 78-95, 106
Copper acetate-benzidine test, 72
Copper sulfate, in detection of ammonia,
        109

in test for water, 70
Coupling reactions, 130-131, 148
Cresols, 90, 331
Cresyl ethers, 297
Cryoscopic constants, 63
Crystallization, 11-26
    fractional, 194, 195, 202, 207
    procedures for, 19-23
    solvents in, 11-17
Crystals, drying of, 23-26
    properties of, 3
    transfer of, 22
Cyanide ion, 72
Cyclic ethers, 183
Cyclic ketones, 181
Cyclization, of aldehyde methones, 263,
        264
Cycloalkanes, derivatization of, 309, 310,
        311-312
    detection of, 143, 179
    separation from mixtures, 200
Cycloalkenes, 312
Cycloalkyl halides, 301-304
Cyclobutane, 179
Cyclobutanone, 181
Cycloheptanone, 181
Cyclohexane, derivatization of, 311-312
    infrared spectrum of, 177-178, 179
    separation from mixtures, 206
Cyclohexanols, derivatization of, 253-254
    detection of, 120
Cyclohexanone, 191, 258, 320
Cyclohexene, 312
    infrared spectrum of, 177-178
Cyclohexylamine, 187-188, 196
Cyclohexyl-α-naphthylurethan, 253-254
Cyclopentane, 179
Cyclopentanone, 181
Cyclopropane, 179
Cystine, 234

D

Data coordination, 215-216
Davidson's indicator systems, 99-100
Decolorizing agents, 18
Density, determination of, 58-61
Derivatives, preparation of, 225-299,
        301-342
    selection criteria for, 225-227
Desiccants, 24, 25-26
Desiccators, 24-26
Diatomaceous earth, 18
Diamines, 132
Diazo compounds, 107, 185
Diazonium salts, in amine detection, 130-
        131

Index of Tables

451

E